The Oxford Dictionary of
Statistical Terms

THE
OXFORD
DICTIONARY OF
STATISTICAL
TERMS

EDITOR

YADOLAH DODGE
University of Neuchâtel, Switzerland

OXFORD
UNIVERSITY PRESS

OXFORD

UNIVERSITY PRESS

Great Clarendon Street, Oxford OX2 6DP

Oxford University Press is a department of the University of Oxford.
It furthers the University's objective of excellence in research, scholarship,
and education by publishing worldwide in

Oxford New York

Auckland Bangkok Buenos Aires Cape Town Chennai
Dar es Salaam Delhi Hong Kong Istanbul Karachi Kolkata
Kuala Lumpur Madrid Melbourne Mexico City Mumbai Nairobi
São Paulo Shanghai Taipei Tokyo Toronto

Oxford is a registered trade mark of Oxford University Press
in the UK and certain other countries

Published in the United States
by Oxford University Press Inc., New York

© International Statistical Institute

The moral rights of the author have been asserted
Database right Oxford University Press (maker)

First published 2003
First published in paperback 2006

British Library Cataloguing in Publication Data

Data available

Library of Congress Cataloging in Publication Data

Data available

Printed in Great Britain
on acid-free paper by
Biddles Ltd, King's Lynn

ISBN 0-19-850994-4   978-0-19-850994-3
ISBN 0-19-920613-9 (Pbk.)   978-0-19-920613-1 (Pbk.)

1 3 5 7 9 10 8 6 4 2

PREFACE

One possible objective of a dictionary is, by definition, to list all words of a target language in alphabetical order and then provide their meaning, whether the words are intrinsically relevant or not. An alternative perspective regarding the function of a dictionary is to provide an explanation of only the terms that are in current use. With an estimated 20,000 statistical terms in our language, and the rapid and continuous growth of the subject, we restricted ourselves to the later definition, limiting our output to some 3,540 entries.

Looking at the antecedents of this present dictionary, in 1949, with financial support from UNESCO, and in collaboration with the International Statistical Institute (ISI), an international programme for teaching statistics was established. Special importance was given to the creation of a statistical dictionary in different languages along with a glossary of statistical terms. In 1951 the ISI invited the late Sir Maurice Kendall to prepare such a dictionary.

A description of the problems which he encountered, along with the methods by which he tried to solve them, is given in a paper read before the ISI Rome meeting in September of 1953 (M.G. Kendall (1954)—*The Projected Dictionary of Statistical Terms. Bulletin of the International Statistical Institute*, 34, Part 2, 629).

With the editorial help of the late Dr. W.R. Buckland, a list of terms along with their definitions was prepared and a draft was submitted to a group of experts in different fields. The draft was examined by, and help in preparing the definitions, was provided by the following persons: R.D.G. Allen; G.A. Barnard; C.I. Bliss; H.S. Booker; D.R. Cox; G.Darmois; J. Durbin; C.E. Eisenhart; D.J. Finney; F.G. Foster; R.C. Geary; C.Gini; P.M. Grundy; H.Hotelling; V. Lahiri; D.V. Lindley; P.C. Mahalanobis; C.A. Moser; E.S. Pearson; J.R.N. Stone; A.Stuart; E.Van Rest; P.E.Vernon; S.S. Wilks; J.Wishart and H.O. Wold.

This first edition of the dictionary with definitions of 1,700 terms was published in 1957 by Oliver and Boyd in London, the second edition in 1960, and the third edition, revised and extended with 2,500 terms, appeared in 1971. There was a further increase to nearly 3,000 terms in the fourth edition which was published by The Longman Group in 1982.

The ISI, which initiated the first dictionary in 1957, invited Dr. F.H.C. Marriott to prepare a revised fifth edition thirty years later. Dr. Marriott retained the same format for the dictionary as Kendall and Buckland and most of the entries remained unchanged apart from minor typographical corrections. Based on the study of literature from 1980 to 1990 Dr. Marriott added 400 new terms to the fifth edition. He also omitted the Italian terms originally provided by Professor Gini and some other terms that seemed by him to be 'ephemeral and obsolete'.

The fifth edition was published in 1990 and reprinted in 1991 and 1992, with 3,166 entries.

In 1998 I was invited by the ISI to undertake the task of revising the *Dictionary of Statistical Terms* for a sixth edition. Just as for the previous editions, an Editorial and Advisory Board team was selected to suggest additional terms for inclusion and at the same time to propose the elimination of some terms that are no longer in current use.

Headed by Sir David Cox of Nuffield College Oxford, the Editorial Board comprised:

Daniel Commenges from the Institute of Public Health and Development, Bordeaux, France;

Anthony Davison of the Federal Institute of Technology in Lausanne, Switzerland;

Patty Solomon of the University of Adelaide; and

Susan Wilson of the Australian National University, Australia.

The Advisory Board was formed by Yuri Belayaev of the University of Umeå, Sweden, Valeri Federov of GlaxoSmithKline, USA, Zhi Geng of Peking University, China, Jane Gentleman of the National Center for Health Statistics, USA and Jana Jureckova of the Charles University, Prague, Czech Republic.

In addition, a website was constructed in which all the terms of the fifth edition were listed and the statistical community was asked to suggest new entries for inclusion with their respective definitions. A total of 320 terms were thus suggested for elimination, and 1,067 terms for addition. In fact we eliminated only 265 entries and added 640 new entries and so the current edition contains 3,540 terms.

For some entries a reference to the literature has been given, in some cases to an early source, and in others to a more recent publication. The editors hope that these and the resulting bibliography add to the value of the dictionary. While some care has been taken in the choice of references, establishing historical priorities is notoriously difficult and the historical assignments are not to be regarded as authoritative; when recent references are given there are often numerous other choices possible. Short expository articles on many of the entries in the dictionary may be found in one or more of the following Encyclopaedias:

International Encyclopaedia of Statistics, eds William Kruskal and Judith M. Tanur (The Free Press, 1978).

Encyclopaedia of Statistical Sciences, eds Samuel Kotz, Norman L. Johnson and Cambell B. Reed (New York: John Wiley and Sons, 1982).

The Encyclopaedia of Biostatistics, eds Peter Armitage and Ted Colton (Chichester: John Wiley and Sons, 1998).

The Encyclopaedia of Envirometrics, ed A.H. El-Sharaawi and W.W. Piergorsch (New York: John Wiley and Sons, 2001).

Statistique: Dictionnaire Encyclopédique, Yadolah Dodge (Paris: Dunod, 1993).

A future goal should be to implement the first definition of a dictionary given at the beginning of this Preface and provide definitions of all terms, going back to the historical beginnings of statistics. In my opinion no statistical term should be treated as ephemeral or obsolete.

I would be very grateful to readers for any comments regarding inaccuracies, corrections, and suggestions for inclusion of new terms, or any matter that could potentially improve the next edition. Please send all your suggestions to the International Statistical Institute via:

http://www.cbs.nl/isi/dictionarysubmitForm.htm.

I am grateful to members of the Advisory Board for helpful comments at the preliminary planning phase of the work as well as providing new definitions.

I would like to thank my assistants S. Vancolen and T. Kondylis for checking the cross-references and for finding many of the historical references; Daniel Berze of the ISI Permanent Office for his valuable help in providing the initial TeX version of the manuscript; Lars Lyberg of Statistics Sweden; Marc Hallin of the ULB; Marie Huskova of Charles University, Carol Blumberg of Winona State University for providing definitions for some new terms; the Swiss National Science Foundation (FN-21-65147-01) for their financial support. Their financial help allowed me to have the assistance of Giuseppe Melfi for valuable help in the initial clean up of the fifth edition and for providing definitions for some new terms. I thank the Oxford University Press, especially Rebecca Allen, for their meticulous care in the production of the dictionary. Finally, I wish to express my sincere appreciation to Sir David Cox whose helpful suggestions and criticisms contributed greatly to this dictionary. Without his guidance and encouragement this dictionary could not have been completed.

Neuchâtel, Switzerland
June 2003

Yadolah Dodge

A

α-error In the theory of testing hypotheses an error incurred by rejecting a hypothesis when it is actually true; more usually known as **error of the first kind** or **type I error** introduced by Neyman and Pearson (1928). In quality control it is equivalent to **producer's risk**.

α-resolvability See **resolvable balanced incomplete block design**.

Abbe, Ernst (1840–1905) Abbe studied at the University of Jena and the University of Göttingen, receiving his doctorate from Göttingen in 1861 with a dissertation on thermodynamics. In 1863 he joined the teaching staff at the University of Jena and he presented the paper *Über die Gesetzmässigkeit in der Vertheilung bei Beobachtungsreihen* for his teaching qualification, in which he derived the **chi-squared distribution** and the distribution of the **serial correlation** coefficient (Kendall, 1971).

Abbe criterion A test of randomness in a time series based upon the fact that in a random series autocorrelation coefficients of order k, for all k greater than zero, should vanish (Abbe, 1906).

Abelson–Tukey score test A significance test in analysis of variance with a minimax property against ordered alternatives (Abelson and Tukey, 1963).

abrupt change Change point problem when the change of the model is sudden (Shaban, 1980). If the change point is known it reduces to a two-sample problem.

absolute deviation The difference, taken without regard to sign, of a variable value from some given value which may be a constant, a population value or a sample value. This concept enters into the **mean deviation**.

absolute difference The difference, taken without regard to sign, between the values of two variables, and in particular of two random variables. This concept is related to the **mean difference**.

absolute error The absolute error of an observation x is the absolute value of deviation of x from its 'true' value.

absolute frequency The actual frequency of a variable, i.e., the number of its occurrences, as distinct from the relative frequency, namely the ratio of the frequency to the total number of occurrences.

absolute measure This term is occasionally used to describe a measure of variable values which is independent of origin and scale; for example, the moment ratio μ_4/μ_2^2 or the corresponding sample statistic m_4/m_2^2. The usage is not entirely satisfactory owing to possible confusion with measures, such as the mean deviation, which are based on absolute quantities, i.e., quantities taken regardless of sign.

absolute moments The moments of a frequency distribution in which the deviations about a fixed point are taken without regard to sign; that is to say, the rth absolute moment about a value a is

$$v'_r = \int_{-\infty}^{+\infty} |x - a|^r dF(x).$$

As for ordinary moments, the absolute moments about an arbitrary value are usually denoted by a prime, those about the **mean** without a prime, e.g.,

$$v_r = \int_{-\infty}^{+\infty} |x - \mu_1|^r dF(x).$$

For even values of r the ordinary and absolute moments are identical. [See also **moment**.]

absolute risk Probability, usually conditionally on a given risk factor, that a disease occurs in a given period of time. Contrasted to **relative risk**.

absolutely unbiased estimator See **unbiased estimator**.

absorbing barrier Certain additive or **random walk** processes represent the motion of a particle in one or more dimensions; limitations may be imposed on the motion in the form of barriers at which there is a probability that the particle will be 'absorbed' and its motion ended (as distinct from being reflected). The boundary lines terminating a sequential sampling process are of this type. [See also **reflecting barrier**.]

absorbing Markov chain An important class of non-ergodic Markov chains; namely, those that possess **absorbing states** (Feller, 1968).

absorbing region A generalization of the idea of an **absorbing barrier**.

absorbing state A state in a stochastic process which, once reached, cannot be quitted. It is also known as an **absorbing barrier**, but this term is generally used where the process terminates when the barrier is reached.

accelerated failure time model A representation of the effect of an explanatory variable on a distribution of failure (survival) time in which the time scale is inflated or deflated by a constant factor depending on the explanatory variable. It was introduced originally in an industrial context but is now widely used in medical and other fields (Kalbfleisch and Prentice, 2002).

accelerated life testing A form of industrial reliability testing in which items are tested in environments more extreme than those normally encountered. The object is to obtain useful results much more quickly than by testing under normal conditions.

accelerated stochastic approximation A method proposed by Kesten (1958) for increasing the speed of convergence in a **Robbins–Munro, Kiefer–Wolfowitz** or other type of stochastic approximation. The method depends upon the number of changes of sign, in relation to the size of the step from one approximation to the succeeding one.

accelerated test Usually a test of reliability of some elements or devices, under conditions much more unfavourable and implying earlier failures. The obtained data have to be recalculated to estimate reliability characteristics under recommended conditions of exploitation.

acceleration by powering Certain arithmetic processes for extracting the characteristic roots of a matrix by iterative methods can be shortened (i.e., the number of iterations required can be reduced) by operating on a power of the matrix. The calculations and the convergence of the iterative process are then said to be accelerated by powering (Wilkinson, 1965).

acceptable quality level (AQL) The proportion of effective units in a batch which is regarded as desirable by the consumer of the batch (Brownlee, 1965); the complement of the proportion of defectives which one is willing to tolerate.

acceptable reliability level A concept, analogous to **acceptable quality level**, in which the characteristic is expressed in terms of failure rate per unit time, e.g., 1000 hours.

acceptance boundary A generalization or alternative name for **acceptance line**.

acceptance control chart A simple form of quality control chart in which lots are accepted or rejected according to whether plotted values of the mean or range of the lot or batch characteristics fall inside or outside the specified acceptance control limits (Duncan, 1986).

acceptance error An alternative name for β-**error**.

acceptance inspection The inspection of items to determine whether they are acceptable, that is to say, conform to standards required by the intending user.

acceptance line In sequential analysis the graph of the **acceptance number** as ordinate against the sample number as abscissa. It is also known as the **acceptance boundary**. There is a corresponding rejection line.

acceptance number In acceptance inspection schemes, and in sequential analysis generally, the number of non-defective items (dependent on the sample number) which, if attained, requires the acceptance of the batch under examination (Duncan, 1986). It is usually accompanied by a rejection number, which is the number of defectives requiring the rejection of the batch. If the number of defectives at any one stage is above the acceptance and below the rejection number, sampling is continued.

acceptance region In the theory of testing hypotheses, a region in the **sample space** such that, if a sample point falls within it, the hypothesis under test is retained.

acceptance sampling Sampling plans specifying criteria used to accept or reject the batch, based on dichotomous classification (e.g., conformal or non-conformal quality characteristics) or on statistics about the shape and the nature of the sample distribution (Dodge, 1970).

accumulated deviation The fitting of a grouped frequency distribution provides, for each grouping interval, a 'theoretical' or 'expected' frequency. The difference of this quantity and the observed quantity, taken with regard to sign and cumulated for increasing values of the variables starting from the least, is called the accumulated deviation. It is equivalent to the difference between the observed and theoretical distribution functions of the grouped distributions.

accumulated process A stochastic process derived from the cumulated sum of random variables. For example, the total damage during time t of a sequence of thunderstorms forms an accumulated process.

accuracy Accuracy in the general statistical sense denotes the closeness of computations or estimates to the exact or true values as contrasted with precision, which refers to reproducibility.

actuarial estimator An estimated survival distribution obtained from right-censored data grouped into the form of a life table. The **Kaplan–Meier estimator** (Kaplan and Meier, 1958) is essentially the limiting case when the grouping is very fine.

actuarial statistics Statistics used in insurance and pensions, mainly concerning the rate of mortality and morbidity (life actuarial statistic) to compute long-term risk (Mitchell, 1974). Non-life actuarial statistics concern short-term risk related to insurance and are subject to more wide fluctuations.

adaptive inference When data are collected over a period of time, the character of the inferences made at any one point may depend on the nature of previous observations and is then said to be adaptive. In a more restricted sense, the observations from a single sample may themselves determine the type of statistic used in drawing inferences about the parent population.

adaptive kernel estimation Methods of kernel density estimation in which kernel width varies according to the local part of data. See **kernel density estimator**.

adaptive methods Appropriate statistical procedures constructed on the basis of characteristics of a sample (Hogg, 1967).

adaptive optimization A system of process control for physical processes proposed by Box and Jenkins (1962), based upon statistical adaptation as a series of observations proceeds through time. It has a particular application in the field of statistical quality control.

adaptive regression A convex combination of two regression methods (Dodge and Jurečková, 2000) adapted in the sense that the optimal value of the convex combination coefficient, minimizing the resulting asymptotic variance, could be estimated from the observations. The resulting estimator attains a minimum asymptotic variance over all estimators of this kind and for any general distribution with a nuisance scale. See also Harter (1983).

adaptive rejection sampling Rejection sampling with a proposal distribution constructed adaptively using information from previous rejected values.

added variable plot A plot used to detect the need to add an explanatory variable in regression analysis.

additive hazard model A model for survival data in which the hazard function is the sum of a baseline hazard function and a function representing the effect of explanatory variables. It is to be contrasted with the more widely used proportional or multiplicative hazards model in which the functions multiply (Kalbfleisch and Prentice, 2002).

additive model A regression model in which the explanatory variables or functions of them appear additively.

additive property of chi squared The property that if two independent variables U_1 and U_2 are distributed as χ^2 with v_1 and v_2 degrees of freedom respectively, their sum $U_1 + U_2$ is distributed as χ^2 with $v_1 + v_2$ degrees of freedom.

additive (random walk) process A stochastic process with independent increments, that is to say, a process x_t is additive if, for $t_1 < t_2 < \ldots < t_n$, the differences $x_{t_2} - x_{t_1}$, $x_{t_3} - x_{t_2}$, etc., are independent. The expressions 'differential process', 'process with independent increments' are usually confined to the case when the time parameter t is continuous. When the increments are discrete and finite, the process may also be said to be additive; a synonym in this case is **random walk**.

additivity of means The representation of the effect of a series of factors acting separately so that the expected value of a response variable is the sum of separate contributions from each factor, there thus being no interaction.

adequate subset In multiple regression, a subset of the regressors that, on the basis of a suitable significance test, contains as much information about the dependent variable as the complete set (Aitkin, 1974). A minimum adequate subset contains no smaller adequate subset.

Adès distributions A family of discrete distributions introduced by Perry and Taylor (1985) and based on the logarithm of a gamma random variable, raised to a power and then discretized. The variance and mean are approximately related by a power law.

adjusted profile likelihood See **modified profile likelihood**.

admissible decision function In the general theory of statistical decision functions an admissible decision function is one for which it can be shown that there is no decision function which is uniformly better. The criterion defining 'better' is usually stated in terms of **risk functions**.

admissible estimator See **strictly dominated estimator**.

admissible hypothesis Generally, a hypothesis is said to be admissible when it is possible within the conditions of the problem. More specifically, if there is a distribution of known mathematical form depending on k unknown parameters θ then a statistical hypothesis can be stated by specifying any set of θ which is a priori possible. Such hypotheses form a set of admissible hypotheses.

admissible strategy A strategy such that no other strategy exists with at least as good an outcome for each possible state of nature and with a better one for some particular state.

admissible test A test of a hypothesis is said to be admissible if, in respect of a particular class of alternatives, there is no other test of given size with uniformly greater power; for example, if a uniformly most powerful test exists, no other test is admissible.

affine α-resolvability See **resolvable balanced incomplete block design**.

affinity A concept introduced by Bhattacharya (1943) as a measure of divergence or distance between two statistical populations. It provides, for example, a generalization of **Mahalanobis distance**.

age-adjusted rate To make comparisons involving death or disease rates in different situations it is essential to adjust for possible differences in age distribution between the groups under comparison. This can be done either by finding, say, the mortality that would apply if the age-specific rates in the study population applied to some standard population with a known age distribution; this leads to the standardized mortality ratio. Alternatively a direct comparison of the study populations in effect by age may be used.

age-dependent birth and death process A **birth and death process** where the birth (fission) and death (extinction) rates are not constant over time but change in a manner which is dependent upon the age of the individual or homogeneous group concerned. This concept is closely related to the **hazard rate distribution** or **force of mortality**.

age-dependent branching process See **branching process**.

age–period–cohort effects In a study with several birth cohorts, with data at several time periods and on individuals of different ages, all three aspects may affect the response. Disentangling them is impossible without special assumptions because of the intrinsic linear relation between the three features (Clayton and Schifflers, 1987).

age–sex-adjusted rate See **age-adjusted rate**.

age-specific death rate A term generally used in life analysis of human or biological populations to denote the probability of dying in the next unit time period of an entity that had survived for a stated period of time.

age-specific mortality rate An actuarial term for the **hazard function** when applied to human populations.

age-specific rate Age-specific rates are rates of an event for a set of age groups into which the ages in an age range of interest have been partitioned. For example, age-specific rates might be calculated for 5-year age groups 0–4, 5–9, 10–14, and so on.

aggregation A word used to denote the compounding of primary data into an aggregate, usually for the purpose of expressing them in a summary form (Cliff and Ord, 1981). For example, national income and price index numbers are aggregative, as contrasted with the income of an individual or the price of a single commodity.

aggregative cluster analysis A form of hierarchical cluster analysis which starts from the individual observations and generates clusters by combining them by joining the individuals or groups judged most similar.

aggregative index An index number which is constructed by aggregating a number of items (as distinct, for example, from picking out a representative item). In price indices, if p_0 typifies the prices in the base period and p_n those for the current period and q_0 and q_n typify the quantities in the base and current periods respectively, then the two principal forms of aggregative index are

$$\frac{\sum p_n q_0}{\sum p_0 q_0} \quad \text{and} \quad \frac{\sum p_n q_n}{\sum p_0 q_n}$$

where the summation takes place over the commodities to be aggregated. [See **Laspeyres' index**, **Paasche's index**.]

aggregative model The statistical study of an economic system usually involves setting up a model expressing known relations between, or hypotheses concerning, the variables under study. When these 'variables' are themselves constructed from groups of individual variables, as when a price index number is substituted for a set of prices, the model is said to be aggregative.

aging distribution A distribution of times of undesirable events, such as failures of equipment or deaths in a population, when risk of such events grows with time of exploitation or with age respectively. Several families of aging distributions are often used in statistical analysis of reliability and survival data.

agreement, coefficient of Suppose m observers provide paired comparisons for n objects. A coefficient of agreement between the verdicts of the m observers is given by

$$u = \frac{8\Sigma}{m(m-1)(n-1)n} - 1$$

where Σ is the sum of the number of agreements between pairs of judges. The coefficient of agreement is a generalization of Kendall's coefficient of rank correlation (τ), to which it reduces when $m = 2$. [See **Kendall's tau**.]

Aitken estimator An estimator obtained by the method of **generalized least squares**. The vector of expected values is specified by a linear model and the covariance matrix of the observations is proportional to a known matrix **V**, say. There are various extensions in which **V** itself has to be estimated and in which nonlinear relations are specified (Aitken, 1948).

Ajne's A_n test A non-parametric test for the uniformity of a circular distribution, proposed by Ajne (1968). Values close to zero of the test statistic

$$A_n = \int_0^{2\pi} \left(N(\theta) - \frac{n}{2} \right)^2 d\theta$$

lead to acceptance of the hypothesis of uniformity, where $N(\theta)$ is the number of the sample observations $\theta_1, \theta_2, \ldots, \theta_n$ that lie in the semicircle $[\theta, \theta + \pi]$.

Akaike's information criterion (AIC) When a model involving q parameters is fitted to data, the criterion is defined as $-2L_q + 2q$, where L_q is the maximized log-likelihood. Akaike (1969) suggested maximizing the criterion to choose between models with different numbers of parameters. It was originally proposed for time-series models, but has also been used in regression.

aleatory variable See **random variable**.

algorithm A procedure which specifies how an algebraic or numerical quantity is to be found, usually numerically, and usually by iterative application of the procedure, rather than by an explicit formula. Algorithms can be characterized in various ways: for example, by whether they produce the exact answer in a finite number of steps and, more generally, by their speed of convergence and indeed an assurance of convergence. The word algorithm derives from Al-Khwarizmi's (Algorithmi in Latin) name, the inventor of algebra, a ninth-century mathematician.

alias This term is used in two branches of statistical analysis. In connection with design of experiments, when a factorial design is only fractionally replicated, certain comparisons do not distinguish between some of the treatment contrasts; these are said to be aliases. In connection with the spectrum analysis of time series, an analogous concept arises because a single width of the interval of observation does not permit a distinction between certain angular frequencies.

alienation, coefficient of If r is the **product moment correlation** between two random variables the coefficient of alienation is $\sqrt{1 - r^2}$. It is equal to the square root of the **coefficient of non-determination**. The term occurs mainly in psychology and is apparently due to Kelley (1919).

all-possible-subsets regression Methods of variable selection that depend on the examination of all possible subsets of regressor variables (in contrast to stepwise methods).

allele One of two or more gene-types that may occur at a given location in the genes of an individual.

allocation of a sample The way in which sample numbers are assigned to various parts of a population by the sampling plan; for example, for a stratified population it may be decided to allocate the total sample number to the strata in proportion to the numbers of individuals in those strata.

allocation rule In discriminant analysis, a rule for assigning a new observation to one of the populations.

allometric growth Changes in shape associated with different growth rates of parts of an organism. If a set of variables x_j corresponds to the logarithms of measurements of length, the first principal component (the major axis of the ellipsoid defined by the covariance matrix) corresponds to uniform growth, and axes orthogonal to it correspond to changes in shape.

allometry The quantitative study of relationship between variation in size and variation in shape when comparing different species from the same family or from the same species (Gould, 1966).

allowable defects In quality control, the number of allowable defects in a sample is the critical number (designated by the particular scheme of sampling inspection) such that, if this number is exceeded, the whole of the remainder of the inspection-lot must be examined or the lot rejected out of hand. [See also **acceptance number**.]

almost certain event See **almost everywhere**.

almost certainly See **almost everywhere**.

almost everywhere Except on a set of measure zero. In a stochastic context the measure will be a probability measure.

almost stationary A term proposed by Granger and Hatanaka (1964) to cover the situation where the series resulting from wide interval sampling of a non-stationary time series is itself stationary.

almost surely See **almost everywhere**.

alphabet A term used in information theory instead of 'sample space' where the outcome of a trial can be one of a finite number of results denoted by the letters $A, B, C, \ldots$. For example, if the outcome is success or failure, denoted by A and B respectively (or 0 and 1), the experience is drawn from a space or 'alphabet' consisting of two letters.

alternating process A process with two states appearing alternatively, e.g., a technical system can be in working state and in repairing. Lengths of working intervals and repairs are used to characterize reliability of the system (Cox, 1962).

alternating renewal process A special case of **alternating process** where sojourn times are independent random variables. It can be described as a two-state **semi-Markov process** in which the independent intervals of sojourn times appear alternatively.

alternative hypothesis In the theory of testing hypotheses, any admissible hypothesis alternative to the one under test.

amount of information See **information**.

amount of inspection In quality control, the number of items (size of sample) taken from each lot and inspected according to the particular sampling plan being employed.

amplitude For a sinusoidal variation the maximum deviation from the mean. The same definition is sometimes used for any approximately stationary series, although the term may then be used more vaguely to describe typical fluctuations from the mean.

amplitude ratio Some time series exhibit seasonal movements which are regular in phase but vary in amplitude from year to year. The actual amplitude in any year, expressed as a proportion of the average amplitude taken over a long period, is called the amplitude ratio. It affords a measure of departure from normal seasonal variation.

analogue computer A device which simulates physically some mathematical process or relationship, and hence one in which the results of the process can be observed as physical quantities, such as voltage or current. Made largely obsolete by modern developments.

analysis of covariance Extension of **analysis of variance** to include decompositions of sums of products of variables. Applications include calculations of least squares adjustments for effect of concomitant variables or treatment contrasts in experiment design (Huitema, 1980).

analysis of variance The total variation displayed by a set of observations, as measured by the sum of squares of deviation from the mean, may in certain circumstances be separated into components associated with defined sources of variation used as criteria of classification for the observations. Such an analysis is called an analysis of variance, although in the strict sense it is an analysis of sums of squares. Many standard situations can be reduced to the variance analysis form (Scheffé, 1959). [See also **ANOVA table**, **variance component**.]

analytic survey A (sample) survey where the primary purpose of the design is the comparison between sectors or subgroups of the population sampled.

ancillary information This phrase is mostly used in the customary sense of information which is additional or supplementary to the main body of information available. It also occurs in the specialized sense of 'information' conveyed by **ancillary statistics**. [See **supplementary information**.]

ancillary statistic A term used by R.A. Fisher (1925c) to denote a statistic used to recover information about a parameter supplementing the maximum likelihood estimator in cases where the latter does not recover all the information contained in the likelihood function. Later work has emphasized the property of having a distribution independent of parameters.

Anderson, Oskar (1887–1960) A leading member of the Continental School of Statistics, Anderson contributed to a broad range of topics including correlation, time-series analysis, non-parametric methods and sample surveys, econometrics and various other applications in the social sciences (Anderson, 1935).

Anderson–Darling statistic A modified version (W_n^2) of the **Cramér–von Mises test** statistic that compares the fit of the observed cumulative distribution function with the expected one (Anderson and Darling, 1954).

Anderson's classification statistic In the two-population discrimination problem, it is required to allocate an object to one of two populations, π_1 or π_2, based on the observed value of a random vector $\boldsymbol{X}$. Anderson's classification statistic is defined by (Anderson, 1951)

$$W = \left\{ \boldsymbol{x} - \frac{1}{2}\left(\overline{\boldsymbol{x}}_1 + \overline{\boldsymbol{x}}_2\right) \right\}' \Sigma^{-1} \left(\overline{\boldsymbol{x}}_1 - \overline{\boldsymbol{x}}_2\right)$$

where $\overline{\boldsymbol{x}}_1$ and $\overline{\boldsymbol{x}}_2$ denote the sample means and Σ is the pooled covariance matrix based on two samples of observations from π_1, π_2. The observed value $\boldsymbol{x}$ is regarded as coming from π_1 or π_2 if $W > c$ or $< c$ respectively, where c is a constant.

Andrews' Fourier-type plot A means of representing multivariate observations (n observations in p dimensions) in such a way that important qualitative features in the data might be revealed, due to Andrews (1972). He suggests that $x_j = (x_{1j}, x_{2j}, \ldots, x_{pj})$, $j = 1, 2, \ldots, n$, should be represented by the function

$$f_{xj}(t) = \frac{x_{1j}}{\sqrt{2}} + x_{2j}\sin t + x_{3j}\cos t + x_{4j}\sin 2t + x_{5j}\cos 2t + \ldots$$

over the range $(-\pi, \pi)$ for t, so that each point in p-space appears as a curve over this range of t. The method has been shown by Andrews, Gnanadesikan and Warner (1973) to have possibilities in the detection of outliers.

angular transformation A variable transformation expressing a variable Y in terms of a variable X by a trigonometrical formula such as the **arc sine transformation**.

angular variables Variables whose values are expressed as angles.

anisotropic distribution See **isotropic distribution**.

annual (block) maximum method Analysis of statistical extremes based on fitting the generalized extreme-value distribution to maxima of blocks of consecutive time-series data.

ANOVA table A summary of results of an **analysis of variance** displayed in a table. Usually such a table contains the degrees of freedom of the model, the sum of squares, and the mean squares, both for the model and for residuals. It may contain also columns for the **F-tests** and the **p-values**.

Ansari–Bradley dispersion test A distribution-free rank test introduced by Ansari and Bradley (1960) of the equality of the scale parameters of two distributions known to be of identical shape and differing at most in their location and scale parameters. The test is restricted to the case of a common median unless the difference between the medians of the distributions is known.

Anscombe residual If y has a non-normal distribution, Anscombe suggested transforming y to $A(y)$ to give an approximately normal distribution, and calculating $A(y) - A(\hat{y})$, where y is the predicted value (Anscombe and Tukey, 1963). In **generalized linear models**, if the variance is given in terms of the mean by $V = V(\mu)$, the appropriate transformation is $A(\cdot) = \int d\mu/V^{1/3}(\mu)$. The residual may then be standardized by dividing by the square root of the variance of $A(y)$.

ante-dependence model For repeated measurements this model defines a family of covariance structures that are intermediate in complexity between the very simple models and the unstructured model, and the particular structure can be developed in several ways. For example, for the conditional independence structure, the observation at time t conditional on the immediate preceding r observations is independent of the earlier observations.

antimode The value taken by a random variable, if any, for which a frequency distribution has a minimum. The term is usually confined to the case where the minimum is not zero and is a true local minimum.

antithetic transforms Transformations of variables which will bring them into the scope of mutually compensated variations as arise in **antithetic variables**.

antithetic variables A term introduced by Hammersley and Morton (1956) used for pairs of variables negatively correlated and which therefore on averaging produce a mean of relatively high precision. They are used in simulation as a variance-reduction technique.

antitonic regression function See **isotonic regression function**.

aperiodic state See **period of a state**.

approximate degrees of freedom A device for approximating the distribution of statistics (usually essentially positive) with a multiple of a chi-squared variable choosing the multiple and degrees of freedom to match the first two moments of the originating distribution. It leads to the approximate applicability of various standard normal-theory procedures (Satterthwaite, 1941).

approximation error In general, an error due to approximation in numerical calculations as distinct, for example, from an error of observation. More particularly, a rounding error. [See **rounding**.]

Aranda-Ordaz model A generalization of the Box–Cox transformation to deal with survival and some discrete data problems (Aranda-Ordaz, 1983).

arbitrary origin In the calculation of the moments of a frequency distribution, it is often desirable to calculate the moments about some convenient, though arbitrary, origin before transforming them to moments about the arithmetic mean as the origin. Moments about an arbitrary origin are usually written with a prime, μ_r', as distinct from those about the mean which are written without a prime, μ_r.

arc sine distribution A distribution function, occurring in the theory of recurrent events (Feller, 1990), of the form

$$F(x) = \frac{2}{\pi} \arcsin \sqrt{x}, \qquad 0 \leqslant x \leqslant 1.$$

arc sine transformation Any transformation involving the arc sine function, particularly the transformation $y = \arcsin \sqrt{(x/n)}$, and modifications thereof, designed to stabilize the variance of a variable x with a binomial distribution in n trials.

ARCH model See **autoregressive conditional heteroscedasticity model**.

area comparability factor In the analysis of vital statistics it sometimes occurs that, whereas the population and deaths of each age are known for the whole country and its localities for the census year, in subsequent years only the total deaths and populations of the localities are known. The problem of adjusting the local crude **death rates** in these later years for comparisons between localities is met by using an area comparability factor. A common form of this factor is obtained by dividing the average **age-specific death rate** for the whole country in the census year by a similar average for the locality. The corrected death rate for any given locality is obtained by multiplying the crude death rate by the area comparability factor.

area sampling A method of sampling used when no complete frame of reference is available. The total area under investigation is divided into small sub-areas which are sampled at random or by some restricted random process. Each of the chosen sub-areas is then fully inspected and enumerated, and may form a frame for further sampling if desired. The term may also be used (but is not to be recommended) as meaning the sampling of a domain to determine area, e.g., under a crop.

Arfwedson distribution A discrete probability distribution proposed by Arfwedson (1951) for an urn sampling problem where drawings are random with replacement. An urn containing N numbered balls is sampled n times; the probability of achieving v different balls $(v = 1, \ldots, n)$ is

$$F(N, n, v) = \binom{N}{v} \frac{f(n, v)}{N^n},$$

where $f(n, v)$ is the number of series containing v selected numbers.

ARIMA See **autoregressive integrated moving average process.**

arithmetic distribution A discontinuous distribution for which the variable values are equidistant and can be represented as $a \pm mb$, $m = 0, 1, 2, \ldots$. In the multivariate analogue the expression **lattice distribution** is used.

arithmetic mean The arithmetic mean of a set of values $x_1, x_2, \ldots, x_n$ is their sum divided by their number, namely $\sum x_j / n$. In current English usage the word 'arithmetic' is frequently omitted so that where a 'mean' is mentioned the arithmetic mean is to be understood, as contrasted with, say, the **geometric mean.**

ARMA See **mixed autoregressive–moving average process.**

Armitage's chi-squared test for trend In a $2 \times r$ contingency table, if the r columns are ordered and indexed by a score x, a component of χ^2 with 1 degree of freedom can be extracted corresponding to a linear trend in the proportions as x increases. The test was given by Armitage (1957).

Arnold distribution An angular distribution first suggested by Arnold (1941), with probability density function

$$f(\theta; \kappa) = \frac{e^{\kappa \cos 2\theta}}{\pi I(\kappa)}, \qquad 0 < \theta \leqslant \pi,$$

where $I(\kappa)$ is the modified **Bessel function** of the first kind and zero order. The distribution is a modification of a special case of the **von Mises distribution.**

array Usually a rectangular arrangement of numbers effectively in a matrix. In older usage a frequency array was simply a frequency distribution.

arrival distribution The probability distribution of the number of items of a defined type which arrive at a given point per unit time. The concept is particularly appropriate to service situations where congestion can lead to queueing: the arrivals can be at a service point or the end of the queue. [See also **Poisson distribution.**]

ascertainment error See **non-sampling error**.

Aspin–Welch test A test (Aspin and Welch, 1949) for the equality of the means of two normal distributions when the variances of the two distributions may be different, making the Student t-test inapplicable. It is constructed to have a close approximation to the nominal size and is to be contrasted with the **Behrens–Fisher test** which stems from the theory of fiducial probability.

assay See **bioassay**.

assignable variation That part of the variation in, for example, an industrial process which can be attributed to specific causes, such as poorly trained operators, faulty machine settings, substandard raw material, etc. (Duncan, 1986).

associable design A concept proposed by Shah (1960b) that, if there are s **partially balanced incomplete block designs** each with v treatments and b blocks, these will be deemed associable designs if the orthogonal matrix $\mathbf{L}$ is also **canonical**, i.e., that $\mathbf{L}'\mathbf{N}_i\mathbf{N}_j'\mathbf{L}$ is diagonal for all $i, j = 1, 2, \ldots, s$, where $\mathbf{N}$ is the **incidence matrix of design**.

associate class A term in the theory of incomplete block designs in which the treatments under comparison are grouped into classes in such a way that the incidence matrix of the design, specifying the number of times each pair of treatments occurs together in the same block, has a simple form based on the class structure.

association A broad term for the relation between two or more variables treated on an equal footing and as such to be contrasted with the dependence of one variable as a response on others considered as explanatory. The term is applied particularly but not exclusively to qualitative variables.

association analysis An expression sometimes used in surveys of human individuals concerning the association of their responses. It is not the analysis of association in the general sense, see **association**. In particular, if association exists, replies to certain questions may permit prediction of responses to other (usually attitudinal) questions.

association, coefficient of A measure of the degree of association between two attributes. In the notation of the previous item, one such coefficient due to Yule (1900) is

$$Q = \frac{ad - bc}{ad + bc}.$$

Another coefficient is

$$V = \frac{ad - bc}{\left|(a+b)(a+c)(b+d)(c+d)\right|^{1/2}},$$

also referred to as the **Phi-coefficient**. A further coefficient of a similar kind,

$$Y = \frac{1 - \sqrt{bc/ad}}{1 + \sqrt{bc/ad}},$$

is called the coefficient of colligation. The concept of the coefficient of association has been developed by Goodman and Kruskal (1954) especially for the more general case of the $m \times n$ table. [See also **contingency**.]

association scheme In the design of **partially balanced incomplete block** experiments, the arrangement which shows which treatments do, or do not, appear together in a block is called the association scheme.

asymmetrical distribution A distribution which is not symmetrical; that is to say, for which there is no central value a such that $f(x - a) = f(a - x)$, $f(x)$ being the frequency function. [See **skewness**.]

asymmetrical factorial design See **symmetrical factorial design**.

asymmetrical test See **one-sided test**.

asymptotic Bayes procedure A term proposed by Whittle (1965) to define a procedure d such that

$$\lim_{c \to 0} \frac{F(S, d)}{F(S)} = 1$$

where $F(S, d)$ is the expected future loss using sequential procedure d from a base of information S and c is the cost of experimentation.

asymptotic distribution The limiting form of a probability distribution dependent on a parameter, such as sample number or time, as that parameter tends to a limit, usually infinity.

asymptotic efficiency The efficiency of an estimator in the limit as the sample size increases. There are different definitions of asymptotic efficiency, see for example **asymptotic relative efficiency.**

asymptotic normality A distribution dependent on a parameter n, usually a sample number, is said to be asymptotically normal if, as n tends to infinity, the distribution tends to the normal form (Esseen, 1945).

asymptotic relative efficiency An asymptotic measure of relative test efficiency usually considers the limiting efficiency as n increases against alternatives that approach the null hypothesis as n increases. The problem was first investigated by Pitman (1936) for estimators with asymptotically normal distributions, and has since been generalized.

asymptotic standard error An estimate that can be used in conjunction with a point estimate of a parameter of interest to provide approximate confidence intervals based on the normal distribution. More formally, if for each n, often a sample size, t_n estimates θ and s_n is such that $(t_n - \theta)/s_n$ converges in distribution to the standard normal distribution as n tends to ∞, then s_n is an asymptotic standard error. It is not necessarily an approximation to the standard deviation of t_n.

asymptotic test A significance test for which the probability distribution under the null hypothesis is derived via the limit laws of probability theory. Often this involves use of either the normal or chi-squared distributions.

asymptotically $\sqrt{n}$-unbiased estimator An estimator with bias tending to zero quicker than $1/\sqrt{n}$ where n is the sample size. This property is essential if the distribution of the estimator is asymptotically normal. See **asymptotic normality**.

asymptotically locally optimal design A term relating to the design of **bioassay** experiments (Andrews and Chernoff, 1955) where the doses cannot be known exactly and it is desirable to minimize the asymptotic variance of the response parameter.

asymptotically most powerful test A test of significance which remains **uniformly most powerful** as the sample size tends to infinity.

asymptotically stationary A concept of limiting probability associated with certain stochastic processes, in particular Markov chains with stationary transition probabilities.

asymptotically subminimax Robbins (1951) introduced the term asymptotically subminimax for a procedure that performs almost as well as the minimax procedure at the value of maximum loss, and much better elsewhere, when n is large.

asymptotically unbiased estimator See **unbiased estimator**.

atom A distribution is said to have an atom at x if $X = x$ has finite probability.

attack rate In medical statistics, the ratio between the number of new cases of sickness and the population at risk in a unit time period.

attenuation The phenomenon whereby when the explanatory variable in a simple regression relation is measured with a random error independent of the true value, the regression of response on the measured value is flatter than the regression on the true value of the explanatory variable. There is a similar effect for correlation and a more complicated one for multiple and nonlinear regression.

attributable fraction The maximum proportion of a disease in a population that can be attributed to a characteristic or risk factor. It is a measure of association that indicates the proportional change in the incidence of a disease given a change in a linked risk factor. Mathematically, the attributable fraction (expressed as a percentage) is characterized as

$$\frac{P(RR-1)}{P(RR-1)+1}100,$$

where P = proportion of the total population that has the characteristic, and RR = relative risk.

attributable risk The proportion of cases of a disease that can be attributed to an exposure factor. In medical statistics, if I_p is the probability of a disease in the population, and I_n the probability in those not exposed to some contributory factor, the attributable risk associated with the factor is

$$\frac{I_p - I_n}{I_p}.$$

attribute A qualitative characteristic of an individual, usually employed in distinction to a variable or quantitative characteristic. Thus, for human beings sex is an attribute but age is a variable. Often attributes are dichotomous, each member of a population being allotted to one of two groups according to whether he or she does or does not possess some specified attribute; but manifold classification can also be carried out on the basis of attributes, as when individuals are classified as belonging to various blood-groups.

auto-catalytic curve See **growth curve**.

autocorrelation The feature that in a stationary time series values at two time points are correlated, the strength of the correlation depending on the separation of the two points, called the **lag**. It is also called **serial correlation**.

autocorrelation coefficient If ξ_t is a stationary stochastic process with mean m and variance σ^2 the autocorrelation coefficient of order k is defined by

$$\rho_k = \rho_{-k} = \frac{1}{\sigma^2} E(\xi_t - m)(\xi_{t+k} - m)$$

where the expectation relates to the joint distribution of ξ_t, and ξ_{t+k}. In a slightly more limited sense, if x_t is the **realization** of a stationary process with mean m and variance σ^2 the autocorrelation coefficients are given by a similar formula where the expectation is to be interpreted as

$$\lim_{n_2 - n_1 \to \infty} \frac{1}{(n_2 - n_1)} \sum_{j=n_1}^{n_2} (x_{t+j} - m)(x_{t+j+k} - m).$$

In a more limited sense still, the expression is applied to the correlations of a finite length of the realization of a series. Terminology on the subject is not standardized and some writers refer to the latter concept as **serial correlation**, preferring to denote the sample value by the Latin derivative 'serial' and retaining the Greek derivative 'auto' for the whole realization of infinite extent. Analogous expressions, omitting division by σ^2, provide autocovariances. [See also **correlogram**.]

autocorrelation function The correlation in a stationary time series between values time h apart considered as a function of h. The series may be defined in discrete or in continuous time and there is an analogous definition for spatial systems in which h is replaced by the vector separation (h_1, h_2) of the points or in isotropic systems by the distance between them. For a time series $\{x_t\}$ defined for $t = 1, \ldots, n$ the population autocorrelation function is estimated from

$$\frac{\Sigma_{t=1}^{n-h}(x_{t+h} - \bar{x})(x_t - \bar{x})/(n - h)}{\Sigma_{t=1}^{n}(x_t - \bar{x})^2/n}$$

or some minor variant thereof.

autocovariance See **autocorrelation**.

autocovariance function For any **stationary process** the function $\gamma(k) = C(x_{t+k}, x_t)$ where C represents the covariance of the terms in the bracket is known as the autocovariance function. It has an obvious generalization to continuous processes when, subject to existence, k may be continuous.

autocovariance generating function A function of a variable z which, when expanded as a power series in z, yields the autocovariance of a **stationary process**.

automatic interaction detection A method of preliminary data analysis for vectors consisting of one dependent variable and one, or more, categorized predictor variable(s). The method was proposed by Belson (1959) and developed by Morgan and Sonquist (1963).

automatic model selection Model selection procedures which do not need human intervention.

autonomous equations In econometrics, an equation which describes the behaviour of one particular group or sector of the economy, e.g., a demand equation describing only buyer's behaviour is said to be autonomous if it is affected only by changes in the behaviour of that particular group or sector. Equations of this kind are sometimes termed 'structural equations' but this is not be recommended, since such equations may include variables from the whole system.

autoregression The generation of a series of observations whereby the value of each observation is partially dependent upon the values of those which have immediately preceded it, i.e., each observation stands in a regression relationship with one or more of the immediately preceding terms. A scheme of autoregression may be regarded as a **stochastic process**.

autoregression quantile An extension of the regression quantile to the linear autoregressive model, introduced by Koul and Saleh (1995); see **regression quantile**.

autoregression rank scores See **autoregression quantile**.

autoregressive conditional heteroscedasticity (ARCH) model A model for financial time series which generates volatility clustering by allowing the innovation variance to depend on the sum of squares of past observations. Generalized ARCH (GARCH) models allow dependence also on past variance (Shephard, 1996).

autoregressive integrated moving average process; ARIMA process A stochastic process u_t in discrete time that is driven according to the relationship

$$(1 - \phi_1 B - \phi_2 B^2 - \ldots - \phi_p B^p)(1 - B)^d u_t = (1 - \theta_1 B - \theta_2 B^2 - \ldots - \theta_q B^q)\varepsilon_{t'},$$

i.e.,

$$\phi(B)\nabla^d u_t = \theta(B)\varepsilon_t$$

where B is the unit time backward shift operator, $\nabla \equiv 1 - B$ and $\{\varepsilon_t\}$ is a sequence of independent and identically distributed shocks or disturbances with expected value zero (Box and Jenkins, 1976). The non-negative integers (p, d, q) are jointly referred to as the *order* of the ARIMA process. The ARIMA model is a powerful method for describing both stationary and non-stationary time series. A special case is the integrated moving average (IMA) process, where d and q are positive integers and $p = 0$, or the **autoregressive process** when d and q are zero.

autoregressive moving average (ARMA) process See **mixed auto-regressive–moving average (ARMA) process**.

autoregressive process A stochastic process suggested by Yule (1921) for the representation of a system oscillating under its own internal forces which, being damped, are regenerated by a stream of random external shocks. The realization of such a scheme in the form of a series defined at equidistant points of time may be expressed as

$$u_{t+j} = f(u_t, u_{t+1}, \ldots, u_{t+j-1}) + \varepsilon_{t+j}, \tag{1}$$

where ε is a random variable and f represents a functional relationship. In most practical applications this is taken as linear, e.g.,

$$u_{t+2} = \alpha u_{t+1} + \beta u_t + \varepsilon_{t+2}, \tag{2}$$

and the name derives from the fact that this may be regarded as a regression of u_{t+2} on u_{t+1} and u_t. The expression is now used to denote any process of type (2).

autoregressive series A series generated by an **autoregressive process**; the realization of an autoregressive process.

autoregressive transformation If there is autocorrelation in the error term of an autoregressive process it is sometimes possible to transform the original variables to new variables such that the autoregressive scheme in the transformed variables has an uncorrelated error term. This procedure is known as an autoregressive transformation.

auto-spectrum A spectrum resulting from the analysis of a single time series. [See also **cross spectrum**.]

average A familiar but elusive concept. Generally, an 'average' value purports to represent or to summarize the relevant features of a set of values; and in this sense the term would include the median and the mode. In a more limited sense an average compounds all the values of the set, e.g., in the case of the arithmetic or geometric means. In ordinary usage 'the average' is often understood to refer to the **arithmetic mean**.

average amount of inspection In quality control, the average number of items inspected per lot. The expression is used either where the sample size is not fixed, as in sequential analysis, or where all lots not accepted with fixed sample size are separately inspected *in toto* and rectified.

average article run length The **average run length** is the average number of samples taken before a signal showing the need for action is given. The average article run length is the average number of *items* sampled before action is taken.

average corrections (for grouping) Corrections to moments for grouping can be regarded in two ways, according as the end-points of the grouping mesh are treated as located at random on the variable scale or not. If they are located at random, correction terms can be derived as the deviations of 'grouped' moments from the true values averaged over all possible positions of the grouping mesh. These are known as average corrections. There is apt to be confusion with the case where the mesh is not treated as randomly located, owing to the fact that the average corrections have the same form as **Sheppard's corrections** which relate to a fixed grouping but require for their validity conditions of high contact on the terminals of the distribution.

average critical value method A method for assessing the relative efficiency of statistical tests in regression analysis of time series due to Geary (1966). Broadly speaking, the average critical value is equivalent to a power function value of one-half for large sample sizes.

average deviation A synonym of **mean deviation**, not to be recommended.

average extra defectives limit A method of ensuring the effectiveness of continuous sampling plans by adjusting a process that has gone out of control (Hillier, 1964). This effectiveness increases as the number of defectives that slip through uninspected decreases before the plan adjusts to the deterioration in quality.

average inaccuracy A measure proposed by Theil (1965) in a regression context, defined as the expected sum of squares of the squared estimation errors as a fraction of the expected sum of squares of the disturbances estimated.

average of relatives An average, usually in the form of an index number, of a set of relatives, that is to say values obtained as the ratio of a magnitude in the given period to the corresponding magnitude in the base period. In price index numbers, the price relatives are usually weighted by the values either of the base period or of the given period. Where the weights used are the values in the base year the formula reduces to that of **Laspeyres.** If values in the given period are used with a harmonic average the formula reduces to that of **Paasche.**

average outgoing quality level See **average quality protection.**

average outgoing quality limit (AOQL) The greatest percentage of defective items that can be found as an average of outgoing quality (Dodge, 1943). A sampling inspection plan may also be designed with the AOQL as a parameter in the sense of it being the maximum percentage defective that can be accepted.

average quality protection In quality control, a procedure which aims at keeping the proportion of defective items in deliveries of a manufactured product (after inspection and rectification if necessary) at or below some specified limit. This limit, usually expressed as a percentage, is called the **average outgoing quality limit**; it may be given in terms of defective or effective units, e.g., as either 5 per cent defective or 95 per cent effective. The actual proportion is called the average outgoing quality level. In cases where the intention is to control the proportion of defectives in each delivered lot (as distinct from the average of a number of lots) the procedure is called lot quality protection, and the limit is the lot tolerance limit or lot tolerance per cent defective. [See also **consumer's risk, rectifying inspection.**]

average run length (ARL) The average run length of a sampling inspection scheme at a given level of quality is the average number of samples taken between the time when the process commences to run at the stated level and that at which the scheme indicates a change from acceptable to rejectable quality level is likely to have occurred. This concept must not be confused with **average article run length.**

average sample number curve The graph of the **average sample number function,** with the function as ordinate against the process parameter as abscissa.

average sample number (ASN) function In sequential analysis, the expected or average sample number required to reach a decision, considered as a function of the parameter concerning which the decision is to be made; for example, in quality control for defective items, the average number inspected per batch, for acceptance of a batch, as a function of the proportion of defects produced by the manufacturing process (Wald, 1947).

average sample run length An alternative name sometimes used for **average run length** in order to distinguish it from the concept of **average article run length**.

average shifted histograms Non-parametric density estimation by averaging histograms with bins of equal width but offset relative to each other (Scott, 1992).

axial distribution A **spherical distribution** that describes the relative frequencies of the possible orientations in three dimensions of a random axis. Since an axis is a direction without a sense, such a distribution can be defined entirely on a unit hemisphere, or conventionally on a unit sphere by assigning the same probability density to diametrically opposite points.

axis, random A random variable defined on the unit hemisphere (or hemi-hypersphere in two or more dimensions). The most important cases are distributions of planes or undirected lines. [See **Dimroth–Watson's distribution**, **Bingham's distribution**.]

axonometric chart A chart devised for the purpose of representing a solid on a plane surface; a **stereogram**.

B

β-error; beta-error An alternative term for **error of the second kind** or **type II error**.

Bachelier process See **Brownian motion process**.

backcalculation A method proposed by Brookmeyer and Gail (1988) to estimate the incidence of past HIV infection from that of observed AIDS incidence data, assuming the distribution of the incubation period is known. The estimated infection curve is then used to predict future AIDS incidence. The method was developed independently by Isham (1988) who used the term **backprojection**. This has a link with **deconvolution**.

backprojection See **backcalculation**.

backward equations See **Kolmogorov equations**.

backward process A representation of an **autoregressive integrated moving average process** in which the current value of $w_t = \nabla d u_t$ is expressed entirely in terms of future values of the series $\{w_t\}$ and the current and future values of the random disturbance sequence $\{\varepsilon_t\}$. Such a representation is useful in estimating values of the series which occurred before the first observation was made.

BACON algorithm (Blocked Adaptive Computationally-efficient Outlier Nominators). A method for the detection of outliers in multivariate and regression data proposed by Billor, Hadi and Velleman (2000) based on a forward search (as opposed to backward search) method. The method starts with a small basic subset, which is likely to be outlier-free, then continues increasing the subset until a stopping criterion is met. Observations that are not in the final basic subset, if any, are declared outliers.

Bagai's Y_1 statistic A variant of **Wilks's criterion** for testing hypotheses involving multivariate normal distribution proposed by Bagai (1962) in the form

$$Y = \frac{|\mathbf{A}|}{|\mathbf{C}|},$$

where $\mathbf{A}$ and $\mathbf{C}$ are independent sums of product matrices based upon sample observations with degrees of freedom n_1 and n_2 respectively.

bagging The process of creating a discriminant function by averaging discriminant functions computed from bootstrap samples; the name is a contraction of bootstrap aggregation (Breiman, 1994).

bagplot The bagplot is a bivariate generalization of the **box plot**. The graphical representation visualizes the location, spread, correlation, skewness and tails of a bivariate data set.

Bahadur efficiency An asymptotic measure of asymptotic relative performance for sequences of test statistics proposed by Bahadur (1960*a*).

balanced bootstrap Bootstrap resampling in which balance is imposed on the sampling scheme, akin to balanced incomplete block designs (Davison and Hinkley, 1997).

balanced confounding In the design of factorial experiments it is sometimes possible to arrange for different components of interaction to be partially confounded to the same extent. The confounding of these components of interaction is then said to be balanced. [See **partial confounding**.]

balanced differences In a systematic sample of an ordered series the selected units are not located at random and therefore no fully valid estimate of sampling error is possible in general. One method of overcoming this difficulty is to construct artificial strata by dividing the series into 'blocks' of equal length and to regard the members falling within a block as having been chosen at random within that block. If there are only two members in the block an estimate of error is based on their difference. If there are more than two it may be based on more complicated linear functions designed to eliminate systematic error, e.g., for seven members there would be used 'balanced differences' of the form:

$$d = \frac{1}{2}y_1 - y_2 + y_3 - y_4 + y_5 - y_6 + \frac{1}{2}y_7.$$

The number of terms is arbitrary but seven or nine will eliminate most of the systematic component of variation. Similar ideas are applicable to systematic sampling in more than one dimension.

balanced factorial experimental design A generalization by Shah (1960*a*) of work by Bose (1947). The following conditions must be satisfied: (*i*) each treatment is replicated the same number of times; (*ii*) each block has the same number of plots; (*iii*) estimates of contrasts for different interactions are uncorrelated; and (*iv*) complete balance is achieved over each interaction. The last is so only if all normalized contrasts in an interaction are estimated with the same variance.

balanced incomplete block See **incomplete block**.

balanced lattice square See **lattice square**.

balanced repeated replication A form of balanced sampling scheme used for variance estimation in sample surveys.

balanced sample If the mean value of some characteristic is known for a population and the value of the characteristic can be ascertained for each member of a sample, it is possible to choose the sample so that the mean value of the characteristic in it approximates to the parent mean. Such a sample is said to be balanced. The object of balancing is to obtain a sample which is representative of the parent in respect of some other characteristic for which the parental value is not known.

Banach's match problem Matches are taken, one at a time, with equal probability from either of two boxes initially containing N matches (Feller, 1968). The problem is that of determining the distribution of the random variable R denoting the number of matches in the remaining box when one box is found to be empty. One gets

$$\Pr(R = r) = \binom{2N - r}{N} 2^{-2N+r}.$$

band chart When a complex quantity, that is to say a magnitude which is the sum of certain component parts, is recorded for successive intervals of time it is often convenient to show the movements of the total on a chart which also shows, for each point of observation, partition into components. The movement of the intervals representing the components describes bands across the chart which may be coloured or cross-hatched to assist the visual interpretation.

bandit problems Problems in the field of **game theory**. A **two-armed bandit** is a gambling machine giving rewards at random at different, unknown, rates according to the arm chosen, and it is required to find the strategy to maximize the rewards for a series of trials.

bandwidth A tuning constant used in smoothing procedures to control the amount of smoothing induced. It is a generalization of the notion of grouping interval.

bar chart The graphical representation of frequencies or magnitudes by rectangles drawn with lengths proportional to the frequencies or magnitudes concerned. There are various complications which can be incorporated into this simple concept. For example, component parts of a total can be shown by subdividing the length of the bar. Two or three kinds of information can be compared by groups of bars each one of which is shaded or coloured to aid identification. Figures involving increases and decreases can be shown by using bars drawn in opposite directions, above and below a zero line.

Barndorff-Nielsen's formula (p^* formula) A higher-order asymptotic formula for the conditional density of a **maximum likelihood estimator** given an ancillary statistic, involving likelihood quantities. In many cases it is exact or nearly so (Barndorff-Nielsen, 1983).

Bartlett, Maurice Stevenson (1910–2001) British statistician. After reading mathematics at Cambridge, Bartlett worked for some years in agricultural research. He subsequently held professorships at Manchester, University College London and Oxford. His interests in statistical theory and application, including time series, and in stochastic processes were wide-ranging, and just a few of his many influential contributions are listed below.

Bartlett adjustment A correction factor applied to a likelihood ratio test statistic to make its distribution under the null hypothesis closer to the asymptotic chi-squared form. It was proposed in special cases by Bartlett (1937) and its properties more recently studied from various viewpoints as part of the broader interest in higher-order asymptotic theory. The correction factor is calculated by forcing agreement between the mean of the test statistic and the mean of the approximating chi-squared distribution and has the remarkable property of improving agreement of the whole distribution function.

Bartlett correction See **Bartlett adjustment**.

Bartlett–Diananda test An extension (Bartlett and Diananda, 1950) of **Quenouille's test** for the fitting of autoregression schemes to time series.

Bartlett–Lewis model A stochastic point process cluster model introduced by Bartlett (1963) and Lewis (1964). In this model the cluster-member process is assumed to be a segment of a renewal process, initiated by the cluster 'centre', and terminating after some finite number of 'renewals'.

Bartlett relation A differential equation relationship proposed by Bartlett (1949) for a class of stochastic processes (including the **Markov process**) between the characteristic function of a vector variable and the derivative of the cumulant function. The term is also used for identities satisfied by the expected values of higher-order log likelihood derivatives.

Bartlett's collinearity test A test for direction and collinearity of latent roots and vectors in principal component analysis. It was proposed by Bartlett (1951) and developed by Kshirsagar and Gupta (1965). The question at issue is whether some of the roots and associated latent vectors are distinguishable. [See also **multicollinearity**.]

Bartlett's decomposition The **Wishart distribution** was shown by Bartlett (Wishart and Bartlett, 1933) to be comprised of k chi-squared variables and $\frac{1}{2}k(k-1)$ standard normal variables which are independent within as well as between these groups. The decomposition depends upon expressing the matrix form of the Wishart distribution as the product of a (lower) triangular matrix and its transpose.

Bartlett's test An approximate test for the homogeneity of a set of variances from a number of independent normal samples, given by Bartlett (1937).

Bartlett's test of second-order interaction A test of significance proposed by Bartlett (1935a) for the presence of a second-order interaction in a $2 \times 2 \times 2$ contingency table. It is based upon the ratio of the cross-product ratios used to determine the first-order interaction.

base A number or magnitude used as a standard of reference. It may occur as a denominator in a ratio or percentage calculation. It may also be the magnitude of a particular time series from which a start is to be made in the calculation of a new relative series, an **index number**, which will show the observations as they accrue in the future in relation to that of the **base period**.

base period The period of time for which data used as the base of an index number, or other ratio, have been collected. This period is frequently one of a year but it may be as short as one day or as long as the average of a group of years. The length of the base period is governed by the nature of the material under review, the purpose for which the index number (or ratio) is being compiled and the desire to use a period as free as possible from abnormal influences in order to avoid bias.

base reversal test See **time reversal test**.

base weight The weights of a weighting system for an index number computed according to the information relating to the base period instead, for example, of the current period. It is usual in writing formulae to denote the information from the base period with a suffix 0 and that for the given period with a suffix n. A price index weighted according to prices and quantities of the base period, i.e., base weights, might be

$$I_{0n} = \frac{\sum p_0 q_0 \left(P_n / P_0 \right)}{\sum p_0 q_0},$$

where p_0, q_0 are the base weights and summation takes place over the commodities composing the index.

baseline hazard A **hazard function** under some standard conditions, comparison with which is used to assess the effect of explanatory variables on survival time.

basic cell A term proposed by Mahalanobis (1944) to denote the smallest area for which a random variable may be considered to have a sufficiently precise meaning. [See also **quad**.]

Basu's theorem The theorem (Basu, 1955) that a statistic which is sufficient and complete for a parameter θ is distributed independently of any statistic whose distribution does not involve θ.

batch variation In quality control, the variations in a product which is made or examined in batches, as distinct from one which is produced or examined continuously. The batch variation may be made up of variation within each batch, due to the ordinary process of manufacture, and variation between batches, which may also be due to the quality of raw materials used. [See also **interclass variance, intraclass variance**.]

Bates–Neyman model A model proposed for the study in time of slight and severe accidents in which the statistical form is known as the **multivariate negative binomial distribution** (Bates and Neyman, 1952).

bathtub curve Frequency distributions of lifetime that have a high ordinate initially, then a minimum followed by a slow increase (Bury, 1975).

battery of tests A term used in applied psychology for a group of tests to which subjects are submitted. The usual objects of subsequent analysis are either to provide predictions of each subject's aptitude for one or more occupations on the basis of a weighted combination of series in the tests, or else to provide variables whose correlations may subsequently be analysed into group or general factors common to the tests.

Baule's equation A generalization of the Mitscherlich equation, used to obtain estimates for two elements in the soil simultaneously

$$Y = A(1 - b_1 R_1^{x_1})(1 - b_2 R_2^{x_2})$$

where Y represents yield, x_1 and x_2 represent amounts of fertilizers applied, the remaining parameters being qualities of the system (Patterson, 1969).

Bayes, Thomas (1701–1761) The problem of passing from a population to the properties of a sample was one of the first studied in probability. Thomas Bayes, an English non-conformist minister, was the first to solve the inverse problem of passage from the sample to population, using ideas that are widely used today (Bayes, 1763).

Bayes' decision rule The decision rule that minimizes **Bayes' risk** for a given prior distribution of unknown parameters. A decision rule d_0 is said to be a generalized Bayes' decision rule if there exists a measure τ on the parameter space Θ (or a non-decreasing function on Θ if Θ is real), such that

$$\int L(\theta, d) f_x(x|\theta) d\tau(\theta)$$

takes on a finite minimum value when $d = d_0$, where $L(\theta, d)$ is the **loss function**. A rule d_0 is said to be an extended Bayes' decision rule (DeGroot, 1970) if for every $\varepsilon > 0$ there is a prior distribution τ such that the **Bayes' risk**

$$r(\tau, d_0) < \inf_d r(\tau, d) + \varepsilon.$$

Bayes' estimation The estimation of population parameters by the use of methods of inverse probability and in particular of **Bayes' theorem**. If $\Pr(\theta|H)$ denotes the prior probability of θ then the posterior probability of θ is given by

$$\Pr(\theta|x_1, x_2, \ldots, x_n, H) = \Pr(\theta|H) \Pr(x_1, x_2, \ldots, x_n|\theta, H).$$

The parameter θ is estimated by choosing that value which maximizes the posterior probability. If **Bayes' postulate** is invoked, $\Pr(\theta|H)$ is constant and the method is equivalent to the maximization of the likelihood $\Pr(x_1, x_2, \ldots, x_n | \theta, H)$.

Bayes' factor In Bayesian statistics, two rival models may be compared on the basis of the ratio of marginal likelihoods, known as the Bayes' factor. It is a multiple of the ordinary likelihood ratio.

Bayes information criterion (BIC) An information criterion of the form $-2\hat{\ell} + p \log n$, where $\hat{\ell}$ is the maximized log-likelihood for a model with p parameters fitted to n observations. It may be derived using Laplace approximation to the marginal distribution of the data and is used for model choice.

Bayes' postulate The postulate that in determining the prior probabilities for use in Bayesian inference, in the absence of information distinct possibilities should be assigned equal probability. The notion has been controversial since early criticisms by Boole. It is rejected in the personalistic approaches to Bayesian inference but some version is needed in the objectivistic approaches and is currently expressed via a notion of reference priors.

Bayes' risk The minimized value of the expected risk in a **Bayes' solution**.

Bayes' solution In the terminology of statistical decision functions, a decision function which minimizes the average risk relative to some prior probability distribution.

Bayes' strategy A strategy for which the available decision rules can be linearly ordered in terms of the **Bayes' risk**.

Bayes' theorem This theorem (Bayes, 1763) states that if $q_1, q_2, \ldots, q_n$ are a set of mutually exclusive events, the probability of q_r, conditional on prior information H and on some further event p, varies as the probability of q_r, on H alone times the probability of p given q_r and H, namely

$$\Pr(q_r | pH) \propto \Pr(q_r | H) \Pr(p | q_r H).$$

If $q_1, q_2, \ldots, q_n$ are exhaustive the constant of proportionality is

$$\left(\sum_{r=1}^{n} \Pr(q_r | H) \Pr(p | q_r H) \right)^{-1}.$$

In the main application of the theorem, p is an observed event and the q's are hypotheses explaining the event. The three terms in the above expression are then called, in order, the **posterior probability**, the **prior probability** and the **likelihood**. The theorem enables the probabilities of the explaining hypotheses to be determined; this use is called the method of **inverse probability**. The principal difficulty lies in determining, or even defining, the prior probabilities and the resolution of the difficulty by **Bayes' postulate** has occasioned much controversy. This does not, however, affect the theorem, which is a simple consequence of the product law of probability.

Bayesian confidence intervals See **Bayesian intervals**.

Bayesian inference Approaches to statistical inference in which the unknowns to be estimated have a **prior probability** distribution which combined with the information from data produces a **posterior probability** distribution for the target quantities. It is uncontroversial when the prior distribution is empirically based on data and the notion of probability thus ultimately frequency-based. Otherwise there are two broad approaches. In one the prior probabilities and thus also the posterior distribution express personalistic degree of belief. In the other it is usual to choose prior distributions which in some sense are non-committal about the target, thus hopefully leading to answers representing reasoned degrees of belief which will be broadly compelling as summaries of the information provided by the data under analysis.

Bayesian inference using Gibbs sampling (BUGS) A software package allowing Bayesian inference using **Markov chain Monte Carlo** simulation. A related set of routines (CODA) enables checking for convergence.

Bayesian interval Inference for an unknown parameter or for a future observation obtained by selecting from a posterior distribution an interval (or more generally a region) of specified posterior probability. Typically this will be a highest posterior density interval. The term credible interval is sometimes used.

Bayesian network Also known as a belief network, this is a graphically based model that makes the specification of the joint probability distribution of a large set of variables, $X = \{X_1, \ldots, X_n\}$, and the propagation of uncertainty efficient because it takes into consideration the conditional independence information in the joint probability distribution (Castillo, Gutiérrez and Hadi, 1997).

Bayesian survival model A model for survival data in which background information, e.g., in the form of expert clinical opinion or evidence from external studies, is formally included in the model in terms of prior distributions for model parameters.

BCa confidence interval A bootstrap percentile confidence interval making allowance for bias and skewness of the estimator distribution, from which the ABC interval is derived by analytical approximation.

Beall–Rescia generalization of Neyman's distribution The probability generating function of Neyman's **types A, B and C distribution** with parameters λ_1 and λ_2 suggest the general form

$$G(z) = \exp\left\{\lambda_1 \Gamma(\beta + 1) \sum_{i=0}^{\infty} \frac{\lambda_2^{i+1}(z-1)^{i+1}}{\Gamma(\beta + i + 2)}\right\}.$$

This is actually a probability generating function and defines Beall–Rescia's generalization of Neyman's distribution (Beall and Rescia, 1953) with parameters β, λ_1, λ_2, $0 < \beta < \infty$, $0 < \lambda_1 < \infty$, $0 < \lambda_2 < \infty$, over the range $\{0, 1, 2, \ldots\}$.

Bechhofer's indifference zone method A procedure proposed by Bechhofer (1954) for selecting from among k normal populations the one with the highest value of the single unknown parameter, with preassigned probability P of being correct whenever the unknown parameter values are outside a zone of indifference.

Behrens–Fisher problem The problem of comparing the means of samples from normal distributions with both variances unknown (Fisher, 1935b). [See **Behrens–Fisher test**, **Welch's test**, **Satterthwaite's test**.]

Behrens–Fisher test A test of significance for the difference between the means of random samples from two independent normal populations with unequal variances. It is based on the concept of **fiducial inference** and has been the subject of considerable controversy. The test uses the statistic

$$d = \frac{\overline{x}_1 - \overline{x}_2}{(s_1^2/n_1 + s_2^2/n_2)^{1/2}}.$$

The fiducial distribution of this statistic is also a **Bayes' solution**, assuming the usual vague priors for means and variances (Behrens, 1929).

Behrens' method A method for estimating the **median effective dose** of a stimulus based upon **quantal responses**. It is closely allied to the **Reed–Münch method**. It is of restricted validity.

belief network See **Bayesian network**.

bell-shaped curve A symmetrical frequency curve, usually of a continuous frequency distribution, which shows a marked similarity to a vertical section through a bell.

Bellman–Harris process A branching process evolving from an initial individual, in which each individual lives for a random length of time and at the end of its life produces a random number of offspring of the same type. The process was first studied by Bellman and Harris in collaboration (1948, 1952).

Beran's tests A class of tests for uniformity of a circular distribution, proposed by Beran (1968, 1969). If $x_1, x_2, \ldots, x_n$ are a random sample from a continuous circular population, and $f(x)$ is any probability density function on the circle, then the null hypothesis of uniformity is rejected if Beran's statistic

$$B_n = 2 \int_0^{2\pi} \left(\frac{n}{2\pi} - \sum_{i=1}^{n} f(x + x_i) \right)^2 dx$$

is sufficiently large. Special cases of the test include **Watson's U_N^2 test**, **Ajne's A_n test** and the circular **Rayleigh test**. [See **circular distribution**.]

Berge's inequality An inequality of the Tchebychev type for two correlated variables proposed by Berge (1938) as follows:

$$\Pr\left\{|x_1| \geqslant k\sigma_1 \text{ or } |x_2| \geqslant k\sigma_2\right\} \leqslant \frac{1 - (1 - \rho^2)^{1/2}}{k^2}$$

for all $k > 0$ and where ρ is the correlation parameter.

Berkeley Madonna A computer package designed to allow easy modelling of dynamical systems (in biology, chemistry, and so on). It quickly integrates coupled differential equations, and can be used to simulate both deterministic and stochastic models. Models are defined using a simple programming language or a flowchart editor. Parameter space is explored graphically using sliders or in a more systematic fashion using a range of sensitivity analysis tools. Models are fitted to data using least squares, or by optimizing a user-defined function. Further details can be obtained from http://www.berkeleymadonna.com.

Berkson's error model A model for errors of measurement of explanatory variables in which the true values are more dispersed than the observed values. It was motivated originally in an experimental setting but is also sometimes relevant in observational contexts (Berkson, 1950).

Berkson's fallacy See **Yule–Simpson paradox**.

Berksonian line A term, derived from a method proposed by Berkson, to decide the regression line between two variables where the values of the independent variables are set at preassigned levels instead of being measured.

Bernoulli, Jakob (1655–1705) In his most famous monograph, the *Ars conjectandi* (1713), which was only published after his death, Jakob Bernoulli, together with his brother Johann, one of the pioneers of the Leibnizian form of the calculus, transformed Huygens' calculus of expectations to make probability its main concept. He formulated and proved the weak law of large numbers, the cornerstone of modern probability and statistics. In 1690, he introduced the term 'integral'.

Bernoulli distribution Another name for the **binomial distribution**.

Bernoulli trials Sequences of events, such as those given by coin tossing or dice throwing, in which successive trials are independent and at each trial the probability of appearance of a 'successful' event remains constant; the distribution of the total number of successes is then given by the **binomial** or **Bernoulli distribution**.

Bernoulli variation A sampling situation in which members are chosen from a population of attributes such that the probability of occurrence is constant; hence the sampling distribution of occurrences in samples of fixed size is binomial. The term is used in contradistinction to **Lexis** and **Poisson variation**.

Bernoulli walk A random walk on the integers, in which at each step the conditional probabilities of moving up or down by one unit are p and $1 - p$ respectively.

Bernoulli's theorem A theorem propounded by Jakob Bernoulli (1713). Effectively it is a proposition in pure mathematics to the effect that the observed proportional frequency in random drawings of individuals from a population of attributes with constant probability p converges to p in probability; or, to put it another way, the proportional frequency in the **binomial distribution** $B(n, p)$ lying within a range $\pm \varepsilon \sigma$ of the mean value p tends to unity with increasing n however small ε may be, σ being the standard deviation $\sqrt{pq}$. Bernoulli himself seems to have regarded the theorem as something beyond a mathematical proposition, perhaps a justification of a frequency theory of probability.

Bernstein's inequality An inequality of the Bienaymé–Tchebychev type (Savage, 1961). If a distribution has mean a and variance σ^2 and if the absolute moment of order r, ν_r, exists and obeys the inequality

$$\nu_r \leqslant \frac{1}{2} \sigma^2 r! h^{r-2},$$

where h is some constant, then

$$\Pr \{|X - a| > t\sigma\} \leqslant 2 \exp \left(\frac{-t^2 \sigma^2}{2\sigma^2 + 2ht\sigma} \right).$$

[See **Bienaymé–Tchebychev inequality**.]

Bernstein's theorem A form of the **central limit theorem** for dependent variables, given by Bernstein (1927).

Berry's inequality This inequality (Berry, 1941) states that

$$\Pr(S \geqslant t\sigma) < 1 - \Phi(t) + 1.88 \frac{M}{\sigma},$$

where $\Phi(t)$ is the distribution function for a unit normal random variable and S is the sum of n random variables X_i, independent and identically distributed of variance σ_i^2. The X_i are bounded so that

$$|X_i - E(X_i)| \leqslant M_i \text{ and } \max(M_i) = M.$$

Bessel function distribution A frequency distribution which involves Bessel functions. For example, the distribution of the covariance of two normal correlated variables involves a Bessel function of the second kind with imaginary argument (Bhattacharyya, 1942).

Bessel functions Various kinds of analytic functions which appear in some statistical distribution formulae. Bessel function of the first, second, third kind of order ν are respectively defined by

$$J_\nu(z) = \sum_{r=0}^{\infty} \frac{(-1)^r (z/2)^{\nu+2r}}{r!\,\Gamma(\nu+r+1)},$$

$$Y_\nu(z) = \frac{(\cos \nu\pi) J_\nu(z) - J_{-\nu}(z)}{\sin \nu\pi},$$

$$H_\nu(z) = J_\nu(z) \pm Y_\nu(z).$$

Complex linear combinations of $J_\nu(z)$ give modified Bessel functions used in **circular normal distributions** and in certain distributions studied by McKay (1932).

Bessel's correction See **unbiased estimator**.

best asymptotically normal estimator The original name given by Neyman (1949*b*) to a class of estimator which later became known as **regular best asymptotically normal estimator**.

best critical region See **critical region**.

best estimator A method of estimating a population parameter so as to satisfy some formal optimality criterion. Examples of such criteria are of maximizing the likelihood, of minimizing mean square error subject to some constraint such as unbiasedness, and so on. In relatively simple problems the criteria often lead to broadly similar or even identical answers but this is not necessarily the case.

best linear unbiased estimator A linear function of the order statistics available in the sample of information of the variable under analysis for which the variance is minimum and which is also unbiased (David and Neyman, 1938).

beta-binomial distribution A discrete probability distribution of the form

$$p(r) = \binom{n}{r} \frac{B(\alpha+r, n+\beta-r)}{B(\alpha, \beta)} \qquad r = 0, 1, 2, \ldots, n,$$

where $B(a, b) = \int_0^1 t^{a-1} (1-t)^{b-1} dt$ is the beta function. It is the distribution obtained by averaging the probability parameter of a binomial distribution with index n over a beta distribution with parameters α and β.

beta coefficient See **moment ratio**.

beta distribution The term beta distribution is usually applied to the form

$$f(x) = \frac{x^{\alpha-1}(1-x)^{\beta-1}}{B(\alpha, \beta)}, \qquad 0 \leqslant x \leqslant 1, \qquad \alpha, \beta > 0,$$

where $B(\alpha, \beta) = \int_0^1 t^{\alpha-1}(1-t)^{\beta-1} dt$ is the beta function. A second form, sometimes known as a beta distribution of the second kind, is

$$f(y) = \frac{y^{\alpha-1}}{B(\alpha, \beta)(1+y)^{\alpha+\beta}}, \qquad 0 \leqslant y < \infty, \qquad \alpha, \beta > 0.$$

This is easily transformed into the first type by putting $x = 1/(1+y)$. It has also been referred to as an 'inverted' beta distribution. Distributions of the first kind are a special case of the Pearson **type I distribution** and those of the second kind are a special case of the Pearson **type VI distribution** (Johnson, Kotz and Balakrishnan, 1995).

beta probability plot The graph of cumulated distribution function as ordinate against the variable as abscissa for beta (type I or type VI) distributions. For type I the variable has finite range. For type VI it is infinite and is used to test a variance ratio. It does not involve specially ruled paper. The plots can be of two kinds: (i) ordered quantities in the unit range against quantities of the appropriate beta distribution of the first kind (Pearson type I) or (ii) ordered ratios of mean squares against quantities of the appropriate beta distribution of the second kind (Pearson type VI).

between-groups variance See **interclass variance**.

Bhattacharyya bounds A system of lower bounding values to the variance of an estimator where no minimum value exists, due to Bhattacharyya (1946). The bounds depend on a series of derivatives of the likelihood function.

Bhattacharyya's distance A measure of the 'distance' between two populations. If the frequency functions are $f(x)$ and $g(x)$ the distance is given by

$$\arccos \int_{-\infty}^{\infty} \{f(x)g(x)\}^{1/2}\, dx.$$

Analogous expressions can be given for discontinuous and multivariate distributions.

bias Generally, an effect which deprives a statistical result of representativeness by *systematically* distorting it, as distinct from a random error which may distort on any one occasion but balances out on the average. For bias in estimation see **unbiased estimator**.

bias–variance trade-off A term particularly used in non-parametric estimation or classification to signify the choice of a **bandwidth** or other **smoothing** parameter by minimizing an estimated **mean square error**. The trade-off is between estimators or classifiers with low bias but high variance, or low variance but high bias, and the usual goal is to choose the smoothing parameter to minimize the variance plus squared bias.

biased coin design A randomized design in which allocation probabilities depend on past outcomes.

biased estimator See **unbiased estimator**.

biased sample A sample obtained by a biased sampling process; that is to say, a process which induces a systematic component of error, as distinct from random error which balances out on the average. Non-random sampling is often, though not inevitably, subject to bias, particularly when entrusted to subjective judgement on the part of human beings.

biased test A test is said to be biased if it gives a lower probability of rejecting the hypothesis under test H_0 when some alternative hypothesis H_1 is true than when H_0 is true. Expressed in another way, if the hypothesis under test is $\theta = \theta_0$ and the **power function** of the test has a minimum value at a point $\theta \neq \theta_0$ then the test is biased.

Bickel–Hodges estimator See **Hodges–Lehmann one-sample estimator**.

Bienaymé, Irénée-Jules (1796–1878) A disciple of Laplace, Bienaymé proved the **Bienaymé–Tchebychev inequality** some years before Tchebychev, and stated the criticality theorem of **branching processes** completely correctly in 1845. His work on correcting the use of the Duvillard life table is perhaps his greatest achievement as a statistician in the public domain.

Bienaymé–Tchebychev inequality An inequality derived by Bienaymé (1853) and rediscovered by Tchebychev (1867); it is a special case of the more general **Tchebychev inequality**. The inequality is generally stated in the form

$$\Pr\{|X - a| > t\sigma\} \leqslant \frac{1}{t^2},$$

where $E(X) = a$ and $E(X - a)^2 = \sigma^2$ exist and $t > 1$. That is to say, the probability that a variable will differ from its mean by more than t times its standard deviation is at most $1/t^2$ for any probability distribution. The limits so placed are in general rather crude but the inequality is a valuable one in the theory of stochastic convergence (Heyde and Seneta, 1977).

bifactor model A model of factor structure, due to Holzinger (1944). It is supposed that a battery of tests can be analysed into a general factor and a number of mutually exclusive group factors, e.g.,

Test	a	b	c	d
1	x	x		
2	x	x		
3	x		x	
4	x		x	
5	x		x	
6	x			x
7	x			x

The factor 'a' is the general factor with factors 'b', 'c' and 'd' associated with mutually exclusive groups of tests; '*b*', for example, occurs in tests 1 and 2 but not elsewhere.

bilateral exponential This distribution is the convolution of an exponential distribution and its mirrored distribution, i.e., the distribution of $X_1 - X_2$ when X_1 and X_2 are independent and have a common exponential distribution.

bilinear model Representation of a function as a ratio of two linear functions. It is the base in particular of one of the simplest nonlinear time-series models.

bimodal distribution A frequency distribution with two **modes**.

binary data Data in which the response variable takes two values, such as success/failure, dead/alive, off/on (Cox and Snell, 1989).

binary experiment An experiment E for which the possible outcomes x fall into one of two distributions is termed a binary experiment (Birnbaum, 1961b).

binary logistic regression A model for binary responses in which the probability of success is a logistic function typically in the form $\exp(\beta^\top x)/\{1 + \exp(\beta^\top x)\}$ for explanatory variables x and parameters β (Cox and Snell, 1989).

binary longitudinal data Longitudinal data with binary responses.

binary sequence In general, any sequence each number of which can take one of two possible values. In a probabilistic context a series of numbers each of which represents the outcome of an (independent or dependent) Bernoulli trial and hence where the two possible outcomes can be represented as 0 or 1 (Cox, 1970).

Bingham's distribution A spherical distribution introduced by Bingham (1974). It is the conditional distribution of a trivariate normal vector with zero mean and arbitrary covariance matrix given that the length of the vector is unity. If the random vector is denoted by $\mathbf{L} = (l, m, n)$, the probability density function of $\mathbf{L}$ is of the form

$$f(\mathbf{L}; \mu, \mathbf{k}) = \frac{1}{4\pi d(\mathbf{k})} \exp\left\{\mathbf{k}_1(\mathbf{L}'\mu_1)^2 + \mathbf{k}_2(\mathbf{L}'\mu_2)^2 + \mathbf{k}_3(\mathbf{L}'\mu_3)^2\right\}$$

where $\mathbf{k} = (\mathbf{k}_1, \mathbf{k}_2, \mathbf{k}_3)$ is a matrix of constants, and μ_1, μ_2 and μ_3 are three orthogonal normalized vectors; $\mu = (\mu_1, \mu_2, \mu_3)$ and $d(\mathbf{k})$ is a constant which depends only on $\mathbf{k}_1, \mathbf{k}_2$ and $\mathbf{k}_3$.

binomial distribution If an event has probability p of appearing at any one trial, the probability of r appearing in n independent trials is

$$\binom{n}{r} q^{n-r} p^r,$$

where $q = 1 - p$. This is the term involving p^r in the binomial expansion of $(q + p)^n$, which, since it arrays the various probabilities for $r = 0, 1, \ldots, n$, is known as the binomial distribution and is often denoted by $B(n, p)$. It is also known as the Bernoulli distribution after Jakob Bernoulli who gave it in his posthumous *Ars conjectandi* (1713).

binomial index of dispersion A statistic for testing whether a set of samples is homogeneous with respect to some common attribute. If there are k samples of sizes $n_1, \ldots, n_k$ with proportions $p_1, \ldots, p_k$ and p is the mean proportion for all members together, i.e.,

$$p = \frac{\sum_{i=1}^{k} n_i p_i}{\sum_{i=1}^{k} n_i},$$

the index of dispersion is

$$\frac{1}{p(1-p)} \sum_{i=1}^{k} n_i (p_i - p)^2.$$

The significance of the index, as denoting departure from homogeneity, may be tested in the χ^2 distribution with $k - 1$ degrees of freedom. The index is a particular case of the **Lexis ratio**.

binomial probability paper A graph paper with a grid which is specially designed to facilitate the analysis of enumeration data, i.e., data in the form of proportions or percentages from a binomial population (Mosteller and Tukey, 1949). Both the rectangular coordinate axes are graduated in terms of the square root of the variable.

binomial variation Another name for **Bernoulli variation**.

binomial waiting time distribution An alternative name for the **negative binomial distribution**. It refers to the realization of a success for the kth time after $k + x - 1$ independent binomial trials.

bioassay A technique for estimating the potency of drugs, poisons, etc., from their effect on biological material (Finney, 1978).

bioequivalence test A comparative test of pharmaceutical products designed to show that they have essentially the same biological activity, as contrasted with a trial designed to uncover differences.

bioinformatics The field of science growing from the application of mathematics, statistics and computer techniques to the analysis of extremely large biological databases, especially those for the human and other genomes.

biometric functions Functions related to the study of lifetime data. Let X be a lifetime random variable and $F(x)$ its distribution function. The function

$$S(x) = \Pr(X > x) = 1 - F(x)$$

is the probability that an individual attains age x. Other related functions are

$$e(x) = E(X - x \mid X > x) = \int_x^{\infty} \frac{S(t)}{S(x)} dt$$

which represents the expected remaining lifetime of an individual who has attained age x,

$$\mu(x) = \frac{F'(x)}{S(x)} \qquad \text{and} \qquad q(x,y) = \frac{S(x) - S(y)}{S(x)}$$

which respectively represent the force of mortality and the conditional probability of dying in age interval (x, y) given that the individual is alive at age x.

biostatistics Statistics applied to biological problems. In the United States, particularly, it has come to mean primarily statistics in medicine and human biology (Armitage and Colton, 1998).

biplot A scatter diagram of a two-dimensional data array in which both rows and columns are represented by points (Gabriel, 1971). The main applications are in **correspondence analysis** and **principal components** analysis.

bipolar factor In factor analysis, a factor which is positively correlated with some variables (tests) but negatively correlated with others. When such a factor is identified with some recognizable quality, before or after rotation, it is regarded as expressing a property which may have a negative as well as a positive intensity, e.g., cowardice as opposed to bravery, cowardice being regarded as a quality in itself and not as mere absence of bravery.

bipolykays The bivariate k-statistics of a sample; an extension of the concept of **polykays** given by Hooke (1956).

Birnbaum–Raymond–Zuckerman inequality An inequality in multivariate analysis giving the probability that deviations from means lie within a hyperellipse

$$\Pr\left\{ \sum \frac{(X_i - \mu_i)^2}{\sigma_i^2 k_i^2} \geqslant t^2 \right\} \leqslant \frac{1}{t^2} \sum \frac{1}{k_i^2}.$$

If the σ_k's are all equal the surface becomes a hypersphere (Birnbaum, Raymond and Zuckerman, 1947). Another inequality of such a type is due to Berge (1938) who made use of the correlations between variables.

Birnbaum–Saunders distribution A family of life distribution developed (Birnbaum and Saunders, 1969) to describe probabilistically the number of stress cycles that a structure will sustain before a fatigue crack grows beyond a critical value. It has the general form

$$f(x) = \frac{1}{2(2\pi)^{1/2}\alpha^2\beta x^2} \cdot \frac{x^2 - \beta^2}{(x/\beta)^{1/2} - (\beta/x)^{1/2}} \exp\left[-\frac{1}{2\alpha^2} \left(\frac{x}{\beta} + \frac{\beta}{x} - 2 \right) \right]$$

$$x > 0, \quad \alpha, \beta > 0.$$

More succinctly,

$$\frac{1}{\alpha} \left[\left(\frac{x}{\beta} \right)^{1/2} - \left(\frac{\beta}{x} \right)^{1/2} \right]$$

is a unit normal variable.

Birnbaum–Tingey distribution See **Smirnov–Birnbaum–Tingey distribution**.

Birnbaum's inequality A 'Student'-type inequality based on order statistics, derived by Birnbaum (1969). It states that if the random variable X has a probability density function $f(x)$ for which

$$f(\mu - x) = f(\mu + x) \text{ for } x \geqslant 0,$$

and $f(\mu + x)$ is non-increasing for $x \geqslant 0$, then in a sample of $2m + 1$ independent realizations of X,

$$\Pr\{|S| > \lambda\} \leqslant \binom{2m+1}{m-r}\binom{2r}{r}[\lambda(\lambda - 1)]^{-r}\, 2^{-(m+r)}$$

for $\lambda > 1$, where

$$S = \frac{X_{(m+1)} - \mu}{X_{(m+1+r)} - X_{(m+1-r)}}.$$

birth and death process A stochastic process describing the growth and decay of a population the members of which may die or give birth to new individuals (Bartlett, 1955). The types mainly studied have relatively simple laws of reproduction and mortality.

birth, death and immigration process A simple extension of **birth and death process** which takes into account immigration into the system from outside. Emigration may be treated synonymously with 'death'.

birth process A stochastic process describing the size of a population in which individual members may give birth to new members. The expression is often confined to the case where the variable (population) increases only by jumps of amount $+1$, the probability of a jump from n to $n + 1$ in time dt being asymptotically $\lambda_n dt$. Here λ_n may also depend on t and in the simplest case $\lambda_n = n\lambda$. [See also **Poisson process**, **branching process**.]

birth rate The crude birth rate of an area is the number of births actually occurring in that area in a given time period, divided by the population of the area as estimated at the middle of the particular time period. The rate is usually expressed in terms of 'per 1000 of population'. This crude birth rate is capable of considerable refinement according to the various specific viewpoints adopted in the analysis of vital statistics. For example, it may be adjusted to allow for changes in the proportion of the female population in the child-bearing age groups. [See **fertility rate**.]

biserial correlation Originally, a coefficient designed to measure the correlation of two qualities, one of which is represented by a measurable variable, the other a simple dichotomy according to the presence or absence of an attribute. The coefficient usually employed is Pearson's biserial η. Later Pearson (1909) extended the connotation of 'biserial' to the case where both characteristics were dichotomies and proposed a coefficient known as biserial r. This has not come into use, being replaced by **tetrachoric correlation**.

bispectrum The Fourier transform of the third-order moment function of a stationary random process. It is the method of analysing quadratic effects as the spectrum is used for linear problems.

Bissinger distributions Systems of distributions constructed by Bissinger (1963) in the form of summations from a parent distribution. The 'Bissinger system' of distributions is defined by

$$\Pr\{X = k\} = \frac{1}{1 - P_0} \sum_{j=k+1}^{\infty} \frac{1}{j} \Pr\{X = j\}, \ k = 0, 1, 2, \dots.$$

Such distributions are known also as **STER distributions** because the summations are of Sums of Truncated forms of the Expected value of the Reciprocal of a variable having the parent distribution.

bit An abbreviation for binary digit, which is a digit of a number written in the scale of two; for example, the number 2 is expressed as 10, 4 as 100 and 8 as 1000. In modern communication theory the term also refers to a single piece (bit) of information conveyed by an electrical impulse.

bivariate beta distribution A bivariate continuous distribution due to Filon and Isserlis (Pearson, 1923), having the joint probability density

$$h(x, y) = \left\{ \frac{\Gamma(p_1 + p_2 + p_3)}{\Gamma(p_1)\Gamma(p_2)\Gamma(p_3)} \right\} x^{p_1 - 1} y^{p_2 - 1} (1 - x - y)^{p_3 - 1}$$

where $x \geqslant 0, y \geqslant 0, x + y \leqslant 1$ and $p_1, p_2, p_3 > 0$. The distribution can be derived as the joint distribution of

$$X = X_1/(X_1 + X_2 + X_3),$$

$$Y = X_2/(X_1 + X_2 + X_3)$$

where X_1, X_2, X_3 are independent random variables having gamma distributions with index parameters p_1, p_2 and p_3.

bivariate binomial distribution An extension of the binomial distribution to the case where a member can exhibit success or failure in each of two attributes (Aitken and Gonin, 1935).

bivariate Cauchy distribution A bivariate probability distribution with density function

$$f(x_1, x_2) = \frac{a}{2\pi(a^2 + x_1^2 + x_2^2)^{3/2}}, \quad -\infty < x_1, x_2 < \infty, \quad a > 0.$$

bivariate distribution The distribution of a pair of variables x_1 and x_2, written for the continuous case as a density $f(x_1, x_2)$. For discontinuous or grouped data the distribution may be set out in a rectangular array known as a bivariate or correlation table.

bivariate exponential distribution Any bivariate distribution that has exponential marginal distributions. Some are special cases of **bivariate gamma distributions** obtained by setting the index parameters to unity. Others include Gumbel's (1960) bivariate exponential distribution with probability density function

$$h(x, y) = \{(1 + ax)(1 + ay) - a\} e^{-x-y-axy}, \qquad x \geqslant 0, \quad y \geqslant 0, \quad a \geqslant 0,$$

and the bivariate case of the **Marshall–Olkin distribution**.

bivariate F-distribution A bivariate generalization of the F-distribution which arises naturally in the simultaneous testing of linear hypotheses in the analysis of variance. The distribution has probability density function

$$h(x, y) = \Gamma\left(\frac{\nu}{2}\right) \nu_0^{-(1/2)\nu} \prod_{i=0}^{2} \left\{ \nu_i^{\nu_i/2} \middle/ \Gamma\left(\frac{1}{2}\nu_i\right) \right\} \frac{x^{(\nu_1-1)/2} y^{(\nu_2-1)/2}}{\{1 + (\nu_1 x + \nu_2 y)/\nu_0\}^{\nu/2}},$$
$$0 \leqslant x, y < \infty$$

where $\nu = \nu_0 + \nu_1 + \nu_2$, which can be shown to be the joint density of

$$X = \frac{X_1/\nu_1}{X_0/\nu_0} \quad \text{and} \quad Y = \frac{X_2/\nu_2}{X_0/\nu_0}$$

where X_0, X_1 and X_2 are independent central chi-squared variables with ν_0, ν_1 and ν_2 degrees of freedom respectively (Ghosh, 1955).

bivariate gamma distribution Any bivariate distribution with gamma marginal distributions. Examples of such distributions are given by Wicksell (1933), McKay (1934) and Cherian (1941).

bivariate hypergeometric distribution A discrete bivariate distribution, introduced by Isserlis (1914), with joint probability mass function

$$\Pr(x, y) = \frac{\dbinom{Np_1}{x} \dbinom{Np_2}{y} \dbinom{N - Np_1 - Np_2}{n - x - y}}{\dbinom{N}{n}},$$

where $x, y = 0, 1, \ldots, n$ such that $x \leqslant Np_1, y \leqslant Np_2$ and $n - x - y \leqslant N(1 - p_1 - p_2); N, n$ are integers with $n \leqslant N$, $0 < p_1, p_2 < 1$ and $p_1 + p_2 \leqslant 1$.

bivariate logarithmic distribution A discrete bivariate distribution introduced by Khatri (1959). It has joint probability mass function

$$\Pr(x, y) = \frac{(x + y - 1)!}{x! y!} \cdot \frac{p_1^x p_2^y}{-\log(1 - p_1 - p_2)}$$

where $x, y = 0, 1, \ldots$, $x + y \geqslant 1$, $0 < p_1, p_2 < 1$ and $p_1 + p_2 < 1$, and arises as the limit of the $(0, 0)$-truncated **bivariate negative binomial distribution**. Alternatively termed bivariate logarithmic series distribution.

bivariate logarithmic series distribution See **bivariate logarithmic distribution**.

bivariate multinomial distribution See **bivector multinomial distribution**.

bivariate negative binomial distribution A discrete bivariate distribution with joint probability mass function

$$\Pr(x, y) = \frac{(x + y + k - 1)!}{x! y! k!} p_1^x p_2^y (1 - p_1 - p_2)^k,$$

where $x, y = 0, 1, \ldots$, $k > 0$, $0 < p_1, p_2 < 1$ and $p_1 + p_2 < 1$. The conditional distributions $\Pr(x|y)$ and $\Pr(y|x)$ are **negative binomial**. Guldberg (1934) introduced this distribution, and Lundberg (1940) first used it in connection with problems of accident proneness.

bivariate normal distribution Two variables X_1 and X_2 with means μ_1 and μ_2 and variances σ_1^2 and σ_2^2 are distributed in the bivariate normal form if their density function is given by

$$f(x_1, x_2) = \frac{1}{2\pi\sigma_1\sigma_2(1 - \rho^2)^{1/2}} \exp\left[-\frac{1}{2(1 - \rho^2)}\left\{\left(\frac{x_1 - \mu_1}{\sigma_1}\right)^2 - \frac{2\rho(x_1 - \mu_1)(x_2 - \mu_2)}{\sigma_1\sigma_2} + \left(\frac{x_2 - \mu_2}{\sigma_2}\right)^2\right\}\right],$$

where ρ is a parameter not greater than unity in absolute value. For any fixed x_1 (or x_2) the variable X_2 (or X_1) is normally distributed. The parameter ρ is called the correlation parameter (Johnson, Kotz and Balakrishnan, 1995).

bivariate Pareto distribution A distribution, in two different forms, where both marginal distributions are univariate Pareto distributions (Mardia, 1962).

bivariate Pascal distribution Consider an experiment of observing individuals one after another for two characters A_1 and A_2 until exactly k individuals possessing both the characters are observed. Let x_i denote the number of individuals possessing character A_i observed in the experiment, $i = 1, 2$. Then (x_1, x_2) has the bivariate Pascal distribution. This distribution is a special case of the bivariate negative binomial distribution.

bivariate Poisson distribution The joint distribution of two random variables (X, Y) such that $X = U + W, Y = V + W$, where U, V, W are independently Poisson distributed. The marginal distributions of X and Y are of Poisson form; the explicit formula for the joint distribution can be written down but is unenlightening.

bivariate sign test A bivariate analogue of the **sign test** proposed by Hodges (1955) which can be regarded as a two-dimensional walk (Klotz, 1959).

bivariate 'Student' distribution A bivariate generalization of Student's t-**distribution**, introduced by Pearson (1923) for graduation purposes. The distribution has the probability density function

$$h(x, y) = \left\{ 2\pi\sigma_1\sigma_2\sqrt{(1 - \rho^2)} \right\}^{-1} \frac{f}{f - 2}$$
$$\times \left[1 + \frac{1}{(f-2)(1-\rho^2)} \left\{ \frac{x^2}{\sigma_i^2} - \frac{2\rho xy}{\sigma_1\sigma_2} + \frac{y}{\sigma_2^2} \right\} \right]^{-(f+2)/2}$$

where $-\infty < x, y < \infty$, $j > 0$. The random variables X/σ_1 and Y/σ_2 have marginal t-distributions with f degrees of freedom, and X and Y have **product moment correlation** coefficient ρ.

bivariate type II distribution A continuous bivariate distribution with probability density function

$$h(x, y) = \left\{ 2\pi\sigma_1\sigma_2\sqrt{(1 - \rho^2)} \right\}^{-1} \frac{n + 1}{n + 2}$$
$$\times \left[1 - \frac{1}{2(n + 2)(1 - \rho^2)} \times \left\{ \frac{x}{\sigma_1^2} - \frac{2\rho xy}{\rho_1\rho_2} + \frac{y}{\sigma_2^2} \right\} \right]^{n}$$

where $-c\sigma_1 < x < c\sigma_1, -\sigma_2 < y < c\sigma_2, c^2 = 2(n + 2)$ and $n > 0$, introduced by Pearson (1923). The marginal distributions of both X and Y are **type II distributions**.

bivariate uniform distribution Any bivariate probability distribution that has uniform marginal distributions. Examples are the contingency-type uniform distribution, otherwise known as Plackett's uniform distribution, and Morgenstern's uniform distribution. [See **contingency-type distributions** and **Morgenstern distributions**.]

bivector multinomial distribution Suppose an individual can exhibit one set of s_1 attributes and also one of another set of s_2 attributes. Let p_{ij} be the probability that an individual exhibits the ith attribute of the first set and the jth attribute of the second set where $\sum_{i,j} p_{ij} = 1$. In a sample of size n the probabilities of the various possible combinations are given by the expansion of

$$\left(\sum p_{ij} \right)^n.$$

This is the bivector of bivariate multinomial distribution. It may be regarded as a univariate multinomial in the $s_1 s_2$ combinations.

Black–Scholes formula A formula used for the pricing of European options, based on the assumption that stock prices follow a geometric Brownian motion (Black and Scholes, 1973).

Blackwell's theorem A theorem (Blackwell, 1948) in renewal theory which states that, provided the renewal distribution is not a **lattice distribution**, then for every fixed $h > 0$ the mean renewal rate in the interval $[t - h, t]$ tends to the reciprocal of the mean of the renewal distribution as $t \to \infty$. The theorem is also true for **general renewal processes**.

Blakeman's criterion In the regression of y on x the correlation ratio (η^2) measures the variation of means of arrays: the total sum of squares of deviations of means of arrays from the hypothetical regression line is given by $(\eta^2 - R^2)/(1 - R^2)$ where R is the correlation coefficient between the variable y and its value Y yielded by the regression line. A comparison of this quantity with its standard error is called Blakeman's criterion. A test of nonlinearity of regression is nowadays usually carried out by analysis of variance.

blinding A procedure used in randomized clinical trials to obscure the treatment to which the patient has been allocated; see **single-blind** and **double-blind**. Although blinding is considered to be a key principle of good clinical trial design, it is not always feasible, as for example in a trial comparing a surgical procedure with non-invasive therapy.

block A grouping of experimental units in a randomized experimental design chosen sometimes for convenience of implementation but more commonly so that the uncontrolled variation between two units in the same block is likely to be less than that between two units in different blocks. By choosing a suitable design in which treatments are assigned in a way balanced with respect to blocks and using an appropriate analysis precision of the estimated treatment contrasts is enhanced.

Blom's method A general method (Blom, 1958) for constructing unbiased estimators of location and scale parameters when the standardized parent distribution is known together with the exact expectations of its order statistics. The estimators are linear functions of the sample order statistics, and are highly efficient for most parent distributions even with small sample sizes, despite their coefficients being based on the asymptotic covariances of the standardized population's order statistics.

Blum approximation A multi-dimensional extension proposed by Blum (1954) of the **Robbins–Munro process** and **Kiefer–Wolfowitz process**.

Blum–Kiefer–Rosenblatt independence test A test (Blum, Kiefer and Rosenblatt, 1961) slightly different from, but asymptotically equivalent to, **Hoeffding's independence test**.

BMDP An acronym for BioMeDical Package. Historically, it was a well-regarded statistical package offering a range of data analysis programs including basic summary statistics and analysis of variance, as well as more advanced programs for multivariate statistics, survival analysis and time series. Its analytical modules were state of the art in the early 1990s, but it has undergone relatively little statistical development since that time. BMDP was popular in the pharmaceutical industry partly because it offered SAS compatible programs for more advanced statistical techniques unavailable in the original base **SAS** package. BMDP is currently available from Statistical Solutions which also maintains SYSTAT and SOLAS. Further details can be obtained from http://www.statsol.ie/bmdp/bmdp.htm.

Bock's three-component model A model proposed by Bock (1958) for allowing for differences between judges in the method of **paired comparisons**. The observed merits of object A_i in its comparison with A_j by judge k could be represented by

$$Y_{ik(j)} = V_i + \omega_{ik} + Z_{ik(j)}, \quad i,j,=1,2,\ldots,t, \quad i \neq j, \quad k = 1,2,\ldots,n,$$

where the brackets around j indicate that this subscript serves merely as a label. V_i represents the merits of object A_i and ω_{ik} and $Z_{ik(j)}$ are random variables, the former being components peculiar to specific objects and judges.

body mass index (BMI) A derived measure of body size defined as an individual's weight divided by the square of their height (i.e., weight/height^2); it is also known as Quetelet's index.

Bonferroni, Carlo Emilio (1892–1960) His name is attached to the **Bonferroni inequalities** which facilitate the treatment of statistical dependence. Improvements to the inequalities have generated a large literature. One reason for his current lack of recognition (he received honours from within his own country, Italy, but the only one from outside was from the Hungarian Statistical Society) may be the fact that his book (Bonferroni, 1927) was never properly disseminated.

Bonferroni inequality Bonferroni inequalities are of use in setting up simultaneous confidence limits (Galambos, 1975). For example, if E_i is the event that the ith interval covers the ith parameter, and $E_i{'}$ is the complementary event of E_i, then the first Bonferroni inequality states that the intersection of the E's obeys the probability relation

$$\Pr\left\{\bigcap_{i=1}^{k} E_i\right\} > 1 - \sum_{i=1}^{k} \Pr\left\{E_i'\right\}.$$

See **Boole–Bonferroni–Fréchet inequality**.

Bonferroni rule The Bonferroni rule is applied to correct the type I error when several tests have been performed (a **multiplicity** problem). It uses the

Bonferroni inequality to show that if each test is done at a size α/n then the global type I error for the n tests is lower than α.

Boole, George (1815–1864) Boole approached logic in a way reducing it to a simple algebra, incorporating logic into mathematics. He also worked on differential equations, the calculus of finite differences and general methods in probability. His name in probability theory and statistical inference is preserved by **Boole's inequality**, an important tool for dealing with presence of statistical dependence. In the design of electronic circuitry, Boolean algebra is widely used.

Boole–Bonferroni–Fréchet inequality An inequality derived in the greatest generality by Fréchet. If i is a discrete random variable, $0 \leqslant i \leqslant N$ say, with frequency function $p(i)$ and jth factorial moment $\mu_{[j]}$, then the inequality states that if the series on the right-hand side of the well-known identity

$$p(1) \equiv \frac{1}{i!} \sum_{j=0}^{N-i} \mu[i+j]' \frac{(-1)^j}{j!}, \qquad i = 0, 1, \ldots, N$$

is truncated, the resulting error is less in absolute magnitude than the first term omitted and of the same sign. **Bonferroni inequalities** follow as the special case when $i = 0$, and may be written

$$0 \leqslant (-1)^t p(0) - \sum_{j=0}^{t} (-1)^j \frac{\mu_{[j]}}{j!} \leqslant \frac{\mu_{[t]}}{t!}, \qquad t = 1, 2, \ldots, N.$$

Boolean factor analysis A form of factor analysis proposed for fitting factor models to **binary data**.

Boole's inequality An inequality developed by Boole (1854) to give the limits to the frequencies in certain logically defined classes in terms of the frequencies in other classes. It has application in probability theory. One simple form states that if $A_1, A_2, \ldots, A_k$ are compatible events, the probability that at least one occurs is not greater than the sum of the probabilities that each occurs independently of the occurrence of the others:

$$1 - \Pr(\text{Not-}A_1, \text{Not-}A_2, \ldots, \text{Not-}A_k) \leqslant \Pr(A_1) + \Pr(A_2) + \ldots + \Pr(A_k).$$

boosting The iterative improvement of a discriminant function by upweighting observations misclassified at earlier stages.

bootstrap methods Simulation procedures for frequentist inference, in the simplest non-parametric case by sampling from a multinomial distribution putting equal probabilities on the observed data, but encompassing also fully parametric and semi-parametric simulation. The name is derived from the phrase 'to lift oneself by one's bootstraps' (Efron, 1979; Davison and Hinkley, 1997).

bootstrap test A test based on data sets simulated from a fitted null model, typically non-parametric. See **bootstrap methods**.

Borel–Cantelli lemmas Two lemmas, given in a simple case by Borel (1909) and more generally by Cantelli (1917), occurring in the theory of probability. The first lemma, in brief, says that if we have a series of events $A_1, A_2, \ldots$, with probabilities $p_1, p_2, \ldots$ (not necessarily independent) and $\sum_{i=1}^{\infty} p_i$ converges, then from some point onwards it is virtually certain that only a finite number of any A_k will occur. The second says that if the events are independent and $\sum p_i$ diverges, then it is virtually certain that an infinite number will occur. The lemmas are useful in providing proofs for the **law of large numbers**.

Borel–Tanner distribution In queueing theory the distribution, under certain conditions, of the number of customers served before the queue vanishes for the first time. The two effective conditions are r customers in the queue at zero time and constant service time and probability of new customers. The original distribution ($r = 1$) is due to Borel (1942) and was generalized by Tanner (1953) for $r > 1$.

Borges' approximation An approximation to the binomial distribution developed by Borges (1970) using a complex expression for calculating the appropriate percentage point of the **normal distribution**.

Boscovich, Rogerius Josephus (1711–1787) A Jesuit priest from Dubrovnik, Dalmatia, who in 1760 developed a simple geometrical method of fitting a straight line to a set of observations on two variables using (constrained) least sum of absolute deviations (Boscovich, 1757). This procedure was subsequently popularized in an algebraic form by Laplace. [See **least absolute deviation (LAD) regression**.]

Bose, R. C. (1901–1987) Statistician born in India who, after working with **Mahalanobis**, spent much of his life at the University of North Carolina. He made pioneering contributions of elegance, in particular to the mathematical theory of the design of experiments and to coding theory.

Bose distribution The distribution of the variance ratio when the variables are correlated was given by Bose (1934).

Bose–Einstein statistics In statistical mechanics one possible basic assumption concerning states and energy levels is equivalent to supposing that r distinguishable particles are distributed among n cells ($r < n$) in such a way that each of the n^r arrangements is equally probable. This gives rise to Maxwell–Boltzmann statistics. If the particles are indistinguishable there are

$$\binom{n + r - 1}{r}$$

distinguishable arrangements and if these are taken as equally probable there result Bose–Einstein statistics. As a particular case, if not more than one particle may appear in any cell there are

$$\binom{n}{r}$$

equally probable arrangements, and these form the basis of Fermi–Dirac statistics (Feller, 1968).

bounded completeness If, in the definition of completeness $E(h(x)) = 0$ implies that $h(x) = 0$ only for all bounded $h(x)$, $f(x|\theta)$ is called boundedly complete.

Bowley index See **Marshall–Edgeworth–Bowley index**.

box and whisker plot See **box plot**.

Box–Cox transformation A technique (Box and Cox, 1964) for estimating a transformation of a continuous response variable so that the requirements of some simple family of models are more nearly satisfied. In particular, power law transformations may be used to induce agreement with a normal-theory linear model. The transformation is estimated either by maximum likelihood or by an almost equivalent Bayesian analysis using a data-dependent prior. It aims to achieve, if possible, the requisite linear structure with constant variance and normally distributed errors but in general some implicit compromise between these three requirements is involved.

Box–Jenkins model A model for predicting non-stationary time series with or without seasonal variations (Box and Jenkins, 1970). It depends on autoregression of the mth difference of the series, with a moving average residual.

Box–Muller transformation If U_1 and U_2 are random uniform $(0, 1)$ variables,

$$X = (-2 \log U_1)^{1/2} \cos 2\pi U_2$$

and

$$Y = (-2 \log U_1)^{1/2} \sin 2\pi U_2$$

are independent standard normal variables (Box and Muller, 1958). The transformation is sometimes used to generate normal variables for **simulation models**.

box plot A graphical method of displaying data, particularly the distribution of one variable for different values of another, due to Tukey (1977). The 'box' covers the interquartile range, with the median dividing it; an extended region covers a predefined quantile range; outlier data are plotted according to represented quantiles.

Box's test An approximate test for the equality of the variances of a number of populations, proposed by Box (1953). The important feature of the test is its robustness under departures from normality in the populations.

Bradford distribution A distribution with density function

$$f(x) = \{\beta / \log(1 + \beta)\} (1 + \beta x)^{-1}, \quad 0 \leqslant x \leqslant 1, \quad \beta > -1$$

used by Bradford (1948) to represent the distribution of references among different sources (in the field of documentation), x denoting a proportion. It can be regarded as a particular case of Pearson's **type VI** or the **beta distribution** of the second kind.

Bradford Hill's guidelines General conditions under which associations established in an observational study are more likely to be causal. They give general guidance, not criteria. The conditions include monotonicity of effect with level of exposure, magnitude and repeatability of effect, subject-matter plausibility and perhaps most controversially specificity of effect. While usually formulated in an epidemiological setting, they apply more generally (Bradford Hill, 1965).

Bradley–Terry model A model for paired comparison (Bradley and Terry, 1952); it postulates that, if X_i and X_j are the responses to the treatments i and j respectively, then $\Pr(X_i > X_j) = \pi_i / (\pi_i + \pi_j)$ in the comparison of treatments i and j. One interprets $X_i > X_j$ as indicative of preference for treatment i over treatment j and π_i is a parameter indicating relative preference of the ith treatment.

branch and bound methods Computer techniques for solving certain discrete optimization problems. If the possible combinations can be arranged in a tree, in which changes from the root outwards are monotonic, optimization may be possible by evaluating only a small proportion of the full range of possibilities. Statistical applications are concerned with the selection of subsets of variables in regression or multivariate analysis (Land and Doig, 1960).

branching Markov process See **branching process**.

branching process A stochastic process describing the growth of a population in which the individual members may have offspring, the lines of descent 'branching out' as new members are born (Athreya and Ney, 1972). Sometimes referred to as a chain-reaction process. If the lifetime distribution of the individual members, i.e., the waiting time in a particular state, follows the negative exponential form, the (continuous time) process is called a branching Markov process. For a general form of lifetime distribution the term age-dependent branching process is used. [See also **Bellman–Harris process**, **Galton–Watson process**.]

branching renewal process A **renewal process** in which each of a series of primary events generates a subsidiary series of events. The complete process is the superposition of the events in the primary and subsidiary processes.

Brandt–Snedecor method A name sometimes given to one of the formulae for calculating χ^2 from a $2 \times n$ table. If the frequencies in the ith column of the table are a_i and b_i and $p_i = a_i/(a_i + b_i), q_i = 1 - p_i, \bar{p} = \sum a_i /\sum(a_i + b_i) , \bar{q} = 1 - \bar{p}$, the summation taking place over the n columns,

$$\chi^2 = \frac{1}{\bar{p}\,\bar{q}} \left\{ \sum(a_i p_i) - \frac{(\sum a_i)^2}{\sum(a_i + b_i)} \right\}.$$

break point See **change point**.

Breslow–Day test A method based on weighted least squares for assessing the homogeneity of odds ratios across a series of 2×2 tables (Breslow and Day, 1980).

Breslow estimator An estimator of the baseline hazard function in a **Cox proportional hazards model**. This is an extension of the **Nelson–Aalen estimator** to models with explanatory variables.

Brownian bridge The part of a Brownian motion process between two points at which it takes known values.

Brownian motion process An additive **stochastic process** in a real variable X_t defined at time t such that $X_t - X_s$ is normally distributed with zero mean and variances $\sigma^2|t - s|$, where σ^2 is a constant (Freedman, 1971).

Brown–Mood procedure For the usual regression model, Brown and Mood (1951) proposed an estimate of the median line $\alpha + \beta x$ which does not make assumptions of normality and common variance of the errors. Given that $y - \alpha - \beta x$ has the same distribution for all $x, \hat{\alpha}$ and $\hat{\beta}$ are such that

$$\text{median } (y_i - \hat{\alpha} - \hat{\beta}x_i) = 0, \quad x_i \leqslant \text{median } x_i$$
$$\text{median } (y_i - \hat{\alpha} - \hat{\beta}x_i) = 0, \quad x_i > \text{median } x_i.$$

Modifications to this procedure were made by Hogg (1975).

Brown's method A method of predicting time series, using the basic idea of exponential weights, proposed by Brown (1959) and further modified (Brown, 1963). The method is designed to allow for first- and second-order polynomial trends.

Bruceton method An alternative designation for the **'up and down' method** or the **staircase method**.

Brunk's test A non-parametric test of the null hypothesis that a random sample of n independent values $x_1, x_2, \ldots, x_n$ comes from a population with distribution function $F(x)$ (Brunk, 1962). The test is based on the range of the statistics $F(x_{(i)}) - i/(n+1), i = 1, 2, \ldots, n$, where $x_{(i)}$ is the ith **order statistic** of the sample.

brushing scatterplots A technique for studying multivariate data, using interactive computer graphics, introduced by Becker and Cleveland (1987). All possible two-variable scatterplots are displayed, and those points falling in a defined area of one of them (the 'brush') are highlighted in all.

Buffon's needle Buffon (1777) showed that if a needle of length $l < 1$ falls at random on a surface ruled with parallels at unit distance, the probability of intersection is $2l/\pi$. For the three-dimensional case, with planes at unit distance, the probability is $l/2$. These results are important in **stereology**.

bulk sampling The sampling of materials which are available in bulk form. That is to say, it is the population which is in bulk; the term does not mean the drawing of a sample in bulk. Examples of such sampling would be the sampling of a shipment of coal for ash content, or of tobacco for moisture content (Duncan, 1962).

bunch map analysis A graphical technique connected with the now rarely used **confluence analysis** of Frisch (1934). This aims to clarify the choice of explanatory variables in multiple regression by first fitting all possible subsets and representing the outcomes graphically.

Burke's theorem A theorem (Burke, 1956) in queueing theory stating that if a single-server queueing process with exponential interarrival and service-time distributions is stationary, then it follows that the departure process is a homogeneous Poisson process with rate equal to the reciprocal of the mean interarrival time.

Burkholder approximation An extension by Burkholder (1956) to cover certain asymptotic properties of stochastic approximation procedures due to Robbins and Munro and to Kiefer and Wolfowitz. [See also **Robbins–Munro process** and **Kiefer–Wolfowitz process**.]

Burr's distribution A frequency distribution introduced by Burr (1942) where the cumulative distribution has a simple algebraic form

$$F(x) = 1 - 1/(1 + x^c)^k, \qquad x \geqslant 0, \qquad c, k > 0.$$

It is one of a class wherein the distribution function, not the frequency function, is defined in analytical terms.

busy period In queueing theory a maximal closed time interval such that all servers in a queueing system are occupied.

Butler–Smirnov test A test proposed by Smirnov and rediscovered by Butler (1969) to decide whether the unknown **cumulative distribution function** of a data set is symmetric about a given median.

Buys Ballot table A method of tabular presentation of a time series used by Buys Ballot (1847), in his meteorological investigations, for the purpose of investigating periodicities. If, for example, a series is suspected of containing a systematic element with period p then the data are arranged as follows:

$$
\begin{array}{ccccc}
u_1 & u_2 & u_3 & \ldots & u_p \\
u_{p+1} & u_{p+2} & u_{p+3} & \ldots & u_{2p}
\end{array}
$$

for as many rows (m) as there are terms in the series, any terms at the end being neglected. The column totals will emphasize the systematic effect of period p.

byte A group of eight **bits** processed as a unit by computers, usually to represent an alphanumeric character.

C

calibration A procedure for converting the results from a measuring instrument into those that would have been obtained by some standardized 'ideal' instrument. More specifically a regression-like relation between the 'ideal' measurement x as explanatory variable and the first device y as response is used to convert a measurement of y into an estimate of the corresponding x.

call-back The inability of an investigator to make contact with a particular designated sample unit at the first attempt raises certain problems of bias due to **non-response**. One method of dealing with this is for the investigator to 'call back' on one or more occasions in order to establish contact.

Camp–Meidell inequality An inequality of the **Bienaymé–Tchebychev** type in which the limits are tighter, the gain being obtained by imposing additional conditions on the probability distribution. For distributions which are continuous and unimodal the inequality states that

$$\Pr\{|X - \mu_0| > \lambda\tau\} \leqslant \frac{4}{9\lambda^2},$$

where μ_0 is the mode and τ is defined by $\tau^2 = \sigma^2 + (\mu - \mu_0)^2$, σ^2 being the variance and μ the mean. It is a particular case of the **Gauss–Winckler inequality**.

Camp–Paulson approximation An approximation (Camp, 1951), based on a general result by Paulson (1942), to the sum of the first $t + 1$ terms of the point binomial. The sum is expressed as the normal integral for a variable value ξ dependent on t and the parameter p of the binomial.

Campbell's theorem An evaluation of the asymptotic distribution of the sum of the effects of random impulses acting with given intensity on a damped system (Kingman, 1993). The impulses are assumed to occur in a Poisson process with rate λ. Each impulse has a given intensity α and has an effect $\alpha\psi(t)$ after time t has passed. Let $\theta(t)$ be the sum of the effects of all impulses occurring prior to time t. Then for $t \to \infty$ the mean and variance of $\theta(t)$ are respectively

$$\lambda\alpha \int_0^\infty \psi(t)dt, \qquad \lambda\alpha^2 \int_0^\infty (\psi(t))^2 dt.$$

Canberra metric A measure of dissimilarity based on p multivariate observations, introduced by Lance and Williams (1967):

$$d_{ij} = \sum_{k=1}^{p} \frac{|x_{ik} - x_{jk}|}{(x_{ik} + x_{jk})}.$$

It is useful only when the variables are non-negative and have a naturally defined zero.

canonical Represented in some preferably simple standard form.

canonical correlation Given two sets of random variables $x_1, \ldots, x_p$ and $y_1, \ldots, y_q$ the first canonical variables are those linear combinations, one of the x's and one of the y's, with unit variance that have maximal correlation, the value of which is the first canonical correlation. Then the second canonical variables are defined similarly subject to being uncorrelated with the corresponding first canonical variable. And so on, leading in general to $\min(p, q)$ pairs in all. The same formal procedure can be applied if the x variables are fixed explanatory variables rather than random but is then more properly called canonical regression. The analysis depends on the linear algebra of the corresponding population or estimated covariance matrices.

canonical variable A variable derived in **canonical correlation** analysis.

Cantelli's inequality An inequality based upon moments proposed by Cantelli (1910). It is of the **Tchebychev** type.

Cantor-type distributions A thoroughly pathological type of probability distribution for which the distribution function F exists and is continuous but for which the density function does not exist.

Capon test A **normal scores test** for a difference between the scale parameters of two otherwise identical populations, proposed by Capon (1961). The ranks R_i, $i = 1, 2, \ldots, m$, of the members of a random sample $x_1, x_2, \ldots, x_m$ from one of the populations, after combination with a random sample $y_1, y_2, \ldots, y_n$ from the other, are transformed to scores which are defined as the expectations of the squares of the **order statistics** of rank R_i in a random sample of $m + n$ realizations from a unit normal distribution; the sum of these scores is the test statistic.

capture–recapture sampling A method of sampling involving the capture, marking and releasing of a number of animals. At each subsequent trapping, a record is made of all previously tagged animals and the untagged animals are marked. This method was practised by Lincoln (1930). When applied to human populations this sampling method is termed 'multiple-record system' (Seber, 1982).

cardioid distribution A **circular distribution** for which the density function has a cardioid curve as its polar representation. It was introduced by Jeffreys (1948), and has the density function

$$f(\theta) = \frac{1}{2\pi}[1 + 2\rho\cos(\theta - \theta_0)], \qquad 0 < \theta \leqslant 2\pi, \qquad |\rho| < \frac{1}{2}.$$

For $\rho = 0$, it reduces to the **circular uniform distribution**.

Carleman's criterion Sometimes two different distributions have the same set of moments. Criteria are therefore desirable to decide when a set of moments determines a distribution uniquely. One such, advanced by Carleman (1926), states that a set of moments $\mu_1, \mu_2, \ldots$, not necessarily about the mean, determines a distribution uniquely if

$$\sum_{j=0}^{\infty} \frac{1}{(\mu_{2j})^{i/2j}} \qquad \text{(for distributions ranging from } -\infty \text{ to } \infty)$$

or

$$\sum_{j=0}^{\infty} \frac{1}{(\mu_j)^{i/2j}} \qquad \text{(for distributions ranging from } 0 \text{ to } \infty)$$

diverges. This was generalized by Cramér and Wold (1936) to the multivariate case.

Carli's index A simple index number of prices proposed by Carli in 1764 (Diewert, 1987). If the prices of a set of k commodities in the base and given period are respectively $p_{0,1}, p_{0,2}, p_{0,3}, \ldots, p_{0,k}$ and $p_{n,1}, p_{n,2}, p_{n,3}, \ldots, p_{n,k}$, Carli's index number is given by

$$I_{0n} = \frac{1}{k} \sum_{i=1}^{k} \frac{p_{n,i}}{p_{0,i}}.$$

It is thus an unweighted index of **price relatives**.

carrier variable A name given to some numerical quantity for which the different values provide the levels of the corresponding factor in a **factorial experiment**. For example, in an agricultural experiment, one carrier variable might be the quantity of fertilizer applied per acre of land and the levels of the factor might be zero and one, two, and three times the standard dressing.

carry-over effect See **residual treatment effect**.

CART An abbreviation for classification and regression trees. This approach partitions the variable space into hyper-rectangles by building a decision tree on one variable at a time (Breiman *et al.*, 1984).

cartogram A device for displaying statistical information of a descriptive nature by means of a symbol on a map. The symbolism may take various forms according to taste, e.g., dots or circles of varying density, or shading in black and white, or use of a full range of colours. The various forms of cartogram are particularly convenient for portraying data according to geographical distribution.

cascade process A general class of stochastic process arising, *inter alia*, in the study of cosmic rays. In general, the collision between a primary electron and some material substance gives rise to a cascade of secondary electrons, which may generate further cascades, and so on. The process is a member of the class known as **birth and death processes** (Heitler, 1937).

case–cohort design A method of sampling from an assembled epidemiological cohort study (or clinical trial) in which a random sample of the cohort is used as a comparison group for all cases that occur in that cohort (Prentice, 1986).

case–control study Design for the retrospective study of rare events by finding cases, say of a disease, and sampling for comparable controls with retrospective determination of explanatory features. There are typically one, two or three controls per case and the controls may be matched individually with cases or sampled from some relevant population. In econometrics the technique is called choice-based sampling (Breslow and Day, 1980).

case fatality rate The risk of dying during a defined period of time for those individuals who have a specific disease. The case fatality rate is often confused and incorrectly termed the mortality rate in clinical literature. The rate at which individuals with a specific disease die during a period of time. This is a disease-specific death rate and should not be confused with a mortality rate. The case fatality rate is $(I(d)/I(p)) \cdot 100$ (during a specified period of time), where $I(d)$ is the number of individuals whose death is attributed to a specific disease, and $I(p)$ is the number of individuals who have been clinically diagnosed with the same disease.

case–referent study Another name for **case–control study**.

catastrophe theory The mathematical theory of situations in which continuous variation in one variable leads to discontinuities in another. The mathematical formulation (Thom, 1975) has led to some statistical and probabilistic studies, particularly in economics.

categorical data Data consisting of counts of observations falling in different categories. The categories may be purely descriptive, or may have a natural order (Stevens, 1968).

categorical distribution A distribution is said to be categorical if the data are sorted into categories according to some qualitative description rather than by a numerical variable.

categorical variable A variable the possible values of which are qualitatively different levels.

category A homogeneous class or group of a population of objects or measurements. The category may be styled after one of the finite characteristics of the population or according to the limits of measurement for which observations are to be allocated to that category of frequency group. For example, people may be categorized according to 'sex' (male or female) or 'age next birthday' (1–5 years; 6–10; 11–15; 16–20; and so on).

category theory Branch of abstract algebra concerned with mappings between sets. It was used by McCullagh (2000) to elucidate the notion of factor.

Cauchy, Augustin-Louis (1789–1857) A French mathematician and engineer, Cauchy pioneered the study of analysis, both real and complex, and the theory of permutation groups. He also researched in convergence and divergence of infinite series, differential equations, determinants, probability and mathematical physics. Numerous terms in mathematics bear Cauchy's name, e.g., the Cauchy integral theorem, in the theory of complex functions, the Cauchy–Kovalevskaya existence theorem for the solution of partial differential equations, the Cauchy–Riemann equations, Cauchy sequences and Cauchy distribution.

Cauchy distribution The distribution on the real line with density

$$\frac{1}{\pi(1 + x^2)}.$$

Its moments do not exist. Student's t-distribution with 1 degree of freedom is of this form.

Cauchy–Schwarz inequality An inequality proposed by Cauchy (1821) stating that, if a_i, b_i are real numbers, then

$$\left(\sum_{i=1}^{n} a_i^2\right)\left(\sum_{i=1}^{n} b_i^2\right) \geqslant \left(\sum_{i=1}^{n} a_i b_i\right)^2.$$

The corresponding result for integrals was given by Schwarz (1885). Many other inequalities can be derived, e.g., **Hölder's inequality**.

causality Philosophically difficult notion of relation between an explanatory variable and a response. Older discussions were non-statistical and involved some notion of necessary and sufficient condition for the response. There are a number of variants of a statistical definition of causality (Holland, 1986). In one the cause must be in some sense prior to the response and alternative allowable explanations of the statistical independence involved must be excluded. In another there is a notion that the possible cause can conceptually be manipulated with a consequent systematic effect on the response.

cause-specific hazard function In situations of competing risks, the hazard function for failure due to particular causes.

cell frequency When a frequency distribution is classified into categories, univariate or multivariate, the subcategories are sometimes known as cells; the frequency with which observations fall into a particular cell is the cell frequency.

cell model A class of models, in particular, as proposed for crystal growth. Randomly distributed 'seeds' grow at the same rate until the circles or spheres make contact. Growth continues where they are not in contact until the space is filled by a convex system of polygons or polyhedra giving the Dirichlet tessellation of the space corresponding to the random process. The statistical properties of the model, and of the **Johnson–Mehl model**, were studied by Meijering (1953).

censoring The failure to observe a variable totally, its value being replaced by a lower limit (right censoring), an upper limit (left censoring) or an interval of values (interval censoring). Sometimes censoring is by design but more commonly it arises in virtue of limitations of the observational process. Provided censoring is uninformative its presence can usually be incorporated into a statistical analysis.

census The complete enumeration of a population or groups at a point in time with respect to well-defined characteristics: for example, population, production, traffic on particular roads. In some connection the term is associated with the data collected rather than the extent of the collection so that the term **sample census** has a distinct meaning. The partial enumeration resulting from a failure to cover the whole population, as distinct from a designed sample enquiry, may be referred to as an 'incomplete census'.

census distribution A term proposed by Skellam and Shenton (1957) for two event-counting distributions arising in the analyses of renewal processes. In the discrete case they decompose into the sum of **Pascal distributions** and **Poisson distributions** for the continuous case.

centile An abbreviated form of **percentile** not in general use, but frequently found in the statistical literature of psychological and educational testing.

central confidence interval A confidence interval for a parameter θ with lower and upper limits t_1 and t_2 is said to be central if

$$\Pr\{(\theta - t_1) < 0\} = \Pr\{(t_2 - \theta) < 0\}.$$

Such intervals may be called equi-tailed.

central factorial moments Strictly speaking, this expression ought to mean the **factorial moments** about the centre or some central value. Actually the term is sometimes applied to the factorial moments calculated about some point near to a central value as distinct from the ends of the range.

central limit theorem A key theorem in the theory of probability concerning, in its simplest form, the limiting normal form of the sum of a large number of independent and identically distributed random variables of finite variance. There are many generalizations and developments. Its use in statistics is two-fold. Sometimes it provides some justification for the tentative assumption that primary data have a normal distribution. Quite separately it provides directly or indirectly valuable approximations to some of the distributions arising in statistical inference.

central moment Strictly speaking this expression ought to mean (and occasionally does mean) a moment taken about the centre of a distribution, i.e., the mid-point of its range. More usually it signifies a moment about the mean. When the distribution is symmetrical the meanings coincide.

central place theory A theory in economic geography (Christaller, 1933) predicting the positions of cities and smaller communities in an approximately uniform area.

central tendency The tendency of quantitative data to cluster around some random variable value. The position of the central value is usually determined by one of the measures of location such as the **mean, median** or **mode**. The closeness with which values cluster round the central value is measured by one of the measures of dispersion such as the **mean deviation** or **standard deviation**.

centre of location When parameters of location and scale are simultaneously under estimate it is possible to choose an origin such that the maximum likelihood estimators are asymptotically uncorrelated. The origin so defined has been called by Fisher (1922b) the centre of location.

centroid method In factor analysis, a method developed by Burt (1940) and Thurstone (1947) for the extraction of factors. It relies on the idea that, if the random variables (tests) are represented as a set of vectors, a common factor may be represented by a vector which passes through the centroid (centre of gravity) of the terminal points of the set. Such methods are now obsolete.

cepstrum A method proposed by Bogert and Healy (1963) for analysing time series believed to contain lags due to 'echo' effects. A high-resolution spectrum is converted to logarithms and any trend or slow-moving waves eliminated. The spectrum of this smoothed series is termed the cepstrum (kepstrum).

certainty equivalence A principle stated by Simon (1956), subsequently generalized by Theil (1957), concerning prediction and regulation through control rules. If there is no future uncertainty concerning the nature of input series, then in a wide class of cases equivalent control rules are obtained by minimizing a statistically or a time-averaged criterion function.

chain binomial model A model introduced by Greenwood (1931) to describe the development of an epidemic through several generations of cases. The model was developed by Greenwood (1949) introducing the concept of a variable probability of infection and by Bailey (1953, 1956) in specifying probability distribution of p, the chain of infection of the individual.

chain block design A form of partially balanced incomplete block design introduced by Connor and Youden (1954) to deal with situations where the number of treatments considerably exceeds a limited block size. Comparisons within blocks are of high precision so that only one or two replications are needed.

chain graph A sequence of terms such that each term depends in some defined way upon the previous term or terms in the series; for example, the chain relative used in the calculation of index numbers upon the chain-base method. The term chain is also used in connection with stochastic processes where the value at one point is determined by values at previous points apart from a random element; or more exactly, the probability distribution at any point, conditional on certain previous values, is otherwise independent of past history. The most common case is the **Markov chain**.

chain graph model A graphical model which can be represented as a directed acyclic graph between blocks of variables, variables within blocks having only undirected edges (Wermuth and Lauritzen, 1990).

chain index An index number in which the value at any given period is related to a base in the previous period, as distinct from one which is related to a fixed base. The comparison of non-adjacent periods is usually made by multiplying consecutive values of the index numbers, which form a 'chain' from one period to another. For example, if the value of the index for period 2 based on period 1 is I_{12} and that for period 1 based on period 0 is I_{01} the chain index for period 2 based on period 0 is $I_{01} \times I_{12}$ (divided by 100 if the index numbers are based on 100 as the standard). In practice chain index numbers are usually formed from a weighted average of link relatives, i.e. the values of magnitudes for a given period divided by the corresponding values in the previous period.

chain relative See **chain index**. The term is synonymous with link relative.

Champernowne distributions A three-parameter system of distributions suggested by Champernowne (1952) in connection with the distribution of incomes:

$$F(x) = 1 - (1/\theta) \arctan \left[\sin \theta \left/ \left\{ \cos \theta + \left(\frac{x_1}{m} \right)^\alpha \right\} \right. \right]$$

where m is the median of x; θ and α are constants.

changeover trial An alternative name for an experiment or trial using a **crossover design**.

change point A change point is the point in a statistical model at which some change of structure occurs (Shaban, 1980).

change point estimator Estimator of the location of the change of the statistical model (Hinkley, 1970).

change point models Models in which a change in parameters or functional form occurs at some point in a sequence of observations.

change point problem A class of statistical procedures (tests and estimators) for analysis of series of observations obtained at time-ordered points when the model for single observations may or need not change at unknown time point(s) (Carlstein, Muller and Siegmund, 1994). The main task is to decide whether the statistical model changes or does not during the observational period and in case of a change to identify its time location. In other words, the main interests are to test the null hypothesis: 'no change' against an alternative: 'there is a change at some unknown point', and to obtain a point or interval estimator of the change point when its presence is suspected. Usually, the change concerns the distributions of the single observations (change in parameters) or the type of dependency among subsequent observations.

channel degrees of freedom An extension of the concept of **degrees of freedom** to the area of information theory in connection with the number of independent passages through a message channel. The acronym 'cdf' should be resisted in view of its obvious clash with terms like cumulative, or continuous, distribution function.

Chapman–Kolmogorov equations Equations determining the development in time of a Markov stochastic process. Typical special cases determine the probability $p_{ij}(t)$ given that if a process is in state i at time zero it is in state j at time t. There are two forms, forward and backward, corresponding to the notions of decomposing respectively the last and the first steps of the process.

characteristic function The characteristic function of a random variable X is the expected value $E(e^{i\lambda X})$ where t is a real number. This may also be expressed as

$$\phi(t) = \int_{-\infty}^{\infty} e^{i\lambda x} f(x) dx$$

or similar formulae for discontinuous random variable. The expression is often abbreviated to c.f. Similarly, the characteristic function of several random variables $X_1, \ldots, X_n$ is the expected value of $\exp(it_1 X_1 + \ldots + it_n X_n)$. The form of the characteristic function for general n, or a continuous range of t, is the characteristic functional.

characteristic functional A generalization of the **characteristic function**, useful in the study of stochastic processes. It is defined as $E(e^{is})$, where

$$s = \int x(t)\eta(t)dt,$$

$x(t)$ is a stochastic process, and $\eta(t)$ is an arbitrary function of t.

characterization A distribution, or family of distributions $F(x)$, is said to be characterized by a property if the property implies $F(x)$. For example, the independence of mean and variance in random samples characterizes the family of normal distributions.

Charlier distribution A little-used term denoting the family of frequency distributions generated by a Gram–Charlier series. [See also **Gram–Charlier series – type A, B and C**.]

Charlier polynomials The name given to a class of polynomial derived by Charlier (1905b, 1905a) in connection with the **Gram–Charlier series type**

B. If $\gamma(m, x)$ is the Poisson term $e^{-m}m^x/x!$ and ∇ is the operator (backward difference) defined by

$$\nabla\gamma(m, x - 1) = \gamma(m, x) - \gamma(m, x - 1)$$

the polynomial G_r is defined by

$$G_r(m, x) = \frac{(-\nabla)'\gamma(m, x)}{\gamma(m, x)}$$

or equivalently

$$G_r(m, x) = \frac{\frac{d^r}{dm^r}\gamma(m, x)}{\gamma(m, x)}.$$

Chauvenet's criterion A test for an **outlier** observation (Chauvenet, 1864) whereby it is to be rejected if it lies outside the lower and upper $1/(4n)$ points of the null distribution. The chance of wrongly rejecting a non-discordant value is approximately 40 per cent in a large sample.

Chebyshev, Pafnutii Lvovich (1821–1894) See **Tchebychev**.

Chebyshev inequality See **Tchebychev inequality**.

Chernoff–Savage theorems Chernoff and Savage (1958) proved that the asymptotic relative efficiency (ARE), under any density f with finite Fisher information for location, of normal-score rank-based tests for the two-sample location problem with respect to the traditional Student test, is larger than or equal to one, with equality under Gaussian f only. The same result immediately extends to linear models (regression, analysis of variance) with independent observations and error density f. A time-series version was obtained by Hallin (1994), who showed, in the context of ARMA models with innovation density f (finite variance and finite Fisher information for location), that the ARE of tests based on residual rank autocorrelations of the van der Waerden or normal-score type with respect to their classical counterparts based on traditional residual autocorrelations is larger than or equal to one, with equality under Gaussian f only. Everyday-practice correlogram-based methods thus are not admissible.

Chernoff's faces Chernoff (1973) suggested representing multivariate observations by computer-generated faces, each variable controlling the shape or size of some feature. The observations could then be arranged or grouped according to their similarity, with the underlying idea that similarities among faces are more easily recognized than among sets of numbers.

chi distribution The distribution of the positive square root of a random variable having the **chi-squared distribution**.

chi-plot A graphical method for investigation of association between two variables (Fisher and Switzer, 1985). The value plotted is the correlation between binary variables defined by the double dichotomy at each point of the data; the diagram may show relationships not apparent in a **scatterplot**.

chi-squared correction See **Yates's correction**.

chi-squared distribution A distribution first given, apparently, by Abbe and rediscovered by Helmert and Pearson (Lancaster, 1969). Its probability density function is

$$f(x) = \frac{1}{2^{\nu/2}\Gamma\left(\frac{1}{2}\nu\right)} e^{-x/2} x^{\nu/2-1}, \qquad 0 \leqslant x < \infty$$

and the distribution function is an incomplete gamma function $\Gamma_{x/2}(\frac{1}{2}\nu)$. It is a particular case of the Pearson type III distribution. The distribution is that of the sum of squares of ν independent normal random variables in standard form. The parameter ν is known as the number of **degrees of freedom**.

chi-squared metric In a two-way table, the χ^2 distance between rows A and B may be defined as

$$d^2(x_A, x_B) = \sum_{j=1}^{n} \frac{a_{..}}{a_{.j}} \left(\frac{a_{Aj}}{a_{A.}} - \frac{a_{Bj}}{a_{B.}} \right)^2$$

where the dot represents summation and j ranges over the n columns. In a contingency table where a_{ij} are counts of independent events, this is a component of chi-squared. It is also the squared **generalized (Mahalanobis) distance** if rows are represented by a vector of dummy variables. If a_{ij} are non-negative scores (other than counts) it may still be regarded as a suitable **dissimilarity** metric. If so, it justifies the **correspondence analysis** of the table, and the representation of the rows and columns as a **biplot**, using this metric.

chi-squared statistic Originally a statistic for comparing observed frequencies with those fitted under some model via the sum (observed minus fitted frequency)^2/ fitted frequency and referring the test statistic to the **chi-squared distribution**. It is sometimes used more broadly for any test statistic with an approximate chi-squared distribution. Note that in the older literature fitted frequencies are often called expected frequencies (Lancaster, 1969).

chi-squared test A test of significance based upon the chi-squared statistic (Pearson, 1900). Such tests occur in many ways, the most prominent being: (i) an overall **goodness-of-fit** comparison of observed with hypothetical frequencies falling into specified classes; (ii) comparison of an observed with a hypothetical variance in normal samples; (iii) combination of probabilities from a number of tests of significance [see **combination of tests**].

chi-statistic The square root of the more familiar **chi-squared statistic**.

choice-based sampling See **case–control study**.

Cholesky decomposition If $\mathbf{A}$ is a symmetric positive definite matrix, it may be expressed as $\mathbf{A}=\mathbf{LL}'$, where $\mathbf{L}$ is a real non-singular lower triangular matrix. This is the Cholesky decomposition (Benoît, 1924), widely used for solving linear equations and matrix inversion.

Chung–Fuchs theorem A theorem on **martingales** due to Chung and Fuchs (1951). It states that if $Y_1, Y_2, \ldots$ are independent and identically distributed variables with $E(Y) = 0$ and $\Pr\{Y = 0\} < 1$ and if

$$S_n = \sum_{j=1}^{n} Y_i,$$

then $\Pr\{\lim\sup_n S_n = \infty, \lim\inf_n S_n = -\infty\} = 1$.

circular chart See **pie chart**.

circular distribution Distribution of a variable considered as an angle and therefore having range $(0, 2\pi)$ and typically having a density $f(\theta)$ such that $f(0) = f(2\pi)$. The circular uniform distribution is the simple form

$$f(\theta) = 1/(2\pi), \qquad 0 \leqslant \theta \leqslant 2\pi.$$

There are many other special forms (Mardia, 1972). [See **circular normal distribution**.]

circular formula The application of some operations to the terms of an ordered series may present difficulties owing to the fact that end terms have no preceding or succeeding terms. For example, in a series of six terms there are only five first differences but if, for reasons of analytical convenience, it is desired to have six differences then this can be secured by reproducing the first term as a 'pseudo' seventh term. This is equivalent to regarding the series as 'circular' and, hence, any resultant formula in the analysis may be said to be of circular type. The device is used in **serial correlation** analysis and also for proving the arithmetic of the **moving average**. More generally, the same device may be used in a stationary stochastic process by regarding successive elements as arranged in a circle. The process is then called circular.

circular histogram A form of diagrammatic representation for grouped angular data. In contrast to the linear **histogram** where the rectangular blocks representing frequencies are erected on a horizontal base, the circular histogram has the blocks attached to the appropriate intervals on the circumference of a circle. [See **polar-wedge diagram**.]

circular lattice distribution A discrete distribution on the circle with

$$\Pr(\theta = \nu + 2\pi r/m) = p_r, \qquad r = 0, 1, \ldots, m - 1$$

and

$$p_r \geqslant 0, \qquad \sum p_r = 1.$$

The points $\nu + 2\pi r/m$ are equidistant on the circle, and the distribution may be considered to be concentrated on the vertices of an m-sided regular polygon. In particular, if all the weights are equal, i.e.,

$$p_r = \frac{1}{m}, \qquad r = 0, 1, \ldots, m - 1$$

then the distribution is called a discrete circular uniform distribution.

circular mean deviation A measure of angular dispersion. For a sample of directions $\theta_1, \theta_2, \ldots, \theta_n$, the circular mean deviation about a central direction α is defined to be

$$d_0 = \pi - \frac{1}{n} \sum_{i=1}^{n} |\{\pi - |\theta_i - \alpha|\}|.$$

Similarly, the population circular mean deviation is given by

$$S_0 = \pi - E\left(|(\pi - |\theta - \alpha|)|\right)$$

which may be alternatively expressed as

$$S = \int_0^\pi \theta dF(\theta + \alpha) + \int_\pi^{2\pi} (2\pi - \theta) dF(\theta + \alpha).$$

circular mean difference For a sample of n angular observations $\theta_1, \theta_2, \ldots, \theta_n$ a measure of angular dispersion defined as

$$\overline{D}_0 = \frac{1}{n^2} \sum_{i=1}^{n} \sum_{j=1}^{n} \{\pi - |(\pi - |\theta_i - \theta_j|)|\}.$$

For a population with distribution function $F(\theta)$, the circular mean difference is correspondingly defined to be

$$\Delta_0 = \pi - E\{|(\pi - |\theta_1 - \theta_2|)|\}$$

where θ_1 and θ_2 are independently distributed as $F(\theta)$. This expression has the alternative form

$$\Delta_0 = \pi - \int_0^{2\pi} \int_0^{2\pi} |\pi - \theta| \, dF(\xi) dF(\theta + \xi).$$

circular normal distribution This has two quite different meanings. One is the bivariate normal distribution with zero correlation and equal variances, so that the contours of equal density are circles. The other is the special distribution defined for variables considered as angles and having density proportional to $\exp(\kappa \cos \theta)$, the normalizing constant being the Bessel function $I_0(\kappa)$. The distribution is sometimes named after von Mises (1919); R.A. Fisher suggested it because of the simple form of sufficient statistics, this accounting for the apparently unusual naming.

circular quartile deviation A measure of dispersion for angular data. The first and third sample quartile directions Q_1^0 and Q_3^0 are defined as the values of the $(n/4)$th observations from the **median direction** m_0 in the clockwise and the anti-clockwise senses respectively. Then the sample circular quartile deviation is defined as the arc length on the unit circle from Q_1^0 to Q_3^0 which contains M_0. The approach can be extended to define other circular quantiles.

circular range For a sample of angular observations, the length of the smallest arc on the unit circle that encompasses them all.

circular serial correlation coefficient Some of the distributional problems connected with estimated serial correlations are eased by regarding the data as spaced around a circle so that the last observation is next to the first, etc., the definition of the estimates being appropriately modified. Unless the circular form has substantive meaning the use of these coefficients seems unwise.

circular test In the construction of index numbers a decision has to be made as to the period upon which to base the index. If an index for period A based upon period B is I_{BA} and for period B based upon period C is I_{CB}, the circular test, derived by Irving Fisher, requires that the index for period A based upon period C, i.e., I_{CA}, should be the same as if it were compounded of two stages, the calculation of A on B and that of B on C. That is to say, we should have

$$I_{CA} = I_{CB}I_{BA}.$$

A similar argument is applied to comparisons between places. Few index numbers in current use satisfy this test.

circular triads In **paired comparisons** concerning three objects X, Y and Z, if X is preferred to Y, Y to Z and also Z to X the triad XYZ is said (in the terminology of Kendall) to be circular. The circular triad shows inconsistent preferences, in the sense that preferences are not being made consistently on a linear scale, and no ranking is possible.

circular uniform distribution See **circular distribution**.

classification A term used rather loosely, often synonymous with **discriminant analysis** or **cluster analysis**. It may also refer to experimental design in such phrases as **crossed classification**.

classification and regression trees (CART) Classification algorithms based on binary splits of variables, often followed by pruning to reduce the complexity of the resulting decision tree (Breiman *et al.*, 1984).

classification statistic In general, a statistic calculated from a sample for the purpose of assigning the population from which the sample emanated to one of a number of classes. The term is practically synonymous with discriminant function. [See **discriminant analysis**.]

Cliff–Ord tests Procedures proposed by Cliff and Ord (1981) for testing for autocorrelation in processes on a regular lattice.

clinical trials In medical statistics, an experiment comparing the effect of treatments or drugs. [See **single-blind, double-blind, protocol**.]

clipped time series A rather unfortunate expression which does not mean that a series is truncated in time or that its values are truncated according to a random variable. It has been applied to a series where the observations on a continuous random variable are in some way approximated by a discontinuous random variable. Loss of efficiency in parameter estimation is usually compensated by computational saving. An infinitely clipped series is one where all positive values are set at $+1$ and all negative values at -1. [See also **hard clipping**.]

closed-ended question See **open-ended question**.

closed sequential scheme In sequential analysis the sampling usually continues until either an acceptance or a rejection boundary is reached. The sample size is not fixed but in order to avoid having (although perhaps, only rarely) to draw large samples before reaching a boundary it may be desirable to fix an upper limit to the sample size. The scheme is then called 'closed'. In the contrary case it is called 'open'.

closed sequential *t*-tests A system of sequential procedures proposed by Schneiderman and Armitage (1962), and developed in later papers (Myers, Schneiderman and Armitage, 1966), which represent an improvement on the **restricted sequential procedures** and the set of sampling plans by Sobel and Wald (1949). This particular type of plan is sometimes referred to as 'wedge' from the plotted shape of the boundaries.

closeness, in estimation In a sense defined by Pitman (1937*a*), given two estimators, x and y, of a parameter θ, if $\Pr\{|x - \theta| < |y - \theta|\} > \frac{1}{2}$, x is a 'closer' estimator of θ than y. It has been shown (Geary, 1944) that where joint distribution of x and y is normal the criterion of 'closeness' is equivalent to that of '**efficiency**', in the sense that if x is closer than y, $\mathrm{var}(x) < \mathrm{var}(y)$.

cluster A group of contiguous elements of a statistical population, e.g., a group of people living in a single house, a consecutive run of observations in an ordered series, or a set of adjacent plots in one part of a field.

cluster analysis A general approach to multivariate problems in which the aim is to see whether the individuals fall into groups or clusters. There are several methods of procedure; most depend on setting up a metric to define the 'closeness' of individuals.

cluster (point) process A point process exhibiting clustering, as a result of heterogeneity (**Cox process**) or 'contagion', a tendency for events to cluster unrelated to the environment.

cluster randomized trial A trial in which whole clusters of individuals or units, rather than the individuals or units themselves, are randomized to different interventions. Cluster randomized trials are also known as group randomization trials; see the series of articles edited by Donner and Klar (2000).

cluster sampling When the basic sampling unit in the population is to be found in groups or clusters, e.g., human beings in households, the sampling is sometimes carried out by selecting a sample of clusters and observing all the members of each selected cluster. This is known as cluster sampling. If the elements are closely grouped they are said to be compact. If they are almost equivalent to a geographically compact group from the point of view of investigational convenience they are said to be quasi-compact. [See also **elementary unit**.]

clustering (*i*) Clustering of events in space or time, as a result of heterogeneity or interdependence. (*ii*) See **space–time clustering**. (*iii*) Clustering of objects in the space defined by multivariate observations. [See **cluster analysis**.]

coalescent process A stochastic model describing the ancestry of a collection of genes. The model fits into the framework of classical population genetics models in a diffusion time scale (Kingman, 1982).

coarsening An extension of missing data proposed by Heitjan and Rubin (1991) aimed at unifying missing and interval censored data.

Cochran's criterion A criterion proposed by Cochran (1950) to compare proportions in matched samples. In the general case the data are arranged in an $r \times c$ table with each row a matched group and each column a sample. The test criterion is

$$Q = \frac{c(c-1)\sum(T_j - \overline{T})^2}{c(\sum u_i) - (\sum u_i^2)}$$

where T_j is total success in the jth sample (column) and u_i the total number of successes in the ith matched group (row). Q has a limiting distribution of χ^2 with $(c-1)$ degrees of freedom if the true probability of success is the sample in all samples.

Cochran's Q-test See **McNemar's test**.

Cochran's rule A rule proposed by Cochran (1941) for rejecting an 'outlier' sample from k samples with m observations in each: the samples can be groups from one set of data. The assumption is that all are from a single normal population and the rejection criterion is based upon the sample variances. [See also **Grubbs' rule, Thompson's rule, Dixon's statistics**.]

Cochran's test A test due to Cochran (1941), for homogeneity of a set of independent estimates of variance. It is based on the ratio of the largest estimate of variance to the total of all the estimates.

Cochran's theorem A theorem on quadratic forms stated by Cochran (1934). If X_i $(i = 1, 2, \ldots, n)$ are independent standardized normal random variables and Q_j $(j = 1, 2, \ldots, k)$ are quadratic forms in the random variables X_i with ranks n_j $(j = 1, 2, \ldots, k)$ and if

$$\sum_{j=1}^{k} Q_j = \sum_{i=1}^{n} X_i^2,$$

then the necessary and sufficient condition for the Q_j to be independent χ^2 random variables with n_j degrees of freedom respectively is that

$$\sum_{i=1}^{k} n_j = n.$$

coefficient Generally this word has the same meaning as in mathematics, but occasionally it is used to denote a dimensionless statistic, e.g., the moment ratio β_2 as a coefficient of **kurtosis** or the coefficient of product moment correlation. In this sense the word 'index' is also used. For a particular coefficient see under the appropriate name, e.g., for 'coefficient of agreement' see 'agreement, coefficient of'.

coefficient of agreement See **agreement, coefficient of**.

coefficient of alienation See **alienation, coefficient of**.

coefficient of association See **association, coefficient of**.

coefficient of concentration See **concentration, coefficient of**.

coefficient of concordance See **concordance, coefficient of**.

coefficient of consistence See **consistence, coefficient of**.

coefficient of contingency See **contingency, coefficient of**.

coefficient of correlation See **correlation, coefficient of**.

coefficient of determination See **determination, coefficient of**.

coefficient of divergence See **divergence, coefficient of**.

coefficient of excess See **excess, coefficient of**.

coefficient of multiple correlation See **multiple correlation, coefficient of**.

coefficient of multiple partial correlation See **multiple partial correlation, coefficient of**.

coefficient of non-determination See **non-determination, coefficient of**.

coefficient of racial likeness See **racial likeness, coefficient of**.

coefficient of self-similarity See **self-similarity, coefficient of**.

coefficient of total determination See **total determination, coefficient of.**

coefficient of variation See **variation, coefficient of.**

coherence Property of self-consistency, especially as applied to personalistic probability. There it leads to the notion that probabilities elicited via hypothetical gambles should obey the usual rules of probability theory in order to be coherent.

coherency The relationship, in the form of a measure of correlation, between the frequency components of two stochastic processes, or time series. The coefficient of coherence at ω is

$$C(\omega) = \frac{c^2(\omega) + q^2(\omega)}{f_x(\omega) f_y(\omega)}$$

where $c(\omega)$ is the **cospectrum** and $q(\omega)$ the **quadrature spectrum** and $f_x(\omega), f_y(\omega)$ are the spectral densities of the series x and y. The concept is analogous to the square of the correlation coefficient (r). The functions in the expression for $C(\omega)$ always obey the 'coherence inequality'

$$c^2(\omega) + q^2(\omega) \leqslant f_x(\omega) f_y(\omega).$$

coherency principle In Bayesian decision theory, the principle that subjective probabilities obey the ordinary laws of probability and that consistent decisions should be based on these probabilities. The probabilities are often described in terms of the assessment of betting odds.

coherent structure A term proposed by Birnbaum, Esary and Saunders (1961) in connection with reliability of multi-component structures whereby functioning components do not interfere with the functioning of the structure. The components, and the structure, in this model take only one of two states: functioning or failure.

cohort A specific subpopulation whose members are typically linked together through a time factor. However, cohort membership can also be defined by other characteristics. Two examples are a group of individuals who have the same birth date and a group of individuals who work in the same building.

cohort life table In contrast to **current life table**, a table based on a cohort of people born in the same period. Thus for a cohort born in 1900, the death rates at 20 and 80 would be based on death rates observed about 1920 and 1980.

cohort study In medical statistics, a study in which a group of individuals, selected according to occupation, age or geographical area, is followed over a period of time. [See **prospective study.**]

cold deck method See **hot deck method.**

collapsed stratum method A method for estimating the variance of the mean of a stratified sample where the sample is based upon two random selections from each of a number of strata of equal size and the number of non-respondents results in too few completed strata. Two or more strata are amalgamated (collapsed) to form one stratum.

collapsibility A statistical inference, a parameter or a model is said to be collapsible if the results obtained in the joint distribution are the same as those obtained in the marginal one.

collective marks, method of A method, introduced by Dantzig (1957), by which generating functions may be derived directly by probabilistic argument. In particular, some very elegant derivations of important results in queueing theory have been obtained by hypothetically marking each customer independently with probability $1 - x$ and not marking him or her with probability x, and then considering the probability that all customers satisfying a certain criterion are unmarked.

collinearity The situation in regression where columns of a design matrix are collinear or nearly so.

collinearity-influential observations Observations with large relative change in the condition number when they are deleted from the design matrix, X, of a regression model. Collinearity-influential observations can either hide or create collinearity.

colouring Addition of structure to a time series, as distinct from its removal by pre-whitening.

combination of tests The combination of a number of probabilities obtained from tests on different groups of data, undertaken so as to assess the probability of the test as a whole. One such test is based on the fact that if k tests give probabilities $p_1, \ldots, p_k$ the statistic

$$-2 \sum_{i=1}^{k} \log p_i$$

is distributed as χ^2 with $2k$ degrees of freedom, provided that the random variables giving rise to the p's are independent and continuous. In the case of discontinuity various modifications are required.

combinatorial methods Statistical methods based on the structure of the data rather than on distributional assumptions. They include **randomization** techniques, **rank-randomization tests** and the study of **runs**, and among newer techniques, **jackknife** and **bootstrap methods**, and many forms of **robust estimation**.

combinatorial test A test of significance in which the sampling distribution of the test statistic is obtained by the algebra of combinatorial analysis. For example, the nature of the process governing the partitioning of N units into k groups may be tested by counting the number of zero groups. On the hypothesis of equiprobability of occurrence of a unit in any group the probability of r empty groups is

$$\Pr\{r|k, N\} = \frac{\binom{k}{t}}{k^N} \Delta^{k-t} 0^N$$

where $\Delta^{k-t} 0^N$ is the $(k-t)$th leading difference of the Nth power of the natural numbers. [See also **difference of zero**.]

commingling analysis A term arising in genetics for a mixture distribution having two or more components.

common factor In factor analysis the factor(s) are classified according to the way in which they contribute to the variables under analysis. Any factor which appears in two or more random variables is called a common factor. If the factor appears in all the random variables it is called a general factor. If it is common to a group of random variables it is called a group factor. A factor appearing in only one random variable is said to be specific. [See also **general factor**.]

common factor space In one geometrical representation of a multivariate situation the variation is regarded as taking place in a space of which each factor represents a dimension. When there are, say, m common factors and s specific factors the whole factor space is one of $n = m + s$ dimensions. Of this the m-dimensional subspace is the common factor space.

common factor variance In factor analysis, that part of the variance of a random variable which is attributable to the factor or factors which it has in common with other random variables, the remainder being due to specific factors or error terms. It is also known as the communality when expressed as a proportion of the total variance.

communality See **common factor variance**.

communicate Two states, j and k, in a **Markov chain** are said to communicate if j is accessible with positive probability from k and k is accessible with positive probability from j.

communicating class Given a state j of a **Markov chain**, its communicating class $C(j)$ is defined as the set of all states k in the chain which communicates with j.

compact (serial) cluster See **cluster, serial cluster**.

comparative mortality figure The ratio of the standardized death rate to the crude death rate in a standard population. It may also be regarded as a form of **Laspeyres' index**. [See also **standardized mortality ratio**.]

comparative mortality index This is a variant of the **comparative mortality figure** and is a weighted average death rate, where the weights are the mean of the actual (current) population and the standard population both expressed in proportions on a common basis of absolute size. In this sense it is a form of **Marshall–Edgeworth–Bowley index**.

compartment models Models, originally used in biology, in which an ingested substance is taken up in several organs or 'compartments', from which it is excreted at different rates. This type of model, in which the intermediate stages are unobservable, has since been used in other disciplines.

compensating error In general, any error which compensates for other errors. More specifically, a class of error with zero mean (unbiased) and subject to the central limit effect, so that the occurrence of several errors will tend to cancel out and their effect become reduced as the errors cumulate. In this sense the term is not to be recommended.

competing risks This term refers to situations in which failure may occur from one of two or more causes. Applications include problems in industrial statistics, and in actuarial, demographic and medical statistics.

competition process A stochastic process, a two-dimensional extension of the **birth and death process**, proposed by Reuter (1961). Two fields of application are competition between species and the spread of an epidemic. A multivariate generalization was covered by Iglehart (1964).

complete case analysis Treatment of missing data by analysing only cases for which data are complete. Very often power is lost thereby, and the inference may be biased.

complete class (of tests) A class of tests is complete if for every test outside the class there is one within the class which is uniformly better in the sense of a measurable decision function.

complete correlation matrix See **correlation matrix**.

complete linkage clustering A method of cluster analysis in which the distance between two clusters is defined as the greatest distance between a pair of items in the respective clusters.

complete set of Latin squares A set of $q - 1$ **Latin squares** each of size $q \times q$ and in $q - 1$ different alphabets and such that each combination of two letters from different alphabets occurs together just once. They exist when q is a prime power. Their main use is in the construction of other designs.

complete sufficient statistic A sufficient statistic essentially of the same dimension as the unknown parameter. This is expressed technically via the condition that if a function of the sufficient statistic has expectation zero for all values of the unknown parameter then the function is identically zero. The statistical implication is that certain estimation and testing procedures are unique (Lehmann, 1986).

complete system of equations A term used mainly in econometrics to denote the equations determining the behaviour of an economic system or part of such a system. The set of equations is said to be complete when it includes all the determining equations governing the system, or a set from which all equations can be deduced. The point of emphasizing completeness is that in some cases it is possible to estimate parameters occurring in an incomplete set of equations, but the estimators may then be biased (Haavelmo, 1943).

completely balanced lattice square See **lattice square**.

completely randomized design A simple form of experimental design in which the treatments are allocated to the experimental units purely on a chance basis.

completeness If, for a parametric family of univariate or multivariate distributions $f(x|\theta)$, depending on the value of a vector of parameters θ, $h(X)$ is any statistic independent of θ and, for all θ,

$$E\left\{h(X)\right\} = \int h(x)f(x|\theta)dx = 0$$

implies that $h(x) = 0$ identically (save possibly on a set measure zero), then the family $f(x|\theta)$ is called complete.

completeness (of a class of decision functions) A class C belonging to the total space D of decision rules is said to be complete if, given any rule $d \in$ D not in C, there exists a rule $d_0 \in$ C that is better than d. A class C is said to be essentially complete if, given any rule d not in C, there exists a rule $d_0 \in$ C that is as good as d. A class of decision rules is said to be minimal complete if it is complete and no proper subclass of it is complete. A class is said to be minimal essentially complete if it is essentially complete and no proper subclass of it is essentially complete.

complex demodulation A variant of the technique of **demodulation** where the basic series or stochastic process is multiplied by a complex non-random function.

complex Gaussian distribution A univariate complex Gaussian distribution is based upon a random variable $z = x + iy$ where the real and imaginary parts are each in a p random variable Gaussian (i.e., normal) form (Goodman, 1963). This concept is important in the spectrum analysis of multiple time series and in some signal processing problems.

complex table A table which shows the classification of a set of data according to more than two different features, as distinct from the one or two features of the simple table. For example, a human population might be tabulated in a complex table according to age, civilian status and sex. The complexity lies not only in the manifold nature of the classification, but in the difficulty of printing the results in a convenient form. This method of tabulation is sometimes referred to as multiple cross-classification.

complex Wishart distribution A form of the **Wishart distribution** required in the analysis of the **complex Gaussian distribution**.

component analysis See **principal components**.

component bar chart A **bar chart** which shows the component parts of the aggregate represented by the total length of the bar. These component parts are shown as sections of the bar with lengths in proportion to their relative size. Visual presentation can be aided by devices of cross-shading or colours.

component of interaction See **interaction**.

component of variance See **variance component**.

composed Poisson distribution A term introduced by Janossy, Renyi and Aczel (1950) to describe the class of distributions now more commonly known as **Hermite distributions**.

composite estimator A weighted combination of two (or more) component estimators. In sample surveys, when appropriate weights are used, its mean square error is smaller that that for either component estimator.

composite hypothesis A composite statistical hypothesis may be defined as any statistical hypothesis which is not simple. This is not entirely satisfactory and the expression often refers to a hypothesis which is a simple hypothesis about one parameter in the presence of **nuisance parameters**.

composite index number A rather vaguely defined term relating to an index number for which the component series are from groups which are different in nature. The definition is somewhat arbitrary in practice since much depends upon the point of view of both the compiler and user of the index. For example, an index number of retail prices would not be regarded as composite from the point of view of a general analysis of the national economy in which 'price' was a single element but it would be regarded as composite by, say, a trade organization operating in only one retail market. In a slightly different sense a national index of production or of business activity is said to be composite at the national level and also composite geographically at a regional level. To be logical, any index number compiled from more than one homogeneous commodity should be called composite, but the expression has its practical uses.

composite sampling scheme A scheme in which different parts of the sample are drawn by different methods; for example, a sample of a national population might be taken by some form of area sampling in rural districts and by a random or systematic method in urban districts.

compositional data Multivariate observations constrained to add to unity, usually the proportions of all the constituents of a mixture. The constraint implies linear dependence and induces negative correlations, and special techniques are needed. Similar problems arise when the sum is less than or equal to unity – when not all possible constituents are measured (Aitchinson, 1986).

compound distribution A distribution specified in terms of another distribution dependent on a parameter that is itself a random variable with a specified distribution. For example, the negative binomial distribution can be specified as a Poisson distribution in which the mean follows a gamma distribution. Compound distributions can be regarded as a special class of **mixture distributions**.

compound hypergeometric distribution A distribution that is produced by averaging the hypergeometric distribution for given x over all possible values of the population characteristic X according to a prior distribution. A number of well-known distributions may be considered as special cases of the compound hypergeometric distribution. For example, the Pólya, the binomial and the rectangular.

compound negative multinomial distribution If, in the context of a **negative multinomial distribution**, we consider distinguishable sets of trials with the probabilities fixed within a set but randomly varying between sets, the resultant distribution will be a compound version of the negative multinomial. When the random variation is a **multivariate beta distribution** the resultant compound form was given by Mosimann (1963).

compound Poisson distribution A distribution resulting from a Poisson distribution of parameter λ where λ itself has a distribution. If the distribution of λ's is represented by $f(\lambda)$ the probability of observing the number k is

$$\Pr(X = k) = \frac{1}{k!} \int_0^\infty e^{-\lambda} \lambda^k f(\lambda) d\lambda.$$

compressed limits In quality control, limits which are more stringent than necessary in the sense that items falling outside them may still be within limits acceptable to the consumer. The object of setting compressed limits is to reveal departure from a controlled state sooner, or with smaller sample size, than might be exhibited by wider limits.

computer-intensive methods Statistical methods such as bootstrap resampling, non-parametric regression and Markov chain Monte Carlo simulation that rely totally on electronic computers for their implementation.

concentration, coefficient of A coefficient advanced by Gini (1912) as a measure of dispersion. It may be defined in terms of the mean difference (Δ_1), also due to Gini, as

$$G = \frac{\Delta_1}{2\mu_1'}$$

where μ_1' is the arithmetic mean. [See **mean difference**.]

concentration, ellipse of For a bivariate normal population with means μ_1 and μ_2, variances σ_1^2 and σ_2^2 and correlation ρ, the ellipse of concentration is given by

$$\frac{1}{1-\rho^2}\left\{\left(\frac{x-\mu_1}{\sigma_1}\right)^2 - \frac{2\rho(x-\mu_1)(y-\mu_2)}{\sigma_1\sigma_2} + \left(\frac{y-\mu_2}{\sigma_2}\right)^2\right\} = 4.$$

It is such that a uniform distribution bounded by the ellipse has the same first and second moments as the normal population. If the ellipse of concentration of one distribution lies wholly inside that of another, the former distribution is said to be more concentrated.

concentration, index of A descriptive index proposed by Gini (1910) to measure the extent to which a quantitative characteristic is concentrated in a few units. If a variable X can take values $x_1, x_2, \ldots, x_n$ (in that order) with frequencies $f_1, f_2, \ldots, f_n$, the sum of the last m units as compared with the total sum obeys the inequality

$$\frac{\sum_{i=n-m+1}^{n} f_i x_i}{\sum_{i=1}^{n} f_i x_i} > \frac{\sum_{i=n-m-1}^{n} f_i}{\sum_{i=1}^{n} f_i}$$

and the extent to which these two expressions depart from equality is taken as a measure of concentration. For incomes and several other characteristics the descriptive index of concentration is the number δ such that

$$\left(\frac{\sum_{i=n-m+1}^{n} f_i x_i}{\sum_{i=1}^{n} f_i x_i}\right)^{\delta} = \frac{\sum_{i=n-m-1}^{n} f_i}{\sum_{i=1}^{n} f_i}$$

concentration matrix The inverse of a covariance matrix; also called precision matrix. The reciprocals of the diagonal elements are conditional variances and minus the standardized off-diagonal elements are partial correlations (Cox and Wermuth, 1996).

concentration parameter See **von Mises distribution**.

concomitant variable In analysis of experiments, this term refers to factors that are detectable and quantifiable and affect the results of the experiment but are not accommodated in the design. In a randomized experiment it is a variable measured on an experimental unit before randomization and used as a basis for error reduction or interaction detection. The rather more general term covariate is sometimes used equivalently. In an observational investigation the term may be used for a variable intended to serve a similar purpose.

concordance, coefficient of In ranking theory, a coefficient measuring the agreement among a set of rankings. If m rankings of n objects are arranged one under another and the rankings summed for each of the n objects, and if

S is the sum of squares of deviations of these sums from their common mean $\frac{1}{2}m(n+1)$, the coefficient of concordance W is

$$W = \frac{12S}{m^2(n^3 - n)}.$$

Complete agreement between the rankings gives $W = 1$ and lack of agreement results in W being zero or very close to it.

concordant sample This concept was introduced by Pitman (1937a, 1938) in connection with his distribution-free test of the difference between two samples. Given two sets of observations $a_1, \ldots, a_n$ and $b_1, \ldots, b_m$, with means $\bar{a}$ and $\bar{b}$, there are

$$\binom{m+n}{n}$$

equiprobable ways of separating these $m + n = N$ observations into two sets of which the available set is one. If $\bar{a} - \bar{b}$ is defined as the 'spread' of a given separation we choose certain separations (with small spreads) as acceptable in the sense that their occurrence does not lead us to infer a real difference in parent populations. A set of a's and b's which forms one of these separations is 'concordant'. In the contrary case they are 'discordant'.

conditional distribution The distribution of one random variable or collection of random variables given the values of other random variables. It may be found from the joint distribution of all the random variables by applying the laws of probability theory. There is a technical mathematical complication if the variables held fixed are continuous, the conditioning event then having probability zero; this difficulty was resolved by Kolmogorov.

conditional expectation A key concept in probability. The conditional expectation of a random variable relative to a sigma-algebra F is the random variable $Y = E(X|F)$ which is F-measurable and which has the property

$$E[Y; B] = E[X; B]$$

for all $B \in F$. Put differently, for two random variables (X, Y) the conditional expected value of Y is the expected value of Y evaluated in its conditional distribution given $X = x$, typically regarded as a function of x. If X is a continuous random variable there is the technical complication that the conditional distribution is taken conditionally on an event of probability zero, namely $X = x$, and therefore is not uniquely defined. This difficulty can be surmounted, but in a statistical context is not a serious problem in that one can take a 'smooth' version regarding the conditioning as in fact on $x < X \leqslant x + \delta$ for some small but nonzero δ.

conditional failure rate An alternative term to **hazard** used in life analysis of physical systems or components.

conditional Gaussian (CG) distribution A joint distribution of mixed discrete and continuous variables. It is defined by the following construction: the marginal distribution of the discrete variables is multinomial and, conditional on these discrete variables, the continuous variables have a multivariate normal distribution (Whittaker, 1990).

conditional independence Independence between random variables conditional on the values of others.

conditional inference The notion that the distributions used in the theory of statistical inference should be conditional on certain observed statistics. This may be done to ensure relevance to the data at hand or may be used to achieve 'exact' probability properties. In Bayesian theory, once an initial specification is set out, conditioning is on all the data.

conditional likelihood The likelihood calculated not from a full probability model but from the model taken conditionally on the values of certain statistics. The objective is usually either to eliminate or reduce dependency on nuisance parameters or to improve relevance of the inference to the specific data under analysis.

conditional logistic regression When data involving a binary response are recorded across several strata, it is often useful to consider logistic regression models in which the relationship between the response and predictors is constant across strata and with separate stratum effects. The simplest example is that of a common odds ratio across a series of 2×2 contingency tables. If the number of strata is large but the number of observations in each stratum is small, as is typically the case with matched data, then maximum likelihood estimates of the parameters of interest can be severely biased owing to the large number of nuisance parameters associated with the stratum effects. In conditional logistic regression, the nuisance parameters are eliminated by conditioning on suitable statistics, e.g., the stratum totals, and likelihood analysis is then applied to the resulting conditional likelihood.

conditional maximum likelihood Inference by maximization of a conditional likelihood.

conditional power function A concept introduced by F.N. David (1947) in connection with the power of tests of randomness in a sequence of alternative events. As in **conditional tests**, the actual sample observed is used to define a sample subspace and the power function considered in this subspace.

conditional probability The probability of an event given that another event is known to have occurred.

conditional regression A regression estimated under certain conditions known a priori to apply to some of the parameters concerned; for example, in estimating price and cross elasticities from time-series data the income elasticities involved can sometimes be assumed to be known from cross-section data. It might be better to find an alternative name, owing to the intimate connection between regression and **conditional distributions**.

conditional statistic   A statistic whose distribution is conditional; that is to say, depends upon some quantity which is held constant. The quantity in question is usually itself some function of the random variables entering into the statistic.

conditional survivor function   The ordinary survivor function is the complement of the distribution function, i.e., $S(x) = 1 - F(x)$. The conditional expression for $X > x_0$ is

$$S(y|x) = \Pr(X - x_0 > y|X > x_0) = \frac{S(y + x_0)}{S(x_0)}.$$

The interpretation of this function, together with the **hazard**, is of importance in life testing and reliability assessment. [See **biometric functions**.]

conditional test   A test of significance is sometimes difficult to apply because the distribution of the test statistic involves unknown parameters of the parent population. This difficulty may sometimes be avoided by introducing restrictions on the sampling distribution, e.g., by considering only samples which have the same mean as that of the observed sample. This is equivalent to making the inference in a subpopulation of samples which have a fixed mean. The distribution and the inference based on it are then said to be conditional.

conditionality principle   The notion that in frequentist inference the probability distribution used for inference should be conditional on aspects of the data that are uninformative about the issues of concern. The principle seems compelling from a Fisherian perspective to achieve relevance to the data under analysis, although there are difficulties in formulating the principle mathematically. From a Neyman–Pearson viewpoint there are apparent clashes with the notion of power.

conditionally unbiased estimator   An estimator t of a parameter θ is said to be conditionally unbiased with respect to statistics $u_1, \ldots, u_n$ if the expectation of t for constant $u_1, \ldots, u_n$ is equal to θ. Symbolically

$$E(t|u_1, u_2, \ldots, u_n) = \theta.$$

confidence band   The region between the upper and lower **confidence limits**.

confidence belt   See **confidence band**.

confidence coefficient   See **confidence interval**.

confidence curves   A unified concept proposed by Birnbaum (1961a) to include point, confidence limit and interval estimation.

confidence distribution   A collection of confidence intervals at various levels subject to natural monotonicity requirements (Cox, 1958b).

confidence interval If it is possible to define two statistics t_1 and t_2 (functions of sample values only) such that, θ being a parameter under estimate,

$$\Pr(t_1 \leqslant \theta \leqslant t_2) = \alpha$$

where α is some fixed probability called the confidence coefficient, the interval between t_1 and t_2 is called a confidence interval. The assertion that θ lies in this interval will be true, on the average, in a proportion α of the cases when the assertion is made.

confidence level An alternative term for **confidence coefficient**.

confidence limits The values t_1 and t_2 which form the upper and lower limits to the **confidence interval**.

confidence region When several parameters are being estimated it may be possible to define regions in the parameter space such that there will be assigned confidence α that the parameters lie within them. This is the generalization of the confidence interval to the case of more than one parameter and the domain so determined is called the confidence region.

confidence set A random subset of a parameter space which has a specified probability of containing an unknown parameter under repeated sampling. The best-known example is a confidence interval, but a confidence set may comprise disjoint subsets or may lie in a set of dimension greater than one.

configuration A set of n observations on a random variable may be represented as a vector in n-dimensional space. A number k of vectors in the space may be regarded as having geometrical properties of interrelationship independently of the coordinate system. In factor analysis, the arrangement of random variable vectors among themselves, and without regard to any frame of reference, is called their configuration. The expression also occurs in a different sense. In a location problem the set of differences among the sample values is called the configuration of the sample. 'Configurational sampling' is sometimes used as a synonym for **grid sampling**.

confirmatory factor analysis A procedure to test a factor model partially specified beforehand, usually in terms of the number of factors and the positions of their zero loadings.

confluence analysis A method of analysis introduced by Frisch (1934) in an attempt to overcome difficulties in regression analysis when there may be linear relations between the explanatory variables or errors of observation introduce 'nearly' linear relations in the observed independent (predicated) variables. The technique is also known as **bunch map analysis**. A relation between the variables which results in the indeterminacy of the coefficients of a regression equation, or approximate indeterminacy where observational errors exist, is called a confluent relation. The technique is now rarely used.

confounder An explanatory variable, measured or unmeasured, distorting the effect of an explanatory variable of interest; see **confounding (ii)**.

confounding A term used in two somewhat different senses: (i) the device in the study of factorial experiments of sacrificing information about contrasts of little interest (typically high-order interactions) in order to improve the precision with which more interesting contrasts can be estimated; (ii) the notion much more broadly that estimation of an effect of direct interest may be distorted by the effect of other features, called confounders, not the direct focus of concern.

conjugate distribution If $F(x)$ is a probability distribution function for which

$$M(t) = \int_{-\infty}^{\infty} e^{tx} dF(x)$$

is finite in an interval I which contains the origin $t = 0$, then for any value of t in this interval, the function

$$G(x, t) = \frac{1}{M(t)} \int_{-\infty}^{x} e^{tu} dF(u)$$

is a probability distribution function which is said to be 'conjugate' to $F(x)$.

conjugate Latin squares Two **Latin squares** are conjugate if the rows of one are the columns of the other.

conjugate prior A prior distribution for a parameter for which there is a sufficient statistic is said to be conjugate if the prior and posterior distributions have the same functional form. An example is the beta prior for the probability in a binomial distribution.

conjugate ranking Given two rankings of n objects, if one is arranged in the natural order and the other (correspondingly rearranged) designated by A, and then the latter is arranged in the natural order and the first (correspondingly rearranged) designated by B, then A and B are said to be conjugate.

connectedness An experimental design for which, given any two treatments θ and ϕ, it is possible to construct a chain of treatments $\theta = \theta_0, \theta_1, \theta_2, \ldots, \theta_n = \phi$ such that every consecutive pair of treatments in the chain occurs together in a block. A design for which this is not true is called disconnected. The important property of a connected design is that all elementary contrasts of treatment effects are **estimable**. For an additive two-way classification model Bose (1949) introduced the notion of connectedness.

conservative confidence interval A **confidence interval** or region is 'conservative' when the actual confidence coefficient exceeds the nominal or stated value.

conservative process A stochastic process governing the behaviour of a population which has constant total size but the members of which can assume independently one of a finite number of states, the variation consisting of transfer from one state to another.

consistence, coefficient of In the analysis of **paired comparisons** the fundamental inconsistency or consistency of preferences may be expressed in terms of **circular triads**. The coefficient of consistence may be defined (Kendall and Babington Smith, 1940) as $1-\left\{24d/(n^3-n)\right\}$ for n odd and $1-\left\{24d/(n^3-4n)\right\}$ for n even, where d is the observed number of circular triads and n is the number of objects being compared.

consistency An estimator is called consistent if it converges in probability to its estimand as sample size increases, and strongly consistent if it converges almost surely.

consistent estimator An estimator which converges in probability, as the sample size increases, to the parameter of which it is an estimator. An example of an inconsistent estimator is the sample mean as estimator of θ in the distribution

$$f(x) = \frac{1}{\pi\left\{1+(x-\theta)^2\right\}}, \qquad -\infty < x < \infty.$$

consistent test A test of a hypothesis is consistent with respect to a particular alternative hypothesis if the power of the test tends to unity as the sample size tends to infinity; and, similarly, it is consistent with respect to a class of alternatives if it is consistent with respect to each member of the class.

constraint A constraint in a set of data is a limitation imposed by external conditions, e.g., that a number of variable values shall have zero mean, or that the sum of frequencies in a set of classes shall be a prescribed constant. There is another sense in which statistical data may be said to be constrained. This is the case of **subnormal dispersion** discussed by Lexis.

consumer price index A price index designed to measure changes in the cost of some specified standard of living. [See **Laspeyres' index**.]

consumer's risk In acceptance inspection the risk which a consumer takes that a lot of a certain quality q will be accepted by a sampling plan. It is usually expressed as a probability of acceptance and depends, of course, on q as well as the sampling plan itself. It is equivalent to the probability of an **error of the second kind** in the theory of testing hypotheses in the sense of corresponding to the acceptance of a hypothesis when an alternative is true. [See also **producer's risk**.]

contagious distribution A class of probability distribution of a compound kind, usually derived from probability distributions dependent on parameters by regarding those parameters as themselves having probability distributions. The name derives from the use of such compound distributions in the study of contagious events such as accidents, occurrences of disease or 'persistence' in weather. [See also **compound distributions**.]

contaminated distribution In the investigation of robustness, observations are often assumed to be drawn from a specified (usually normal) distribution with high probability p, and from some other distribution with probability $1-p$. The resulting **mixture distribution** is referred to as a contaminated (normal) distribution.

content validity In factor analysis, the ability of a set of test items to provide an estimate of a latent trait.

contingency The difference in the cells of the **contingency table** between the actual frequency and the estimated frequency on the assumption that the two characteristics are independent in the probabilistic sense. If f_{ij} is the frequency in the ith row and jth column and $f_{i.}, f_{.j}$ are the respective row and column totals, and if the total frequency is n, the difference in question is

$$f_{ij} - \frac{f_{i.} f_{.j}}{n}.$$

The square contingency is given by

$$\chi^2 = \sum_{i,j} \frac{n(f_{ij} - f_{i.} f_{.j}/n)^2}{f_{i.} f_{.j}}.$$

The mean square contingency, usually denoted by ϕ^2, is given by

$$\phi^2 = \frac{\chi^2}{n}.$$

contingency, coefficient of Any coefficient purporting to measure the strength of dependence between two characteristics on the basis of a contingency table. In the notation of the previous term, K. Pearson's coefficient is defined by

$$C = \left(\frac{\chi^2}{n + \chi^2} \right)^{1/2} = \left(\frac{\phi^2}{1 + \phi^2} \right)^{1/2}.$$

Tschuprov's coefficient is defined as

$$T = \left(\frac{\chi^2}{n\sqrt{(p-1)(q-1)}} \right)^{1/2}$$

where p and q are the number of rows and columns in the **contingency table**.

contingency table The members of an aggregate may be classified according to qualitative or quantitative characteristics. Where the characteristics are qualitative a classification according to two or more of them may be set out in a two-way or multiway table known as a contingency table. For example, if the characteristic A is p-fold and a characteristic B is q-fold then the contingency table will be one of p rows and q columns. The cell corresponding to A_j and B_k contains the number of individuals bearing both of those characteristics. In general, the order of the rows and columns is arbitrary. There are corresponding definitions with more than two characteristics. [See also **contingency, coefficient of.**]

contingency-type distributions A class of bivariate distributions proposed by Plackett (1965). For given marginal distribution functions F and G with unknown bivariate distribution function H, he sets the **odds ratio**

$$\frac{H(1 - F - G + H)}{(F - H)(G - H)}$$

equal to an arbitrary positive constant regardless of the point of dichotomization (x, y), which leads to the equation

$$H = \frac{S - \left\{S^2 - 4\psi(\psi - 1)FG\right\}^{1/2}}{2(\psi - 1)} \tag{1}$$

where $S = 1 + (F + G)(\psi - 1)$. Contingency-type distributions are so called because of the connection, through the odds ratio, with 2×2 **contingency tables**. Writing $F = x$, $G = y$ in (1) we get the contingency-type uniform distribution, which has probability density function

$$h(x, y; \psi) = \frac{\psi \left\{(\psi - 1)(x + y - 2xy) + 1\right\}}{\left[\left\{1 + (\psi - 1)(x + y)\right\}^2 - 4\psi(\psi - 1)xy\right]^{3/2}}, \qquad 0 \leqslant x, y \leqslant 1.$$

continuity correction Correction applied when approximating a cumulative distribution function (tail area) of a discrete variable by a continuous distribution such as the normal. When the discrete values are equally spaced the correction involves half the spacing interval. There is a close connection with quadrature formulae in numerical analysis.

continuous process A name sometimes employed to denote a stochastic process $[x_t]$ which depends on a continuous parameter t. It is apt to lead to confusion with the continuity of x and is, perhaps, better avoided. The same applies to 'discontinuous' or 'discrete' process.

continuous random variable A random variable has a continuous distribution if all point values have zero probability. Most interesting continuous distributions have the stronger property of absolute continuity, meaning that the probability of any set is obtained by integrating a probability density function over that set. In statistical analysis variables are called continuous if reasonably regarded as derived by rounding a continuous variable to a specified level; the ultimately discrete character of such data is occasionally important.

continuous sampling plans A type of sampling inspection plan for use where a production process is continuous in the sense that batching is not a rational procedure. It may also be applicable where immediate process control is required and the rectification of outgoing products. The plans may be single-level (Dodge, 1943) or multi-level.

contour level See **patch**.

contrast A contrast among the parameters in analysis of variance, or among the treatment and interaction effects in an experiment, is a linear function of these quantities with known constant coefficients which sum to zero. It is said to be an elementary contrast if it has only two nonzero coefficients $+1$ and -1.

control There are two principal ways in which this term is used in statistics. If a process produces a set of data under what are essentially the same conditions and the internal variations are found to be random, then the process is said to be statistically under control. The separate observations are, in fact, equivalent to random drawings from a population distributed according to some fixed probability law. The second usage concerns experimentation for the testing of a new method, process or factor against an accepted standard. That part of the test which involves the standard of comparison is known as the control.

control chart A graphical device proposed by Shewhart (1931) used to show the results of small-scale repeated sampling of a manufacturing process. It usually consists of a central horizontal line corresponding to the average value of the characteristic under investigation, quantitative or qualitative, together with upper and lower control limits between which a stated proportion of the sample statistics should fall. Any marked divergence above or below these control limits will tend to indicate that new causes are at work beyond those responsible for the random variations inherent in large-scale production. Points outside the limits will signal the need for special enquiries to identify the new factor(s) at work. An alternative approach is through using cumulative results for which the **cumulative sum chart** was devised.

control limits See **control chart**.

control method A method for reduction of sampling error by correlating the variable of interest with another, the control random variable, amenable to theoretical analysis.

control of substrata A term used in sampling enquiries to denote the employment of prior knowledge of the population cell values in an n-way table formed according to the n factors which are being used in a scheme of **multiple stratification**. For example, if a population is stratified by age and sex, knowledge of the number of individuals of each sex in each age group enables the sample to be controlled by these substrata. If only the marginal frequencies were known the sample could be controlled by strata but not by substrata, as, for example, if the numbers of each sex in the population and the age distribution of the population were known, but not the numbers of each sex in each age group.

controlled process An industrial process is said to be controlled when the mean and variability of the product remain stable. The variation is then due to random effects or the combination of small factors of a non-cumulative kind. The expression 'under statistical control' is to be avoided in favour of 'statistically in control' or 'statistically stable'.

convergence in measure See **stochastic convergence**.

convergence in probability See **stochastic convergence**.

convolution Let $F_1(x), F_2(x), \ldots, F_n(x)$ be a sequence of distribution functions, with densities $f_1(x), f_2(x), \ldots, f_n(x)$. The distribution F whose density is

$$f(x) = \int_{\mathbb{R}^{n-1}} f_1(x_1) f_2(x_2) \ldots f_{n-1}(x_{n-1}) f_n(x - \sum_{i=1}^{n-1} x_i) dx_1 dx_2 \ldots dx_{n-1}$$

is called the convolution of the distributions. The relationship is sometimes written

$$F(x) = F_1(x) * F_2(x) * \ldots * F_n(x).$$

If the associated random variables are independent, $F(x)$ is the distribution function of their sum.

Cook's statistic In **regression diagnostics**, a measure of **influence** (Cook, 1979). The statistic is

$$D_i = \frac{1}{ps^2} \left(\hat{\beta}_{(i)} - \hat{\beta} \right)' X'X \left(\hat{\beta}_{(i)} - \hat{\beta} \right)$$

where $\hat{\beta}_{(i)}$ is the vector of regression coefficients calculated omitting the ith point, and s^2 is the estimated residual variance after fitting the p regressors $X_1, \ldots, X_p$.

copula A joint distribution for two or more variables, each being marginally standard uniform.

corner test See **medial test**.

Cornish–Fisher expansion A reversion of the Edgeworth expansion of a frequency function used by Cornish and Fisher (1937) to tabulate the significant points of certain probability integrals.

corpuscle problem The oldest problem in stereology, of inferring the diameter distribution of spheres from two-dimensional sections.

corrected moment A moment of a set of observations which has been adjusted for some effect such as the bias arising from its being calculated from a grouped frequency distribution rather than from the original data. [See also **correction for grouping**, **Sheppard's corrections**.]

correction factor The quantity $(\sum x)^2/n$ subtracted from $\sum x^2$ to give $\sum(x - \bar{x})^2$, or the corresponding correction for a sum of products.

correction for continuity See **continuity correction**.

correction for grouping When data are grouped into frequency distributions the approximation which becomes necessary by reason of having to regard frequencies as being concentrated at the mid-points of class intervals may impart a bias to the calculations of the moments of the distribution. Under certain conditions it is possible to correct for this effect, the best-known corrections being due to Sheppard. Other corrections have been advanced for distributions which are abrupt at one, or both, terminals. The problem of **average corrections** has been shown to lead to expressions similar to those for **Sheppard's corrections**.

corrections for abruptness A system of corrections to the moments of frequency distributions which do not have high-order contact at the limits of the range. Such corrections were devised by Pairman and Pearson (1919) and were proposed as suitable for use in cases where **Sheppard's corrections** do not apply.

correlation Sometimes used broadly to mean some kind of statistical relation between variables, the word is sometimes restricted to broadly linear relations and even to relations represented by bivariate or multivariate normal distributions. In other contexts the word correlation is to be contrasted with regression. In the former the variables concerned are regarded on an equal footing whereas regression usually implies that one of the variables is regarded as a response, the other variable or variables being explanatory.

correlation, coefficient of A measure of the interdependence between two variables. It is usually a pure number which varies between -1 and 1 with the intermediate value of zero indicating the absence of correlation, but not necessarily the independence of the variables. The limiting values indicate perfect negative or positive correlation. If there are two sets of observations $x_1, \ldots, x_n$ and $y_1, \ldots, y_n$ and a score is allotted to each pair of individuals, say a_{ij} for the x-group and b_{ij} for the y-group, a generalized coefficient of correlation may be defined as

$$\Gamma = \frac{\sum a_{ij} b_{ij}}{\sqrt{\sum a_{ij}^2 \sum b_{ij}^2}}$$

where $\sum$ is a summation over all values of i and j $(i \neq j)$, from 1 to n. This general coefficient includes **Kendall's** τ, **Spearman's** ρ and Pearson's **product moment correlation** r as special cases according to the method of scoring adopted. In the last case, for example, the scoring is based on variable values with $a_{ij} = x_i - x_j$, $b_{ij} = y_i - y_j$. If positive values of one variable are associated with positive values of the other (measured from their means) the correlation is sometimes said to be direct or positive, as contrasted with the contrary case, when it is said to be inverse or negative. There are numerous other correlation coefficients of a different character. [See **cube of correlation**.]

correlation matrix For a set of random variables $X_1, \ldots, X_n$ with correlations between X_i and X_j denoted by r_{ij}, the correlation matrix is the square matrix of values (r_{ij}). Its determinant is the correlation determinant. The matrix is symmetric since $r_{ij} = r_{ji}$. The diagonal elements r_{ii} are unity.

correlation ratio In a bivariate frequency table with random variables X and Y the correlation ratio of X on Y is defined by

$$\eta_{xy}^2 = \frac{\sum (\bar{x}_i - \bar{x})^2}{\sum (x - \bar{x})^2}$$

where the summation in the numerator takes place over y-arrays, $\bar{x}_i$ is the mean of the ith array and $\bar{x}$ is the mean of x in the whole distribution, and the summation in the denominator takes place over all the values. It may be regarded as the ratio of the variance between arrays to the total variance. There is an analogous expression for the correlation ratio of Y on X.

correlation table The frequency table of a bivariate distribution. The difference between a correlation table and a **contingency table** is that the former usually denotes a grouped frequency distribution with intervals defined in terms of the variable values and therefore possessing a natural order and clearly defined width.

correlogram In time-series analysis, the graph of the serial correlation of order k as ordinate against k as abscissa. It is generally presented only for non-negative values of k since $r_k = r_{-k}$ (conventionally). This term was introduced by Wold (1938) but the diagram was earlier used by Yule (1921), who called it a 'correlation diagram'.

correspondence analysis A method proposed by Hirschfeld (1935) for the analysis of multivariate discrete data which is equivalent to a special case of **canonical correlation** and also to a scale-free variant of **principal components** analysis (Guttman, 1959).

cospectrum The covariance between the two cosine components and between the two sine terms in a spectrum analysis of the relationship between two time series. The cospectrum measures the covariance of the components which are in phase. [See also **quadrature spectrum, coherency**.]

cost function In sampling theory, a function giving the cost of obtaining the sample as a function of the relevant factors affecting cost. It may relate to only a part of the entire cost, e.g., by providing for the cost of collecting the sample but not for the cost of tabulation.

counter model, type I A stochastic process related to the physical behaviour of Geiger–Müller counters. A model of type I is one which records a count at the first arrival not covered by the pulse of the previous count.

counter model, type II A model of type II (see entry for type I for background information) is one in which a count is recorded as the first arrival not covered by the pulse of any previous arrival.

counterfactual model A model for causal inference contains two potential response variables for exposing and not exposing each unit to the action of a cause. For each unit, only one of the potential responses is observed, and the other cannot be observed and hence is called counterfactual (Holland, 1986).

counting distribution A probability distribution formed by the number of events occurring in a fixed period of time. If the counting period begins just after an event the distribution may be termed a synchronous distribution (Haight, 1965).

counting process A general name for a stochastic process $(N(t), t \geqslant 0)$ in which $N(t)$ represents the total number of point events which have occurred up to time t.

counting process methods For studying a group of individuals each of whom is moving between a finite number of states, such as survival levels. The exact transition times (in continuous time) form the modelling base, although these times are often incompletely observed due to, say, censoring (Andersen, Hansen and Keiding, 1991).

covariance The first product moment of two random variables about their mean values. The term is also used for the estimator from a sample of a parent covariance.

covariance, analysis of See **analysis of covariance**.

covariance function Equivalent to autocovariance function. [See **autocorrelation coefficient** and **function**.]

covariance kernel See **mean value** function.

covariance matrix For n random variables $X_1, \ldots, X_n$ for which the covariance of X_i and X_j is c_{ij}, the square matrix (c_{ij}) is called the covariance matrix. The diagonal terms are the variances: $\mathrm{var}(X_i)$. Alternative names are the dispersion matrix and the variance–covariance matrix. If the random variables are standardized so as to have unit variances the covariances become correlations and the matrix becomes the **correlation matrix**.

covariance stationary process A stochastic process $x(t)$ is stationary in the covariance or covariance stationary if the covariance function

$$R(v) = E[x(t)x(t + v)]$$

exists and is independent of t for all integral v.

covariate See **concomitant variable**.

covariation The joint variation of two or more random variables.

covarimin See **factor rotation**.

coverage A term used in sampling in two senses: (i) to denote the scope of the material collected from the sample members (as distinct from the extent of the survey, which refers to the number of units included); (ii) to mean the extent or area covered by the sampling as in expressions such as '50 per cent coverage', which means that one-half of the population under discussion has been examined. The word was also introduced by Tukey (1947) to mean the probability content of a **statistical tolerance region** based on independent observations from a probability distribution.

coverage probability The probability that a confidence interval or region contains its parameter as assessed by long-run frequency.

coverage problems In geometrical probability, problems concerned with the proportion of an area or length covered by randomly placed, possibly overlapping, objects. There are important applications in stereology.

Cox, Gertrude Mary (1900–1978) Gertrude Cox is principally known as a gifted statistical administrator and entrepreneur, as well as for her contributions to psychological statistics and experimental design (Cochran and Cox, 1957). She was a founding member of the International Biometric Society in 1947, served as editor of its journal, *Biometrics*, from 1947 to 1955, and was president from 1968 to 1969. She was president of the American Statistical Association (ASA) in 1956.

Cox and Stuart's tests Two classes of **sign test** (Cox and Stuart, 1955) of the hypothesis that the distribution of a continuously distributed random variable does not change through time, one being especially sensitive to monotonic changes in the location of the distribution, the other being especially sensitive to monotonic changes in dispersion.

Cox process See **doubly stochastic Poisson process**.

Cox proportional hazards model Model for the analysis of censored survival data in which the hazard function for failure is the product of a function of time, the baseline hazard and a function of explanatory variables. The term proportional hazards is a slight misnomer in that the explanatory variables may be time-dependent (Cox, 1972).

Cox's regression model A model for the effects of various covariates in survival analysis (Cox, 1972). The hazard function is assumed to be an unknown function of time multiplied by a factor involving the covariates, and the parameters are estimated by maximizing a likelihood function conditional on failure times and the number at risk at those times. However, this likelihood is not a conditional likelihood but a partial likelihood.

Cox's theorem A theorem which states specific conditions under which a sequential test of a mean of a normal distribution against a **composite hypoth-**

esis can be constructed (Cox, 1952*a*; Cox, 1952*b*) . The test is based upon the application of the **sequential probability ratio test** to transformed values of the original observations.

Craig's theorem A theorem, stated by Craig (1933), on the independence of quadratic forms in the analysis of variance. If x_i $(i = 1, \dots, n)$ are n independent standardized normal random variables, then the quadratic forms $x'Ax$ and $x'Bx$ are statistically independent if and only if the product of the symmetric matrices A and B, namely AB, is zero.

Cramér, Harald (1893–1985) Harald Cramér, a Swedish mathematician, contributed pathbreaking research in probability, statistics and insurance mathematics, and to the illumination of statistics as a coherent mathematical discipline (Cramér, 1946).

Cramér–Lévy theorem A theorem conjectured by Lévy and proved by Cramér (1936). It states that if X and Y are independent random variables whose sum is normally distributed then both X and Y have normal distributions.

Cramér–Rao bound In its simplest form, this gives the reciprocal of the Fisher information as the lower bound for the variance of an unbiased estimate of a parameter in a regular parametric estimation problem. There are many generalizations (Cramér, 1946).

Cramér–Rao efficiency The efficiency of an estimator $\tilde{\theta}$ which will permit the condition of equality to exist in the **Cramér–Rao inequality** (Katz, 1950).

Cramér–Rao inequality An inequality giving a lower bound to the variance of an estimator of a parameter. If t is an estimator of θ in a distribution with frequency function $f(x, \theta)$ and if the bias $b(\theta)$ is given by

$$b(\theta) = E(t) - \theta,$$

the inequality states that

$$\mathrm{var}(t) \geqslant \frac{(1 + db/d\theta)^2}{E\left(\partial \log f / \partial \theta\right)^2}.$$

Results along these lines have been given by many authors and the question of priority is unsettled. In English writings the inequality is almost invariably known by the names of Cramér (1946) and Rao (1945), singly or in conjunction.

Cramér–Tchebychev inequality An inequality of the **Bienaymé–Tchebychev** type depending on the second and fourth moments, namely

$$\Pr\left\{|X - a| > t\sigma\right\} \leqslant \frac{\mu_4 - \sigma^4}{\mu_4 - 2t^2\sigma^4 + t^4\sigma^4}, \qquad t > 1,$$

where σ^2 is the variance and μ_4 the fourth moment of the distribution and a is the mean. Like many inequalities of this type, it has several names (Berge, 1932).

Cramér–von Mises test A test for the difference between an observed distribution function and a hypothetical distribution function. It was proposed by Cramér (1928) and, independently, by von Mises (1931). If $F_n(x)$ is the observed distribution function and $F(x)$ its hypothetical counterpart, the criterion is

$$\omega^2 = \int_{-\infty}^{\infty} (F_n(x) - F(x))^2 dx.$$

The sampling distribution of ω is not known. To meet this difficulty Smirnov (1936) considered an alternative form

$$\omega_n^2 = \int_{-\infty}^{\infty} (F_n(x) - F(x))^2 f(x) dx$$

which is independent of F and therefore provides a distribution-free test. The test was modified still further by Anderson and Darling (1952, 1954) in the form

$$W_n^2 = n \int_{-\infty}^{\infty} (F_n(x) - F(x))^2 \omega(F(x)) f(x) dx$$

where $\omega(t)$ is some real non-negative function defined for $0 \leqslant t \leqslant 1$.

Cramér–Wold device The result following from the continuity theorem that weak convergence of a random vector is equivalent to weak convergence of its linear combinations.

credible interval See **Bayesian interval**.

criterion This word is used in statistics in its everyday sense in a number of contexts, e.g., the likelihood criterion for testing hypotheses. In earlier literature the phrase 'the criterion', otherwise unqualified, is found to denote a function which distinguishes between various types of **Pearson curves**. The criterion is

$$\kappa = \frac{\beta_1(\beta_2 + 3)^2}{4(2\beta_2 - 3\beta_1 - 6)(4\beta_2 - 3\beta_1)}$$

where β_1 and β_2 are the Pearson measures of **skewness** and **kurtosis**.

critical quotient In the analysis of extreme values from unlimited distributions, Gumbel (1958) defined a critical quotient

$$Q(x) = \frac{-f^2(x)}{f'(x)\{1 - F(x)\}}$$

where $f'(x)$ is the first derivative of the frequency function.

critical region A formal test of a statistical hypothesis is made on the basis of a division of the **sample space** into two mutually exclusive regions. If the sample point falls into one (the region of acceptance) the hypothesis is accepted; if in the other region (the region of rejection) it is rejected. Both regions are, in a sense, critical, but it is customary to denote the second by the term critical region. If, among critical regions of fixed size, there is one which minimizes the probability of **error of the second kind** it is called the best critical region. If, for a set of alternative hypotheses, the probability of an error of the second kind is less than the probability of an error of the first kind (or equivalently, the power is greater than the size) the region is said to be unbiased (Lehmann, 1986).

critical value The value of a statistic corresponding to a given **level of significance** as determined from its sampling distribution, e.g., if $\Pr(t > t_0) = 0.05$, t_0 is the critical value of t at the 5 per cent level.

Crofton's theorem A basic theorem of **geometric probability**, given by Crofton (1869). 'The measure of the number of random lines which meet a given closed convex plane boundary is the length of the boundary.'

Cronbach's alpha A method to assess the internal consistency or reliability of a multiple-item instrument (or scale). The underlying construct, or latent variable, is internally consistent if its items are highly intercorrelated (Cronbach, 1951).

cross amplitude spectrum A procedure used in spectrum analysis of time series for describing the covariance between the two series which form a bivariate time series. It shows whether and how the amplitude of a component at a particular frequency in one series is associated with a large or small amplitude at the same frequency in the other series.

cross-correlation See **cross-covariance**.

cross-covariance For the stationary stochastic processes $X(t)$ and $Y(t)$, the covariance $\gamma_{xy}(w)$ between $X(t)$ and $Y(t+w)$ for given w, where w may take negative, zero or positive values; the function $\gamma_{xy}(w)$ is called the cross-covariance function. Similarly, the dimensionless quantity

$$\rho_{xy}(w) = \frac{\gamma_{xy}(w)}{\sigma_x \sigma_y},$$

where σ_x^2 and σ_y^2 are the variances of the X and Y processes respectively, is for given w called the cross-correlation coefficient and for general w called the cross-correlation function. $\gamma_{xy}(w)$ and $\rho_{xy}(w)$ are the generalizations to two series of the **autocovariance** and **autocorrelation functions**.

cross intensity function In a point process events of several types might be occurring along the one-dimensional time axis. The cross intensity function is a method proposed by Cox and Lewis (1966) for dealing with the relationship between these different types of events considered jointly against time.

cross-over design In its original sense, a design involving two treatments which could be applied more than once to the same set of subjects. The subjects would be divided into pairs and each pair treated first with the treatments A and B and then with the treatments B and A, 'crossed over'. More recent usage has extended the meaning to cases where the pairs of subjects are divided into two sets and each pair consists of one where the response is expected to be better and one where it is expected to be worse. In the first set A is applied to the 'better' members and in the second set to the 'worse' members. The method can be extended to cases where there are more than two treatments but if the number is large other designs are usually preferable (Kershner and Federer, 1981).

cross range The name given to a criterion introduced by Hyrenius (1953) in testing for identity of two rectangular distributions with respect to changes in the range of variation. If, in samples of n', n'', the smallest observations in the two distributions are denoted by L', L'' and the largest by U', U'' then the cross ranges are $U'' - L', U' - L''$. For $V = U'' - L'$, Hyrenius obtained the probability density function:

$$\frac{(n' - 1)n''}{n' + n'' - 1} v^{n'' - 1} \text{ for } 0 \leqslant v \leqslant 1$$

$$\frac{(n' - 1)n''}{n' + n'' - 1} v^{-n'} \quad \text{for } v \geqslant 1.$$

cross-sectional survey A method of data collection whereby a battery of questions is asked of participants at one single point or in a relatively small interval in time. Inferences about a population must be anchored to the time period in which the sample was taken. Data from cross-sectional surveys are typically unable to be used to prove the existence of cause-and-effect relationships. See **causality**, **longitudinal survey**.

cross spectrum In connection with the spectrum analysis of bivariate time series, the product of the amplitude spectrum and the **phase spectrum** is known as the cross spectrum. It is equivalent to the **Fourier transform** of the **cross-covariance** function. If the **cross-correlations** between series 1 and 2 are $\rho_{(12)s}$, $s = -\infty, \ldots, \infty$, the cross-spectral density is

$$\omega_{12}(\alpha) = \sum_{s=-\infty}^{\infty} \rho_{(12)s} e^{is\alpha} = c(\alpha) + iq(\alpha)$$

where $c(\alpha)$ is the **cospectrum** or co-spectral density and $q(\alpha)$ the **quadrature spectrum** or quadrature spectral density. The sum of squares $c^2 + q^2$ is called the **amplitude** and, if $\omega_1(\alpha)$ and $\omega_2(\alpha)$ are spectral densities of the two series, the quantity

$$C(\alpha) = \frac{c^2(\alpha) + q^2(\alpha)}{\omega_1(\alpha) + \omega_2(\alpha)}$$

is called the **coherence**. The quantity, plotted against α as abscissa,

$$\arctan\left(\frac{q(\alpha)}{c(\alpha)}\right)$$

is the phase diagram and the quantity

$$\frac{\omega_1(\alpha)}{\omega_2(\alpha)}C(\alpha),$$

similarly plotted, is called the gain diagram.

cross-validation Assessment of the quality of an estimator or predictor formed using one part of a sample according to its performance on the other part. The simplest version is leave-one-out cross-validation, which sums the errors obtained by predicting each observation from the remainder, while in k-fold cross-validation the sample is split into k parts of equal size. There is a connection with information criteria such as **Akaike's information criterion**. An older term is the predictive sample reuse method.

cross-validation criterion This criterion consists in the controlled or uncontrolled division of a data sample into two subsamples. The choice of a statistical predictor, including any necessary estimation, is made on one subsample and then the assessment of its performance is made by measuring its predictions against the other subsample. This procedure has also been termed the predictive sample reuse method.

crossed classification A feature of the largest of nested designs. If, say, there are two factors A and C then where every level of C appears with every level of A this two-way layout is termed completely crossed. Anything less than this is deemed to be partially crossed. [See also **hierarchical group divisible design**.]

crossed factors See **nested design**.

crossed weight index number An index number is said to have crossed weights if it results from two subsidiary index numbers, with different weights, after the application of some process of averaging. The most commonly quoted crossed weight formula is that of Fisher's **'ideal' index number** which is the result of geometrically crossing (averaging) the index number formulae attributed to **Paasche** and **Laspeyres**. It may be written as a price index:

$$I_{0n} = \left(\frac{\sum p_n q_0}{\sum p_0 q_0}\frac{\sum p_n q_0}{\sum p_n q_n}\right)$$

where 0 and n are subscripts relating to the base year and current year respectively. The **Marshall–Edgeworth–Bowley index** also has crossed weights in this sense.

crude moment See **raw moment**.

crypto-deterministic process A particular kind of stochastic process, due to Sir Edmund Whittaker (1943), where the initial conditions contain all the uncertainty. Apart from this uncertainty the development of the process in time is of a completely determinate character.

CSM test A test of significance developed by Barnard (1947) for data, in the form of a 2×2 table, arising from comparative trials; for example, where prearranged numbers are taken from each of two sources to compare the proportions of some attribute in the two sets. The name CSM derives from the three conditions of Convexity, Symmetry and Maximum sensitivity which determine the **critical regions** of the test (and indirectly from the military rank Company Sergeant Major).

cube of correlation The square ρ_{XY}^2 of the correlation coefficient between two random variables X and Y may be interpreted as the percentage of variance of one of the variables which is 'linearly explained' by the other. On the other hand, under the assumptions of a linear model

$$Y = \alpha + \beta X + \varepsilon,$$

where α and β are some constants and ε is an error term which is symmetric and independent of X, the cube of the correlation coefficient may be expressed as

$$\rho_{XY}^3 = \frac{\gamma_Y}{\gamma_X},$$

where γ_Y and γ_X are the coefficients of skewness of Y and X, respectively (as long as $\gamma_X \neq 0$). More generally, if the error term ε is normally distributed, the rth ($r \geqslant 3$) power of the correlation coefficient is equal to the ratio of the rth standardized cumulants of Y and X. Thus, the cube of correlation may be interpreted as the percentage of skewness which is 'preserved' by a linear model. From this result, Dodge and Rousson (2001) derived a criterion for choosing the response variable in a simple regression problem. Since ρ_{XY}^3 is necessarily smaller than or equal to one (in absolute value), the response variable should have a smaller skewness than the explanatory variable (in absolute value).

cubic lattice An extension of the **square lattice** in which the number of treatments is a perfect cube and they are regarded as arranged on the points of a cubic lattice. From the point of view of factorial experiments the treatments are regarded as the combination of three factors each at k levels. [See **quasi-factorial design**.]

cuboidal lattice design This experimental design is a development of the **cubic lattice** design in much the same way as the rectangular lattice is a development of the square lattice. The cubic lattice design is suitable for numbers of treatments which are perfect cubes (k^3) whereas the cuboidal lattice design is appropriate to a number of treatments of the form $k^2(k + 1)$.

cumulant The cumulants are constants of a frequency distribution defined in terms of the moments by the identity in t

$$\exp \left(\sum_{r=0}^{\infty} \frac{\kappa_r t^r}{r!} \right) = \sum_{r=0}^{\infty} \frac{\mu_r' t^r}{r!} .$$

They are thus given by the coefficients in the expansion of a power series formed from the logarithm of the characteristic function of a variable, if such an expansion exists. The earlier name for the quantities was semi-variant or half-invariant, a term introduced by Thiele (1889). The word cumulant is due to Cornish and Fisher (1937).

cumulant generating function A function of a variable t which, when expanded in powers of t, has the cumulants of a distribution (or numerical multiples of them) as the coefficients in the expansion. The only cumulant generating function in common use is the logarithm of the **characteristic function**, which results in

$$K(t) = \log \phi(t) = \sum_{r=0}^{\infty} \kappa_r \frac{(it)^r}{r!}$$

where κ_r is the rth cumulant.

cumulative chi-squared statistic A statistic proposed by Taguchi and developed by Hirotsu (1986) for testing against ordered alternatives to the null hypothesis. Variants of the test are used for **contingency tables** and for **analysis of variance**.

cumulative distribution function A synonym for the **distribution function**.

cumulative error An error which, in the course of the cumulation of a set of observations, does not tend to zero. The relative magnitude of the error does not then decrease as the number of observations increases.

cumulative frequency (probability) curve See **distribution function**.

cumulative frequency (probability) function A synonym for the **distribution function**.

cumulative incidence In the context of **competing risks** the probability that a particular failure has occurred as a function of time. This term is somewhat misleading in that it is closer to the cumulative distribution function than to the cumulative risk. It can be considered as the cumulative distribution of the time of occurrence of a particular failure.

cumulative normal distribution The cumulative frequency function (distribution function) of the **normal distribution**.

cumulative process A development of the regenerative process, and a generalization of the **renewal process**, introduced by Smith (1955a). As its name implies it is concerned with accumulation of regeneration points, or some attribute occurring at such a point, with the passage of time. An alternative formulation by Kendall (1948) presented this process as a variant of the standard **birth and death process**.

cumulative risk (or hazard) The integral of the **hazard function**.

cumulative sum chart These control charts are intended to replace the standard form; the aim is to make it unlikely that 'lack of control' will be indicated when a process is 'in control' or that a marked change in population mean will remain undetected. In cumulative sum control charts cumulative totals are plotted against the number of observations. If $\bar{x}_j$ denotes the mean of the jth sample, and σ the known standard deviation of $\bar{x}_j$, then it is convenient to consider the points on the control chart as having coordinates (m, X_m), where

$$X_m = \sigma^{-1} \left[(\bar{x}_1 - \mu) + (\bar{x}_2 - \mu) + \ldots + (\bar{x}_m - \mu) \right]$$

and μ is the target mean. In practice this can be effected by calculating the cumulative sum

$$\sum_{j=1}^{m} (\bar{x}_j - \mu)$$

and using an appropriate scale.

cumulative sum distribution Generally, if $X_1, \ldots, X_n$ are a set of random variables the distribution of

$$X \equiv \sum_{i=1}^{n} X_i$$

is the cumulative sum distribution.

current life table A life table in which each survival probability is based on current experience (say, deaths in the age group in the preceding year). The corresponding life expectations do not relate to any real population; contrast **cohort life table**.

current status data A type of censored data in which it is observed whether an event has occurred or not at a particular time of observation. This comes from a particular case of discrete time observation where there is only one time of observation and the data are a case of interval-censored data where the time axis is divided in two intervals. [See **interval censoring**.]

curtailed inspection In quality control, inspection is said to be curtailed if it is stopped at some point otherwise than is provided for by the sampling inspection plan. Usage, however, is not uniform and the expression is also found to denote the stoppage of inspection provided by the plan itself, e.g., in 'cutting off' before acceptance or rejection boundaries are reached in **open sequential schemes**. [See also **cut-off, truncation**.]

curtate A word used in vital statistics to denote the integral number of years, as distinct from the nearest number of years, for which a given state has existed. For example, if an assurance matures in 3 years 9 months, the curtate duration is 3 years. An analysis which uses curtate periods will, in most cases, be subject to downward bias.

curve fitting An expression used in two rather different senses in statistics: (*i*) to denote the fitting of a mathematically specified frequency curve to a frequency distribution (see **Pearson curves**); (*ii*) to denote the fitting of a mathematical curve to any statistical data capable of being plotted against a time or space variable, e.g., regression data or time series.

curved exponential family A representation of a distribution in exponential family form in which the canonical (natural) parameters are expressed in terms of a smaller number of further parameters via nonlinear functions. The canonical statistic (minimal sufficient statistic) has dimension greater than that of the parameter space.

curvilinear regression See **nonlinear regression**.

curvilinear trend A trend that is not linear. It may be expressed by a polynomial, a more complicated mathematical expression such as a logistic curve, or some smoothing process such as a moving average or kernel function.

CUSUM procedures Statistical procedures for detection of changes based on the cumulative sums of residuals or equivalently based on the partial sums of residuals; recursive version is based on partial sums of recursive residuals, non-recursive version is based on partial sums of non-recursive residuals.

cut-off The artificial truncation of a sampling process at a point when it becomes apparent that enough data have been collected for the purpose in view. [See also **sequential analysis**.]

cycle In time series, any periodic variation may be described as a cycle. Often, however, the term is reserved for cycles generated by the autoregressive structure of the series, as opposed to **seasonal variation**, caused by outside influences. A disturbance to the series may affect the phase of the cycle in this sense, while a seasonal variation has always the same phase.

cyclic design A class of balanced and partially balanced experimental design in which the blocks are constructed by cyclic permutations of the treatments.

cyclic order An arrangement of n permutations of n objects such that if the first is denoted by $1, \ldots, n$ the second is $2, \ldots, n, 1$ and the third $3, \ldots, n, 1, 2$, etc. The process provides some of the basic **Latin square** designs; for example, with four treatments

$$
\begin{array}{cccc}
A & B & C & D \\
B & C & D & A \\
C & D & A & B \\
D & A & B & C \\
\end{array}
$$

cylindrically rotatable design A development of the **rotatable (response surface) design** proposed by Herzberg (1966) in which the variances of the estimated responses at points on the same $(k - 1)$-dimensional hypersphere centred on a specified axis are equal.

D

D-optimal design An experimental design chosen to minimize the generalized variance of the estimated parameters. In a D_s-optimal design the generalized variance of a subset of the parameters is considered (Kiefer and Wolfowitz, 1960). [See **design optimality**.]

D^2 statistic A quantity introduced by Mahalanobis (1930) as a measure of the **distance** between two populations with different means but identical dispersion matrices. The distance between the populations is

$$\Delta^2 = \sum \alpha_{jk}(\mu_{1j} - \mu_{2j})(\mu_{1k} - \mu_{2k})$$

where (α_{jk}) is the inverse of the dispersion matrix; Δ^2 is estimated by D^2.

D_n^+ statistic A statistic introduced by Wald and Wolfowitz (1939) in connection with setting confidence limits to continuous distribution functions:

$$D_n^+ = \sup_{-\infty < x < \infty} \{F_n(x) - F(x)\}$$

$$= \max_{1 < i < n} \left(\frac{i}{n} - U_i\right)$$

where U_i is an ordered sample from a uniform distribution $(0, 1)$. This statistic may also be used as a **goodness-of-fit** test (see **Kolmogorov–Smirnov test**) and may be generalized as

$$D_n^+(\gamma) = \sup \{F_n(x) - \gamma F(x)\}.$$

δ-index (Gini) See **concentration, index of**.

d-separation theorem For variables whose distribution is specified by a directed acyclic graph (DAG), a result due to Pearl (1998) gives for three sets of variables A, B, C the conditions under which A is conditionally independent of B given C. The letter 'd' stands for directed (Lauritzen, Thiesson and Spiegelhalter, 1994).

D'Agostino's test A test for normality (D'Agostino, 1971) based on the ratio of a linear unbiased estimate of the standard deviation (using order statistics) to the usual mean square estimate.

Dalenius, Tore (1917–2002) A Swedish statistician who graduated from the University of Uppsala. Dalenius founded the Survey Research Center at Statistics Sweden and eventually became professor in statistics at the University of Stockholm. Dalenius's main contributions include work on sampling schemes, optimal stratification, non-sampling errors in surveys, and cognitive methods in questionnaire design. He was one of the founders of the International Association of Survey Statisticians (IASS) and was its first scientific secretary (1973–1977).

Dalenius's theorem If a continuous distribution is subdivided into groups, the mean square deviation of a point from the mean of its group has a stationary value when the divisions are equidistant from the means of the groups they divide. The result was given by Dalenius (1950) in the context of optimal stratification.

damped oscillation In an oscillatory time series, if the amplitude from peak to trough progressively decreases along the series it is said to be subject to a damped oscillation. Certain derived series of a time series, such as the **correlogram**, are also said to be subject to a damped oscillation when they exhibit this effect.

damping factor If, in a damped oscillation, the amplitude (peak to trough) diminishes at a constant rate along the series, the ratio of one amplitude to the preceding amplitude is called the damping factor. For example, the correlogram of a second-order autoregressive scheme may be written in the form

$$r_k = \frac{\gamma^k \sin(k\theta + \psi)}{\sin \psi}, \qquad k \geqslant 0,$$

and the damping factor is γ.

Dandekar's correction An adjustment proposed by Dandekar (Rao, 1973) in the calculation of χ^2 for a 2×2 table. [See also **Yates's correction**.]

Daniel, Cuthbert (1905–1997) An American statistician. He received both a BS and MS in chemical engineering (1925, 1926). An interest in physics led him to the University of Berlin for a year but he soon abandoned physics to return to the United States as an instructor at Cambridge School, Kendall Green, MA. Later, for four years, he taught physics and the sciences to teachers in New York City. The journal *Technometrics* began publication in 1959 and Cuthbert Daniel was one of its early associate editors. In volume 1 of the journal he published his paper 'Use of half normal plots in interpreting factorial two-level experiments' and in volume 3 the paper 'Locating outliers in factorial experiments'. Both papers are motivated by a practical concern for the problems faced by experimenters dealing with data.

Daniels, Henry Ellis (1912–2000) Henry Ellis Daniels was born in London. He graduated from the University of Edinburgh. After working in textile research, he went to Cambridge as a lecturer, and became professor in mathematical statistics at the University of Birmingham. His major contributions are in the theories of saddle point approximations, the strength of composite materials, and epidemics.

Daniels' rank correlation coefficient A general formulation of rank correlation based on scores which includes virtually all other rank correlation coefficients as special cases (Daniels, 1944).

Daniels' test A distribution-free test for trend (Daniels, 1950), the basis of which is to treat a time series

$$(t_i, X(t_i)), \qquad i = 1, 2, \ldots, n$$

as a sample of n observations from a bivariate population, and test for trend by means of **Spearman's** ρ.

Darmois–Koopman's distributions An exponential family of distributions introduced by Darmois (1953), Koopman (1936) and Pitman (1936) of the form

$$f(x) = \exp\left\{ \sum_{j=1}^{k} A_j(\theta)B_j(x) + C(x) + D(\theta) \right\}.$$

It is the most general form possessing a set of k jointly sufficient statistics for its k parameters. [See **sufficiency**.]

Darmois–Koopman–Pitman theorem This theorem states that under certain regularity conditions on the probability density, a necessary and sufficient condition for the existence of a sufficient statistic of fixed dimension is that the probability density belongs to the exponential family. [See **Darmois–Koopman's distribution**.]

Darmois–Skitovich theorem A theorem due to Darmois (1951) and Skitovich (1954) stating that a sufficient condition for independent, but not necessarily identically distributed, variables to have normal distributions is the independence of any two linear functions of the form

$$a_1X_1 + \ldots + a_nX_n, \qquad b_1X_1 + \ldots + b_nX_n, \qquad \text{with } a_ib_i \neq 0 \ \ (i = 1, \ldots, n).$$

data; datum Data are characteristics or information, usually numerical, that are collected through observation ('data' is plural; 'datum' is singular, although the former term is often used as if it were singular). The word can also be used to describe statistics (i.e., aggregations or transformations of raw data).

data analysis Usually carries the imputation **exploratory data analysis**, or is a translation of the equivalent French *analyse des données*.

data augmentation algorithm A version of multiple imputation used in Bayesian latent variable models, a precursor of Gibbs sampling.

data depth Approaches to ordering multivariate data towards a centre analogous to the median of a univariate sample (Liu, 1990).

data sharpening A simple data-perturbation method that can be used to reduce bias (Choi and Hall, 1999). Alternatively, the preprocessing of data input to a standard estimation procedure in order to improve its performance.

database In computer technology, a file for storing data with indexing facilities for extracting specified groups of entries.

David–Barton test See **Siegel–Tukey test**.

David's empty cell test A distribution-free test proposed by David (1950) of the null hypothesis that a continuous population has a particular completely specified probability density function against the general alternative hypothesis that it has not. The idea behind the test is that if the area under the hypothesized density function is divided into a number of vertical strips of equal area, the bases of which are considered as cells, then a simple random sample from the actual population would tend to be more evenly dispersed among the cells if the hypothesis were true than if it were false. The number of empty cells should thus provide a reasonable index of the truth or falsity of the hypothesis.

de Finetti's theorem The result (de Finetti, 1937) that an infinitely exchangeable sequence of binary random variables can be regarded as a sequence of independent Bernoulli trials with probability p, where p itself has a probability distribution. The result is in particular important in the development of de Finetti's theory of personalistic probability.

de Moivre, Abraham (1667–1754) The first textbook of a calculus of probabilities to contain a form of local central limit theorem grew out of the activities of the lonely Huguenot de Moivre who was forced up to old age to make his living by solving problems of games of chance and annuities on lives for his clients whom he used to meet in a London coffeehouse. Author of *Doctrine of chance* (1718).

death process A stochastic process in continuous time defined on the state space of non-negative integers, representing the decay of a population in which only deaths may occur. The expression is usually confined to processes for which instantaneous transitions to the next lower integer only are possible, the probability of a transition from n to $n - 1$ in time dt being asymptotically $\mu_n dt$; μ_n may be time-dependent.

death rate The number of deaths in a given period divided by the population exposed to risk of death in that period. For human populations the period is usually one year and if the population is changing in size over the year the divisor is taken as the population at the mid-year. The death rate as so defined is called 'crude'. If some refinement is introduced by relating mortality to the age and sex constitution (or other factors) for comparative purposes the rate is said to be standardized.

decapitated distribution A negative binomial distribution from which the zero class is missing. The term 'decapitated' was employed by Yule (1944) to characterize a truncated distribution from which the zero class only had been removed.

decile One of the nine values of a random variable which divide the total frequency into 10 equal parts. [See also **quantile**.]

decision function In its most general form a decision function is a rule of conduct which, at any stage of a sampling investigation, specifies whether to take further observations or whether enough information has been collected, and in the latter case, what decision to make upon it. At each stage beyond the first the decision function is a function of the preceding observations. Until the development of sequential methods, decision functions were mostly of a simple type, based on a fixed sample size, which enjoined the acceptance or rejection of a hypothesis or set limits to a parameter under estimate (Wald, 1950). The above definition provides for a sequential situation wherein the investigator may not reach a decision about the hypothesis but proceeds to take further observations. The class of all decision rules which are admissible in the circumstances of a particular case is called a **complete class**. The **risk function** $R(\theta, d)$ permits a natural ordering of decision rules. Let Θ be the parameter space. A decision rule d_1 is said to be *as good as* a rule d_2 if $R(\theta, d_1) \leqslant R(\theta, d_2)$ for all $\theta \in \Theta$. A rule d_1 is said to be *better than* a rule d_2 if $R(\theta, d_1) \leqslant R(\theta, d_2)$ for all $\theta \in \Theta$, and $R(\theta, d_1) < R(\theta, d_2)$ for at least one $\theta \in \Theta$. A rule d_1 is said to be *equivalent to* a rule d_2 if $R(\theta, d_1) = R(\theta, d_2)$ for all $\theta \in \Theta$.

decision rule An alternative name for a **decision function**.

decision space In analysis of decisions, the set of all possible decisions.

decision theory See **Bayes' decision rule**, **minimax strategy**.

decomposable models A subset of log-linear models which have closed form of MLEs. They also are a subset of **graphical models** for categorical variables, and they can be defined recursively: a decomposable model can be decomposed into two decomposable models, and a complete graph represents a decomposable model (Whittaker, 1990).

decomposition The act of splitting a time series or other system into its constituent parts. A typical time series is often regarded as composed of four parts: (*i*) a long-term movement or trend; (*ii*) oscillations of more or less regular period and amplitude about this trend; (*iii*) a seasonal component; (*iv*) a random, or irregular, component. Any particular series need not exhibit all of these but those which are present are usually presumed to act in an additive fashion, i.e., are superimposed, and the process of determining them separately is one of decomposition. A more modern approach (Wold, 1938) seeks to decompose the series into deterministic and non-deterministic elements. This is known as Wold's decomposition or predictive decomposition.

deconvolution In a sense the inverse of convolution but in a statistical framework. If $f = h * g$ the problem is to find g knowing f and h. Most often f

is observed with noise and this is an **ill-posed problem** so that some regularization techniques are very useful. Applications in signal restoration and in epidemiology. [See also **backcalculation**.]

decreasing failure rate A distribution in which the hazard rate decreases in time. The summary description old better than new applies.

decreasing hazard rate Conventionally, a hazard rate that does not increase; it can, therefore, be constant (Barlow, Marshall and Proschan, 1963). [See also **increasing failure rate**.]

defective probability distribution A probability distribution normally assigns a probability measure over the appropriate interval in such a way that the sum of all the assigned probabilities is unity. However, probability measures are sometimes encountered which assign a total mass, say p, which is less than unity. Such a measure is called a defective probability measure with defect $1-p$; in a sense, the missing probability may be regarded as at infinity.

defective sample A sample resulting from an enquiry which has been incompletely carried out, e.g., because certain assigned individuals have not been examined, because records have been lost, or because (in plant or animal experiments) certain members have died.

defective unit In quality control, a unit which does not reach some prescribed standard and is therefore to be rejected, in contrast to an effective or acceptable unit.

defining contrast In the analysis of fractional–factorial designs the comparisons between those treatment combinations which are used with those that have not been used is called the defining contrast (Finney, 1945). They define the sectors into which the whole replicate is divided.

degree of belief Term used in connection with a view of probability emphasizing assessment of uncertainty of events or propositions. There are broadly two approaches, one based on the notion of reasonable degree of belief given suitable evidence and the other stressing the notion of 'your' degree of belief. Special arguments are needed to show that the usual mathematical properties of probability hold. In the personalistic theory these are requirements of internal consistency or coherence.

degrees of freedom Term borrowed from dynamics for number of independent components or in a narrower sense for the rank of a quadratic form. In **analysis of variance** the degrees of freedom associated with a particular set of parameters are the number of linearly independent effects that can be estimated and a similar definition applies to the associated sums of squares, or in **generalized linear models** to the log-**likelihood ratio test**. A particularly important use of the term is in connection with the **chi-squared distribution** with d degrees of freedom defined as the distribution of the sum of squares of

d independent random variables each with a normal distribution of zero mean and unit variance.

DeGroot, Morris H. (1931–1989) An American statistician. He wrote three books, *Optimal statistical decisions* (1970), *Probability and statistics* (1975) and *Bayesian analysis and uncertainty in economic theory* (1987). He was an elected fellow of the American Statistical Association, the Institute of Mathematical Statistics, the International Statistical Institute and the Econometric Society. Most of his research was on the theory of rational decision making under uncertainty.

Delaunay triangulation A tessellation associated with a point process (Delaunay, 1963). Those points of the process are joined for which the cells given by the **Dirichlet tessellation** have a side, or face, in common.

delta distribution A lognormal distribution with a positive probability mass at zero (Aitchinson, 1955).

delta method A procedure for obtaining an approximate variance by Taylor series expansion of a statistic, usually for use with a normal approximation.

Delthiel polygon See **Dirichlet tessellation**.

Deming, W. Edwards (1900–1993) Long recognized as a leading proponent of statistical sampling and approaches to quality improvement, Deming led major efforts to bring statistical approaches and methods to bear on problems in government, science and industry and was deeply concerned with the broader aspects of achieving high quality (Deming, 1950).

demodulation A technique used in spectrum analysis of time series for the purpose of detecting and describing the more unusual kinds of non-stationary behaviour. The method consists of multiplying a series by a suitable function and applying a **filter** to leave only the frequency in a narrow band.

demography A broad social science discipline concerned with the study of human populations. Demographers deal with the collection, presentation and analysis of data relating to the basic life-cycle events and experiences of people: birth, marriage, divorce, household and family formation, employment, aging, migration and death. The discipline emphasizes empirical investigation of population processes, including the conceptualization and measurement of these processes and the study of their determinants and consequences. Practitioners frequently draw on related disciplinary areas – sociology, economics, political science, anthropology, psychology, public health and ecology – to illuminate their analyses. They may explore biological and biosocial aspects of fertility and mortality in areas such as reproductive health and epidemiology.

dendrogram A diagram of a hierarchical kind exhibiting the relation of subsets of a structure, like a pedigree chart; so called because the diagram (usually read downwards from the main trunk) branches like a tree.

density estimation Methods for estimating **density functions** from data without assuming any functional form. **Histograms** and **frequency polygons** are classical methods: more recently **kernel density estimator** methods and **spline** techniques for smoothing histograms have been introduced.

density function See **probability distribution**.

departure process The stochastic process formed by the intervals between the departure times from the service counter or station of a queueing situation. This departure process is of importance in the study of tandem queues (Jackson, 1957).

dependence Quantities are dependent when they are not independent. [See **independence**.] For dependence in regression analysis, see **regression**.

dependent variable See **regression**.

descriptive statistics A term used to denote statistical data of a descriptive kind or the methods of handling such data; more broadly, methods of analysis, graphical or tabular, without any probabilistic formulation. When the data are determined by national authorities they are referred to as official statistics.

descriptive survey A (sample) survey where the principal objective is to estimate the basic statistical parameters (means, totals, ratios) of the population or its subdivisions.

design of experiments The design of investigations in which the investigator has essential control over the system, including the allocation of '**treatments**' to '**experimental units**'.

design matrix A matrix which specifies the levels of factors to be used in an experimental design, i.e., an $n \times p$ matrix $\mathbf{X}$ that specifies observations by treatment combinations in a linear model $E(\mathbf{Y}) = \mathbf{X}\theta$ related to a given design of experiment or a classification model (Dodge, 1985).

design optimality The study of designs that give optimal estimates of the parameters. Kiefer and Wolfowitz (1959) considered a generalization of the covariance matrix of the least squares estimates to a moment matrix for a general linear model, and discussed minimization of functions of this matrix. Particular choices of this function give A-optimality, D-optimality and E-optimality. Further work relates to optimal designs for autocorrelated observations (Pukelsheim, 1993).

destructive test Sometimes it is possible that the carrying out of an inspection test on a manufactured product will result in destruction of the particular test specimen. This is called a destructive test. Under such conditions there is every incentive to design sampling schemes which minimize the number of items to be tested.

determinant correlation See **vector alienation coefficient**.

111

determination, coefficient of The square of the product moment correlation between two variables, R^2, so called because it expresses the proportion of the variance of one variable, Y, given by the other, X, when Y is expressed as a linear regression on X. More generally, if a dependent variable has **multiple correlation R** with a set of independent variables, R^2 is known as the coefficient of determination. The quantity is also known as the index of determination.

deterministic model A deterministic model, as opposed to a stochastic model, is one which contains no random elements and for which, therefore, the future course of the system is determined by its position, velocities, etc., at some fixed point of time.

deterministic process A stochastic process with a zero error of prediction; one in which the past completely determines the future of the system.

detrimental variable In confluence analysis, when a new variable is added to a set of explanatory variables the fit may be made worse. On the 'bunch map' the pencil of beams becomes less compact than before. The new variable is then regarded as detrimental.

deviance A term proposed by Kendall to denote the sum of squares of observations about their mean. A measure proposed by Nelder and Wedderburn (1972) for judging the degree of matching to the data of the model when the parameter estimation is carried out by maximizing the likelihood. The quantity $-2L_{\max}$, the deviance, is, for example, $\sum (x_i - \mu)^2/\sigma^2$ for the normal distribution where x_i are the observations.

deviance residual A residual constructed as the signed square root of a deviance contribution in a **generalized linear model** (McCullagh and Nelder, 1989).

deviate The value of a random variable measured from some standard point of location, usually the mean. It is often understood that the value is expressed in standard measure, i.e., as a proportion of the parent standard deviation.

diagonal regression A suggestion of Frisch (1934) that in fitting a straight line to data with both variables subject to error the line should pass through the mean with slope the ratio of the standard deviations of the two variables. It is not a regression in the sense of defining a conditional mean or other location measure.

dichotomous variable See **binary data**.

dichotomy A division of the members of a population, or sample, into two groups. The definition of the groups may be in terms of a measurable variable but is more often based on quantitative characteristics or attributes.

difference of zero A set of numbers represented in a convenient way to provide certain probabilities. Usually, the notation

$$\Delta^r 0^s = \sum_{j=0}^{r} (-1)^j \binom{r}{j} j^s$$

is termed a difference of zero, or a leading difference.

difference sign test When considering observations on a phenomenon which is moving through time and generates an ordered set, i.e., a time series, one useful test against trend etc. is the difference sign test. This consists of counting the number of positive first differences of the series; that is to say, the number of points where the series increases. It is used especially as a test against linear trend and as such is superior to the **turning point** test but inferior to tests based on rank order.

differential–difference equation A type of equation often arising in stochastic processes, in which both derivatives and differences appear. It can often be reduced to a differential equation by the use of generating functions or integral transforms.

differential process See **additive process**.

diffuse prior See under **informative prior**.

diffusion index A term proposed by Burns and Mitchell (1946) and Moore (1950) to denote the proportion of a set of time series in a given collection of series which are increasing at a given point of time.

diffusion process A type of **additive process** describing certain kinds of diffusion. The process is such that the 'displacement' of the random variable (its increment) in time dt follows a normal distribution with variance proportional to dt.

digamma function The derivative of the logarithm of the gamma function, also sometimes known as the psi function.

dimension reduction In multivariate analysis, **ordination** methods, in which p-variate observations are represented approximately in fewer dimensions.

Dimroth–Watson's distribution A **spherical distribution** of the symmetric girdle type, introduced independently by Dimroth (1963) and Watson (1965); see also Dryden and Mardia (1991). Its probability density function is

$$f(l, m, n) = \frac{b(K)}{2\pi} \exp\left\{ -K(l\lambda + m\mu + n\nu)^2 \right\}, \qquad -\infty < K < \infty$$

where (l, m, n) and (λ, μ, ν) are the vectors of direction cosines of a three-dimensional directional random variable and its mean direction respectively, and

$$b(K) = \frac{1}{2 \int_0^1 e^{-Kt^2} dt}.$$

direct correlation See **correlation, coefficient of**.

direct probability An expression which is supposed to be antithetical to **inverse probability**, although neither expression is very logical. It usually denotes probability when used to proceed from the given probabilities of prior events to the probabilities of contingent events; for example, if it is given that the probability of throwing each number with an ordinary six-faced die is $\frac{1}{6}$ the probability of throwing a score of 15 in three throws is directly ascertainable. The relation is analogous to the deductive relations of logic.

direct sampling A term used when the sample units are the actual members of the population and not, for instance, some kind of record relating to such numbers, such as census form, ticket or registration card. The term relates to the directness of the observation of the units which enter into the sample, not to the process by which they are selected. [See also **indirect sampling**.]

direct standardization It is often required to compare death rates (or morbidity rates etc.) after allowing for differences in the population structure in terms of age or sex. The direct method of standardization weights the rate in each group p_i by the number N_i in a standard population to give a standardized death rate $\sum N_i p_i / \sum N_i$. The indirect method calculates the expected deaths that would occur if standard rates P_i applied to the numbers n_i in each group, and gives a standardized death rate $(d / \sum n_i p_i) \times S$, where d is the total observed deaths and S is a standard death rate. This is S times the **standardized mortality ratio**. The indirect method does not require knowledge of p_i.

directed acyclic graph (DAG) A representation of the dependencies among a set of variables in which each variable is represented by a node and in which some but in general not all pairs of nodes are joined by directed edges in such a way that no directed cycle is present. A missing edge represents a particular kind of conditional independence. DAGs have developed from Sewall Wright's path analysis and are widely used in Bayesian belief and knowledge systems and in empirical statistical analysis (Pearl, 1998).

directional data Data representing points on a hypersphere; sets of observations $x_1, \ldots, x_p$ satisfying $\sum x_i^2 = 1$.

Dirichlet, Johann Peter Gustav Lejeune (1805–1859) Dirichlet was educated at the University of Göttingen, where Carl Friedrich Gauss was one of his mentors. He is known for a theorem in which the existence of an infinite number of primes in any arithmetic series is proven. He gave the first set of conditions sufficient to guarantee the convergence of a Fourier series under the so-called Dirichlet conditions.

Dirichlet distribution The multivariate analogue of the **beta distribution**.

Dirichlet series distribution A power series distribution defined by Siromoney (1964) by the formula

$$\Pr(X = k) = \frac{a_k e^{-\lambda_k \theta}}{f(\theta)}, \qquad k = 0, 1, 2, \ldots$$

where

$$f(\theta) = \sum_{j=0}^{\infty} a_j e^{-\lambda_j \theta}$$

(supposing the series to converge).

Dirichlet tessellation Associated with any realization of a point process in space is a system of polygons in two dimensions, or polyhedra in three dimensions, containing those points closest to each point of the process. This is the Dirichlet tessellation. The polygons are also sometimes referred to as Voronoi, Delthiel or Thiessen polygons (Sibson, 1980).

dirty data Data with missing, unusual or incorrect values. A term used in data management strategies.

disability adjusted life year (DALY) Somewhat arbitrary measure used in international and other comparisons of life expectancy in which years of life subject to various handicaps are downweighted. Similar to **quality adjusted life year (QALY)** which tends to be used more in a health economics context of allocation of resources for health care.

discontinuous process See **continuous process**.

discontinuous variable A random variable which can take only a discontinuous set of values.

discordant sample See **concordant sample**.

discounted least squares method An extension by D'Esopo (1961) of the basic least squares analysis incorporating weights to the squared deviations in the form of an exponentially decaying series.

discrete circular uniform distribution See **circular lattice distribution**.

discrete distribution Probability distribution of a **discrete random variable**.

discrete lognormal distribution A distribution obtained as the compound of a Poisson distribution with the mean parameter λ distributed lognormally.

discrete random variable A **random variable** with a countable number of possible values is called a discrete random variable. That is, X is discrete if there is a finite or countable sequence of numbers $x_1, x_2, \ldots$ such that $\Pr(X = x_1) + \Pr(X = x_2) + \ldots = 1$ and $x_1, x_2, \ldots$ are called possible values of X.

discrete wavelet transformation A class of fast orthogonal decompositions widely used in signal processing, image compression and related areas of statistics owing to its ability to approximate complex signals using small numbers of coefficients, these being estimated by penalized least squares or similar criteria.

discriminant analysis Given a set of multivariate observations on samples, known with certainty to come from two or more populations, the problem is to set up some rule which will allocate further individuals to the correct population of origin with minimal probability of misclassification. This problem and sundry elaborations of it give rise to discriminant analysis.

dishonest process A process, in particular a stochastic process, which does not satisfy some physically 'natural' properties and is thus in a sense pathological. See **honest process**.

disorder problem Older notion for **change point problem**.

dispersion, analysis of An obsolete term for **multivariate analysis of variance**.

dispersion index A statistic used to compare observed variability with that to be expected under some simple model and often forming the basis of a significance test of agreement with that model. Particularly important cases are statistics that compare observed variance with that expected under a Poisson distribution, under a binomial distribution and under an exponential distribution, in all of which the variance is a known function of the mean. Then the index of dispersion is the ratio of the estimated variance to the appropriate function of the sample mean.

dispersion matrix See **covariance matrix**.

dispersion stabilizing transformation A term proposed by Ruben (1966) for a generalization of the variance stabilizing transformation. The generalization covers the stabilization of covariance matrices.

displaced Poisson distribution A situation in which the number of events in excess of a threshold value r, when it is assumed that at least the r events do occur, has a displaced Poisson distribution with parameters λ and r

$$\Pr(x) = \frac{e^{-\lambda}\lambda^{x+r}}{I(r,\lambda)(x+r)!},$$

where

$$I(r,\lambda) = \sum_{\gamma=r}^{\infty}(e^{-\lambda}\lambda^{\gamma})/\gamma!.$$

disproportionate subclass numbers In the **analysis of variance**, except in the case of analysis by reference to a single classification, the arithmetic and estimation of class effects are greatly simplified when there are equal numbers of observations in the subclasses or when the numbers are proportionate. In the contrary case, the subclass numbers are said to be disproportionate and the analysis, though theoretically straightforward, is much more complicated.

dissimilarity Any measure of dissimilarity between individuals, usually defined to be symmetrical ($d_{ij} = d_{ji}$), non-negative, and zero only for identical individuals.

distance This word is used in many statistical contexts in its ordinary sense, e.g., the 'distance' of a value x from some origin a is $x - a$. A specialized use occurs in the notion of 'distance' between two random variables, X and Y, which may be defined as the **expected value** of $X - Y$; or the 'distance' between two populations, which may be defined as the difference of their means (though other definitions are possible). [See also **Bhattacharyya's distance**, D^2 **statistic**.]

distance measures Measures of distance between two points in multi-dimensional space, such as Euclidean distance or city-block metric. [See also L_2-**metric** and L_1-**metric**.]

distributed lag A term introduced by Irving Fisher (1925a) in connection with the analysis of correlation between time series. It is based on the assumption that a given cause occurring at one point of time will exert its effect at various future points and will thus be 'distributed' over terms which lag behind the original cause.

distribution The theoretical or empirical distribution of a random variable may be expressed in classical terms, as a set of probabilities or probability densities, as a frequency distribution, or as a percentage distribution. The classical way to define or summarize a distribution is by providing theoretical or observed probabilities (or probability densities) at each value that the random variable can achieve, or for each of a set of mutually exclusive, exhaustive categories of values that the random variable can achieve. These probabilities add (or densities integrate) to one. A frequency distribution (for observed data) presents the frequency (count) of observations in each of a set of mutually exclusive, exhaustive categories of values that the random variable can achieve. The frequencies add to the total number of observations. A percentage distribution (for observed data) presents the percentage of observations in each of the categories. The percentages add to 100 per cent.

distribution-constant statistic A statistic whose distribution does not depend on the parameters of an underlying model. Ancillary statistics are special cases.

distribution curve See **distribution function**.

distribution-free method A method, for example, of testing a hypothesis or of setting up a confidence interval, which does not depend on the form of the underlying distribution; for example, confidence intervals may be obtained for the median, based on binomial variation, which are valid for any continuous distribution. Distribution-free inference or distribution-free tests are sometimes known as non-parametric but this usage is confusing and should be avoided. It is better to confine the word 'non-parametric' to the description of hypotheses which do not explicitly make an assertion about a parameter.

distribution-free sufficiency A concept introduced by Godambe (1966) and related to **linear sufficiency**. An estimation is said to be distribution-free sufficient if any other estimator is independent, in terms of the accumulated prior knowledge, of the original estimator and the population value under estimation.

distribution function The distribution function $F(x)$ of a random variable X is the total frequency of members with values less than or equal to x. As a general rule the total frequency is taken to be unity, in which case the distribution function is the proportion of members bearing values $\leqslant x$. Similarly, for p variables $X_1, X_2, \ldots, X_p$ the distribution function $F(x_1, x_2, \ldots, x_p)$ is the frequency of values less than or equal to x_1 for the first variable, x_2 for the second, and so on.

distribution of run lengths The distribution of lengths of runs of attributes in a series. A run is defined as a sequence of items of the same kind terminated by one or more items of another kind.

disturbed harmonic process This form of stochastic process was proposed by Yule (1927) to explain the continual change of amplitude and shift of phase which appears to be typical of time series from the economic and meteorological sciences. If a series is observed at points $t = 1, 2, \ldots$ an ordinary harmonic movement may be expressed by the equation

$$u_{t+2} - 2u_{t+1} + u_t = 0.$$

If we replace the zero by a random term ε_{t+2} on the right-hand side the motion is said to be that of a disturbed harmonic. It is a limiting case of an **autoregressive process** but is not stationary.

disturbed oscillation A time series which exhibits a continual shift of period and amplitude in its oscillation is said to possess a disturbed oscillation.

divergence A measure of distance between two probability distributions, defined as

$$J = \int [f_1(x) - f_2(x)] \log \frac{f_2(x)}{f_1(x)} dx.$$

It is invariant under non-singular transformations. This is linked to the **Kullback–Leibler information**.

divergence, coefficient of A coefficient introduced by Lexis to measure departure from simple or Bernoullian sampling in the sampling of attributes. [See **Lexis ratio**.]

diversity, index of Various measures of the spread of an ecological community over species or other groups. The **Shannon–Wiener index** is the most widely used.

Divisia–Roy index An index number of Divisia's type for prices, constructed as a **Konyus index number** by taking the quantities to be optimal for a fixed nominal income of the consumer.

Divisia's index An index number due to Divisia (1925). It is in the form of a chain index. If prices p_i and quantities q_i are regarded as functions of the time, the price index is defined as

$$I_p = \exp\left(\int_c \frac{\sum q_i dp_i}{\sum q_i p_i}\right)$$

where c denotes the path of the prices. There is a similar form for a quantity index, I_q. The indices have the property that changes in total expenditure are proportional to the product of I_p and I_q.

Dixon's statistics A class of statistics for the rejection of the largest observation proposed by Dixon (1950, 1951) of the form

$$r_{ij} = \frac{x_{(n)} - x_{(n-i)}}{x_{(n)} - x_{(j+1)}}, \qquad i = 1, 2, \qquad j = 0, 1, 2$$

where $x_{(k)}$ means the kth value of a sample of n when ranked by size.

Dodge continuous sampling plan The first **continuous sampling plan** was proposed by Dodge (1943). It assumed a knowledge of the production process. This restriction was relaxed by Derman, Johns and Lieberman (1959) who made other developments.

domain of study In sample surveys a certain subgroup of a population may be of particular interest. Such a subgroup of interest is called a 'domain'. Domains frequently cut across the strata and the various stage units of a sample. Formulae for domains cutting across strata have been devised.

dominating strategy A decision strategy which is in no respect worse than all others and in some respect is better.

Donsker's theorem A theorem due to Donsker (1951) concerning the insensitivity of the asymptotic distribution of certain functions of a sequence of independent and identically distributed random variables with zero mean and finite positive variance to changes in the distribution of these variables. The theorem is alternatively known as the **invariance principle** or the **functional central limit theorem**.

Doolittle technique A method of solving the normal equations in least squares proposed by Doolittle (1878) which proceeds by systematic backward elimination of the variables. It is a more compact version of an approach originally proposed by Gauss (1823). It has been modified and extended by others.

dose–response curve A curve of regression type relating response to dose.

dose–response relationship An expression of the result of a dose or exposure as a function of the amount of the dose or exposure. An observed dose–response relationship in observed studies strengthens, but does not prove, a causal hypothesis. [See **causality**.]

dotplot A one-dimensional scatterplot of the observed values of a variable in which the horizontal axis represents the scale of measurement and the observations are stacked vertically according to their values. Dotplots are useful for visualizing the location and spread of a small set of observations, and for comparing the empirical distributions of a number of relatively small data sets. The dotplot provides a simple alternative to the stem and leaf plot.

double binomial distribution A weighted average of two binomials with parameters p_1 and p_2 with $p_1 < p_2$ and weights w_1 and w_2, with $w_1 + w_2 = 1$.

double-blind A **clinical trial** in which both the patient and the doctor who assesses the response are kept unaware of the identity of the treatment, so that subjective biases are avoided, is said to be double-blind. [See **single-blind**.]

double censoring As described by Groeneboom and Wellner (1992), we observe indicators $I_{T \leqslant G_1}$ and $I_{T \leqslant G_2}$ as in case II **interval censoring** but in addition the exact value of T is observed if it falls between G_1 and G_2.

double confounding A **confounding** of two different groups of treatment contrasts with two different sources of variation in an experimental design. Thus in the row and column layout of a **quasi-Latin square** certain interactions are confounded with rows, and certain others with columns.

double dichotomy The division of a set of members by two dichotomies, usually according to attributes; thus a set may be divided into A's and not-A's and each of these two into two subsets according to whether they bear a second attribute B or not.

double exponential distribution The distribution with density function

$$f(x) = \frac{1}{2} b^{-1} e^{-b|x-c|}, \qquad -\infty < x < \infty,$$

b and c being constants, $b > 0$. The distribution may be regarded as an ordinary exponential together with its reflection about the point c. See also **Laplace distribution**.

double exponential regression An expression sometimes used for a regression equation in which the dependent variable y is a linear combination of two simple exponential terms in x. [See also **mixed exponential response law**.]

double hypergeometric distribution If, from a finite population containing two kinds of item, a random sample is drawn without replacement and then a second random sample without replacing the first, the distribution of the (X_1, X_2) successes takes the form of the double hypergeometric (Pearson, 1924a).

double logarithmic chart A chart in which both the horizontal and the vertical axis are scaled in logarithms usually to base 10.

120

double Pareto curve A continuous frequency function whose ordinate is the sum of two functions of the **Pareto curve** type, e.g.,

$$f(x) = \frac{A}{x^{1+\alpha}} + \frac{B}{x^{1+\beta}}, \qquad \alpha, \beta > 0, \quad x \geqslant 0.$$

double Poisson distribution A distribution in which the parameter λ of the Poisson series is itself regarded as distributed in the Poisson form. The distribution arises particularly in ecology where the numbers of offspring of parents which are themselves distributed over space in the Poisson form are also distributed in that form in the neighbourhood of the parent.

double-ratio estimator An estimator built up from four random variables, X_1, X_2, Y_1, Y_2, by using the ratio of the ratios Y_1/X_1 and Y_2/X_2. [See **ratio estimator**.]

double reversal design An extension of the **switchback design** or reversal design to the case of two treatments over four periods $A_1 B_2 A_3 B_4$ for one-half of the experimental units and $B_1 A_2 B_3 A_4$ for the other half; the suffixes are the time periods. This design allows for trends in the responses which are independent of the treatment effects.

double sampling A standard form of sample design for industrial inspection purposes. In accordance with the characteristics of a particular plan, two samples are drawn, n_1 and n_2, and the first sample inspected. The batch can then be accepted or rejected upon the results of this inspection or the second sample be inspected and the decision made upon the combined result. The term has also been used somewhat loosely for what is called **multi-phase sampling** and the two-stage version of **multi-stage sampling**. There is a further usage whereby a first sample provides a preliminary estimate of design parameters which govern the size of the second sample to achieve a desired overall result. [See **Stein's two-sample procedure**.]

double-tailed test See **two-tailed test**.

doubly non-central F-distribution The distribution of the ratio of two **non-central χ^2** distributions. [See also **non-central F-distribution**.]

doubly stochastic matrix See **stochastic matrix**.

doubly stochastic Poisson process A point stochastic process, suggested by Bartlett, where the rate of occurrence in a Poisson process is itself a stochastic process (Cox, 1955).

doubly truncated normal distribution A distribution of normal (Gaussian) form but with the variable not extending to infinity, being truncated below and above at specified points.

down cross The point where a time series, measured about its mean, changes in sign from positive to negative. Correspondingly a point where it changes in sign from negative to positive is called an up cross.

Downton's estimators A class of estimators of moments based on linear combinations of the order statistics (Downton, 1966).

downward bias Bias which tends to reduce a magnitude below its true value. In index number theory the expression frequently occurs but has a rather obscure meaning owing to the somewhat arbitrary nature of the definition of bias in index numbers. In the theory of estimation an estimator t is biased downwards if $E(t) < \theta$, the parameter under estimation. Similarly, upward bias is such as to tend to raise a magnitude above its true value.

Dragstedt–Behrens' method See **Behrens' method**.

dropout Failure of a unit to complete a longitudinal study.

dual process For every queueing process it is possible to create an associated process by interchanging the distributions of interarrival and service times respectively: this associated process was termed the dual process by Takacs (1965).

dual theorem A fundamental theorem of linear programming. The 'primal' problem in the variables x_j for $j = 1, \ldots, n$ is to maximize $z = c_1 x_1 + c_2 x_2 + \ldots + c_n x_n$ subject to linear programming constraints:

$$a_{i1} x_1 + a_{i2} x_2 + \ldots + a_{in} x_n \leqslant b_i, \qquad i = 1, \ldots, m,$$
$$x_j \geqslant 0, \qquad j = 1, \ldots, n.$$

The 'dual' problem in variables u_i for $i = 1, \ldots, m$ is to minimize $v = b_1 u_1 + b_2 u_2 + \ldots + b_m u_m$ subject to

$$a_{1j} u_1 + a_{2j} u_2 + \ldots + a_{mj} u_m \geqslant c_j, \qquad j = 1, \ldots, n,$$
$$u_i \geqslant 0, \qquad i = 1, \ldots, m.$$

The theorem states that for values of x_j $(j = 1, \ldots, n)$ and u_i $(i = 1, \ldots, m)$ satisfying the above inequalities then $z \leqslant v$, and that if z^*, v^* are the solutions of the optimization problems $z^* = v^*$.

duality In mathematical programming any linear maximization problem can be expressed in its dual linear minimization problem (Arthanari and Dodge, 1981).

dummy observation In **analysis of variance** with **disproportionate subclass numbers** it is sometimes possible to obtain good approximations to the correct analysis by adding 'dummy' observations. These observations are generally inserted with values equal to the means of the particular cells concerned. An analogous use of class means is made to replace missing values in the **missing plot technique**.

dummy treatment In order to preserve the symmetry or other features of an experimental design it is sometimes useful to pretend that treatments are applied to certain units when, in fact, no treatments are applied; they are then called dummies. For example, a factorial experiment on the effect of two fertilizers may provide for the application of each in two different concentrations and also for a control (nil) application. One possible design would provide for two factors at three levels, but the third level of each, involving no application of fertilizer, would be a dummy treatment.

dummy variable A quantity written in a mathematical expression in the form of a variable although it represents a constant; for example, in a regression equation

$$y = \beta_0 + \beta_1 X_1 + \beta_2 X_2 + \ldots + \beta_p X_p$$

it may be more convenient to attach to the coefficient β_0 a dummy X_0 which is always unity so that the expression may be written

$$y = \sum_{i=0}^{p} \beta_i X_i.$$

The term is also used, rather laxly, to denote an artificial variable expressing qualitative characteristics; for example, the presence or absence of an attribute may be indicated by attaching the values 0 or 1 to the individuals concerned.

Duncan's test (*i*) Duncan (1952) suggested a modified form of the **Newman–Keuls test** in which the error rate per experiment is $1 - (1 - \alpha)^k$, where there are k degrees of freedom for treatments (k independent comparisons are possible). (*ii*) Duncan (1955) suggested an entirely different test, based on the idea that losses associated with **type I** and **type II errors** are additive. The mathematics are complicated, but in practice it is equivalent to Fisher's least significant difference test, except that the appropriate t-value is replaced by a quantity dependent on the F-value for testing treatment effects. This quantity decreases as F increases, so that the test may be regarded as a smoothed version of Fisher's proposal to examine the significant differences only when F is significant. The test is sometimes referred to as the k-ratio t-test.

Dunnett's test A **multiple comparison** test intended for comparing each of a number of treatments with a control.

Dunn's test A **multiple comparison** test based on the Bonferroni inequality.

duo–trio test A test in which three objects, two of which are alike, are presented to a judge who attempts to select the dissimilar object after being given the identity of two of them. [See also **triangle test**.]

duplicate sample A sample collected concurrently, i.e., in the course of the same sample survey under comparable conditions, with a first sample. It acts in much the same way as a **replication** except that in some cases the only thing which can be replicated is the act of taking a second independent sample. For example, a second independent sample in a survey will afford additional information on the sampling error but nothing has been replicated in the sense of a repeated experiment beyond the act of selecting a duplicate sample. The act of taking several such samples may be called replicated sampling. [See **interpenetrating sample**.]

duplicated sample A sample which is taken up twice for enquiry by two different parties of investigators. Two **interpenetrating samples** assigned to two investigating parties are sometimes so arranged, for purposes of control of field operations, as to have some common sample units; these common units constitute a duplicated (sub)sample. This should be distinguished from **duplicate sample**.

Durbin–Watson statistic A statistic testing the independence of errors in least squares regression (Durbin and Watson, 1950) against the alternative of serial correlation. The statistic is a simple linear transformation of the first serial correlation of residuals and, although its distribution is unknown, it is tested by two bounding statistics which follow Anderson's distribution.

Durbin's multi-stage variance estimator Durbin (1967) suggested a variance estimator for use in a multi-stage sampling procedure where the method of selection is two units per stratum with unequal probabilities and without replacement.

Dvoretzky's stochastic approximation theorem A general theorem on the convergence of transformations with superimposed random errors (Dvoretzky, 1956). It applies to the stochastic approximation procedures of **Robbins–Munro** and **Kiefer–Wolfowitz**. [See also **Sacks' theorem**.]

Dwass–Steel test A distribution-free multiple comparison procedure proposed independently by Steel (1960) and Dwass (1960). The Dwass–Steel k-sample rank statistic is the maximum **Wilcoxon rank sum test** statistic over all possible pairs of populations.

dynamic model In econometrics a model is said to be dynamic if it possesses either or both of these properties: (i) at least one variable occurs in the structural equations with values taken at different points of time or in the form of time derivatives etc.; (ii) at least one equation contains a function of time. If the first property is present the system is sometimes called multitemporal, and if neither is present it is sometimes called unitemporal. These terms are not ideal; they do not mean that several 'time' variables or one 'time' variable are involved. The term could reasonably be used in many other contexts as a contrast with a static model.

dynamic programming A method for securing a sequence of consistently optimum solutions to multi-stage decision problems; sequential decision taking, in fact (Bellman, 1957).

dynamic stochastic process A class of second-order process introduced by Stigum (1963) which combines a wide sense stationary process and a deterministic non-stationary process in a single representation.

dynamic treatment allocation An experiment, in particular a clinical trial, in which the treatment, perhaps a dosage, at each time may depend on intermediate responses observed up to that point.

E

ecologic study An investigation that examines the correlation between disease/event rates and exposure rates. The unit of analysis is groups rather than individuals. Therefore, it is important to note that the exposure for each individual is unknown. An example of such a study would be an investigation into the correlation between the rate of lung cancer in a group of adult males and the rate of smoking in the same group.

ecological fallacy Errors of interpretation arising by assuming effects established over aggregates of individuals (e.g., countries) apply to individuals. Also called 'aggregation bias'.

econometrics Modelling economic systems, usually with stochastic elements (Green, 2000).

edge effects Issues arising in the analysis of spatial or temporal data as the boundary of the region of data availability is approached.

Edgeworth, Francis Ysidro (1845–1926) His writing reflecting the advantage of a classical education, Edgeworth made contributions to the moral sciences, economics, probability and statistics (Edgeworth, 1885). His early and important work on **index numbers** and utility was followed by later statistical work now perhaps seen as 'Bayesian'. In 1891 Edgeworth became the first editor of the *Economic Journal*, and he continued in one or another editorial role until his death.

Edgeworth index See **Marshall–Edgeworth–Bowley index**.

Edgeworth expansion Asymptotic expansion associated with the **central limit theorem** introduced by Edgeworth (1905). For the simplest case of the sum of n independent and identically distributed random variables, the distribution of the standardized sum is given by the limiting normal distribution adjusted by a series in powers of $1/\sqrt{n}$ involving Hermite polynomials. The series is different from the **Gram–Charlier series** in which the ordering of successive terms is via the degree of the polynomial (Barndorff-Nielsen and Cox, 1989).

effect modifier A feature of study individuals such that a treatment or risk factor has different effect at different levels of the feature, i.e., that there is an interaction between the feature and the treatment. The term is mostly used in an epidemiological context.

effective degrees of freedom See **approximate degrees of freedom**.

effective range A group of observations may contain a limited number of outlying observations at either or both ends of the range. A very rough measure of dispersion may be obtained by taking the 'effective' range, i.e., the range after these outlying values have been removed. The removal may have to be a matter of subjective judgement and inferences based on effective range are of somewhat doubtful value; in fact the term itself is not a good one.

effective unit See **defective unit**.

efficiency The concept of efficiency in statistical estimation is due to Fisher (1922*b*) and is an attempt to measure objectively the relative merits of several possible estimators. The criterion adopted by Fisher was that of variance, an estimator being regarded as more 'efficient' than another if it has smaller variance; and if there exists an estimator with minimum variance v the efficiency of another estimator of variance v_1 is defined as the ratio of v/v_1. It was implicit in this development that the estimator should obey certain other criteria such as **consistency**. For small samples, where other considerations such as bias enter, the concept of efficiency may require extension or modification. The word is also used to denote the properties of experimental designs, one design being more efficient than another if it secures the same precision with less expenditure of time or money.

efficiency equivalence A broad extension of the concept of **asymptotic relative efficiency** to the case of regression parameters in second-order processes (Striebel, 1961).

efficiency factor The efficiency factor of an experimental design is usually expressed in terms of a ratio of **variances** of estimation, or, from a slightly different viewpoint, in terms of the size or number of replications needed to attain a given precision. Thus in an agricultural field trial laid out in incomplete blocks, the efficiency factor is the ratio by which the variance of a given treatment estimate would be reduced by ignoring blocks, if this did not alter the variance per plot. In general the efficiency factor will vary according to the standard of comparison.

efficiency index A concept proposed by Armitage (1959) for comparing the asymptotic relative efficiencies of methods of assessing the difference between exponential survival curves.

Efron's self-consistency algorithm Formulation of a method for dealing with missing and censored data involving averaging over fictitious observations. Closely related to the **EM algorithm** and sometimes called 'missing information principle'.

Ehrenfest model A diffusion model in Markov chain form, proposed by Ehrenfest and Ehrenfest (1907), without absorbing states and with the probabilities converging to zero with increasing time.

Eisenhart models In variance analysis it is customary to draw a distinction between type 1, in which the classification variables are fixed, and type 2 in which they are themselves random variables. The distinction has been pointed out by Eisenhart (1947) whose name is frequently attached to the models. [See also **variance components.**]

elementary contrast See **contrast**.

elementary renewal theorem A theorem in renewal theory which states that if $m(t)$ is the expected number of renewals in the interval $(0, t)$ and the renewal times have distribution function $F(t)$, then

$$\lim_{t \to \infty} \frac{m(t)}{t} = \begin{cases} \dfrac{1}{\mu} & \text{if } \mu < \infty \\[2mm] 0 & \text{if } \mu = \infty \end{cases}$$

where

$$\mu = \int_0^\infty t\, dF(t),$$

i.e., events grow at a rate determined by the mean interval.

elementary unit One of the individuals which, in the aggregate, compose a population: the smallest unit yielding information which, by suitable aggregation, leads to the population property under investigation. Cases occur where the term may be ambiguous, e.g., if an age distribution is to be estimated from a sample of households then the person is the elementary unit; but if, at the same time, the size of household is to be estimated, the household is the elementary unit. [See **basic cell.**]

Elfving distribution An approach to the distribution of the range in samples from a normal population proposed by Elfving (1947) making use of the probability integral transformation.

elimination Methods of solving simultaneous equations by eliminating variables to give a single equation, and calculating the rest by substitution. These methods are almost always used for small-scale calculations, but for many variables, using floating-point arithmetic in a computer, they are notoriously liable to a build-up of rounding errors.

ellipsoidal normal distribution See **spherical normal distribution**.

elliptical normal distribution A **bivariate normal distribution** in which $\rho = 0$, but $\sigma_1 \neq \sigma_2$.

elliptical truncation A method of **truncation** of multi-normal distributions described by Tallis (1963) which leads to relatively simple formulae for the truncated distribution.

elliptically symmetric distributions A name given to distributions whose density functions depend only on quadratic functions of the variables so that contours of constant density are elliptical. They include spherically symmetric distributions and multivariate normal distributions as special cases.

ellipticity A term sometimes used to describe a **bivariate normal distribution** in which the variances are not equal, leading to the idea of a 'test of ellipticity', i.e., a test of the inequality of the variances.

EM algorithm An iterative algorithm adapted particularly to maximum likelihood estimation from incomplete data. The two steps, Expectation and Maximization, are alternated, and, under weak conditions, the procedure converges to a local maximum (Dempster, Laird and Rubin, 1977).

empirical Bayes' estimator An estimator derived according to an **empirical Bayes' procedure**. The non-parametric form was given by Johns (1957) and a 'smooth' form for discrete distributions by Maritz (1966).

empirical Bayes' procedure A procedure for estimation or prediction based on Bayes' theorem with the prior distribution estimated from appropriate data. There are two broad types, depending on whether the prior distribution is specified in parametric form usually involving unknown hyperparameters or on whether the prior distribution is specified only non-parametrically (Robbins, 1956, 1964; Maritz, 1966).

empirical distribution function Given an ordered sample of n independent observations

$$x_{(1)} \leqslant x_{(2)} \leqslant x_{(3)} \leqslant \ldots \leqslant x_{(n)},$$

the function $F_n(x)$ defined as

$$F_n(x) = \begin{cases} 0, & x < x_{(1)} \\ \dfrac{k}{n}, & x_{(k)} \leqslant x < x_{(k+1)} \qquad k = 1, 2, ..., n-1 \\ 1, & x_{(n)} \leqslant x \end{cases}$$

is called an empirical distribution function.

empirical exponential family An exponential family obtained by exponential tilting of an **empirical distribution function**, useful as a basis for semi-parametric inference and bootstrap confidence intervals.

empirical likelihood A likelihood obtained by profiling a tilted empirical distribution function subject to constraints, imposed for example by estimating equations (Owen, 1988).

empirical probit The analysis of experimental results concerned with the relationship between levels of a stimulus and the responses thereto is usually made in terms of the percentage of test subjects reacting to the stimulus. The analysis is often facilitated by transforming the percentages into **probits**. The probits corresponding to these observed percentages are known as empirical probits, in contradistinction to **expected probits** and **working probits**, which relate to a fitted regression line.

empirical statistical model Probabilistic model to represent variation in data but having no specific subject-matter foundation.

empty cell test A rather inefficient class of distribution-free tests characterized by the test statistic being the number of cells in a uni- or multivariate classification into which none of the sample members fall. [See **David's empty cell test**, **Wilks's empty cell test**.]

end corrections An expression used in several statistical contexts, to denote corrections made to extreme values. For example, in taking a systematic sample from material which varies continuously it may be possible to obtain some increase in accuracy of estimation by employing corrections to the term at each end of the sample. Again, in fitting a curve to a time series, corrections may be made to end values to improve the fit. The expression also occurs in the calculation of serial correlations where certain alterations are necessary to formulae relating to infinite series to allow for the fact that the observations have finite length. In many such cases the word 'adjustment' is better than 'correction', which carries an implication that the values concerned are in error. [See **finite multiplier**.]

endogenous variable The statistical representation and analysis of multivariate systems generally involve a primary division of variables into those which are endogenous and those which are exogenous. Endogenous variables are those which form an inherent part of the system, as for instance price and demand in an economic model. Exogenous variables are those which impinge on the system from outside, e.g., rainfall or epidemics of disease. It is possible for a variable to be endogenous in one model and exogenous in another: for example, rainfall might be regarded as exogenous to an economic model of the automobile industry but endogenous to a meteorological model describing climatic states.

end-point estimation Estimation of the upper or lower support point of a distribution, important in fields such as reliability and extreme-value statistics. End-point estimators often have non-standard limiting behaviour.

Engset distribution A distribution occurring in congestion theory. It is equivalent to the binomial truncated at the higher values:

$$\Pr(X = x) = \frac{\binom{n}{x} p^x q^{n-x}}{\sum_{j=0}^{k} p^i q^{n-j}}, \qquad k < n.$$

ensemble The infinite set of **realizations** of a single **stochastic process**. The concept is frequently extended to apply to a sample set of realizations.

ensemble average Given a number of independent sample **realizations** from the same generating process, the mean of the process can be estimated by the overall mean of the samples. The estimator is called an 'ensemble average'. If the process is stationary, its mean can be estimated by a simple average over time instead of by the 'ensemble average'.

entrance time An alternative name for **first passage time**.

entropy See **diversity, index of**.

entry plot A plot through which a **serial cluster** is entered or determined.

envelope power function A power function based on a critical region which itself is obtained as the envelope of a number of critical regions. Where no best critical region exists for all alternative hypotheses, such an envelope provides a 'good' test in the sense that it picks out and amalgamates parts of critical regions which are best for particular alternatives.

envelope risk function A concept analogous to the **envelope power function** but applied to the risk functions associated with decision functions.

environmental statistics The study of the variability in environmental sciences.

epidemic The occurrence in a community or region of cases of an illness (or an outbreak) clearly in excess of expectancy. The number of cases indicating presence of an epidemic will vary according to the infectious agent, size, and type of the population exposed, previous experience or lack of exposure to the diseases, and time and place of occurrence. Epidemicity is thus relative to usual frequency of the disease in the same area, among the specified population, at the same season of the year. A single case of communicable disease long absent from a population or the first invasion by a disease not previously recognized in that area requires the immediate reporting and epidemiologic investigation; two cases of such a disease associated in time and place are often sufficient evidence of transmission to be considered an epidemic. An increase in the incidence of disease in a specified geographic area that is greater than that historically expected. In contemporary clinical parlance, an epidemic encompasses outbreaks of disease from 10,000 influenza cases to a single incidence of a disease long absent from a community. Contrary to popular usage, an epidemic is not synonymous with only mass infection.

epidemic process A model for the spread of an epidemic in a population. Models of this type are also used to represent the increase of a beneficial genetic mutation, or the spread of rumours.

epidemiology The quantitative study of health problems in the population; nowadays the term is used without necessary reference to infectious disease.

EPSEM sampling See **equal probability of selection method**.

equal ignorance, principle of In connection with the use of **Bayes' theorem**, the absence of definitive information on prior probabilities generally leads to an assumption of equal ignorance, or a uniform distribution of prior probabilities.

equal probability of selection method Any method of sampling in which the population elements have an equal probability of selection. EPSEM sampling can result either from equal probability selection throughout, or from variable probabilities that compensate each other in multi-stage selection.

equal spacings test A non-parametric test of uniformity for circular distributions that is invariant under rotation. Under the hypothesis of uniformity, the arc lengths $T_i, i = 1, 2, \ldots, n$, subtended on the unit circle by the angles between n independent sample directions each have expected value $2\pi/n$. The test criterion is

$$L = \sum_{i=1}^{n} \left| T_i - \frac{2\pi}{n} \right|,$$

large values of which indicate clustering of the angular observations.

equal-tails test A symmetrical **two-tailed test**.

equalizer decision rule A decision rule d such that the risk function $R(\theta, d) = C$, a constant, for all members θ of the parameter space Θ. The concept is of use in finding minimax decision rules. [See **minimax estimation**.]

equally correlated distribution A multivariate distribution is equally correlated if the correlation between each pair of variables is equal to ρ.

equilibrium distribution A physical, economic or social system which has settled down to a stable statistical behaviour has a stationary, or equilibrium, distribution which specifies the limiting proportions spent in the designated states of the system.

equilibrium renewal process A **general renewal process** in which the starting point is a random time, so that the renewal density is constant.

equi-normal distribution See **modified normal distributions**.

equitable game An alternative name for **fair game**.

equivalence class A term used in the construction of cyclic designs whereby certain permutations of treatments emerge as rearrangements of others (David and Wolock, 1965).

equivalence testing See **bioequivalence test**.

equivalent deviate If P is a probability or proportion and $f(x)dx$ is the frequency element of a distribution, almost always continuous, and usually in a standard form free of parameters, the equivalent deviate of P relative to the distribution is Y where

$$P = \int_{-\infty}^{Y} f(x)dx.$$

Particular examples are the **normal equivalent deviate** and the **logit**.

equivalent dose The dose of a standard preparation having the same effect, i.e., the same expected mean response, as a specified dose of a test preparation. It is of particular importance under conditions of **similar action** when the ratio of any dose of the test preparation and its equivalent is a constant, the **relative potency**.

equivalent samples Any two samples selected by the same sampling scheme which contain the same set of d ($\leqslant n$) population members yield the same information about the parameters, irrespective of whether these d members appear with different frequencies in the two samples.

equivariant estimator See **regression equivariant estimator, scale equivariant estimator, translation equivariant estimator, Pitman estimator**.

ergodic state An alternative name for a state of a **Markov chain** which is both **aperiodic** and **persistent**.

ergodicity Generally, this word denotes a property of certain systems which develop through time according to probabilistic laws. Under certain circumstances a system will tend in probability to a limiting form which is independent of the initial position from which it started. This is the ergodic property. A stationary stochastic process $\{X_t\}$ may be regarded as the set of all realizations possible under the process. Each such realization may have a mean m_r. If the process itself has a mean $E(X_t) = \mu$, the ergodic theorem of Birkhoff and Khintchine states that m_r exists for almost all realizations (Khintchine, 1933). If, in addition, $m_r = \mu$ for almost all realizations the process is said to be ergodic. In this sense ergodicity may be regarded as a form of the law of large numbers applied to stationary processes.

Erlang, Agner Krarup (1878–1929) A Danish probabilist. Erlang graduated from Copenhagen in 1901 and joined the Copenhagen Telephone Company in 1908. He wrote a book entitled *The theory of probabilities and telephone conversations* and was a pioneer in the application of stochastic models.

Erlang distribution A family of gamma-type distributions, with k an integer, of the form

$$f(t) = \frac{\lambda}{\Gamma(k)}(\lambda t)^{k-1}e^{-\lambda t}$$

proposed by Erlang for the interarrival times and service times in queueing problems. It is equivalent to a **chi-squared distribution** with an even number ($2k$) of degrees of freedom.

Erlang's formula An early result in congestion theory given by Erlang (1917) concerning the degree of hindrance experienced by a telephone subscriber who is unable to effect a call because all lines or channels are in use. The assumption is that the call is then lost and the formula is frequently known as Erlang's loss formula. The probability that, of N channels, all are in use is obtained by placing $x = N$ in

$$\Pr(x) = \frac{\lambda^x / x!}{\sum_{j=0}^{N} (\lambda^j / j!)}, \qquad x = 0, 1, 2, \ldots, N.$$

error A term with many uses. Sometimes it has the everyday meaning of mistake but more commonly it represents the random term attached to a statistical model to represent that part of the variation not represented in systematic terms; there is then no necessary implication of the first meaning.

error band In estimation or prediction the estimated or predicted value is bracketed by a range of values determined by standard errors, confidence intervals or similar methods within which the value may be supposed to lie with a certain probability. This is called the error band.

error function A form of the **normal distribution** defined as

$$\mathrm{erf}(x) = \frac{1}{\sqrt{2\pi}} \int_0^x e^{-t^2} dt.$$

It is also sometimes known as the 'Kramp function'. Another form of the **normal distribution**, the complement of the error function, is defined as

$$\mathrm{erfc}(x) = 1 - \mathrm{erf}(x).$$

error in equations An equation in variables may be inexact either because the equation is not a complete representation of the situation, as in a demand–supply equation which omits other factors such as income or employment, or because it is disturbed by extraneous sources of variation as in an autoregression equation. These departures from the relationship expressed by the equation are known as errors in the equation, as distinct from effects such as observational errors in the variables themselves.

error mean square The residual or **error sum of squares** divided by the number of **degrees of freedom** on which the sum is based. It provides an estimator of the residual or error variance.

error of estimation In general, the difference between an estimated value and the true value of a parameter, or, sometimes, of a value to be predicted.

error of observation An error arising from imperfections in the method of observing a quantity, whether due to instrumental or to human factors.

error of the first kind If, as the result of a statistical test, a statistical hypothesis is rejected when it should be accepted, i.e., when it is true, then an error is committed. This class of error is termed an error of the first kind and is fundamental to the theory of testing statistical hypotheses associated with the names of Neyman and Pearson. The frequency of errors of the first kind can be controlled by an appropriate selection of the regions of acceptance and rejection; that is to say, by choice of appropriate **critical regions** it is possible to ensure that the probability of committing an error of the first kind is an assignable constant.

error of the second kind If, as the result of a test, a statistical hypothesis is not rejected when it is false, i.e., when it should have been rejected, then an error is made. This class of error is termed an error of the second kind and, like the **error of the first kind**, it is fundamental to the Neyman–Pearson theory of testing statistical hypotheses. Unlike the error of the first kind, however, it is not, in general, controlled by the simple process of selecting regions of acceptance and rejection. The customary procedure in choosing tests of hypotheses is to fix the frequency of the first kind of error and, with this restriction, to minimize the frequency of the second kind of error. [See also **power function, consumer's risk**.]

error of the third kind A definition proposed by David (1947), perhaps not entirely seriously. She suggested that there was a third kind of error which might be committed in testing statistical hypotheses: that of selecting the test falsely to suit the significance of the particular sample data available. A somewhat different type of error of the third kind was suggested by Mosteller (1948) in proposing a non-parametric test for deciding whether one population, out of k populations characterized by a location parameter, has shifted too far to the right of the others. He defines it as 'the error of correctly rejecting the **null hypothesis** for the wrong reason'.

error reducing power A term used in connection with the smoothing of time series. Each observation is regarded as composed of a true value and an error of observation which is independent of the errors in other observations. The smoothing process is an attempt to approximate to the true values and to reduce the errors. The success of any process is measured by its error reducing power, one common measure being the extent to which the variance of a random series is reduced if the process is applied to it.

error sum of squares In **analysis of variance** it is customary to regard the data as generated by a model (usually linear) consisting of certain class effects plus a stochastic component. When estimates are made of the class effects and subtracted from the observations, the residuals are estimates of the contribution from the stochastic component and the sum of squares of these residuals is known as the 'error sum of squares', though 'residual sum of squares' is preferable from many points of view. [See also **pooling of error**, **variance components**.]

error variance The variance of an error component. Thus, if the generating model of a set of data consists of certain systematic components together with a stochastic component, the variance of the latter is the error variance. The expression can also be understood in a wider sense, as the variance of error in repetitions of an experimental situation, whether the 'error' is due to sampling effects or not. It makes for clarity if expressions such as 'error variance' are eschewed in favour of 'residual variance' but the use of the former type of wording is very widespread.

errors in surveys The errors in a sample survey arise both from sampling effects and from other sources not connected with sampling, i.e., they would also be present for a complete survey. It has become customary to use the word 'error' to cover all these types of departures from representativeness, whereas in some statistical contexts 'error' denotes 'sampling error' and the other effects are called 'biases'.

errors-in-variables model A regression model that takes specific account of errors in the variables (including the regressors) as well as the error in the equation.

Esseen's lemma This result, due to Esseen (1945), gives an estimate of the closeness of two distribution functions F and G in terms of the closeness of the characteristic functions on an arbitrary interval $(-T, T)$. It states that if F and G are distribution functions with respective characteristic functions f and g, and if G has first derivative G', then

$$\sup_{x} |F(x) - G(x)| \leqslant \frac{1}{\pi} \int_{-T}^{T} \left| \frac{f(t) - g(t)}{\pi t} \right| dt + \frac{24}{\pi T} \sup_{x} |G'(x)|.$$

Esseen-type approximation In connection with the normal approximation to the common discrete distributions, Esseen (1945) proposed an improvement to the Lyapunov results concerning conditions for, and speed of, convergence.

essential completeness See **completeness (of a class of decision functions)**.

estimable A parameter which can be estimated in some appropriate sense from the data under analysis.

estimate In the strict sense an estimate is the particular value yielded by an **estimator** in a given set of circumstances. The expression is, however, widely used to denote the rule by which such particular values are calculated. It seems preferable, following Pitman (1939a), to use the word **estimator** for the rule of procedure and 'estimate' for the values to which it leads in particular cases.

estimating equation An equation involving observed quantities and an unknown which serves to estimate the latter; one of a set of such equations involving several unknowns.

estimation Estimation is concerned with inference about the numerical value of unknown population values from incomplete data such as a sample. If a single figure is calculated for each unknown parameter the process is called **point estimation**. If an interval is calculated within which the parameter is likely, in some sense, to lie, the process is called **interval estimation**.

estimation–maximization (EM) algorithm See **EM algorithm**.

estimator A rule or method of estimating a parameter of a population. It is usually expressed as a function of sample values and hence is a variable whose distribution is of great importance in assessing the reliability of the estimate to which it leads.

ethics In a research context the constraints placed on investigators, especially in a social and medical context, by legal and other restrictions aimed to prevent harm to individual participants in the study.

etiologic fraction The proportion of disease cases that are directly linked to a specific exposure. For diseases with multiple causes, the rate of diseases may still be quite high for members of a group not exposed to the factor being studied. The etiologic fraction accounts for cases that are linked to the exposure of interest. This fraction is calculated mathematically as $(C(e) - C(n))/C(e)$, where $C(e)$ = case exposed and $C(n)$ = case not exposed.

Euclidean distance The distance $d_{ij}^2 = \sum(x_{ik} - x_{jk})^2$, used particularly in multivariate analysis in p-dimensional state space.

Euler's conjecture The conjecture that no Graeco-Latin square of side $p = 4n + 2$ existed. It is true for $p = 6$, but was disproved by Bose, Shrikhande and Parker (1960), who found orthogonal squares of side 10, and other values excluded by the conjecture.

even summation In the smoothing of time series, a moving average taken over an even number of terms. The method produces a complication when applied to equally spaced data because it yields a term which is mid-way between the two central observations of the portion of the series being summed and averaged. A second summation of the first results, again based upon an even number of terms, will bring the results back into line with the original data.

event This is often used in its everyday meaning. In probability theory an event is a set of outcomes of interest; in the formal theory the events considered are all sufficiently regular (Borel-measurable) sets in some space of possible outcomes in a random system. Occasionally, especially in some of the older literature, the word event is used to denote a point in a stochastic point process.

event history data Term used especially in sociology for duration (e.g., survival) data and for generalizations in which a sequence of point events is observed on each individual.

event space See **sample space**.

evolutionary operation A technique for the optimization of established full-scale processes (Box and Draper, 1998). It is used for experimenting on production processes. The basic elements are (*i*) provision to introduce a routine of systematic small changes in the levels at which process variables are held; (*ii*) provision to feed back the results derived from making these small changes to the operating supervisor; and (*iii*) an organization which continually reviews the results and suggests new action to be introduced later.

evolutionary process Any non-stationary stochastic process. The probability distributions associated with the process are not independent of the time.

evolutionary spectrum In a non-stationary process or time series, a spectrum will apply strictly only to a limited period of time. Hence, for the whole realization the spectral function will be time dependent or evolutionary.

exact chi-square test See **Fisher–Yates test**.

exact statistical method Used in interval estimation to imply that the probability distributions involved are completely known and that probability levels quoted are exactly achieved.

exact test A significance test in which the distribution of the test statistic under the null hypothesis can be computed exactly, i.e., is independent of any nuisance parameters and does not involve mathematical approximation. Fisher's exact test for the 2×2 table is a well-known example.

exceedances Values, in particular of a time series or spatial process, exceeding a specified threshold. [See **extreme values**.]

exceedances, distribution of In two random samples S_1 and S_2 of size n_1 and n_2, the number of observations in S_2 which exceed in magnitude some specified number of observations in S_1 has the distribution of exceedances (Gumbel and von Schelling, 1950).

exceedances tests A class of distribution-free tests of identity of two populations, characterized by having a test statistic that is the number of observations in a simple random sample from one population that exceed a particular order statistic in a simple random sample from the other population (Chakraborti and van der Laan, 1996). [See **exceedances, distribution of.**]

excess, coefficient of A name given to a measure of **kurtosis**. It is defined as

$$\gamma_2 = \frac{\mu_4}{\mu_2^2} - 3 = \frac{\kappa_4}{\kappa_2^2}$$

where μ's represent **moments** and κ's **cumulants**. The coefficient is simply the excess of the value of κ_4/κ_2^2 over the zero value it takes in the case of a normal distribution.

exchangeability The property of a joint distribution of a sequence of random variables of being unchanged by an arbitrary permutation of the sequence. De Finetti (1974) showed that an infinite exchangeable sequence is equivalent to a mixture of independent and identically distributed random variables.

exchangeable (variables) The random variables $X_1, X_2, \ldots, X_n$ are exchangeable if all the $n!$ permutations $(X_{k_1}, X_{k_2}, \ldots, X_{k_n})$ have the same n-dimensional probability distribution. The variables of an infinite sequence $\{X_n\}$ are exchangeable if $X_1, X_2, \ldots, X_n$ are exchangeable for each n. An alternative term is 'symmetrically dependent'.

exhaustive sampling A term occasionally encountered which indicates that, in order to achieve the sample size necessary for the required precision, the sampling has included all of the population under study, i.e., a complete enumeration rather than a sample has been taken.

exogenous variable The opposite of an **endogenous variable**.

expectation See **expected value**.

expected frequencies See **theoretical frequencies**.

expected normal scores test See **normal scores test**.

expected probit The probit at some experimental dose calculated from the **probit regression line** fitted to the data. The expression is also used for the corresponding value obtained from the provisional line at any stage of iteration.

expected value The mean of a random variable (scalar or vector) defined as an average with respect to its probability distribution. For a discrete random variable this is a sum, and for a continuous random variable an integral, or in general it is the Stieltjes integral

$$\int x \, dF(x),$$

where $F(x)$ is the cumulative distribution function of X. If the integral does not converge the expected value is regarded as not defined.

experiment An investigation in which the investigators have sufficient control of the system under study, in particular to be able to determine the assignment of different units of study to different treatments (conditions or modes of intervention). Experiments are to be contrasted with observational studies in which, even though the investigators may determine which units are studied and the observational process, the assignment of treatments is outside their control.

experimental error In general, any error in an experiment whether due to stochastic variation or bias. More specifically, the expression is used to denote the essential probabilistic variation to be expected under repetition of the experiment, not actual mistakes in design or avoidable imperfections in technique. It is one aim of a good design to provide valid measures of the experimental error in the more restricted sense.

experimental unit See **plot**.

expert system Used rather loosely for computer programs designed to play the role of an expert consultant. Such programs have, in particular, been developed for medical diagnosis and for guidance in statistics.

explanatory variable See **independent variable**.

exploratory data analysis Methods of examining, summarizing and displaying data that involve no statistical inference or modelling; particularly the techniques developed by Tukey (1977).

exploratory survey See **pilot survey**.

explosive process A rather too vivid term to describe a stochastic process which has no bound to the expectation of the mean square: a process whose values may increase without limit (in absolute magnitude, so that oscillations are possible) as time goes on.

explosive stochastic difference equation The difference equation $x_t + \alpha_1 x_{t-1} + \alpha_2 x_{t-2} + \ldots + \alpha_k x_{t-k} = \mu_t$ derived from an **explosive process**. If one or more roots of the individual equation

$$\sum_{j=0}^{k} \alpha_j z^{t-j} = 0$$

lie outside the unit circle the values of x_t increase without limit as t increases. The same expression would be applied to more general difference equations with a variable possessing the same property of not remaining within finite bounds.

exponential curve A series of observations ordered in time which has a constant, or approximately constant, rate of increase can be represented over a long period by the curve:

$$y = ae^{bt}$$

where a and b are constants and t is time. This, or some simple transformation, is called the exponential curve. The fitting of an exponential trend of this form by the **least squares method** is facilitated by transforming into the logarithmic form:

$$\log y = \log a + bt.$$

exponential dispersion model A generalized form of exponential family in which an additional parameter controls its variability (Jorgensen, 1987).

exponential distribution   A distribution of the form

$$f(x) = \frac{1}{\mu} \exp\left(-\frac{x - m}{\mu}\right), \qquad m \leqslant x < \infty.$$

The parameter μ is the standard deviation of the distribution and is also equal to the distance of the mean from m.

exponential family   A distribution with density of the form

$$f(x) = \exp\left\{a(\theta)b(x) + c(\theta) + d(x)\right\},$$

where θ is a parameter and a, b, c, d are known functions, is said to belong to the exponential family. It includes the normal, gamma, binomial and Poisson distributions. The definition can be generalized to vector parameters. [See **Darmois–Koopman's distributions.**]

exponential regression   A term used to denote a relationship of the form

$$y_i = \alpha + \beta e^{\lambda x_i} + \varepsilon$$

where ε is a random residual.

exponential smoothing   A method used in time series to smooth or to predict a series. There are various forms, but all are based on the supposition that more remote history has less importance than more recent history. For instance, in time-series analysis a predictor y_t is often expressed as an average of some variable observed at previous points of time, with weights which diminish in importance as the period becomes more remote

$$y_t = \beta x_t + \beta^2 x_{t-1} + \beta^3 x_{t-2} + \ldots .$$

Formulae of this kind lie at the basis of prediction methods developed by Beverton and Holt (1957), Brown (1959), Harrison (1964) and later writers. Originally designed to deal with a stationary series, they have been extended to cover series containing linear trend and seasonal variation.

extended Bayes' decision rule   See **Bayes' decision rule.**

extended group divisible design   The extension of the ordinary group, divisible, partially balanced incomplete design was for an m-associate class and proposed by Hinkelmann and Kempthorne (1963). The definition is lengthy; see Hinkelmann (1964).

extended hypergeometric distribution   This distribution arises in testing the hypothesis that two proportions p_1 and p_2, derived from binomial random variables x_1 and x_2, are equal. The distribution is that of x_1 conditional on the fixed sum $x_1 + x_2$ and is of the form

$$f(x; t) = \frac{g(x)t^x}{P(t)}$$

where $t = p_1 q_2 / p_2 q_1$ and $P(t)$ is the factorial generating function of the **hypergeometric distribution**.

extended negative binomial distribution A form of the decapitated **negative binomial distribution** introduced by Engen (1974). As the index tends to zero, the distribution approaches the **logarithmic series distribution**, but the distribution, conditional on $x \geqslant 1$, exists for negative values of the index $(0 > n > -1)$ and gives distributions longer tailed than the logarithmic.

extensive sampling A term used to denote sampling where the subject matter, or geographical coverage, of a sample is diffuse or widespread as opposed to intensive, where it is narrowed to a small field. Extensive sampling may refer to a case where a wide variety of topics are covered superficially, rather than a few topics in detail, or a large area is surveyed broadly, rather than a small area studied in detail. The term could also be used with reference to time; that is to say, of sampling covering a long period. It would be convenient to distinguish the cases as space-extensive, item-extensive and time-extensive respectively.

external validity The extent to which the results of a research study can be generalized to individuals and situations beyond those involved in the study (Gall, Borg and Gall, 1996).

extra period changeover design The extension of a changeover design for a further period not contemplated in the original design. A feature of such extra period designs is that, in the analysis, direct and residual effects are orthogonal and hence the new data can be incorporated without difficulty in the analysis.

extra-Poisson variation The situation in which a sequence of count data have a variance greater than the mean and hence show more variability than would a comparable set of data from a Poisson distribution.

extrapolation The practice of predicting or estimating values using projections of data outside the range of the independent variables represented in a sample. For example, an estimated regression line may be extended beyond the observed range of the data and used to predict values outside of this range.

extrapolation forecasting A simple technique for predicting future incidence of disease or infection by extrapolating from empirical curves fitted to observed incidence data. It is useful for short-term predictions when a well-defined end-point, such as AIDS incidence data collected under routine surveillance, is available.

extremal index A measure of the dependence among rare events. Its inverse is the mean number of maxima or minima occurring in a cluster of extrema. See Coles (2001).

extremal intensity The particular values of the **intensity function** or **hazard function** that appear as parameters in asymptotic distributions of extreme values.

extremal process A class of stochastic process proposed by Dwass (1964), derived from three extreme-value distribution functions.

extremal quotient The ratio of the absolute value of the largest observation to the smallest observation in a sample. For continuous variables which are symmetrical and unlimited at both ends of the range the logarithm of the extremal quotient is symmetrically distributed.

extremal statistic Any function which depends upon observations at both extremes of the sample. For example, the **range** and the **extremal quotient**.

extreme rank sum test A non-parametric test proposed by Youden (1963). If I objects are independently ranked by J judges, the individual ranks r_{ij} can be summed for each object. An extreme rank sum derived by this method can be tested using tables in Thompson and Wilke (1963).

extreme studentized deviate A statistic, given by

$$t_n = \frac{(x_{(n)} - \bar{x})}{s} \quad \text{or} \quad \frac{(\bar{x} - x_{(1)})}{s},$$

where s is the sample standard deviation based on $n - 1$ degrees of freedom, n is the sample number, $x_{(1)}$ and $x_{(n)}$ are the smallest and largest members of the sample and $\bar{x}$ is the sample mean.

extreme-value distributions The distribution of the largest (smallest) observation in a sample. The distributions may be exact or asymptotic in form (Gumbel, 1958) and there are three types in accordance with the population from which the sample is drawn. The three main types are associated with the names of Gumbel (1935), Fréchet (1927) and Weibull (1939). [See also **range**, **Gumbel distribution**, **Fréchet distribution**, **Weibull distribution**.]

extreme values The largest or smallest variable values borne by the members of a set. Slightly more generally, the expression signifies values neighbouring the end values.

F

F-distribution The distribution of the ratio of two independent quantities each of which is distributed like a variance in normal samples, i.e., in the **type III**, **chi-squared** or **gamma** form. The distribution, due to R.A. Fisher, may be put in the form

$$\frac{(\nu_1/\nu_2)^{\nu_1/2}\Gamma\left(\frac{1}{2}\left(\nu_1+\nu_2\right)\right)}{\Gamma\left(\frac{1}{2}\nu_1\right)\Gamma\left(\frac{1}{2}\nu_2\right)}\frac{F^{(\nu_1-2)/2}}{\left((\nu_1 F/\nu_2)+1\right)^{(\nu_1+\nu_2)/2}}$$

where ν_1 and ν_2 are the degrees of freedom of the numerator and denominator of the ratio $F = s_1^2/s_2^2$. The distribution was first studied by Fisher in a transformed form (see **z-distribution**), the ratio F being so denominated by Snedecor (1937) from the first letter of the discoverer's name. The distribution is a simple transform of the **type I** or **beta distribution**.

F-test A test based on the ratio of two independent statistics, each of which is distributed as the variance in samples from normal populations with the same parent variance. Usually the statistics themselves are quadratic estimators of the parent variance. The test is employed in **analysis of variance** to test the homogeneity of a set of means.

factor This word occurs in statistical contexts in several senses: (i) in the ordinary mathematical sense, e.g., a factor of an algebraic expression; (ii) to denote a quantity under examination in an experiment as a possible cause of variation, e.g., in a 'factorial' experiment; (iii) (adapted from psychology) in multivariate analysis, an unobserved variable postulated to explain the structure of observed variables; (iv) to denote a constituent item in an average or index number.

factor analysis A branch of multivariate analysis in which the observed values of random variables $x_i, i = 1, 2, \ldots, p$, are supposed to be expressible in terms of a number $m < p$ of unobserved factors f_i together with residual elements. One such model is expressed by

$$X_i = \sum_{j=1}^{m} a_{ij} f_j + b_i s_i + c_i \varepsilon_i$$

where s_i is a factor specific to the ith variable, ε_i is an error variable and the a's, b's and c's are structural constants of the model which it is the object of the analysis to estimate. The coefficients a_{ij} are known as factor loadings. That part of the variance of X_i which is attributable to the f's is called the communality; that attributable to s_i is called the specificity; and that attributable to the ε_i is called the unreliability. The complement of the last is called the **reliability**.

factor loading The use of the word 'loading' rather than 'weighting' in this context is due to the fact that the terminology of factor analysis developed from psychology. The word 'saturation' is used in a similar sense to denote the extent to which a variable is 'saturated' with a common factor. [See **factor analysis**.]

factor matrix The matrix of coefficients (a_{ij}) appearing in the relations between variables and factors in **factor analysis**.

factor pattern In the **factor matrix** certain items may be known or assumed on prior grounds to be zero; for example, if the jth factor f_j does not appear in the ith variable, $a_{ij} = 0$. The pattern of nonzero coefficients as distinct from their actual values is called the factor pattern. It may be regarded as defining the model of structure, in terms of factors, which is under investigation. In oblique factor analysis, i.e., factor analysis where the factors are correlated, it may be necessary to distinguish between the factor pattern, as so defined, and the factor structure which expresses the way in which the factors are dependent among themselves.

factor reversal test A test proposed for index numbers by Irving Fisher (1927). The idea was that, in an index number of price, if the symbols for price and quantity are interchanged, there should result an index of quantity, and that this, multiplied by the original price index, should give an index of changes in total value. The test is obeyed by Fisher's **'ideal' index number** but not by most of those in current use, e.g., those of Laspeyres and Paasche. [See also **time reversal test**.]

factor rotation The final stage of **factor analysis** is, in geometric terms, a rotation of the factor axes in order to achieve some correspondence between the numerically derived factors and explanatory entities. Analytic methods may be divided broadly into those dealing with orthogonal factors, where the loadings are uncorrelated, and oblique factors involving correlated loadings. One group of methods pursues the maximization of the scatter among factor loadings; for example, quartimax and varimax for orthogonal rotation and oblimax for oblique rotation. An alternative approach, using the minimization of cross-products, is termed quartimin and covarimin, which latter method is the inverse of varimax.

factor score In **factor analysis**, the rating of an individual on any of the factors in the model. Once the **factor loadings** have been estimated, the scores can be estimated. Two methods are in common use: Bartlett (1935b) suggested a method that is approximately unbiased; Thomson (1939) gave an alternative that gives approximately minimum mean square error.

factorial cumulant The factorial cumulant of order r, $\kappa_{[r]}$, is defined as the coefficient of $t^r/r!$ in the expansion of the **factorial cumulant generating function** as a power series in t.

factorial cumulant generating function The logarithm of the **factorial moment generating function**, formed by analogy with the cumulant generating function, which is the logarithm of a moment generating function.

factorial design See **factorial experiment**.

factorial distribution A discrete distribution for which the successive frequencies are factorial qualities. Irwin (1963) defined a whole class of such distributions of the form

$$f(r) = \frac{ka_r}{\theta^{[r+1]}}, \qquad \theta > 0, \qquad r = 0, 1, 2, \ldots$$

where $x^{[0]} = 1$, $x^{[r]} = x(x+1)\ldots(x+r-1)$ denotes the ascending factorial. The family is sometimes known as the inverse factorial series distribution. A particular case is Waring's distribution, with $a_r = a^{[r]}$, so that it becomes

$$f(r) = (\theta - a)\frac{a^{[r]}}{\theta^{[r+1]}}.$$

The name was given in recognition of Waring's discovery in the eighteenth century of the expansion of $(\theta - a)^{-1}$ in reciprocals of factorials. In particular, if $a = 1$ and $\theta - a = p$ the distribution becomes

$$f(r) = \frac{pr!}{(p+1)\ldots(p+r+1)}$$

and it is known as the Yule distribution (1925*b*).

factorial experiment An experiment designed to examine the effect of more than one factor, each factor being applied at two or more levels so that differential effects can be observed. The term is frequently used in a slightly narrower sense, as describing an experiment investigating all possible treatment combinations which may be formed from the factors under investigation. The levels may be measured quantitatively, as when fertilizer is applied to plots in a given weight per unit area, or qualitatively, as when patients are considered at two levels 'inoculated' and 'not inoculated'. See Raktoe, Hedayat and Federer (1981).

factorial moment A type of moment used for discontinuous distributions defined at equal variable intervals. If f_r is the frequency at x_r the jth factorial moment about arbitrary origin a is

$$\mu'_{(j)} = \sum_{r=-\infty}^{\infty} (x_r - a)^{(j)} f_r,$$

where

$$(x_r - a)^{(j)} = (x_r - a)(x_r - a - 1)\ldots(x_r - a - j + 1).$$

In most cases the variable intervals are taken as units, the variable values as $0, 1, 2, \ldots$ and a as zero, in which case

$$\mu'_{(j)} = \sum_{r=j}^{\infty} r(r-1)\ldots(r-j+1)f_r.$$

factorial moment generating function A function of a variable t which, when expanded formally as a power series in t, yields the factorial moments as coefficients of the respective powers. It is used almost entirely for discontinuous distributions defined at equal distances of the variable, say at $x = 0, 1, \ldots$. If f_r is the frequency at x_r a factorial moment generating function is given by $\omega(t)$, say, where

$$\omega(t) = \sum_{r=0}^{\infty} f_r(1+t)^r = \sum_{j=0}^{\infty} \mu'_{(j)}\frac{t^j}{j!}$$

where $\mu'_{(j)}$ is the jth **factorial moment** about zero.

factorial multinomial distribution See **multivariate hypergeometric distribution**.

factorial sum A sum entering into the calculation of **factorial moments**. If the frequency of a value of random variable r $(r = 0, 1, \ldots, k)$ is f_r the factorial sum of order j is

$$\sum_{r=0}^{k} r(r-1)\ldots(r-j+1)f_r = \sum_{r=0}^{k} r^{(j)}f_r.$$

failure rate The probability density of failures in **renewal theory**, as a function of age.

fair game In the theory of games, a game consisting of a sequence of trials is deemed to be a 'fair' game if the cost of each trial is equal to the expected value of the gain from each trial. A 'fair' game in this sense may not be fair as between a pair of adversaries with unequal resources: it is well known that at a 'fair' game the player with the larger sum to stake has the better chance of ruining his or her opponent.

false positive; false negative A false positive in a clinical setting is a test result indicating the presence of a condition that is, in fact, not present. A false negative in a clinical setting is a test result indicating the absence of a condition that is, in fact, present. A false-negative result on a screening test will lead to unnecessary treatment or more invasive testing. A false-positive result will lead to undertreatment and may allow a disease to spread. A screening test that is very sensitive will report few false-negatives and a test with high specificity will return few false-positives.

fatigue models Statistical models used to describe the lifetime, X, of materials (such as a piece of chain, wire or cable) which are subject to stress level y until they break. It is known that the higher the stress level y the lower the lifetime X. The lifetime–stress-level curves associated with different pieces lead to non-intersecting percentile curves associated with elements of increasing strength. These curves are known in the engineering literature as Wöhler curves.

feature selection The selection of subsets of variables to describe the main characteristics in a population, with some reduction of redundancy and noise.

feed-forward neural network A neural network in which feedback between hidden layers is not allowed.

Fellegi's method Fellegi (1963) proposed a method of sampling without replacement with probability proportional to size, which allows for rotation of the sample simultaneously with the exact calculation of joint probabilities of selection of sets of units.

Fermi–Dirac statistics See **Bose–Einstein statistics**.

Ferreri distribution A system of distributions described by Ferreri (1964) with probability density functions of form

$$\Pr(x) = \sqrt{b}\frac{1}{G\left(-\frac{1}{2},a\right)} \cdot \frac{1}{e^{a+b(x-\xi)^2} + c}, \qquad c = 1 \text{ or } -1,$$

where

$$G(p,a) = \Gamma(p+1) \sum_{j=1}^{\infty} (-c)^{j-1} \frac{e^{-ja}}{j^{p+1}}.$$

The distribution depends on the four parameters a, b, c and ξ.

fertility gradient If a field, or other stretch of land which is to be used for agricultural experimentation, is divided into plots ready for the experimental treatments, some part of the differences in yield between contiguous plots may be due to inherent variation in the fertility of the soil. If this inherent variation is slowly decreasing, or increasing, from one side to the other there is said to be a fertility gradient. It is one aim of randomization and other devices of experimental design to eliminate bias due to the existence of such gradients.

fertility rate The number of live births in a unit period expressed as a proportion (usually per thousand) of potentially fertile women in the population concerned. 'Potentially fertile' is usually defined by reference to age, e.g., by taking women from 15 to 50 years old. This crude or general fertility rate can be refined or standardized in respect of age, social class, etc.

fiducial distribution A distribution of a parameter required for **fiducial inference** about that parameter. It is not a probability distribution in the customary sense, but is derived from the distribution of estimators containing all the relevant information in the sample. In earlier literature it is sometimes referred to as a 'fiducial probability distribution'.

fiducial inference A type of statistical inference based on **fiducial distribution**, introduced by R.A. Fisher (1930). The object of the inference is to make probabilistic statements about the values of unknown parameters and to some extent it resembles the theory of **confidence intervals**. In simple cases, results from fiducial theory agree numerically with those from the theory of confidence intervals, but this is not so in general. When this type of inference and the related confidence interval type were first propounded they were often confused. The confusion survives to the extent that 'fiducial' is sometimes applied to inference of the confidence interval type.

fiducial limits In **fiducial inference**, limits between which a parameter is considered to lie. The term occasionally occurs as a synonym of **confidence limits**. The two are often the same in commonly occurring cases, but their conceptual genesis is different.

fiducial probability See **fiducial distribution**.

Fieller–Hartley–Pearson measure of association A non-parametric measure of association for paired samples, first studied by Fieller, Hartley and Pearson (1957). It is calculated by replacing each value of random variable in the Pearson sample **product moment correlation** coefficient by its corresponding expected **normal scores test**.

Fieller's theorem A theorem giving limits of the confidence interval type for a ratio, stated in its general form by Fieller (1940).

filter Any method of isolating constituents in a time series; a mathematical analogy of the 'filtering' of a ray of light or sound by removing unsystematic effects and bringing out the constituent harmonics.

finite arc sine distribution A combinatorial distribution occurring in stochastic process analysis; it expresses the probability of return to zero of the winnings in a series of trials, with probability one-half of success at each trial. For $n \to \infty$ this tends to the ordinary **arc sine distribution**.

finite Markov chain A Markov chain with a finite number of states.

finite multiplier See **finite population correction**.

finite population A population of individuals which are finite in number.

finite population correction If a sample of n values is drawn without replacement from a population of limited size N, the sampling variances of the statistics derivable from it depend, in general, on N as well as n. For example, the variance of the sample mean $\bar{x}$ may be written

$$\text{var}\,(\bar{x}) = \frac{\sigma^2}{n}\left(1 - \frac{n}{N}\right)$$

where

$$\sigma^2 = \frac{1}{N-1}\sum_{i=1}^{N}(x_i - \mu)^2 \quad \text{and} \quad \mu = \frac{1}{N}\sum_{i=1}^{N}x_i.$$

The factor $1 - n/N$ is called the finite sampling correction. Formulae for sampling from a finite population are not, in general, expressible as the product of formulae for the infinite case and terms dependent solely on n and N.

finite sample breakdown point The finite sample breakdown point of an estimator $T_n = T_n(x_1, \ldots, x_n)$ was introduced by Donoho and Huber (1983). If we replace m arbitrary components of the sample $\boldsymbol{x} = (x_1, \ldots, x_n)$ by arbitrary (unfavourable) values, then starting with some m, this replacement procedure leads to an infinite value of T_n. Let m be the minimum number with this property; then $\varepsilon_n = m(\boldsymbol{x})/n$ is called the (finite sample) breakdown point corresponding to $\boldsymbol{x}$. Another version of the breakdown point is obtained if m observations are added to the sample, instead of the replacement.

finite sampling correction An alternative term for the **finite population correction**.

first digit distribution The distribution of first digits of numbers in tables of statistical and other data. Under reasonable assumptions, it may be shown that, approximately,

$$p_k = \log_{10}(k+1) - \log_{10}k, \qquad k = 1, 2, \ldots, 9.$$

first limit theorem If a sequence of distribution functions tends to a single distribution function F then the corresponding characteristic functions tend uniformly in any finite interval to the characteristic function of F. This is generally known as the first limit theorem. Although known to earlier writers, it seems to have been proved rigorously for the first time independently by Lévy (1925) and Cramér (1925).

first passage time An important concept in the theory of stochastic processes, being the time random T until the first instant that a system enters a state j given that it starts in state i, including the time taken to leave state i.

first return time A more explicit name for **return time**.

first-stage unit See **multi-stage sampling**.

Fisher, Irving (1867–1947) With his crusades for health, the League of Nations, etc., Irving Fisher used about half of his working time as professor of economics. Otherwise, his 'pillars and arches', as Schumpeter called his contributions to theoretical economics, might have grown to a kind of temple of economics/econometrics (Fisher, 1922a).

Fisher, Ronald Aymler (1890–1962) R.A. Fisher transformed the statistics of his day into a powerful and systematic body of theoretical concepts and practical methods. This achievement was all the more impressive because at the same time he pursued a dual career as a biologist, laying down, together with Sewall Wright and J.B.S. Haldane, the foundations of modern theoretical population genetics. Fisher began to develop an extraordinary range of statistical methods fitted into a theoretical superstructure. These fell into four main categories: statistical tests of significance and distribution theory, contributions to theoretical statistics and estimation theory, fiducial inference, and the design of experiments (Fisher, 1935a). Throughout his life he received many awards and honours.

Fisher–Behrens test See **Behrens–Fisher test**.

Fisher–Hsu–Roy distribution The distribution of the characteristic roots $\lambda_1, \lambda_2, \ldots, \lambda_p$ of the sample covariance matrix in sampling from a p-variate normal distribution, discovered independently by Fisher (1939), Hsu (1939), Roy (1939) and Girshick (1939). The distribution has probability density function

$$\frac{\pi^{1/2p}}{2^{1/2p(n-1)}} \prod_{j=1}^{p} \frac{\lambda_j^{1/2(n-p-2)} \exp\left(-\frac{1}{2}\sum_{i=1}^{p}\lambda_i\right)}{\Gamma\left(\frac{1}{2}(n-j)\right)\Gamma\left(\frac{1}{2}(p+1-j)\right)} \prod_{j<k}(\lambda_j - \lambda_k)$$

where n is the sample size and the λ_j are in descending order.

Fisher information See **information**.

Fisher information matrix See **information matrix**.

Fisher–Irwin test Because of a paper by Irwin (1935) dealing with the same problem, this is an alternative name for the **Fisher–Yates test** which itself is also known as the Fisher exact test for a 2×2 table.

Fisher model A term proposed by Ogawa (1963) to denote the general class of experimental designs without technical errors but based upon Fisher's randomization procedure. [See also **Neyman model**.]

Fisher–Yates test The use of χ^2 as a test of independence in a double dichotomy has limitations if the cell frequencies are small. Yates (1934) proposed a correction for continuity in these circumstances and, following a suggestion by R.A. Fisher, also gave an 'exact' test in the form of a computation for the probability of any observed set of cell frequencies. If the four cell frequencies in a 2×2 table are denoted by a, b, c and d with a and d diagonal, then the probability of this set of frequencies on the hypothesis of independence is

$$\frac{(a+b)!(c+d)!(a+c)!(b+d)!}{(a+b+c+d)!a!b!c!d!}.$$

The test proceeds by calculating the exact probabilities of the frequencies observed and of those deviating more than the observed from the situations of independence, and cumulating the results. It is also known as the exact chi-square test.

Fisher's 'B' distribution The distribution of the square root of a variable following a **non-central χ^2 distribution**. It is a hypergeometric function and was derived by Fisher in a study of the multiple correlation coefficient.

Fisher's distribution See ***F*-distribution**.

Fisher's (spherical normal) distribution An extension of the **von Mises distribution** to the spherical case. Its density may be defined on the surface of the unit sphere with centre at the origin by

$$f(l, m, n) = \frac{K}{4\pi \sinh K} \exp\left\{-K(l\lambda + m\mu + n\nu)\right\}, \qquad K > 0,$$

where (l, m, n) and (λ, μ, ν) are the vectors of direction cosines of a direction random variable in three dimensions and its mean direction respectively. K is described as a concentration parameter, for as K tends to zero the distribution tends to the spherical uniform distribution, whilst as K tends to infinity the density progressively concentrates towards the mean direction. The distribution is also known as the **spherical normal distribution**.

Fisher's transformation (of the correlation coefficient) A transformation of the sample correlation coefficient r according to the formula

$$z = \tanh^{-1} r.$$

The distribution of z for samples from a bivariate normal population approaches normality much more rapidly than does that of r; and its variance, unlike that of r, is little affected by the population correlation coefficient even for samples of moderate size.

fixed base index An index number for which the **base period** for the calculations is selected and remains unchanged during the lifetime of the index.

fixed effects (constants) model An alternative designation for **model I** (or **first kind**) in **analysis of variance**.

fixed variable In regression theory the model under investigation is of the type

$$Y = \beta_0 + \beta_1 X_1 + \ldots + \beta_p X_p + \varepsilon$$

where the β's are constants, the X's are variables in the mathematical sense and ε (and hence Y) are random variables. Methods of estimating the β's are the same whether the X's are selected arbitrarily or are themselves the values of variables, provided that in the latter case the conditional distribution of ε is the same for all values of the X's. The X's are, in such an event, sometimes known as 'fixed variables'.

Flemming–Viot process A measure-valued diffusion process model describing the distribution of gene frequencies when the type space of a gene is complex and may not be countable.

fluctuation A movement up or down between consecutive items of a series of numbers or numerical observations. In a different sense the variation of a statistic from sample to sample is also referred to as a sampling fluctuation.

Fokker–Planck equation An equation originally occurring in the theory of diffusion when drift is taken into account. It may be written in the form

$$\frac{\partial v(x,t)}{\partial t} = -2c\frac{\partial v(x,t)}{\partial x} + D\frac{\partial^2 v(x,t)}{\partial x^2}$$

where $v(x,t)$ is the probability density for displacement x at time t, D is the diffusion coefficient and c represents drift (Kolmogorov, 1931). The equation occurs in the theory of stochastic processes as a limiting case of random walk or additive processes. It can be generalized to state- and time-dependent diffusions, where it is the forward version of the **Chapman–Kolmogorov equations**.

folded contingency table A symmetric (square) contingency table, i.e., one in which the frequency with ith row and jth column is equal to that in the jth row and ith column.

folded distributions A measurement may be recorded without its algebraic sign: as a result the underlying distribution of a measurement is replaced by a distribution of absolute measurements. The resulting 'folded' distributions take the name of the underlying distribution, thus: 'folded normal distribution', 'folded Cauchy distribution', etc. [See also **half-normal distribution**.]

folded median An alternative name for a **Bickel–Hodges estimator**.

folding In empirical data, observations are sometimes recorded without being given a sign, and therefore have to be regarded as all having the same sign (conventionally 'plus'). This affects their distribution since the negative point is 'folded' over and added to the positive part. This situation should be distinguished from that in which a variable is essentially positive.

follow-up A further attempt to obtain information from an individual in a survey or field experiment because the initial attempt has failed or later information is available.

force of mortality A term used in actuarial or other analysis of human life to indicate the **age-specific death rate** or **hazard function**. [See **actuarial statistics**.]

forecasting 'Forecasting' and 'prediction' are often used synonymously in the customary sense of assessing the magnitude which a quantity will assume at some future point of time, as distinct from 'estimation' which attempts to assess the magnitude of an already existent quantity. For example, the final yield of a crop is 'forecast' during the growing period but 'estimated' at harvest.

The errors of estimation involved in prediction from a regression equation are sometimes referred to as 'forecasting errors'.

Foster's criteria A number of criteria proposed by Foster (1953) for deciding whether a state of a **Markov chain** $\{x_n, n = 0, 1, 2, \ldots\}$ with stationary transition probabilities, a denumerable state space and with all states **communicating** and **aperiodic** is **positive recurrent, null recurrent** or **transient**; the criteria are particularly suited to the special type of Markov chains occurring in queueing theory.

four-fold table See **contingency table**.

Fourier analysis The theory of representing functions of a variable t as the sum of a series of sine and cosine terms of type $a_j \cos(2\pi j/\lambda_j), j = 0, 1, \ldots$. The λ's are not necessarily commensurable and hence the analysis is more general than harmonic analysis which considers series of terms such as $\cos(2\pi j/\lambda)$ where λ is some constant.

Fourier transform The Fourier transform of a function $f(x)$ defined on $\mathbb{R}$ is the function

$$\tilde{f}(s) = \int_{-\infty}^{\infty} e^{isx} f(x) dx.$$

The most important statistical applications are the **characteristic function** and the **spectrum** of a time series.

fractal A term introduced by Mandelbrot (1975) for objects having features with a great range of scales, so that the perimeter increases indefinitely as the scale of measurement is reduced (the coastline of a country behaves roughly in this way). Technically, a fractal is defined as an object with Hausdorff dimension greater than its topological dimension. The idea is also relevant to the study of long-range dependence.

fractile See **quantile**.

fractile graphical analysis The name suggested by Mahalanobis (1961) for a distribution-free regression analysis based upon order statistics.

fraction defective In quality control, that proportion of a number of units which are defective.

fractional Brownian motion A Gaussian process $\{X_t\}$ with means zero and correlations $\frac{1}{2}(|t|^{2H} + |s|^{2H} - |t - s|^{2H})$, where $0 < H < 1$ is called the self-similarity or Hurst coefficient; the **Wiener process** arises when $H = 1/2$.

fractional replication Where there are a large number of treatment combinations resulting from a large number of factors to be tested, it is sometimes impracticable to test all the combinations with one experimental layout. In such cases resort may be made to a fractional, i.e. partial, replication. This device is likely to be useful only where certain high-order interactions can be regarded as negligible.

frailty model This refers to a random effects model for time-to-event data such as survival data.

frame A list, map or other specification of the units which define a population to be sampled. The frame may or may not contain information about the **size** or other supplementary information of the units, but should have enough details so that a unit, if included in the sample, may be located and taken up for enquiry. The nature of the frame exerts a considerable influence over the structure of a sample survey. It is rarely perfect, and may be inaccurate, incomplete, inadequately described, out of date, or subject to some degree of duplication. Reasonable reliability in the frame is a desirable condition for the reliability of a sample survey based on it. In multi-stage sampling it is sometimes possible to construct the frame at higher stages during the progress of the sample survey itself. For example, certain first-stage units may be selected in the first instance; and then more detailed lists or maps may be constructed by compilation of available information or by direct observation only of the first-stage units actually selected.

Fréchet distance A distance between two probability distribution functions G and H of two random variables U and V, proposed by Fréchet (1957), is defined by

$$d(G, H) = \left[\min_{U,V} E\|U - V\|^2\right]^{1/2},$$

where the minimization is taken over all values of the random variables U and V having distribution functions G and H, respectively. Dowson and Landau (1982) showed that, if G and H are elliptically symmetric, $d(G, H)$ can be written as

$$d(G, H) = \left[\|\mu_U - \mu_V\|^2 + \text{tr}\left\{\Sigma_U + \Sigma_V - 2\left(\Sigma_U\Sigma_V\right)^{1/2}\right\}\right]^{1/2},$$

where μ_U, μ_V, Σ_U and Σ_V are the means and covariance matrices of the random variables U and V, respectively.

Fréchet distribution An alternative name for the type III, or second asymptotic, distribution of extreme values. The distribution function is

$$F(x) = \exp(-y^{-k})$$

where $y = (x - a)/b$, $b > 0$ and $x \geqslant a$, and k is a positive constant related to the shape. [See also **generalized extreme-value distribution**.]

Fréchet's inequalities A set of inequalities due to Fréchet (1940) which state that if $A_1, A_2, \ldots, A_n$ are arbitrary events, and S_k is defined by

$$S_k = \sum_{1 \leqslant i_1 < i_2 < \ldots < i_k \leqslant n} \Pr(A_{i_1} A_{i_2} \ldots A_{i_k}) \qquad \text{for} \quad k = 1, 2, \ldots, n$$

with $S_0 = 1$, then it follows that

$$\frac{S_{r+1}}{\binom{n}{r+1}} \leqslant \frac{S_r}{\binom{n}{r}} \qquad \text{for } r = 0, 1, \ldots, n-1.$$

A similar set of inequalities, due to Gumbel, states that

$$\frac{\binom{n}{r+1} - S_{r+1}}{\binom{n-1}{r}} \leqslant \frac{\binom{n}{r} - S_r}{\binom{n-1}{r-1}} \qquad \text{for } r = 1, 2, \ldots, n-1.$$

freehand method A method of describing the relationship in a series of data, ordered in time or space, whereby the general trend is estimated by drawing a line freehand through or near the series of plotted observations.

Freeman–Tukey transformation A transformation of the form $\sqrt{x} + \sqrt{x+1}$ to stabilize the variance of Poisson data proposed by Freeman and Tukey (1950).

frequency The number of occurrences of a given type of event, or the number of members of a population falling into a specified class. Where the frequency is expressed as a proportion of the total number of occurrences or the total number of members, it is called the relative or proportional frequency, but where no ambiguity can arise these ratios may simply be called frequencies. [See **histogram**.]

frequency curve The graphical representation of a continuous frequency distribution, the variable being the abscissa and the frequency, or preferably the frequency per unit of measurement, the ordinate. The frequency curve may be viewed as the limiting form of the frequency polygon as the number of observations tends to become infinitely large, and the class intervals indefinitely small.

frequency distribution A specification of the way in which the frequencies of members of a population are distributed according to the values of the variables which they exhibit. For observed data the distribution is usually specified in tabular or graphical form, with some grouping for continuous variables.

frequency domain The analysis of time series in terms of spectral properties, i.e., in terms of a generalization of Fourier analysis. [See **spectrum, time domain**.]

frequency moment See **probability moment**.

frequency polygon A diagram showing the form of a frequency distribution; the frequencies are graphed as ordinates against the values of the variable as abscissae and the tops of the ordinates joined one to the next. The diagram may be used to exhibit the frequencies of a continuous distribution if the frequencies are grouped in variable intervals; it is then customary to erect ordinates at the middle of the intervals.

frequency response function In the analysis of time series a linear weighted average of $u(t)$ may be expressed as:

$$v(t) = \int_0^\infty a(\tau)u(t - \tau)d\tau$$

where $a(\tau)$ is a system of weights known as the impulse response function. The frequency response function, or transfer function, is the **characteristic function** or Fourier transform of $a(\tau)$, namely

$$w(\alpha) = \int_0^\infty e^{i\tau\alpha}a(\tau)d\tau.$$

frequency surface The bivariate analogue of the **frequency curve**.

frequency table A table drawn up to show the distribution of the frequency of occurrence of a given characteristic according to some specified set of class intervals. It may be univariate or multivariate but there are difficulties in presenting data tabulated by more than two variables.

frequency theory of probability The frequency theory of probability regards the probability of an event as the limit of the frequency of occurrence of that event in a series of n trials as n tends to infinity. The existence of this limit is an axiom of the theory as proposed by von Mises (1919), but later axiomatizations (Kolmogorov, 1933) avoid the difficulties associated with it by taking the probability as a measure associated with a set of points (events) and proceeding on the basis of measure theory. This avoids the difficulty only for a mathematician. For the statistician the problem of relating probability to frequency of occurrence remains.

frequentist A user of the **frequency theory of probability**.

Freund–Ansari test See **Siegel–Tukey test**.

Friedman's test A non-parametric test related to two-way **analysis of variance**. To test differences between rows, the items in each column are ranked, and the test depends on the row totals of these ranks (Friedman, 1937).

full information method In econometrics, a method of deriving estimates of parameters in a stochastic model which are subject to all the a priori restrictions of that model. [See also **limited information method**.]

functional central limit theorem An alternative name for **Donsker's theorem**.

functional relationship The problem of estimating an exact relationship among variables when each is observed with random error is known as the functional relationship problem. [See **regression, errors-in-variables model**.]

fundamental probability set A set of objects or events which are basic to a probabilistic situation, in the sense that all other objects or events under consideration are derived from them by compounding. It follows that all probabilities are expressible by the rules of addition, multiplication, etc., of probabilities in terms of the probabilities of the fundamental set. Failure to specify the fundamental set explicitly leads to confusion, occasionally to error, and in particular to a number of paradoxes. The fundamental probability set is sometimes called the reference set. By a slight extension the actual probabilities of the fundamental set are also referred to as the fundamental probability set.

fundamental random process See **Brownian motion process**.

funnel plot A plot of an outcome measure of interest versus its estimated standard error (or sample size) which is used to informally assess publication bias in meta-analyses and overviews. If part of the 'funnel' in the plot is missing, particularly for small estimated effects or sample sizes, this may be indicative of publication bias.

Furry process An early variety of a **birth and death process** (Furry, 1937).

futility index Futility is the conditional probability that a clinical trial will fail to demonstrate the superiority of a new therapy given the accumulated data and projected sample size of the study. It thus provides a probabilistic basis for the early stopping of a clinical trial of a new therapy based on the results already observed (Ware, Muller and Braunwald, 1985).

fuzzy set theory The theory of sets in which some items may be members of two or more sets (Zadeh, 1965). The 'fuzziness' is associated, not with random variation, but with uncertainties about the set boundaries. The theory has been used, in particular, in **decision theory** and **pattern recognition**.

G

g-statistics The sample values of standardized cumulants. See **gamma coefficients**. There are a variety of other meanings in specific applied fields.

Gabriel–Sen statistic An **ANOVA** and **MANOVA** statistic based on rank scores proposed by Gabriel and Sen (1968). The statistic provides a distribution-free procedure to test all subgroups of samples on all subsets of variables, and to set simultaneous confidence bands on all location differences.

Gabriel's test A test of homogeneity of subsets of mean values in **analysis of variance**, proposed by Gabriel (1964). It can be regarded as an extension of the **Scheffé test**.

gain function The function $G(f)$ in a factorization of the **frequency response function** $w(\alpha)$ into the form

$$w(e^{i2\pi f}) = G(f)e^{i2\pi\phi(f)}.$$

$\phi(f)$ is known as the phase function.

Galton, Francis (1822–1911) An English scientist who developed the basic statistical concepts of regression lines and correlation between variables. His pioneering contributions were wide-ranging, including to meteorology and to the use of fingerprints. He contributed to the foundation of the journal *Biometrika* and to the establishment of the Chair of Eugenics subsequently occupied by K. Pearson and R.A. Fisher.

Galton–McAlister distribution A frequency distribution in which the logarithm of the variable, or some simple linear function of it, is normally distributed. The distribution was suggested by McAlister (1879). It is now more generally known as the **lognormal distribution**.

Galton ogive See **ogive**.

Galton–Watson process A Markov-type **branching process** whose earliest application was to the survival of family names. It was first stated by Galton and Watson (1874). The first complete solution was by Steffensen (1930).

Galton's individual difference problem A problem discussed by Galton in the latter half of the nineteenth century. In modern terms it is equivalent to determining the differences of random variable values or their expectations for certain individuals, based on their ranks. For example, Galton considered the problem: how should a prize be divided between the winner and the second and third members in a contest, assuming that the underlying distribution of abilities is normal?

Galton's rank order test A simple application of rank order statistics for testing the difference between two treatments which occurred in some work for Darwin (1876); summarized and developed by Hodges (1955) using results of Chung and Feller (1949).

gambler's ruin The name given to one of the classical topics in probability theory. A game of chance can be related to a series of Bernoulli trials at which a gambler wins a predetermined sum of money for every success and loses a second sum of money for every failure. The play may proceed until the gambler's initial capital is exhausted and the gambler is ruined. The statistical problems involved are concerned with the probability of the ruin of a player, given the stakes, initial capital and chances of success, and with such matters as the distribution of the length of play. There are many variations to this classical problem, which is closely associated with problems of the **random walk**; in particular, of **sequential sampling**.

game theory Generally, that branch of mathematics which deals with the theory of contests between two or more players under specified sets of rules. The subject assumes a statistical aspect when part of the game proceeds under a chance scheme, e.g., by the throw of a die or when strategies are selected at random. [See also **strategy**, **zero-sum game**, **minimax principle**, **fair game**.]

gamma coefficients Ratios, analogous to **moment ratios**, which are based upon the cumulants

$$\gamma_1 = \frac{\kappa_3}{\kappa_2^{3/2}}, \qquad \gamma_2 = \frac{\kappa_4}{\kappa_2^2}$$

and, in general,

$$\gamma_r = \frac{\kappa_r}{\kappa_2^{r/2}}, \qquad r = 3, \ldots.$$

gamma distribution A probability density function of the form

$$f(x) = \frac{e^{-x}x^{\alpha-1}}{\Gamma(\alpha)}, \qquad 0 \leqslant x < \infty,$$

where $\Gamma(\alpha) = \int_0^\infty x^{\alpha-1}e^{-x}dx$. An older term is Pearson's type III or simply **type III distribution**. The distribution function $F(x)$ is an incomplete gamma function: hence the name. Its importance in statistics derives partly from the fact that $c\chi^2$, where c is a numerical constant, is actually or approximately distributed in the gamma form under certain conditions.

gamma-frailty models **Frailty models** in which the frailty is assumed to have a gamma distribution.

gamma function The function defined by

$$\Gamma(x) = \int_0^\infty t^{x-1} e^{-t} dt.$$

In general $\Gamma(x + 1) = x\Gamma(x)$. This integral is convergent whenever $x > 0$.

gamma integral See **gamma function**.

Gantt progress chart An application of the **bar chart** due to Gantt in 1917, of use in industrial statistics; see Clark (1922). An actual performance or output is expressed as a percentage of a quota or planned performance per unit of time. Account may also be taken of the cumulative performance by plotting it, with the planned cumulative performance, as ordinate against time as abscissa.

GARCH model See **autoregressive conditional heteroscedasticity (ARCH) model**.

Gart's test Gart (1969) proposed an exact test for comparing proportions in matched samples where order within pairs may be important. The test is related to the **Fisher–Yates test** for 2×2 tables.

Garwood distribution A distribution proposed by Garwood (1940) for the waiting time of vehicles arriving randomly at vehicle-controlled traffic lights.

Gauss, Karl Friedrich (1777–1855) The German mathematician Gauss is regarded as one of the greatest mathematicians and astronomers of all time. He founded modern number theory and aspects of probability theory. He shaped the treatment of observations into a practical tool. Various principles which he advocated became an integral part of statistics and his theory of errors remained a major focus of probability theory up to the 1930s.

Gauss elimination See **elimination**.

Gauss–Markov theorem A fundamental theorem dealing with an unbiased estimator of a population characteristic, based upon a linear combination of sample observations drawn from that population. The theorem is to the general effect that an unbiased linear estimator of a parameter has minimum variance, when the estimator is obtained by least squares. The theorem can be extended in many directions, e.g., by considering the simultaneous estimation of several parameters or linear functions of them.

Gauss–Poisson distribution Another name for the **Poisson–normal distribution**.

Gauss–Seidel method A classical method for the iterative solution of a set of linear equations, particularly those arising from least squares solutions: an extension by Seidel (1874) of the fundamental method due to Gauss (1823).

Gauss–Winckler inequality An inequality concerning the moments of a continuous distribution about the mode, if ν_r is the absolute moment of order r about the mode, assumed unique:

$$\{(r+1)\nu_r\}^{1/r} \leqslant \{(n+1)\nu_n\}^{1/n}, \qquad r < n.$$

This was given by Gauss for $r = 2$, $n = 4$ in the form $\mu_4/\mu_2^2 \geqslant 1.8$ and generalized by Winckler (1866). The expression is also used to denote an inequality of the **Bienaymé–Tchebychev** type covering limits to the probability of deviations from the mode. There are various forms of the inequality, which is also associated with the names of Camp, Meidell and Narumi. [See **Camp–Meidell inequality**.]

Gaussian distribution An alternative name for the **normal distribution**.

Gaussian process Any stochastic process whose finite-dimensional distributions are all normal.

Geary, Roy (1896–1983) An Irish scientist whose research in mathematical statistics, demography, national accounting and economics is influential. Alongside his theoretical achievements, he contributed enormously to the building of official statistical records during the first 60 years of the State through his work with the Central Statistics Office. He also played a key role in establishing the reputation of The Economic and Social Research Institute in Dublin.

Geary's contiguity ratio A generalization of the **von Neumann's ratio** (1941) for one-dimensional data proposed by Geary (1954). The special case is where the data analysed represent values for irregular areas with a well-defined pattern of contiguity.

Geary's ratio As a test of normality, the moment ratio for **kurtosis** has the drawback that its sampling distribution is skew even for quite high values of n, the sample size. In order to overcome this, Geary (1935) proposed a test in the form of a ratio

$$\frac{\text{mean deviation}}{\text{standard deviation}},$$

which, in samples from a normal distribution, tends to $\sqrt{2/\pi}$ as n tends to infinity. The distribution of Geary's ratio tends to the normal form fairly rapidly and, as a test, the ratio aims at detecting departures from mesokurtosis in the parent population.

Geary's theorem A **characterization** theorem (Geary, 1936) showing that independence of the sample mean and variance implies normality.

Gehan test A rank test for the comparison of two right-censored samples reducing to the Wilcoxon test in the uncensored case (Gehan, 1965).

general factor In component analysis, a component which is common to all the observed variables; a factor which is involved in the variances of all the tests in a **battery of tests** subjected to a **factor analysis**. In the theory due to Spearman the common factor variance reduces to the variance of a single general factor and a psychological significance is attributed to this general factor or g. [See **common factor**.]

general interdependent system An interdependent model where each residual is assumed to be uncorrelated only with explanatory variables of the same equation. The classic **simultaneous equations model** assumes that each residual is uncorrelated with all explanatory variables irrespective of the equation in which they occur *vis-à-vis* the residuals.

general renewal process A class of stochastic process in which the time to the first event has distribution function $F_1(t)$, but thereafter the intervals between events are independently and identically distributed with distribution function $F(t)$. $F(t)$ is called the 'renewal distribution' whilst $F_1(t)$ is referred to as the distribution of the first renewal time.

generalized additive model (GAM) Extension of a **generalized linear model** in which explanatory variables may be replaced by non- or semiparametric smooth terms, typically obtained through **penalized likelihood** fitting of spline terms or through a **scatterplot smoother** (Hastie and Tibshirani, 1990; Green and Silverman, 1994).

generalized Bayes' decision rule See **Bayes' decision rule**.

generalized binomial distribution An alternative name for the **Poisson binomial distribution**.

generalized bivariate exponential distribution A generalization of the bivariate form obtained directly from the **multivariate exponential distribution**. This generalization (Marshall and Olkin, 1967a) is important in connection with the form of the underlying model used in life assessment problems.

generalized contagious distribution A wider generalization of the Neyman **type A, B or C distribution** than the **Beall–Rescia generalization of Neyman's distribution** proposed by Gurland (1958). Its probability generating function is

$$H(z) = \exp(-m_1) \exp\{m_1 \cdot {}_1F_1[\alpha, \alpha + \beta, m_2(z-1)]\}$$

where ${}_1F_1$ is the confluent hypergeometric function. When $\alpha = 1$ this distribution is the Beall–Rescia distribution.

generalized cross-validation A variant of cross-validation modified for computational reasons, typically involving the trace of a smoothing matrix.

generalized distance Often used for the distance defined by

$$D^2 = (x_1 - x_2)'\mathbf{V}^{-1}(x_1 - x_2),$$

where x_1, x_2 are vectors and $\mathbf{V}$ is a dispersion matrix. When x_1 and x_2 are both means, it corresponds to the **Mahalanobis distance** or its estimate, the D^2 **statistic**.

generalized distribution This expression occurs in two senses: (i) as a more complicated form similar to some known distribution; (ii) as obtained from a known distribution when its parameters are themselves random variables. The latter usage should be discarded in favour of **compound distribution**.

generalized estimating equations Method for estimation, often used for regression models with correlated discrete responses, that is based on quasi-likelihood theory and extends generalized least squares. Estimating equations are applied with postulated, so-called working, covariance matrices, often chosen for convenience but designed to reduce the loss of efficiency that would arise if the observations were presumed independent. The estimating equations yield consistent estimators of the regression parameters (Diggle, Liang and Zeger, 1994).

generalized extreme-value distribution A distribution fitted to maximum wave heights, windspeeds, financial returns and so on. If $X_1, \ldots, X_k$ is a random sample from a continuous non-degenerate distribution, then if sequences $\{a_k\}$ and $\{b_k\}$ exist such that $Y = a_k(\max\{X_1, \ldots, X_k\} - b_k)$ has a non-degenerate limiting distribution as $k \to \infty$, it has form

$$H(y) = \exp\left\{-(1 + \xi y)_+^{-1/\xi}\right\}, \quad -\infty < \xi < \infty,$$

apart from location and scale. The **Gumbel distribution** function

$$\exp\{-\exp(-y)\}, \quad -\infty < y < \infty,$$

arises in the limit as $\xi \to 0$. The heavy-tailed **Fréchet distribution** arises for $\xi > 0$ and the short-tailed negative **Weibull distribution** for $\xi < 0$. Distributions for minima arise by taking $-Y$. Extensions give the limiting joint distribution for a fixed number of extreme order statistics (Coles, 2001).

generalized gamma distribution The general form of the gamma distribution has probability density function of the form

$$f(x) = \frac{(x - \gamma)^{\alpha - 1} \exp\{-(x - \gamma)/\beta\}}{\beta^\alpha \Gamma(\alpha)}, \quad x > \gamma,$$

where α, β and γ are positive parameters. The standard **gamma, chi-squared, exponential** and **Weibull distributions** all appear as special cases (Stacy, 1962).

generalized inverse A matrix $\mathbf{X}^-$ satisfying the condition $\mathbf{X}\mathbf{X}^-\mathbf{X} = \mathbf{X}$ is a generalized inverse (g-inverse) of $\mathbf{X}$. A principal motivation for g-inverse is that the system of inhomogeneous simultaneous linear equations $\mathbf{X}\theta = y$ has a solution $\theta = \mathbf{X}^- y$ if and only if $\mathbf{X}^-$ satisfies $\mathbf{X}\mathbf{X}^-\mathbf{X} = \mathbf{X}$. In **generalized linear models** $E(y) = \mathbf{X}\theta$ when the design matrix is not of maximal rank, computation of g-inverse becomes important for estimation of parameters (Rao, 1962; Rao and Mitra, 1967). [See **Moore–Penrose inverse**.]

generalized least squares estimator A method proposed by Aitken (1934) for estimating the k parameters of the vector β in a linear equation when the disturbances v are not independent but their variance–covariance matrix is known up to a scalar constant.

generalized linear model An extension of the linear regression model to settings where responses have an exponential family distribution, such as the binomial, Poisson or gamma distributions, the mean μ of which is related to the linear predictor $\eta = x_1\beta_1 + \ldots + x_p\beta_p$ through a **link function**, $g(\mu) = \eta$, typically nonlinear. The inference procedures are typically likelihood-based, and competing models are compared using the **deviance** statistic. Generalized linear models unify **log-linear models**, **logistic regression** models and **probit** models, in addition to the linear multiple regression model. Not to be confused with the so-called general linear model, a term formerly used for linear regression with a non-diagonal covariance matrix (Nelder and Wedderburn, 1972; McCullagh and Nelder, 1989).

generalized maximum likelihood estimator An approach to parameter estimation involving the concept of asymptotic efficiency proposed by Weiss and Wolfowitz (1966) to remove some of the problems associated with classical maximum likelihood estimators.

generalized mixed model The extension of the **generalized linear model** by the inclusion of additional random terms in the linear predictor.

generalized multinomial distribution An n-dimensional distribution of discrete variables proposed by Tallis (1962) which includes an additional parameter ρ as the correlation coefficient between pairs of variables with common mean and marginal distributions which are multinomial.

generalized Pareto distribution A distribution of the form

$$G(y) = 1 - (1 + \xi y/\tau)_+^{-1/\xi}, \qquad y > 0, \quad \tau > 0, \quad -\infty < \xi < \infty,$$

fitted to exceedances of windspeeds, financial losses, wave heights, etc., over a high threshold. This is the only distribution having the threshold-stability property $\{1 - G(y + u)\}/\{1 - G(u)\} = 1 - G(y/\sigma)$ for some σ. There is a close connection to the use of the generalized extreme-value distribution for maxima (Davison and Smith, 1990).

generalized polykays An extension of the concept of **polykays** and **bipolykays** by Dayhoff (1964) to 'n-way' polykays.

generalized power series distribution A random variable X taking nonnegative integral values with probability

$$\Pr(X = x) = \frac{a_x \theta^x}{f(\theta)}, \qquad x = k, k + 1, \ldots, \quad \theta > 0,$$

is a generalized power series distribution. When $k = 0$ the generalized power series distribution reduces to a power series distribution.

generalized right angular designs A generalization proposed by Tharthare (1965) to produce a further class of four associate **right angular designs**.

generalized sequential probability ratio test An extension by Weiss (1953) of the basic Wald **sequential probability ratio test** where constant acceptance and rejection limits (A, B) are not necessarily used at each stage of sampling: at the ith stage predetermined number A_i B_i may be used.

generalized stable law See **stable law**.

generalized STER distribution A generalization of the **STER distribution** corresponding to a non-negative integer random variable Y is a discrete random variable X which has a frequency distribution of the form

$$\Pr(X = x) = \frac{\sum_{y=x+k}^{\infty} \{1/(y - k + 1)\} \Pr(Y = y)}{\sum_{y=k}^{\infty} \Pr(Y = y)}$$

where $x = 0, 1, 2, \ldots$, and $k = 1, 2, 3, \ldots$. The STER distribution is a special case of this where $k = 1$.

generalized T^2 distribution A generalization to several samples by Lawley (1939) and Hotelling (1951) of the two-sample T^2 test for the equality of mean vectors. The exact form of this distribution is not available but an asymptotic formula was given by Ito (1956, 1960).

generating function A function of a quantity t which, when expanded as a power series in t, yields as the coefficients the values of some quantity of statistical interest such as the probability of events or the moments of a frequency distribution. The **characteristic function** is an important case of a moment generating function, a phrase which is often abbreviated to m.g.f. The theory of probability has made use of the generating function approach since the time of de Moivre in the first half of the eighteenth century.

genetic algorithm An algorithm used specially in combinatorial optimization.

genetic heterogeneity This describes a situation where a disease may be due to different genes in different individuals.

GenStat A powerful statistical package for modelling and data analysis. Its name derives from GENeral STATistics, and it has a long history of applications in agriculture, especially for the design and analysis of variance of experiments. GenStat was originally developed by statisticians at the Rothamsted Experimental Station, UK, and is still the package of choice for many biometricians and statisticians working in agricultural and environmental statistics. It has a wide range of routines for modern statistical techniques including mixed models, spatial statistics and resampling methods, and extensive graphical capabilities. GenStat for Windows is a popular version with a graphical user interface. Further details can be found at http://www.nag.co.uk/stats/tt/tt_tech.html.

geometric distribution A discrete distribution. When an experiment with two possible results, classified as success or failure, is repeated independently until the first success is obtained, the probability function is

$$\Pr(X = r) = p(1 - p)^{r-1}, \qquad r = 1, 2, \dots.$$

geometric mean A measure of location, the geometric mean (G) of n positive quantities is the positive nth root of the product of these quantities. The geometric mean of $x_1, \dots, x_n$ can be calculated by

$$G = \left(\prod_{j=1}^{n} x_i \right)^{1/n}.$$

Where it exists the geometric mean lies between the harmonic mean and the arithmetic mean. The geometric mean of a frequency distribution may be written in terms of relative frequencies in the groups:

$$G = \prod_{j=1}^{k} (x_j^{f_j}),$$

where f_j is the relative frequency at x_j. For a continuous distribution with frequency function $f(x)$ with positive support, it may be defined by the equation

$$\log G = \int_{0}^{\infty} f(x) \log x \, dx.$$

geometric probability The study of problems relating to random structures in space.

geometric range This expression occurs very infrequently in two senses: (i) as the ratio of the extremes of a sample; (ii) as the geometric mean of the extremes of a sample.

geostatistics A term introduced by Matheron (1962) for the study of 'regionalized variables'; that is, variables supposed to follow some spatial stochastic process. The term is used particularly in hydrogeology, oceanography, environmental sciences and in the study of mineral deposits, using **kriging** or similar methods.

Gibbs sampler A **Markov chain Monte Carlo** algorithm widely used for numerical approximation of multivariate Bayesian posterior distributions, in which values of each variable are successively simulated from their conditional distributions given current values of all other variables. It is a particular form of the **Metropolis–Hastings algorithm**.

Gibrat distribution A form of the logarithmic normal or **lognormal distribution**. If the random variable Z has the **normal distribution** with mean zero and variance one, and a, b, c are constants, then $a + \exp(c + bZ)$ has the Gibrat distribution.

Gini, Corrado (1884–1965) From a strictly statistical point of view, Gini's contributions pertain mainly to mean values and variability and association between statistical variables, with influential contributions also to economics, sociology, demography and biology. Between 1926 and 1932 he was president of the Italian Central Institute of Statistics (ISTAT). He had a considerable interest in editorial activity and in 1920 he founded *Metron*, an international journal of statistics, and in 1934 *Genus*, a journal of the Italian Committee for the Study of Population Problems. In 1936 he founded the Faculty of Statistical, Demographic and Actuarial Sciences at the University of Rome of which he was dean until 1954.

Gini's mean distance A measure of dispersion introduced by Gini (1910) in the study of income distributions. The statistic

$$G = \binom{n}{2}^{-1} \sum_{i<j}^{n} 2|x_i - x_j|$$

is an estimate of a linear measure of dispersion which exists for all distributions with a mean value. It had been used earlier (Helmert, 1876) as an unbiased estimate of the **probable error** of the normal distribution.

Gini's ratio The ratio $E(|X - \mu|)/2\mu$, related to **Gini's mean distance**, mainly used as a measure of income inequality.

girdle distribution An **axial distribution** that has its maximum density in a plane around a great circle (rather than at the end-points of an axis). If there is rotational symmetry about the axis perpendicular to the girdle plane, the distribution is called a symmetric girdle distribution.

Gittins index A statistic, based on a dynamic allocation procedure, determining the optimal strategy in **bandit problems** (Gittins and Jones, 1974).

GLIM Acronym for Generalized Linear Interactive Modelling. Release 4 was developed by the GLIM Working Party of the Royal Statistical Society. It was designed to fit generalized linear models and is attractive in this regard, facilitating transformations and model selection, for example. GLIM also contains many standard statistical techniques for data exploration and analysis, has high-resolution graphics and a useful macro facility. The GLIM Macro Library provides techniques that are not available in the main package, including **generalized additive models**, and users can write macros tailored to their own analyses. GLIM is widely respected by statisticians but its use has been largely replaced by larger mainstream statistical packages and systems. Further details can be found at http://www.nag.co.uk/stats/GDGE_soft.asp.

Glivenko–Cantelli lemma This lemma states that for random samples of n from populations with any distribution function $F(x)$,

$$\sup_{-\infty < x < +\infty} |F_n(x) - F(x)| \to 0$$

with probability unity as n increases (Glivenko, 1933); here $F_n(x)$ is the **empirical distribution function**.

Glivenko's theorem Another term for the **Glivenko–Cantelli lemma**.

GLLAMM Acronym for Generalized Linear Latent and Mixed Models (Rabe-Hesketh, Pickles and Skrondal, 2001). It is a specialist **Stata** program for fitting structural equation models, latent class models and generalized linear mixed models among other things by maximum likelihood and Gaussian quadrature, especially adaptive Gaussian quadrature. GLLAMM is highly regarded by statisticians and epidemiological users, and the program and documentation can be downloaded from http://www.iop.kcl.ac.uk/IoP/Departments/BioComp/programs/gllamm.html.

glyph A graphical representation of multivariate data suggested by Anderson (1960). Each observation is represented by a circle, with rays of different length indicating the values of the observed variables.

Gnedenko–Koroljuk distributions See **Gnedenko–Koroljuk theorem**.

Gnedenko–Koroljuk theorem A theorem proved by Gnedenko and Koroljuk (1951) which showed that the problem of determining the distribution of the maximum discrepancy between two empirical distribution functions $F_n^{(1)}$ and $F_n^{(2)}$, each based upon a sample of n independent values from a common continuous distribution F, reduced to a random walk problem with a well-known solution. The distributions of the maximum positive discrepancy and the maximum absolute discrepancy, i.e., of

$$D_n^+ = \sup_{-\infty < x < +\infty} \left[F_n^{(1)}(x) - F_n^{(2)}(x) \right]$$

and

$$D_n = \sup_{-\infty < x < +\infty} \left| F_n^{(1)}(x) - F_n^{(2)}(x) \right|,$$

are known as the Gnedenko–Koroljuk distributions. [See also **Kolmogorov–Smirnov test**.]

Gnedenko's theorem A theorem showing the convergence of probability functions of sums of independent random variables, which take on only values of the form $a + kh$ ($h > 0, k$ an integer), to the density function of the normal distribution. This theorem contains the de Moivre–Laplace theorem as a particular case.

Gompertz curve See **growth curve**.

Gompertz–Makeham distribution A generalization of the Gompertz distribution (Gompertz, 1825) in which a constant is added to the hazard function

(Makeham, 1860). The hazard function for the Gompertz–Makeham distribution is

$$h(t) = \rho_0 + \rho_1 e^{\rho_2 t},$$

where $\rho_0 = 0$ corresponds to the Gompertz form which has a long history in describing mortality curves. The Gompertz–Makeham distribution is useful in actuarial studies and survival analysis for modelling lifetime data via parametric regression models. [See also **growth curve.**]

Goodman–Kruskal G statistic An index of association proposed by Goodman and Kruskal (1954) for the multinomial sampling model in a complete cross-classification where both of the classifications have an intrinsic and relevant order. The sample statistic $\hat{G}$ is defined in terms of probabilities p related to similarly and dissimilarly ordered classifications, or where such classifications have a neutral effect.

Goodman–Kruskal tau A measure of association for cross-classification in contingency tables. It is asymmetric between the two qualities concerned, being based on a probability interpretation involving the notion of predicting one given the other.

goodness of fit In general, the goodness of agreement between an observed set of values and a second set derived from the fitting of a model to the data. The term is used in particular in relation to the fitting of theoretical distributions to observations and the fitting of regression lines. The excellence of the fit is often measured by some criterion depending on the squares of differences between observed and theoretical value, and if the criterion has a minimum value the corresponding fit is said to be 'best'. [See also **Kolmogorov–Smirnov test.**]

Gosset, William Sealy (1876–1937) Better known by his pseudonym, 'Student', Gosset's name is associated with the discovery of the t-distribution and its use. In 1908, he published two papers in *Biometrika*: the first of these derived what is now known as Student's t-distribution and the second dealt with the small-sample distribution of Pearson's correlation coefficient (Gosset, 1908a; Gosset, 1908b). He had a profound effect on the practice of statistics in industry and agriculture.

Gower's similarity coefficient A similarity measure (Gower, 1971) defined in terms of p multivariate observations. It is particularly designed for mixtures of continuous and binary data, including missing observations:

$$S = \frac{\sum_{k=1}^{p} z_{ijk} w_{ijk}}{\sum_{k=1}^{p} w_{ijk}}$$

where

$$z_{ijk} = 1 - \frac{|x_{ik} - x_{jk}|}{r_k}$$

and where r_k is the range of x_k and $w_{ijk} = 0$ if x_{ik} or x_{jk} is unobserved, 1 otherwise.

grade For a continuous population, the grade of an individual variable value is the proportion of the total frequency with values less than or equal to that value; it is thus equivalent to the (cumulated) distribution function of that value. For discontinuous distributions the grade is similarly defined except that, by convention, an individual bearing the specified variable value counts as half an individual for the purpose of calculating proportional frequencies, the other half being regarded as lying in the remaining part of the range to the right of the specified value. The concept of grade was introduced by Galton (1907) for a continuous population to replace that of rank.

grade correlation In a continuous bivariate population, the correlation between the grades of its members. In a bivariate normal population with correlation parameter ρ the grade correlation ρ' is given by a formula $\rho' = 2\sin(\pi\rho/6)$.

gradual changes Change point problems when the model can begin to change gradually at some unknown point.

graduation See **smoothing**.

Graeco-Latin square An extension of the **Latin square**. Formally, it is an arrangement in a square of two sets of letters (say $A, B, \ldots$, and $\alpha, \beta, \ldots$), one of each in each cell of the square, such that no Roman letter occurs more than once in the same row or column, no Greek letter occurs more than once in the same row or column, and no combination of the two occurs more than once anywhere. For example, a 4×4 square of this kind is

$$A\alpha \;\; B\beta \;\; C\gamma \;\; D\delta$$
$$B\gamma \;\; A\delta \;\; D\alpha \;\; C\beta$$
$$C\delta \;\; D\gamma \;\; A\beta \;\; B\alpha$$
$$D\beta \;\; C\alpha \;\; B\delta \;\; A\gamma$$

The arrangement is used in experimental designs to allocate treatments of three factors so that all comparisons are orthogonal. The arrangement also provides four orthogonal classifications of the 16 cells, by rows, columns, Roman and Greek letters.

Gram–Charlier series type A An expression of a frequency function in terms of derivatives of the normal curve. If $H_r(x)$ is the **Tchebychev–Hermite polynomial** of order r the series with zero mean and unit standard deviation is

$$\frac{1}{\sqrt{2\pi}}e^{-\frac{1}{2}x^2}\left\{1 + \frac{1}{2}(\mu_2 - 1)H_2 + \frac{1}{6}\mu_3 H_3 + \frac{1}{24}(\mu_4 - 6\mu_2 + 3)H_4 + \ldots\right\}.$$

The name derives from the work of Gram (1883) and Charlier (1905a) who used the series to approximate to frequency functions. It is to be distinguished from **Edgeworth expansion**.

171

Gram–Charlier series type B   A series proposed by Charlier (1905a) to represent a discontinuous function in terms of differences of a Poisson variable. There are many difficulties in the use of the series and it is rarely employed.

Gram–Charlier series type C   A further series proposed by Charlier (1928) to avoid difficulties due to negative frequencies which can arise with type A. The series expands a frequency function in the form

$$f(x) = \exp\left(\sum \gamma_r H_r\right)$$

where the H_r are polynomials. This series has not come into use.

Gramian   A matrix, or its determinant, based on sums of squares and products and therefore positive semi-definite. Correlation matrices based on coefficients not of product moment type may be non-Gramian, and this can cause difficulties, particularly in **factor analysis**.

Gram's criterion   A criterion which states that for n continuous functions $f_i(x)$ to be linearly independent in the interval $a \leqslant x \leqslant b$ it is necessary and sufficient that $|d_{ik}| = 0$, where $|d_{ik}|$ is the determinant defined by

$$d_{ik} = \int_a^b f_i(x)f_k(x)dx, \qquad i, k = 1, \ldots, n.$$

[See **singular distribution**.]

grand tour   The visualization of multivariate data by plotting sequences of smoothly varying two-dimensional projections.

graphical chain model   A general graphical model is described by a chain of blocks which is a mixture of directed and undirected edges. Any two vertices from the same block are only joined by an undirected edge or a line, and two from different blocks are only jointed by a directed edge or an arrow (Wermuth and Lauritzen, 1990).

graphical estimator   A constant chosen by trial and error from a geometrical representation generally to secure a linear plot of observations on specially ruled paper.

graphical model   A representation of the dependencies among a set of variables in which each variable is represented by a node and in which some, but in general not all, pairs of nodes are joined by edges. The idea has developed from Sewell Wright's path analysis. A missing edge represents some kind of statistical independence between the corresponding variables, the precise nature of which, e.g., the conditioning set, depends on the nature of the graph. For example, edges may be directed or undirected. For contrasting treatments, see Lauritzen (1996) and Cox and Wermuth (1996).

Greenhouse–Geisser test   A modified variance ratio test used in the analysis of **repeated measures designs**. The test assuming equal correlations may give spurious significance when the assumption is false; the Greenhouse–Geisser test is (sometimes extremely) conservative.

Grenander estimators   Non-parametric maximum likelihood estimators of (*i*) a decreasing density, (*ii*) an increasing hazard and (*iii*) an increasing hazard for repeated events (Grenander, 1956).

Grenander's uncertainty principle   The product of a measure of 'resolvability' in the estimation of spectral densities (Δ_1) and a measure of the reliability, in the sense of variability, of the estimator (Δ_2) was shown by Grenander (1951) always to have a positive lower bound. The product $\Delta_1\Delta_2$ is a measure of 'uncertainty' in the estimator.

grid   A rectangular mesh on a plane formed by two sets of lines orthogonal to each other, each line of each set being at a constant interval from the adjacent lines. It is used in some forms of area sampling.

grid sampling   A form of cluster sampling, the clusters being individual areas of a grid and hence consisting of groups of basic cells arranged in some standard geometrical pattern. The term 'configurational sampling' is also used in the same sense.

group comparison   A comparison between groups of individuals, usually on the basis of a representative value (such as a mean) from each.

group divisible design   A class of experimental design studied extensively by Bose. The parameters for treatments (v), blocks (b), replicates (r), groups (m), group size (n) and λ_1, λ_2 allocation factors are related as follows:

$$v = mn,$$
$$bk = vr,$$
$$\lambda_1(n-1) + \lambda_2 n(m-1) = r(k-1),$$
$$Q = r - \lambda_1 \geqslant 0,$$
$$P = rk - v\lambda_2 \geqslant 0.$$

Where $Q = 0$ the group divisible (GD) design is singular. For $Q > 0$ and $P = 0$ the GD design is semi-regular; regular GD designs are characterized by $Q, P > 0$ (Bose and Connor, 1952).

group divisible incomplete block design   An incomplete block design with v treatments replicated r times in b blocks of size k is group divisible if the treatments can be divided into m groups of n treatments each so that treatments belonging to the same group occur together in λ_1 blocks and for different groups in λ_2 blocks. If $\lambda_1 = \lambda_2$ the design reduces to a balanced **incomplete block**.

group divisible rotatable designs   A class of response surface designs proposed by Das and Dey (1967) in which the factors are divided into two groups. The design is rotatable for each group separately when the factors in the other groups are held constant at some set of levels.

group factor See **common factor**.

group screening methods Screening designs are directed towards finding the few effective factors out of a large list of possibles which can affect the response in the design of experiments. Group screening methods consist of putting the factors in groups, testing these groups and then testing the factors in the significant groups.

group sequential design A sequential design in which progress is monitored, not after each individual result, but on several occasions when groups of results have been collected. Such designs are used particularly in **clinical trials**.

group testing This refers to a situation in which a large number of samples, e.g., of blood, are pooled then assayed as a single sample. If the pooled sample tests positive, then each individual sample is tested; if, however, the pool tests negative, no further testing is done. The rationale for the procedure is that if the disease or infection being detected is relatively rare or of low prevalence in a population, then testing the pooled sample may be cheaper than testing each individual sample.

grouped Poisson distribution In the pure birth stochastic process if we assume that the time between two events has the gamma distribution then the population size at a given time has the grouped Poisson distribution with parameters k and λ with probability function

$$\Pr(x) = \sum_{j=1}^{k} e^{-\lambda} \frac{\lambda^{xk+j-1}}{(xk+j-1)!}, \qquad x = 0, 1, \ldots,$$

where $\lambda > 0$.

grouping lattice A lattice or mesh of equal variable intervals which, superimposed on a variable scale, defines the intervals in which the frequencies are grouped.

growth curve In general, an expression giving the size of a population y as a function of a time variable t, and hence describing the course of its growth. The expression may also be used to denote the growth of an individual. If the relative growth rate declines at a constant rate, i.e.,

$$\frac{1}{y}\frac{dy}{dt} = -b, \qquad b > 0,$$

the curve is exponential. If the asymptotic value of y as t tends to infinity is a positive constant c, then

$$y = c + ae^{-bt}.$$

A growth curve for which

$$\frac{dy}{dt} = by(k - y)$$

is called logistic or autocatalytic. Its explicit form is

$$y = \frac{k}{1 + ae^{-kbt}}.$$

A rather more general form of type

$$y = \frac{k}{1 + e^{c\phi(t)}},$$

where $\phi(t)$ is some function of time, is also called logistic. The analysis of growth curves usually involves the problems of **repeated measures designs**.

Grubbs' estimators In biological and other experimentation, the situation where variations of material under test between successive measures are unavoidable is common. Grubbs (1948) proposed a technique for separating the sources of error. The **maximum likelihood estimators** (for normally distributed readings) of the variances of two instruments or procedures and the population variance are

$$\hat{\sigma}_1^2 = S_x^2 - S_{xy}, \qquad \hat{\sigma}_2^2 = S_y^2 - S_{xy}, \qquad \hat{\sigma}^2 = S_{xy}.$$

Grubbs' rule A criterion for the rejection of outlying large observations proposed by Grubbs (1950) and based upon a studentized maximum residual $R_{(n)}$. It may also be used in the form of $R_{(1)}$ to cover the case of an unduly small observation.

Gumbel distribution An alternative name for the type I, or first asymptotic, distribution of extreme values. The distribution function is

$$F(x) = \exp(-e^{-y})$$

where $y = (x - a)/b$, $b > 0$ and $-\infty < x < \infty$.

Gumbel's inequalities See **Fréchet's inequalities**.

Gupta's subset selection method A procedure proposed by Gupta (1956) for selecting from among k populations of the same family a subset of random size which contains the population with the highest value of the single unknown parameter, with a probability greater than or equal to a preassigned value.

Gupta's symmetry test An asymptotically distribution-free test of the symmetry of a continuous distribution about an unknown median.

Gurland's generalization of Neyman's distribution The mixture of Neyman's type A distribution with two parameters λ_1 and $\lambda_2 p$, where p has the **beta distribution** with parameters α and β, is Gurland's generalization of Neyman's distribution with parameters $\lambda_1, \lambda_2, \alpha$ and β where all four parameters are positive. **Beall–Rescia's generalization** is a special case of this Gurland generalization.

Guttman scaling A scaling technique in psychological statistics for ordering and weighting responses to single questions. It is related to **correspondence analysis**.

H

Hadamard, Jacques Salomon (1865–1963) A French mathematician. He was educated in Paris. After graduating from the École Normale Supérieure in 1888, Hadamard obtained his doctorate in 1892. He made many contributions in different areas of mathematics.

Hadamard matrix A square matrix whose entries are +1 and −1 and all of whose rows are orthogonal, useful for systematic resampling schemes such as **balanced repeated replication** and for fractional factorial experiments.

Haldane's discrepancy measures The set of measures

$$
D_r = \begin{cases} \dfrac{(n+r)!}{n!} \displaystyle\sum_{i=1}^{k} \dfrac{n_i! \pi_i^{r+1}(\theta)}{(n_i+r)!}, & r = \ldots, -3, -2; 1, 2, \ldots, n-1, \\[2ex] -\dfrac{1}{n} \displaystyle\sum_{i=1}^{k} n_i \log \pi_i(\theta), & r = -1, \end{cases}
$$

of discrepancy between the observed frequencies

$$
n_1, n_2, \ldots, n_k, \qquad \sum_{i=1}^{k} n_i = n,
$$

of k mutually exclusive events and the hypothetical expectations $n\pi_1(\theta)$, $n\pi_2(\theta)$, $\ldots$, $n\pi_k(\theta)$, proposed by Haldane (1951). Under suitable regularity conditions, minimization of D, with respect to θ provides consistent and efficient estimators of the parameters θ; in particular, D_1, which is known as Haldane's discrepancy, provides estimators that are sometimes much easier to compute than the corresponding **maximum likelihood** or **minimum chi-squared** estimators.

half-Cauchy distribution A special case of the folded Cauchy distribution which, in standard form, has the density function

$$
f(x) = \frac{2}{\pi(1+x^2)}, \qquad x \geqslant 0.
$$

The distribution is analogous with the **half-normal distribution**. [See **folded distributions**.]

half-drill strip One of the older systematic experimental designs which was commonly used for comparisons between two agricultural factors, e.g., two cereal strains. Owing to the various disadvantages of this kind of design, compared with the more modern designs based upon randomization, it has lost popularity.

half-invariant See **semi-invariant**.

half-normal distribution A special case of the folded **normal distribution**. If the distribution of X is $N(0, \sigma)$ then the distribution of $Z = |X|$ is 'folded' in half and has the density function

$$f(z) = \frac{1}{\sigma}\sqrt{\frac{2}{\pi}}e^{-z^2/2\sigma^2}, \qquad z \geqslant 0.$$

half-normal plot A plot of residuals in a linear model against their expected values assuming normality (the expected values of the normal order statistics). In the half-normal form, absolute values of residuals and expectations are used (Daniel, 1959).

half-normal probability paper A **normal probability paper** with negative abscissae omitted.

half-plaid square A **Latin square** design in which additional treatments are applied to complete rows (or columns) introduced by Yates (1937a). For example, a 3×3 factorial fertilizer trial may be laid out in a 9×9 Latin square, using three different varieties, each in three rows.

half-replicate design An experimental design based upon the principle of **fractional replication**, which employs only one-half of the complete number of treatment combinations in a basic design.

half-width This expression is sometimes used in relation to **central confidence intervals** to denote the upper or lower half of an interval. In fixed interval prediction, such as may be used in a **control chart**, the half-width refers to the distance on the scale of the variable between, say, the process average and the upper or lower control limit.

Hammersley–Clifford theorem A general specification of probability distributions consistent with a given undirected graphical Markov model which determines the conditional independencies among the variables (Besag, 1974). The specification hinges on potential functions defined over the cliques (maximal closed subsets) of the graph.

hamming A procedure for smoothing the spectrum of a time series using weights of $\frac{1}{4}, \frac{1}{2}$ and $\frac{1}{4}$. It was proposed by Hamming and Tukey (1949) and may have had some connection with the earlier work on meteorological series by von Hann. [See also **hanning**.]

hanning A procedure for smoothing the spectrum of a time series associated with the Austrian meteorologist von Hann. He used weights of $\frac{1}{4}, \frac{1}{2}$ and $\frac{1}{4}$ for smoothing meteorological data (not the spectrum) many years before the period of modern spectrum analysis (Tukey, 1977). [See also **hamming**.]

hard clipping (limiting) Certain nonlinear transformations of stochastic signal processes are used in electronic systems which work in real time. This form of processing, called hard clipping, is used in order to reduce the number of information bits which the system has to process. With hard clipping only one binary digit per sample value of an input signal is used. Information concerning the input signal is lost and the spectrum calculated from the clipped signal is a distorted form of the original signal spectrum. By a sine transformation of the sample covariance of the clipped process a consistent estimator is derived for the spectral density of the input signal.

hard-core model A point process in which the points are reference points on solid objects that cannot overlap. Typically, this implies a minimum distance between the points.

Hardy summation method A method of determining the moments of a frequency function defined at equidistant points, or grouped in equal intervals, by repeated summation of the frequencies. The summations give **factorial moments** from which ordinary moments may easily be derived.

Hardy–Weinberg equilibrium If two **alleles** A and a occur in a population in proportions p and q, and the genotypes AA, Aa and aa are in proportions p^2, $2pq$ and q^2, they are said to be in Hardy–Weinberg equilibrium. The equilibrium is achieved after one generation of random mating.

Harley approximation An approximation to the t-distribution proposed by Harley (1957) based upon a transformation of the correlation coefficient.

harmonic analysis The analysis of a series of values into constituent periodic terms. [See also **Fourier analysis, periodogram, spectral function.**]

harmonic dial A method of representing harmonic constituents of a time series introduced in a geophysical context by Bartels (1935b). A harmonic component is represented by a vector with length proportional to its intensity and angular orientation proportional to the phase. A set of components then appear like a number of hands on a clock.

harmonic distribution See **Zipf's law.**

harmonic mean The harmonic mean of a set of observations is the reciprocal of the arithmetic mean of their reciprocals. It may be written in the discrete case for n quantities $x_1, \ldots, x_n$, as

$$\frac{1}{H} = \frac{1}{n} \sum_{i=1}^{n} \frac{1}{x_i},$$

or, in the continuous case, as

$$\frac{1}{H} = \int_{-\infty}^{\infty} \frac{f(x)}{x} dx,$$

where $f(x)$ is the frequency function, provided of course that the integral exists. For frequency distributions where the variable values are non-negative it may be shown that the harmonic mean is less than either the geometric mean or the arithmetic mean.

harmonic regression A term occasionally used for regression on trigonometric functions, used for the detection and estimation of seasonal effects.

Harris recurrence A form of recurrence for Markov chains on a general state space, a subset of which is Harris recurrent if the number of visits to it is almost surely infinite.

Harris walk A state-dependent random walk on the set of non-negative integers, for which the matrix of transition probabilities consists of zeros except for the elements:

$$P_{0,1} = 1,$$
$$P_{j,j+1} = a_j, \qquad 0 < a_j < 1, \qquad 1 \leqslant j,$$
$$P_{j,j-1} = 1 - a_j.$$

The walk was studied by Harris (1952) using **Brownian motion process**.

Hartley, Herman Otto (1912–1980) A German statistician. Hartley earned a PhD degree in mathematics at Berlin University in 1934, and a PhD degree in statistics under John Wishart at Cambridge University in 1940. He taught at University College London, at Iowa State College and at Texas A & M University. Hartley was well known for his work on the foundations of sampling theory, and also made important contributions to mathematical optimization, estimation with incomplete data, estimation of variance components, and establishment of safe doses in carcinogenic experiments. Hartley collaborated with Egon Pearson to produce the classic two-volume *Biometrika 'Tables for Statisticians'*.

Hartley–Rao scheme A sampling scheme by Hartley and Rao (1962) in which sampling is without replacement from a finite population. The scheme requires the random ordering of the elements i of the population, from which a systematic sample of n is selected such that P_i, the probability that the ith element is included, is proportional to the size of the element.

Hartley's test A test proposed by Hartley (1950) for the equality of the variances of a number of normal populations, based on the ratio between the largest and the smallest sample variances.

hat matrix The projection matrix $\mathbf{H} = \mathbf{X}(\mathbf{X'X})^{-1}\mathbf{X'}$ corresponding to least squares estimation in the **linear regression** model with **design matrix X**; it satisfies $\hat{y} = \mathbf{H}y$, where $\hat{y}$ is the vector of fitted values for the response vector y. Introduced by Hoaglin and Welsch (1978), the hat matrix is used in **regression diagnostics**.

Hayashi, Chikio (1918–2002) A Japanese statistician. He received a BA in mathematics from the University of Tokyo in 1942, and a DSc from the same university in 1954. In 1977 he was named Honorary Fellow of the Royal Statistical Society. In 1982 he was awarded the Purple Ribbon Medal from the Japanese government. Hayashi is the author or co-author of 28 books in the Japanese language in different fields of statistics. He is also the author of books in English such as *Data analysis for comparative social research* (1992), *Treatise on behaviourmetrics* (1993), and *Quantification of qualitative data. Theory and method* (1993).

hazard In general, a word implying the existence of chance of risk (from Arabic *az-zahr*, meaning a die). Specialized usage occurs principally in connection with life analysis of systems or components. The hazard is the probability that an individual, functioning at time t_0, will fail in the interval $(t, t + \delta t)$ and this is given by $h(x) = f(x)/\{1 - F(x)\}$, where $f(x)$ and $F(x)$ are the probability density function and the cumulative distribution function. If this **instantaneous (death or failure) rate** is calculated at a number of survival times, the resulting hazard rates may be regarded as a hazard function.

hazard function See **Cox's regression model**.

hazard rate distribution A group of distributions for which the hazard satisfies certain conditions. The group can be broadly divided between distributions with **increasing hazard rate** (IHR distributions) and those with **decreasing hazard rate** (DHR distributions), but a wider grouping is that of distributions with increasing (decreasing) hazard rate *on average* (IHRA, DHRA distributions). The **exponential distribution**, which has a constant hazard rate, is a natural boundary between IHR and DHR distributions.

Hellinger, Ernst (1883–1950) A Polish mathematician. Hellinger obtained his doctorate from the University of Göttingen in 1907. In 1914 he was appointed professor of mathematics at the University of Frankfurt. In 1939 he joined the staff at Northwestern University.

Hellinger distance The Hellinger distance between two distributions with density functions $f(x)$ and $g(x)$ is given by

$$H = \int \left\{ f(x)^{1/2} - g(x)^{1/2} \right\}^2 dx.$$

It takes values between 0 (when $f(x) = g(x)$) and 2 (when $f(x)$ and $g(x)$ do not overlap), and is invariant for all non-singular transformations of x.

Helly–Bray theorem This theorem, sometimes referred to as Helly's second theorem, states that, if $f(x)$ is a continuous function and $\{F_k(x)\}$ a sequence of uniformly bounded non-decreasing functions that converge weakly to $F(x)$ in the interval $[a, b]$, then

$$\lim_{k \to \infty} \int_a^b f(x) dF_k(x) = \int_a^b f(x) dF(x).$$

181

Helly's first theorem A theorem which states that every sequence of distribution functions contains a subsequence which tends to some non-decreasing function, not necessarily a distribution function, at all continuity points of this non-decreasing function. It is sometimes referred to as Helly's lemma.

Helmert, Friedrich Robert (1843–1917) A German mathematical physicist. Helmert's interest in statistics resulted from his research in geodesy. He was professor of geodesy in Aachen in 1872, and in Berlin in 1887. His research included fundamental work on the **chi-squared distribution**.

Helmert criterion See **Abbe criterion**.

Helmert distribution The distribution of the sample standard deviation or, equivalently, of the sample variance in samples from a normal population. It may be written

$$f(x) = \frac{n^{\frac{1}{2}(n-1)}}{\Gamma\left\{\frac{1}{2}(n-1)\right\} 2^{\frac{1}{2}(n-3)}} \left(\frac{x}{\sigma}\right)^{n-2} \frac{e^{-nx^2/2\sigma^2}}{\sigma}, \qquad 0 \leqslant x < \infty,$$

where σ^2 is the population variance and s^2 the sample variance (Helmert, 1876). If ns^2/σ^2 is put equal to χ^2 the distribution becomes that of χ^2 with $n-1$ degrees of freedom. It is a form of the Pearson **type III distribution**.

Helmert transformation An orthogonal linear random variable transformation due to Helmert. If $X_1, X_2, \ldots, X_n$ have zero mean and unit variances the transformation is given by

$$Y_1 = (X_1 - X_2)\frac{1}{\sqrt{2}},$$

$$Y_2 = (X_1 + X_2 - 2X_3)\frac{1}{\sqrt{6}},$$

$$Y_3 = (X_1 + X_2 + X_3 - 3X_4)\frac{1}{\sqrt{12}},$$

$$\vdots$$

$$Y_{n-1} = \{X_1 + \ldots + X_{n-1} - (n-1)X_n\}\frac{1}{\sqrt{n(n-1)}},$$

$$Y_n = (X_1 + \ldots + X_n)\frac{1}{\sqrt{n}}.$$

heritability In general terms, the proportion of the variability of a character attributable to inheritance.

Hermite distribution The probability generating function of a generalized Poisson distribution can be expressed as

$$g(z) = \exp\left\{a_1(z-1) + a_2(z^2-1) + \ldots\right\},$$

where $\sum_{i=1}^{\infty} a_i = \lambda$, the Poisson parameter. This generalized distribution tends to the Poisson distribution with parameter a_1 if $a_2, a_3, \ldots$ become negligible compared with a_1 in some limiting process. If a_2 does not become negligible compared with a_1, then the limiting distribution is the Hermite distribution with parameters a_1 and a_2. See Kemp and Kemp (1965).

Hermite polynomials The family of orthogonal polynomials associated with the standard normal distribution.

heterogeneity Applies to populations for which the individuals do not follow the same model. It has several particular usages: in genetic epidemiology we may encounter **genetic heterogeneity**. In **analysis of variance** heterogeneity of variances may occur; in regression models explanatory variables produce non-identically distributed variables; this can be called explainable heterogeneity. Even after adjusting on explanatory variables there may remain unexplainable heterogeneity, due for instance to **clustering**; this may be represented by **random effects** or **frailty models**.

heterokurtic See **kurtosis**.

heteroscedastic See **scedasticity**.

heterotypic A term used in relation to Pearson distributions. For certain values of the moment ratios β_1 and β_2 the differential equation defining the family of distributions has infinite moments of order 8 or more. In this region the standard error of the sample estimate of β_2 would be infinite and hence the fitting of the distributions by the method of moments would be inappropriate. Distributions of the Pearson family with such values of β_1 and β_2 were called heterotypic.

Heywood case In factor analysis, a **correlation matrix** may correspond to a factor model in which one of the communalities exceeds unity, implying a negative specific variance. This is known as a Heywood case. It implies that certain correlation structures cannot arise from any standard factor model, but a factor model may give sample correlation matrices of this type because of random error.

$Hh_n(x)$ function A function derived by integration and differentiation of the 'normal' function $e^{-x^2/2}$. The function of zero order is defined as

$$Hh_0(x) = \int_x^{\infty} e^{-t^2/2} dt,$$

and for positive n the function is defined by recurrence:

$$Hh_n(x) = \int_x^{\infty} Hh_{n-1}(t) dt.$$

Similarly

$$Hh_{-n}(x) = \left(-\frac{d}{dx}\right)^n Hh_0(x)$$

$$= \left(-\frac{d}{dx}\right) Hh_{-n+1}(x).$$

The Hh_{-n} functions are Hermite functions. [See **Tchebychev–Hermite polynomial.**]

hidden Markov model A model for an observed process involving an unobserved Markov process and a measurement equation that relates the unobserved to the observed process. Also called state-space model (MacDonald and Zucchini, 1997).

hidden periodicity, scheme of A term advanced by Schuster (1898), and later used extensively by other writers, e.g., Wold (1938), to denote a time series, or more generally a stochastic process, which is generated by the addition of a finite number of harmonic terms and a random residual component. One of the objects of analysing such a series is to determine the amplitude, period and phase of each 'hidden' component.

hierarchical birth and death process A type of birth and death process proposed by Blom (1960) in which the states of the process are divided into substates according to given rules. Although only processes with constant transition probabilities were considered the analysis was generalized to include an infinitely denumerable number of states.

hierarchical cluster analysis Any form of cluster analysis that proceeds by either combining or dividing clusters (agglomerative or divisive cluster analysis). These procedures define a **dendrogram** (in contrast with methods that seek an optimal division into c groups without reference to the optima for $c + 1$ or $c - 1$).

hierarchical group divisible design The designation proposed by Roy (1953) for group divisible designs with m associate classes. These designs were later developed by Raghavarao (1964) under the more general title.

hierarchical model A model with more than one level of random variation, particularly in the Bayesian context in which the notion of exchangeability plays a central role.

hierarchy If, in a matrix of **intercorrelations** of a set of variables, the rows and columns can be so arranged to give the highest correlations in the upper left-hand corner and the lowest correlations in the lower right-hand corner and when this is done there is a constant proportional relationship between adjacent columns, except for diagonal terms, the table is called a hierarchy (Spearman, 1904) and the intercorrelations are said to be a hierarchical order. Thus, for two rows denoted by a and b and two columns by c and d, the correlations obey the so-called tetrad relations $r_{ac}r_{bd} = r_{ad}r_{bc}$. Under certain conditions the fact that the correlation matrix is hierarchical is a necessary and sufficient condition that the variation can be accounted for by a single factor common to the variables. The quantities $r_{ac}r_{bd} - r_{ad}r_{bc}$ are called tetrad differences.

high contact In relation to a frequency function $f(x)$, the order of contact of the function with the variable axis, or at infinity if the range is infinite, is said to be high if $x^r f(x)$ or its limit vanishes at the terminals for some high value of r. What constitutes a 'high' value for this purpose is somewhat arbitrary. This property of high contact is one of the conditions for the application of the **corrections for grouping** known as **Sheppard's corrections**.

higher-order asymptotic theory A refinement of asymptotic theory in which the usual normal and chi-square approximating distributions are refined in order to improve the approximation to the 'exact' distributions. More mathematically expressed, errors in the approximation that are of order $1/n$ in the standard theory are improved to order $1/n^{3/2}$ or order n^{-2} (Barndorff-Nielsen and Cox, 1994).

highest posterior density interval See **Bayesian interval**.

Hill estimator An estimator of the shape parameter of a heavy-tailed distribution, based on extreme order statistics; an associated plot is called a Hill plot (Hill, 1975).

histogram A univariate frequency diagram in which rectangles proportional in area to the class frequencies are erected on sections of the horizontal axis, the width of each section representing the corresponding class interval of the variable. [See also **frequency polygon**.]

historical controls In observational studies, a control group chosen from past experience and thus differing from the treatment group in the time of diagnosis and possibly in other features.

hitting point See **waiting time**.

Hodges, Joseph Lawson (1922–2000) An American statistician. Hodges entered the University of California at age 16 and obtained his BA in mathematics in 1942, and his PhD in 1949 under the supervision of Jerzy Neyman, and then joined the statistics faculty where he remained for the rest of his life. He is best known for his work in non-parametric inference. In a technical report with Fix, written in 1951 but published in 1989, he pioneered non-parametric density estimation. He is the inventor of the **Hodges bivariate sign test** and (with Lehmann) of the **Hodges–Lehmann one-sample estimator**. One of the most striking discoveries was the phenomena of superefficiency which exploded long-held beliefs about maximum likelihood estimation and had a profound effect on asymptotic theory.

Hodges–Ajne's test A non-parametric test of the uniformity of a **circular distribution**, first proposed as a bivariate sign test by Hodges (1955), but later studied independently by Ajne (1968) in connection with circular uniformity. Regarding n independent angular observations $\theta_1, \ldots, \theta_n$ as n points on a circle, the test statistic is the maximum possible number of points that can be made to lie to one side of a diameter of the circle by judicious orientation of the diameter. [See also **Hodges' bivariate sign test**.]

Hodges–Lehmann one-sample estimator An estimator of location proposed by Hodges and Lehmann (1963). When the appropriate model for n observations $z_1, \ldots, z_n$ is

$$z_i = \theta + \varepsilon_i, \qquad i = 1, \ldots, n,$$

where the ε's are mutually independent unobservable random variables, and each ε comes from a continuous population (not necessarily the same one) that is symmetric about zero, the Hodges–Lehmann one-sample estimator of θ is defined as

$$\hat{\theta} = \text{median}\left\{ \frac{z_i + z_j}{2}, i < j \right\}.$$

This estimator is relatively insensitive to outliers.

Hodges' bivariate sign test This analogue, proposed by Hodges (1955), of the classical **sign test**, can be shown to have the same null distribution as a test proposed by Daniels (1954a).

Hoeffding, Wassily (1914–1991) A Finnish mathematical statistician who spent most of his career in the United States. After schooling in Denmark and Germany, Hoeffding gained his PhD from the University of Berlin in 1940. In 1946 he moved to the United States and joined the staff at the University of North Carolina, Chapel Hill.

Hoeffding C_1 statistic A test criterion proposed by Hoeffding (1951) for a most powerful rank order test. It was extended by Terry (1952) for specific parametric alternative hypothesis. It formalized an earlier notion of Fisher and Yates.

Hoeffding's independence test A distribution-free test for independence of the elements of a continuous bivariate random variable (Hoeffding, 1948b).

Hoeffding's inequality If $X_1, \ldots, X_n$, are independent random variables not necessarily identically distributed and the X_i are bounded, Hoeffding's (1963) inequality states that, if $0 \leqslant X_i \leqslant 1$, for $0 < t < 1 - \mu$

$$\Pr(\bar{X} - \mu \geqslant t) \leqslant \exp\left\{-nL(\mu, t)\right\} \leqslant \exp(-2nt^2),$$

where

$$L(\mu, t) = (\mu + t) \log\left(1 + \frac{t}{\mu}\right) + (1 - \mu - t) \log\left(1 - \frac{t}{1 - \mu}\right),$$

and $\bar{X} = (X_1 + \ldots + X_n)/n$, $\mu = E(\bar{X})$.

Hoeffding's U-statistics See **U-statistic**.

Hölder's inequality An inequality that states that if $p, q > 1$ and $1/p + 1/q = 1$, then for any non-negative random variables X and Y

$$E(XY) \leqslant (E(X^p))^{1/p} \cdot (E(Y^q))^{1/q}$$

provided the mathematical expectations exist. [See also **Schwarz's inequality**.]

Hollander–Proschan 'new better than used' test A distribution-free test (Hollander and Proschan, 1972) of the null hypothesis that the failure rate of items is age-independent, against the class of **new better than used distributions** alternatives.

Hollander's bivariate symmetry test On the basis of n mutually independent realizations of the bivariate random variable (X, Y) the acceptability of the hypothesis

$$H_0 : \Pr(X < x \text{ and } Y < y) = \Pr(X < y \text{ and } Y < x) \qquad \text{for all } x, y,$$

is examined by this distribution-free test (Hollander, 1971). The hypothesis H_0 is natural when an experimenter is testing for a treatment effect and finds it convenient (or necessary) to have the same subjects receive the treatment and also act as controls. Since (X_i, Y_i) then represent two observations on the same subject, it is unrealistic to assume that X_i and Y_i are independent. The hypothesis of no treatment effect is precisely H_0.

Hollander's parallelism test A distribution-free test (Hollander, 1970) for the parallelism of two regression lines, restricted in application to situations where there are an equal and even number of observations corresponding to each line. It is a special case of the **Wilcoxon signed rank test**.

homogeneity This term is used in statistics in its ordinary sense, but most frequently occurs in connection with samples from different populations which may or may not be identical. If the populations are identical they are said to be homogeneous and, by extension, the sample data are also said to be homogeneous. In a more restricted sense populations may be said to be homogeneous in respect of some of their constants, e.g., k populations with identical means but different dispersions are homogeneous in their means.

homogeneous process A stochastic process is said to be homogeneous in space if the transition probability between any two state values at two given times depends only on the difference between those state values. The process is homogeneous in time if the transition probability between two given state values at any two times depends only on the difference between those times.

homokurtic See **kurtosis**.

homoscedastic See **scedasticity**.

honest process For a generalized Markov birth process, the solution of the differential equations is such that for all t, $\sum p_i(t) = 1$ if $\sum 1/\lambda_i$ is divergent; the process is then designated an 'honest' process. If, however, $\sum 1/\lambda_i$ is convergent then, for some t, $\sum p_i(t) < 1$ and the process is termed 'dishonest' or pathological. The λ_i are the average times spent in states i.

horseshoe effect In **ordination** it often happens that similarities are well defined for objects close together in a one-dimensional array, but near zero whenever they are separated by more than a critical distance. When a two-dimensional representation is found, e.g., by **multi-dimensional scaling**, the points representing the objects tend to lie on a horseshoe-like curve.

Horvitz–Thompson estimator A method of estimating the population total when sampling without replacement from a finite population and when unequal probabilities of selection are used. The estimator is unbiased, linear and can be used with a variety of basic sample designs (Horvitz and Thompson, 1952).

Hosmer–Lemeshow statistic A global goodness-of-fit statistic for testing the fit of a binary logistic regression model (Hosmer and Lemeshow, 1989). Observations are sorted in increasing order of their predicted probabilities and divided arbitrarily into approximately 10 groups; the statistic is then obtained by calculating the Pearson chi-squared statistic from the table of observed and expected frequencies. Large values of the H–L statistic (and small P-values) indicate lack of fit of the model. Simulation studies have shown that the statistic is distributed as chi-squared when there is no replication in subgroups based on patterns in the explanatory variables. However, the H–L statistic suffers several disadvantages, including low power, sensitivity to the choice of interval cut-points, and issues to do with testing the model on the training data. Moreover, the statistic is only suitable for small data sets.

hot deck method A fast procedure for imputing missing values in which for each non-respondent for one or more items, a matching respondent on the basis of the variables observed is found, and the missing items for the non-respondent are replaced by the matched respondent's values. For the cold deck method, the matching respondent is from a past survey.

Hotelling, Harold (1895–1973) A major developer of the foundations of statistics and an important contributor to mathematical economics, Hotelling introduced the multivariate T^2, principal components analysis, and canonical correlations (Hotelling, 1933).

Hotelling's T^2 A generalization by Hotelling (1931) of **Student's distribution** to the multivariate case, and like Student's t, available to test the significance of a broad class of statistics including means and differences of means, regression coefficients and their differences. If, for example, the measurements of p variables on a random sample of n individuals from a multivariate normal distribution of unknown covariance matrix are to be used to test the hypothesis that the respective population means are $\mu_1, \ldots, \mu_p$, then T is defined as the non-negative square root of

$$T^2 = n \sum l_{ij}(\bar{X}_i - \mu_i)(\bar{X}_j - \mu_j),$$

where $\bar{X}_i$ is the sample mean of the ith variable, the summations are from 1 to p, and l_{ij} is a typical element of the p-rowed matrix inverse to that of sample covariances. Tests of significance involving T^2 can be carried out by the variance ratio distribution.

Hotelling's test (dependent correlations) When two correlation coefficients are calculated between variables X_1, X_2 and X_1, X_3, from a single sample from a multivariate normal distribution, Hotelling (1940) suggested a test statistic

$$t = (r_{12} - r_{13})\sqrt{\frac{(n-3)(1+r_{23})}{2D}}$$

where D is the determinant of the sample correlation matrix. Alternative tests have been proposed by Williams (1959) and Dunn and Clarke (1971).

Householder tridiagonalization A symmetric $n \times n$ matrix $\mathbf{A}$ may be reduced to tridiagonal form by a sequence of $n-2$ transformations of the type $\mathbf{A} = \mathbf{PAP}$, where $\mathbf{P}$ is orthogonal. This procedure, due to Householder, is important in calculating eigenvalues and eigenvectors.

Huber estimator A **robust estimator** of location, defined by $\sum \psi(x-T) = 0$, where

$$\psi(x) = \begin{cases} x & \text{if } |x| \leqslant k \\ k \text{ sgn}(x) & \text{if } |x| > k. \end{cases}$$

It is an **M-estimate** with a **minimax** property; for a symmetrically contaminated normal distribution with a known proportion of contamination, it minimizes the variance in the worst possible case (Huber, 1964).

Hudson's R_m An estimate of the minimum number of recombination events in a sample of DNA sequences based on mutation patterns incompatible with the **infinitely-many-sites** assumption of mutation.

Hudson–Kreitman–Aguade test (HKA test) A statistical test for the selective neutrality based on within-species polymorphism and between-species divergence.

Hunt–Stein theorem A theorem concerning critical regions associated with a particular significance test. For example, in the set of **critical regions** of size α which are also invariant, there may be one, say W_0, which is **uniformly most powerful** and **most stringent**.

Hurst coefficient See **fractional Brownian motion**.

hyperbolic secant distribution A frequency distribution, related to the logistic distribution, with probability density function

$$f(x) = \frac{1}{\pi}\text{sech } x$$

and cumulative distribution function

$$F(x) = \frac{1}{2} + \frac{1}{\pi}\tanh^{-1}(\sinh x).$$

hypercube The original term for orthogonal array as introduced by Rao (1946b)

hyperexponential distribution A term proposed by Morse (1958) for a mixture of exponential distributions in the general form

$$b\gamma e^{-2\gamma bt} + b(1-\gamma)e^{-2(1-\gamma)bt}, \qquad t > 0, \quad b > 0, \quad 0 \leqslant \gamma \leqslant 1.$$

hypergeometric distribution A discrete distribution generally associated with sampling from a finite population without replacement. The frequency of r 'successes' and $n - r$ 'failures' in a sample of n so drawn from a population of N in which there are Np 'successes' and Nq 'failures' $(p + q = 1)$ is

$$\frac{1}{N^n}\binom{n}{r}(Np)^{(r)}(Nq)^{(n-r)},$$

where $N^{(r)} = N(N-1)\ldots(N-r+1)$. As N tends to infinity the distribution tends to the ordinary binomial form. The distribution derives its name from the fact that the probability generating function has the form of a hypergeometric series (Pearson, 1924b).

hypergeometric waiting time distribution An alternative name for the **inverse hypergeometric distribution**.

hyper-Graeco-Latin square A generalization of the **Latin** and **Graeco-Latin square** in the form of a $p \times p$ square in which each cell contains one of the characters of each of k types $(k > 2)$, the characters of each type being p in number and constituting a Latin square; and the types being mutually orthogonal so that no combination of the characters of different types occurs more than once anywhere in the design. The maximum value of k never exceeds $p - 1$. For example, the 4×4 square

$$A\alpha 1 \; B\beta 2 \; C\gamma 3 \; D\delta 4$$
$$B\gamma 4 \; A\delta 3 \; D\alpha 2 \; C\beta 1$$
$$C\delta 2 \; D\gamma 1 \; A\beta 4 \; B\alpha 3$$
$$D\beta 3 \; C\alpha 4 \; B\delta 1 \; A\gamma 2$$

shows how comparisons between 16 observations may be considered in five independent sets, corresponding to rows, columns, Roman letters, Greek letters and numerals.

hypernormal dispersion See **Lexis variation**.

hyper-Poisson distribution This discrete distribution has a complicated probability function in two parameters, λ and θ. If $k = 1$ this distribution becomes the **Poisson distribution** with parameter θ. If $\lambda < 1$ it is sometimes referred to as a sub-Poisson distribution and if $\lambda > 1$ as a super-Poisson distribution (Bardwell and Crow, 1964).

hyperspherical normal distribution A p-dimensional distribution $(p > 2)$ with probability density function:

$$p_\theta(t) = c_p \left\{ \sin^{p-1}(t - \theta_0) \right\} \exp \left\{ k \cos(t - \theta_0) \right\}, \qquad 0 \leqslant t \leqslant \pi,$$

where

$$c_p = \left\{ \int_0^\pi (\sin^{p-1} t) \exp(k \cos t) dt \right\}^{-1}.$$

For the case $p = 3$, the distribution is sometimes referred to as the **spherical normal distribution**.

hypothesis, statistical A hypothesis concerning the parameters or form of the probability distribution for a designated population or populations, or, more generally, of a probabilistic mechanism which is supposed to generate the observations.

hypothesis testing The testing of conformity with a probabilistic model for data often with a postulated alternative representation in mind.

hypothetical population A statistical population which has no real existence but is imagined to be generated by repetitions of events of a certain type; for example, the binomial distribution as generated by the throws of a die, or crop yields on a set of plots imagined as all the possible ways in which a set of yields might occur under the conditions of an experiment.

I

'ideal' index number Irving Fisher (1922*a*) advanced certain criteria which should be obeyed by 'good' index numbers. Of the large collection of formulae investigated only a few obeyed his tests. One of these was termed the 'ideal' index. It may be written

$$\left(\frac{\sum p_n q_0}{\sum p_0 q_0} \times \frac{\sum p_n q_n}{\sum p_0 q_n} \right)^{1/2}$$

where p_0, q_0 represent prices and quantities in the base period and p_n, q_n those of the period for which the index is being calculated. The ideal index is the geometric mean of the **Laspeyres** and **Paasche index numbers**. [See also **crossed weight index number, factor reversal test, time reversal test**.]

identical categorizations In the investigation of the relationships between two or more categorized variables, the variables investigated may simply be the same variable observed either on different occasions or on related samples. Such interactions are said to have identical categorization.

identifiability The property of a statistical model or parameter in such a model that it can be estimated consistently from a very large amount of data on that system. Usually models are chosen to be identifiable although this cannot be taken as necessary or sufficient. On the one hand estimation of an identifiable model from realistic amounts of data may be so poor as to be ineffective and on the other hand there may be powerful subject-matter reasons for, say, including non-identifiable parameters and studying their effect by sensitivity analysis. Also in some cases useful bounds can be found for non-identifiable parameters. The concept is emphasized more in econometrics than in mainstream statistics.

idle period In single-server queueing theory, a time period beginning when a customer departs to leave the server free, and lasting until the instant at which the next customer arrives for service.

ill-conditioned A matrix is ill-conditioned if the ratio of the largest to the smallest eigenvalue is large, so that it is nearly singular. A covariance matrix of estimates of this type implies **multicollinearity**.

ill-posed problem A problem in which the solution is not unique, at least numerically. The solution is to add more structure or a priori knowledge to the problem, for instance by regularization techniques based on some kind of penalization. Inverse problems, such as **deconvolution** problems, are often ill-posed.

illness–death model A model which represents the evolution of the health status of subjects and involving typically three states: health, illness and death. This is a generalization of survival models, particularly useful for studying chronic diseases in elderly people where the death rate from some health states is high. This is a particular case of a **multistate model**.

illusory association or correlation A statistical relation between variables that while real does not have a subject-matter interpretation of direct dependency. Nonsense correlation is an alternative term. A traditional example is the time-series relation between the number of storks nesting on roofs in Oslo and the birth rate.

image analysis The extraction of the underlying scene from an image, usually pixellated and subject to degradation by noise. Also known as image enhancement or image reconstruction.

image processing Representation of a photographic or other image in digital form, as a step in automatic **pattern recognition**.

imbedded process A stochastic process in continuous time is considered only at the time points where a change of state occurs. These points of discontinuity can be thought of as forming a new discrete time variable. A new stochastic process can be derived by defining the state of the process at time n to be that immediately following the nth transition in the old process. The new process in discrete time is said to be 'imbedded' in the old process in continuous time. In this sense, the imbedded process can, for example, be Markov derived from a non-Markov process or a renewal process derived from a point process.

immigration In **birth and death processes**, immigration is generally regarded as independent of population size and so differentiated from birth; emigration, on the other hand, can usually be treated in the same way as death.

implicit strata The n groups, zones or subclasses into which, in systematic sampling, the population is divided by the sampling interval $k = N/n$, where N is the population size and n the sample size.

importance sampling Sampling schemes in which the parts of the sampling frame or space making the most important contributions to the estimand are sampled with increased probability, relative to others, and so are down-weighted when the estimate is computed. It is most common in Monte Carlo estimation of integrals, but the idea applies more widely, e.g., to the **Horvitz–Thompson estimator**.

improper distribution A probability distribution with a distribution function $F(x)$ for which $F(-\infty) > 0$ or $F(+\infty) < 1$. Distributions for which $F(-\infty) = 0$ and $F(+\infty) = 1$ are said to be proper, although in most circumstances the adjective is redundant and is omitted.

improper prior A notional type of probability density function $p(\mu)$ of a parameter μ for which

$$\int_{-\infty}^{\infty} p(\mu)d\mu$$

is not finite. Improper prior probability density functions are extensively used in Bayesian inference with the intention of representing the state of knowing little about a parameter, e.g.,

$$p(\mu) \propto \text{constant}, \quad -\infty < \mu < \infty.$$

impulse response function See **frequency response function**.

imputation The process of replacing **missing data** in a large-scale survey.

inadmissible estimator See **strictly dominated estimator**.

incidence The number of new cases of a disease per unit of time in a given population, or the number of new cases divided by the size of the population at risk. Incidence and **prevalence** are the two main measures of the impact of a disease in a population.

incidence matrix of design A **block** experiment may be described by a treatment matrix of order $k \times t$ for each block. If we need only allocation of treatments to blocks without reference to allocation within blocks then the b treatment matrices $\mathbf{T}_j$ can be condensed into an incidence matrix $\mathbf{N}$ of order $t \times b$. The matrix merely records whether or not a particular plot or treatment appears in the experiment, not the outcome of the observations.

incidence rate The incidence of an event (typically a disease) is the number of new occurrences of the event that occur during a specified period of time in a population at risk for the occurrence of the event. The incidence rate is the number of new occurrences of the event divided by the size of the population at risk for the occurrence of the event.

incidental parameters See **partially consistent observations**.

includances tests A class of distribution-free tests of identity of two populations, characterized by having a test statistic that is the number of observations in a simple random sample from one population that are enclosed, i.e., 'included', between a specified pair of order statistics in a simple random sample from the other population. [See also **exceedances tests**.]

inclusion probability In survey sampling, the probability that a member of a population will appear in the sample drawn. Joint inclusion probabilities correspond to appearance in the sample of specified subsets of size two, three, and so forth.

incomplete beta function This function is defined as

$$B_t(s,r) = \int_0^t y^{s-1}(1-y)^{r-1}dy, \qquad s,r > 0, \qquad 0 \leqslant t \leqslant 1.$$

The ratio of the incomplete beta function to the complete beta function is generally written

$$I_t(s,r) = \frac{B_t(s,r)}{B(s,r)} = \frac{\int_0^t y^{s-1}(1-y)^{r-1}dy}{\int_0^1 y^{s-1}(1-y)^{r-1}dy}.$$

[See **beta distribution**.]

incomplete block A basic form of experimental design introduced by Yates (1936). If material is divided into blocks and it is desired to allocate certain treatments to the units of a block, the treatments may be too numerous for them all to appear in each block. When a block contains fewer than a complete replication of the treatments it is called incomplete. Differences between blocks are then discussed by a more elaborate analysis than is required for complete blocks (see **recovery of information**). If each block contains the same number of treatments and they are arranged so that every pair of treatments occurs together in the same number of blocks, the design is said to be balanced.

incomplete census See **census**.

incomplete data When exact values of variables of interest on a sample taken at random from the target population cannot be observed. This includes in particular **censoring**, **truncation** and **coarsening**.

incomplete gamma function A function defined as

$$\Gamma_t(\alpha) = \int_0^t e^{-x}x^{\alpha-1}dx, \qquad \alpha > 0, \qquad 0 \leqslant t < \infty.$$

It is the distribution function of the **gamma distribution** multiplied by $\Gamma(\alpha)$.

incomplete Latin square An alternative name for the **Youden square**.

incomplete moment The ordinary moment of a distribution of density $f(x)$ about an arbitrary origin a is given by

$$\mu_r' = \int_{-\infty}^\infty (x-a)^r f(x)dx.$$

If this is modified to make the limits of integration extend only from $-\infty$ to t, the expression

$$\int_{-\infty}^t (x-a)^r f(x)dx$$

is sometimes called the incomplete moment of order r, provided that it exists.

incomplete multi-response design An experiment design introduced by Srivastava (1964) to deal with a situation where not all the variables are measured for each experimental unit. This position can arise from a state of physical impossibility, e.g., irreparable damage, from inconvenience or from other considerations such as expense.

inconsistent estimator See **consistent estimator**.

increasing failure rate A non-negative random variable X is said to have an increasing failure rate (IFR) distribution if for all $x > 0$ and for all $t \geqslant 0$ such that $\Pr(X \geqslant t) > 0$, the quotient $\Pr(X > x + t)/\Pr(X > t)$ is a decreasing function of t. If the distribution F of X has a probability density function f, then IFR is equivalent to the failure rate

$$q(t) = f(t)/(1 - F(t))$$

being an increasing function of t, for all t for which $1 - F(t) > 0$.

increasing hazard rate See **increasing failure rate**.

independence In the calculus of probabilities, independence is usually defined by reference to the principle of compound probabilities. Two events are independent if the probability of one is the same whether the other is given or not, i.e., $\Pr(A) = \Pr(A \mid B)$ and $\Pr(B) = \Pr(B \mid A)$. From this it follows that the probability of the compound event $\Pr(A \cap B) = \Pr(A)\Pr(B)$ if the events are independent. As a matter of axiomatization it may be preferable to use relations of the type $\Pr(A \cap B) = \Pr(A)\Pr(B)$ as definitions to avoid difficulties arising when $\Pr(A)$ or $\Pr(B)$ is zero. In statistics two random variables X_1 and X_2 are independent if their distribution functions are related by

$$F(x_1, x_2) = F(x_1, \infty)F(\infty, x_2)$$

or equivalently if their density functions, should they exist, are related by $f(x_1, x_2) = f_1(x_1)f_2(x_2)$. Generally, n variables $X_1, \ldots, X_n$ are independent if

$$F(x_1, x_2, \ldots, x_n) = F(x_1, \infty, \infty, \ldots, \infty)F(\infty, x_2, \infty, \ldots, \infty) \ldots$$
$$F(\infty, \infty, \infty, \ldots, x_n).$$

It is not enough that they should be pairwise independent. The word is also applied in the ordinary mathematical sense to describe the independence of two or more variables.

independence frequency In a contingency table, the frequency which would be found in a particular cell if the attributes defining it were independent; for example, if the r rows of the table have total frequencies $A_1, \ldots, A_r$ and the s columns have frequencies $B_1, \ldots, B_s$, and

$$\sum A_i = \sum B_j = N,$$

the independence frequency in the ith row and jth column is $A_i B_j / N$.

independent action Suppose that doses x_1 and x_2 of two stimuli have expected quantal response rates of $P_1(x_1)$ and $P_2(x_2)$. The two stimuli are said to display independent action if, for all x_1 and x_2, the expected response rate to a simultaneous application of both doses is

$$P = P_1 + P_2 - P_1 P_2 = 1 - (1 - P_1)(1 - P_2).$$

The extension to three or more stimuli is obvious.

independent censoring A condition under which a simple partial likelihood can be used, which amounts to ignoring the mechanism leading to censoring. A rigorous definition involves counting process theory such as in Andersen *et al.* (1993). A simple model is when censoring is defined by random variables independent of the failure process.

independent increments, process with See **additive process**.

independent trials The successive trials of an event are said to be independent if the probability of outcome of any trial is independent of the outcome of the others. The expression is usually confined to cases where the probability is the same for all trials. In the sampling of attributes, such a series of trials is often referred to as **Bernoulli trials**. It includes all the classical cases of drawing coloured balls from urns with replacement after each draw, coin tossing, dice rolling and the events associated with other games of chance.

independent variable This term is regularly used in contradistinction to 'dependent variable' in regression analysis. When a variable Y is expressed as a function of variables $X_1, X_2, \ldots$, plus a stochastic term, the X's are known as 'independent variables'. The terminology is rather unfortunate since the concept has no connection with either mathematical or statistical dependence. Modern usage prefers 'explanatory variable', 'covariate' or 'regressor'.

index number A quantity which shows by its variations the changes over time or space of a magnitude which is not susceptible to direct measurement in itself or to direct observation in practice. Examples of these magnitudes are business activity, physical volume of production, and wholesale prices. Important features in the construction of an index number are its coverage, base period, weighting system and method of averaging observations. The above definition relates to the usual meaning of the expression 'index number'. The term can also be applied to a series of values which are standardized by being referred to a basic period or area.

index of response An alternative, and simpler, method than the method of concomitant observations for estimating the effects of treatments in the analysis of experimental data. The method consists of constructing an initial index of response by combining two or more variables and treating this construction as a new variable.

indicator variable A random variable defined to equal one when a particular event occurs and to equal zero otherwise. The term indicator function is used when the event is replaced by a set.

indifference-level index number A synonym for **Konyus index number**.

indirect sampling Sampling from documents, or some record of the characteristics of a population, rather than the recording of information obtained at first hand from units of the population themselves. For example, it is becoming customary to obtain preliminary information on the results of, say, a national census by analysing a sample of the census forms before the full analysis is undertaken; the population is then subject to indirect sampling. [See also **direct sampling**.]

indirect standardization The process of transforming two statistics into a comparable metric by conforming them to the expected rates of a related third variable. The indirect method of adjustment may be used if specific rates of the population for standardization are unknown or unstable. The stratum-specific rates of a larger population are applied to the number of persons within each stratum of the population of interest to obtain the expected number of events. Thus the indirect method of standardization does not require knowledge of the actual event rates in the population being standardized. Indirect standardization is used when specific rates for a group are unknown or unstable. For instance, the stratum-specific rates for a larger population, such as a national entity, are applied to a region. In this instance the national rates, obtained from the more stable data, become the expected rates and are compared with observed events in the region. This is characterized mathematically as $O(e)/E(r)$, where $O(e)$ is the observed rate of the group under consideration and $E(r)$ is the expected rate in a larger, more stable population. See **direct standardization**.

induction The logical model in which general conclusions are developed from specific observations. The epistemological model of developing overarching principles from a collection of events. For instance, consistent observations of the moon's position in the night sky led to hypotheses about the lunar cycle.

inductive behaviour A term introduced by Neyman (1937b) to indicate the adjustment of a course of action based upon a limited amount of information in relation to established ideas or 'permanencies'. In particular, when the relative merits of a number of courses of action depend upon the nature of the frequency function of some observed variates, the rule of inductive behaviour is equivalent to a test of a statistical hypothesis. [See also **decision function**, **hypothesis**.]

inefficient estimator An estimator whose variance is larger than the minimum possible. [See **efficiency**.]

inefficient statistic A statistic with less than the maximum possible efficiency, e.g., by having larger sampling variance. In fact 'efficiency' relates to precision, irrespective of robustness, cost and time; and 'inefficiency' has the corresponding interpretation. [See also **efficiency**.]

inequality coefficient A measure due to Theil (1961) of the absolute difference between the actual and the forecast change in a system. If P_i and A_i are predicted and actual values, the coefficient was originally defined as

$$U = \left\{ \frac{\sum (P_i - A_i)^2}{n} \middle/ \frac{\sum A_i^2}{n} \right\}^{1/2}$$

and more recently modified to

$$U = \left[\frac{\sum (P_i - A_i)^2}{n} \right] \middle/ \left[\left\{ \frac{1}{n} \sum P_i^2 \right\}^{1/2} + \left\{ \frac{1}{n} \sum A_i^2 \right\}^{1/2} \right].$$

The summation is over n values for which forecast and actual values can be compared, usually n consecutive years.

inferential statistics The process of making inferences about a population from findings based on sampled observations. Inferential statistics are used to go beyond the description of the data and to examine hypotheses about underlying research questions.

infinite divisibility A distribution F is infinitely divisible if and only if for each n it can be represented as the distribution of the convolution (sum)

$$S_n = X_{1,n} + \ldots + X_{n,n}$$

of n independent random variables with a common distribution F_n.

infinitely-many-sites model A model of DNA sequence evolution where each mutation is assumed to occur at an entirely new site, never before mutant. There is a unique gene tree which can be deduced as a perfect phylogeny from a sample of such DNA sequences.

infinitesimal jackknife Jackknife in which analytical derivatives replace the numerical derivatives used in the usual jackknife.

inflated distributions A term introduced by Singh (1963) and Pandy (1965) to describe discrete distributions, e.g., Poisson, which have been modified to take account of an observed excess of zeros as compared with the standard form of the distribution with the same parameters. Other names for the modified distributions are 'with zeros', e.g., Poisson with zeros, and 'pseudo-contagious'.

influence The property of an observation of having a large effect on a fitted regression model. This typically arises either because the response is an **outlier** or because the observation is a high **leverage point**.

influence function The basis of an array of tools for assessing the influence on a given estimator of a contaminated observation, proposed by Hampel (1971). If T is the estimator, and it assumes the value t for a sample of n independent observations from the population with distribution function F, and the value

t_c when a further value ξ, the contaminant, is added to produce a sample of size $n + 1$, then the finite sample influence curve is defined to be the array of values over n of

$$\mathrm{IC}_{T,F;n}(\xi) = (n + 1)(t_c - t).$$

The influence curve $\mathrm{IC}_{T,F}(\xi)$ is defined to be the asymptotic value of the finite sample influence curve as n tends to infinity. The influence curve is sometimes called the influence function.

influential observation An observation is said to be influential if its removal from the data has a substantial effect on the inference from the analysis.

information The word 'information' occurs frequently in statistics with its ordinary meaning. In a specialized sense in the theory of estimation, the amount of information about a parameter θ from a sample of n independent observations drawn at random from a population with frequency function $f(x, \theta)$ is defined as

$$nE\left(\frac{\partial \log f}{\partial \theta}\right)^2 \equiv n \int_{-\infty}^{\infty} \left(\frac{\partial \log f(x, \theta)}{\partial \theta}\right)^2 f(x, \theta) dx.$$

Under some general regularity conditions the reciprocal of the information gives a lower bound for the variance of unbiased estimators of θ, so that the greater the variance the less the 'information' and an equivalent expression is

$$-nE\left(\frac{\partial^2 \log f}{\partial \theta^2}\right).$$

The concept generalizes to the case of several parameters. [See **information matrix**.]

information criterion A tool for model comparison that trades off model fit and complexity, usually measured using the number of model parameters. Examples are **Akaike's information criterion** and the **Bayes information criterion**. The best-fitting model is sometimes taken to be that minimizing the criterion.

information matrix In generalization of the definition of **information** for one parameter under estimate, the information matrix of a sample of n drawn independently from a population with frequency function $f(x, \theta_1, \ldots, \theta_p)$ is the matrix for which the element in the ith row and jth column is

$$nE\left(\frac{\partial \log f}{\partial \theta_i} \frac{\partial \log f}{\partial \theta_j}\right) \equiv n \int_{-\infty}^{\infty} \frac{\partial \log f(x; \theta_i)}{\partial \theta_i} \frac{\partial \log f(x; \theta_j)}{\partial \theta_j} f(x; \theta) dx.$$

As in the case of one parameter, certain regularity conditions on f are required; and in many cases an equivalent expression is

$$-nE\left(\frac{\partial^2 \log f}{\partial \theta_i \partial \theta_j}\right).$$

This generalizes to the case of independent but non-identically distributed variables.

information theory In the theory of communication developed by Shannon and Weaver (1949) information is defined by expressions of the form

$$I = -\sum_{i=1}^{s} p_i \log p_i.$$

Various statistical measures, such as the **Kullback–Leibler distance function**, are based on information theory in this sense. It is not to be confused with Fisher **information**. [See **diversity, index of, Shannon–Wiener index**.]

informative censoring Censoring is informative if its occurrence depends conditionally on the failure process of interest, given explanatory variables. For ignoring the mechanism leading to censored data, both **independent** and non-informative censoring must be assumed.

informative prior In Bayesian inference, a prior probability density function that reflects positive empirical or theoretical information regarding the value of an unknown parameter. The contrary case, where little or no information about the parameter is available, is represented by means of a diffuse prior, alternatively known as a vague prior.

inherent bias A rather loosely defined expression which in general means a bias which is due to the nature of the situation and cannot, for example, be removed by increasing the sample size or choosing a different type of estimator. An example of inherent bias is the systematic error of an observer or an instrument; a further example, in the interrogation of human population, is the distortion of truth by the respondent for whatever reasons. It is possible also to speak of the inherent bias of a method of estimation, although in this context the word 'inherent' appears redundant. For example, in the theory of index numbers it may be shown that the standard formulae of **Laspeyres** and **Paasche** possess an inherent bias due to the methods of weighting and averaging the items.

inhomogeneous Poisson process A stochastic process of points in time or space in which the probability of a point in any small element of time or space of small duration (extent) δ is asymptotically a function, the rate function, of time or space multiplied by δ and is independent of occurrences in all non-overlapping elements (Cox and Miller, 1965).

input/output process A broad class of stochastic processes, the state $\xi(t)$ at time t being the resultant of two random variables X, Y

$$\xi(t) = X(t) - Y(t),$$

X and Y being called respectively the input and the output. Examples are queueing processes, provisioning and epidemic processes.

inspection diagram This term is capable of several interpretations. It may describe a diagram of a manufacturing process showing the inspection points as part of the process. It may be applied to the diagrammatic layout of a double or multiple sampling scheme which describes the stages from the inspection of the first sample to the acceptance or rejection of the **inspection lot**. It may refer to the graph upon which are summarized the elements of a sequential sampling plan together with the course of the actual sampling process. Finally, it may be applied to the graph of a sampling inspection plan which shows the **operating characteristic** curve.

inspection lot A **lot** presented for inspection, which may be carried out on each member of the lot or on a sample of members only.

instantaneous death rate A version of the **age-specific death rate** where the 'next unit time period' is reckoned to be short.

instantaneous state A state of a **Markov chain** in continuous time for which the expected **sojourn time** is zero.

instrumental variable In econometrics, and generally in the analysis of the structure of a stochastic situation, an instrumental variable is a predetermined **variable** which is used to derive consistent estimators of the parameters of the system. Its use is inefficient in relation to a complete set of equations in the sense that only a limited amount of information is employed. On the other hand it may be applied to incomplete systems. [See also **limited information methods, reduced form method**.]

integer programming A form of **linear programming** where the solutions must be an integer form.

integrated data A class of statistical data in which the values for short unit intervals can be added together to give a series of values relating to longer intervals; for example, daily rainfall can be integrated into a new series of weekly, monthly or annual rainfall figures each of which will possess a longer time base than the previous series. On the other hand, a series of, say, temperature readings cannot be integrated in this sense and series for longer time intervals must be derived by averaging or the selection of typical values.

integrated moving average (IMA) process See **autoregressive integrated moving average (ARIMA) process**.

integrated spectrum A concept analogous to the cumulative distribution function in the same fashion as the spectral density function is analogous to the probability density function. [See **spectrum**.]

intensity In the harmonic analysis of time series, a measure which provides an estimate of the amplitudes of the constituent harmonics. If the series is $u_1, \ldots, u_n$ the intensity for a period μ is defined as $A^2 + B^2$ where

$$A = \frac{2}{n} \sum_{j=1}^{n} u_j \cos \frac{2\pi j}{\mu}, \qquad B = \frac{2}{n} \sum_{j=1}^{n} u_j \sin \frac{2\pi j}{\mu}.$$

An entirely different use of this term occurs in the theory of life testing etc. Here it is equivalent to the **force of mortality, hazard** or **age-specific death (failure) rate**. The intensity of a point process is the mean number of points per unit measure.

intensity function In an actuarial context, the 'intensity' in this second sense is an alternative term to **hazard, age-specific death rate** or **force of mortality**. If a number of values are calculated, i.e., for a sequence of ages, then these values can be regarded as the intensity function.

intensive sampling Like **extensive sampling** this expression may mean two different things: (i) sampling in a particular area with a dense scatter of sampling points; (ii) sampling wherein information on a restricted range of topics is sought by probing on them very deeply with an intricate schedule of questions.

interaction While sometimes used in the broad sense of effects not operating separately, in statistical discussions it is typically restricted to effects that do not act additively on some response variable. Positive interaction between two explanatory variables on an equal footing corresponds to the biological notion of synergism. When there is interaction between an explanatory feature and a background variable, the latter may be called an effect modifier. Interaction between an explanatory variable and blocking or stratification factors is often used to estimate error. Interactions involving effect reversals may be called qualitative (Cox, 1984).

interblock See **block**.

intercalate Latin square A term proposed by Norton (1939) for any Latin square or rectangle which may be embedded in a larger **Latin square**.

intercept The constant term in a regression equation.

interclass correlation This expression denotes correlation in the ordinary sense; the qualifying adjective 'interclass' is only employed to distinguish ordinary correlation from '**intraclass correlation**'.

interclass variance In the **analysis of variance** of data subject to multiple classification, the sum of squares of all observations about their mean is expressed as the sum of squares of observations about class means plus the sum of squares of class means about the mean of the whole. The former, divided by an appropriate number of degrees of freedom, is sometimes called **intraclass variance**; those of the latter type, again divided by degrees of freedom, are called interclass variances. The expressions are convenient and the quantities in question estimate components of variance in the model generating the observations, but strictly speaking they are not always variances. Sundry synonymous expressions occur such as between-class variance, within-class variance, external and internal variance, and so on.

intercorrelation A term used to denote the correlation of a number of variables among themselves, as distinct from the correlations between them and an 'outside' or dependent variable.

intercropping The growing of two or more crops simultaneously on the same ground. Experiments to compare such combinations involve special problems in design.

interdecile range This term is usually interpreted as the variable range between the first and ninth **deciles**. Like the **interquartile range** it provides an indication of the spread of the frequency but does not appear to have come into use as a measure of dispersion. [See also **semi-interquartile range**.]

internal least squares A name given by Hartley (1948) to a method of dealing with nonlinear regression. He was led to consider a regression of a variable Y on the accumulated sums of Y as well as on independent variables X. This was called internal regression and the estimation of parameters of the regression equation by least squares was called internal least squares.

internal regression See **internal least squares**.

internal validity In experiments, the extent to which extraneous variables have been controlled by the researcher, so that any observed effects can be attributed solely to the treatment variable (Gall, Borg and Gall, 1996).

internal variance See **interclass variance**.

interpenetrating samples (subsamples) When two or more samples are taken from the same population by the same process of selection the samples are called interpenetrating samples. The samples may or may not be drawn independently, linked interpenetrating samples being an example of the latter. There may be different levels of interpenetration corresponding to different stages in a multi-stage sampling scheme. Thus, in a two-stage sampling scheme with village as the primary and household as the second-stage unit, when the sample villages are distributed into two interpenetrating subsamples we have interpenetration at the first stage only; but when the sample of households within every sample village is broken up into two interpenetrating subsamples we have interpenetration at the second stage; and we can have interpenetration of a mixed type, e.g., the four subsamples obtained by combining the two earlier types. Generally, the subsamples are distinguished not merely by the act of separation into subsamples but by definite differences in survey or processing features, e.g., when different parties are assigned to different subsamples, or one subsample is taken up earlier in time than the others.

interpolation The use of a formula to estimate an intermediate data value.

interquartile range (IQR) The variable distance between the upper and lower **quartiles**. This range contains one-half of the total frequency and provides a simple measure of dispersion which is useful in **descriptive statistics**. [See also **semi-interquartile range, quartile deviation, box plot.**]

interval censoring Two definitions of interval censoring can be found in the literature. Groeneboom and Wellner (1992) define case I and case II interval censoring. Case I gives rise to **current status data**. In case II, we observe indicators $I_{T \leqslant G_1}$ and $I_{T \leqslant G_2}$. The situation can be generalized to observing a survival process (or more generally a counting process) at an arbitrary number of discrete times: what is observed is the interval in which the event occurred.

interval distribution A probability distribution formed by the intervals, or gaps, in time or distance between events or occurrences.

interval estimation The estimation of a population parameter by specifying a range of values bounded by an upper and a lower limit, within which the true value is asserted to lie, as distinct from **point estimation** which assigns a single value to the true value of the parameter. The unknown value of the population parameter is presumed to lie within the specified interval on a stated proportion of occasions, under conditions of repeated sampling, or in some other probabilistic sense. The first of these two approaches is that of **confidence intervals** due to Neyman (1937a), which regards the value of the population parameter as fixed and the limits to the intervals as random variables. A second approach is that of **fiducial limits** due to R.A. Fisher (1930) where the population parameter is regarded as having a 'fiducial probability' distribution which determines the limits.

intervention analysis Study of the effect of a policy change on some variable, typically involving the comparison of parts of a time series before and after the intervention (Box and Tiao, 1975).

interviewer bias In surveys of human populations by interview, bias in the responses or recorded information which is the direct result of the action of the interviewer. This bias may be due, among other things, to failure to contact the right persons; to the failure of the interviewer to establish proper relations with the informant, with the result that imperfect or inaccurate information is offered; or to systematic errors in recording the answers received from the respondent.

intrablock See **block**.

intrablock analysis Least squares analysis of an unbalanced block design in which block parameters are treated as fixed, not random.

intrablock subgroup In the orthodox design of symmetrical factorial experiments in incomplete blocks, the treatments in the same block as the control may be considered to form a group in the mathematical sense. This approach to the matter provides a simple way of specifying the treatments in the different blocks. For example, if there are n factors each at two levels, the 2^n treatments form a group in which the square of every element is the identity element. If there are 2^c blocks available $(c < n)$ there will be $2^c - 1$ interactions confounded forming, with the identity, a group of 2^c members. The 2^{n-c} treatments having an even number of letters (identifying treatments) in common with them form the intrablock subgroup.

intraclass correlation A measure of correlation within the members of certain natural groups or 'families'. For example, if a variable X is measured on a number of members, say k, of a family and it is desired to ascertain the correlation between the k members, a correlation table is constructed in which each of the $\frac{1}{2}k(k-1)$ pairs of members is represented twice, according to which member of the pair is taken as providing the first, and which the second, variable. For several families the correlation tables are superimposed, and a **product moment correlation** computed for the resulting bivariate table. This is the intraclass correlation. In practice it is not, in fact, necessary to construct the actual table in order to compute the coefficient. The concept is closely allied to a variance ratio in variance analysis, which has largely superseded it.

intraclass variance In **analysis of variance**, where the data are classified into groups, the total variation (sum of squares about the grand mean) may be expressed as the sum of two components, expressing variation among means of groups and variation within groups. The latter is the sum of squares about group means pooled for the various groups, and an estimate of variance within groups based on it is known as within-group or intraclass variance.

intrinsic accuracy The intrinsic accuracy of a location distribution with frequency function $f(x - \theta)$ is defined as

$$I = \int_{-\infty}^{\infty} \left(\frac{\partial \log f(x)}{\partial x} \right)^2 f(x)dx.$$

Under very general conditions the variance of any unbiased estimator t of θ based on an independent sample of n observations cannot be less than $1/nI$. The intrinsic accuracy thus gives a limit to the variance independent of particular parameters (Fisher, 1922b).

invariance This term denotes a property that is not changed by a particular family of transformations. For example, an orthogonal transformation of a set of independent normal variates leaves the properties of independence and normality unaffected; they are invariant under the transformation. Other important families include origin and scale changes.

invariance method A principle of estimation or hypothesis testing which requires an estimator or hypothesis under test to remain invariant if the data of the problem undergo some transformation.

invariance principle This usually refers to the notion that inferences should be invariant to data transformations such as changes to units of measurement; often these transformations form a mathematical group. Another usage is as an alternative name for **Donsker's theorem.**

inventory problems In **operational research**, the problem of maintaining an inventory or store of goods to meet predicted demands.

inverse correlation See **correlation, coefficient of.**

inverse distribution This expression is used in several senses. For example: (i) to denote the distribution of the reciprocal of a variable; (ii) to denote the distribution of sample numbers where sampling is continued until a predetermined number of successes has been attained, e.g., the 'inverse hypergeometric' is the distribution of sample sizes n required to obtain k successes in sampling without replacement from a finite population; (iii) to denote distributions (mainly discrete) where the frequencies are reciprocal quantities, see, for example, **factorial distribution**; (iv) by extension of (ii) to denote similar expressions in more general situations, see **inverse Pólya distribution**; (v) instead of expressing the distribution function $F(x)$ in terms of the variable x, to denote expansions of x in terms of $F(x)$.

inverse factorial series distribution See **factorial distribution.**

inverse Gaussian distribution The distribution whose density is

$$f(x) = \left(\frac{\lambda}{2\pi x^3}\right)^{1/2} \exp\left\{-\frac{\lambda(x-\mu)^2}{2\mu^2 x}\right\}, \qquad x > 0, \quad \lambda > 0, \quad \mu \geqslant 0.$$

The name is due to Tweedie (1945), who noted that the cumulant generating function

$$\log E(e^{-tX}) = \frac{\lambda}{\mu}\left\{1 - \left(1 + \frac{2\mu^2 t}{\lambda}\right)^{1/2}\right\}$$

is the inverse function of the corresponding function $-\kappa_1 t + \kappa_2 t^2/2$ for the normal distribution. It arises as the first passage time distribution of Brownian motion.

inverse hypergeometric distribution See **inverse distribution.**

inverse moment If the rth moment is defined as $E(X^r)$, then the rth inverse moment is defined as $E(X^{-r})$. A similar relationship exists between the rth **factorial moment** and its inverse.

inverse multinomial sampling A form of sequential analysis in which sampling is terminated when one or other of a specified subset of classes has occurred k times.

inverse normal scores tests See **normal scores tests.**

inverse Pólya distribution If, in a finite population of size $n_1 + n_2$ consisting of two components of size n_1 and n_2, units are drawn at random replacing each type with c additional units of the same kind until k units of the first kind are achieved, then the distribution of the number of units of the second kind has the inverse Pólya distribution.

inverse polynomial A form of polynomial proposed by Nelder (1966) for use in multi-factor response designs. The general form is $x/y = P_n(x)$ where $P_n(x)$ is a polynomial of order n with non-negative coefficients, y is the response and x the stimulus. This form of response function is bounded and its second-order form is not necessarily symmetric. In this way it overcomes two disadvantages of the ordinary polynomial as a response function.

inverse probability The probability approach which endeavours to reason from observed events to the probabilities of the hypotheses which may explain them, as distinct from direct probability, which reasons deductively from given probabilities to the probabilities of contingent events. A principal theorem in this connection is **Bayes' theorem**. The word 'inverse' usually means inverse in some logical relationship but is sometimes used in the sense of 'prior in time'. For example, in connection with the analysis of stochastic processes, it is sometimes desirable to consider the past history of a system rather than its future development. The past probabilities of transition which have given rise to the present state are thus sometimes called 'inverse'. This usage is comparatively rare and not to be recommended. [See **projection**.]

inverse problems These problems occur by inverting an integral transformation and often occur in signal restoration. Inverse problems are often **ill-posed**.

inverse sampling A method of sampling which requires that drawings at random shall be continued until certain specified conditions dependent on the results of those drawings have been fulfilled, e.g., until a given number of individuals of specified type have emerged. In this sense it is allied to **sequential sampling**.

inverse sine transformation See **arc sine transformation**.

inverse tanh transformation See **Fisher's transformation**.

inverse transformation method A method used in simulation for obtaining random numbers for different distributions given a pseudo-random function. Let u be a pseudo-random number and let F be the distribution function of X. Then $x = F^{-1}(u)$ is a random number of X.

inversion An inversion of a sequence of elements is an interchange in the position of a part. The number of inversions in a ranking forms the basis of several tests of independence in series and of certain rank correlation coefficients. There is an entirely different usage of the word 'inversion' in the so-called 'inversion' theorem which proves that a frequency distribution is uniquely determined by its **characteristic function** or more generally relates a transform of function or sequence back to the originating function or sequence.

inverted beta distribution See **beta distribution**.

inverted Dirichlet distribution Just as the **Dirichlet distribution** is the multivariate analogue of the **beta distribution**, the inverted beta distribution has its multivariate analogue in the inverted Dirichlet distribution (Tiao and Guttman, 1965). It may also be obtained from $k + 1$ independent and randomly distributed chi-squared variables $Y_1, \ldots, Y_{k+1}$ with $2\nu_1, \ldots, 2\nu_{k+1}$ degrees of freedom, where the ratios $X_i = Y_i/Y_{k+1}$ $(i = 1, \ldots, k)$ have the k-dimensional inverted Dirichlet distribution.

irreducible Markov chain A **Markov chain** is said to be irreducible if all pairs of states of the chain communicate, so that the chain consists of exactly one **communicating class**.

irregular kollectiv An infinite series of a finite number of characteristics obeying the following laws: (i) The proportion of a given characteristic in the first n terms tends to a limit as n increases. (ii) Any infinite subsequence of the kollectiv designated by some independent rule possesses the same limiting property. The irregular kollectiv was taken by von Mises (1919) as the basis of his frequency theory of probability.

Irwin distribution See **factorial distribution**.

Ising–Stevens distribution If there are n_1 and n_2 objects of two kinds C_1 and C_2 arranged at random in $n = n_1 + n_2$ positions along a circle then the distribution (Stevens, 1939; Ising, 1925) of the number of runs of either object is

$$
\Pr(x) = \binom{n_1}{x}\binom{n_2 - 1}{x - 1} \Big/ \binom{n_1 + n_2 - 1}{n_2}
$$
$$
= \binom{n_1 - 1}{x - 1}\binom{n_2}{x} \Big/ \binom{n_1 + n_2 - 1}{n_1}.
$$

isodynes When contour maps of power are plotted on a β_1, β_2 diagram, where β_1 and β_2 are the Pearson measures of skewness and kurtosis, lines of constant power are known as isodynes.

isokurtosis See **kurtosis**.

isometric chart A chart which attempts to depict three-dimensional material on a plane. It is a form of **axonometric chart** where the distances on the three axes are measured on an equal scale. There are various conventional combinations for the angles at which two out of three axes are drawn *vis-à-vis* the horizontal.

isomorphism The logical equivalence of two systems in the sense that one can be obtained from the other by a translation or reinterpretation of basic notions and symbols. For example, the mathematical part of the theory of probability is isomorphic to a branch of the theory of additive set functions.

isotonic regression function An isotonic function is a real function $\mu(\cdot)$ on a real finite set $\mathcal{X}$ if, for all $x, y \in \mathcal{X}$, $x \leqslant y$ implies $\mu(x) \leqslant \mu(y)$, where $\leqslant$ is a reflexive, transitive, antisymmetric binary relation. If $w(\cdot)$ is a positive function and $g(\cdot)$ a real function on $\mathcal{X}$, then the isotonic regression function $g^*(\cdot)$ of g with weights $w(\cdot)$ is the isotonic function on $\mathcal{X}$ which minimizes

$$\sum_{x \in \mathcal{X}} [g(x) - f(x)]^2 w(x)$$

in the class of isotonic functions f on $\mathcal{X}$. Antitonic functions and antitonic regression functions are similarly defined by replacing '$\leqslant$' with '$\geqslant$'.

isotropic distribution A distribution in two or more dimensions in which the probability density depends only on the Euclidean distance from a 'centre'.

isotropy A contingency table is isotropic when the associations in all tetrads of any four frequencies for two rows and two columns are of the same sign. An isotropic contingency table remains isotropic in whatever way the table may be condensed by grouping adjacent rows or columns, even if it be condensed to a **four-fold table**. The case of complete independence is a special case of isotropy since the association is zero for every tetrad of the **independence frequencies**. The expression was introduced by Yule (1906) in a discussion of contingency tables based on an underlying normal distribution.

isotype method See **pictogram**.

item analysis In psychological testing, analysis based on the responses to individual questions is referred to as item analysis.

iterated bootstrap Bootstrap resampling of samples from a bootstrap population, usually to improve properties of confidence intervals or tests, or for sensitivity analysis. The double bootstrap, involving two levels of resampling, is most common in practice, though further iteration is possible in principle (Hall, 1986; Davison and Hinkley, 1997).

iterated logarithm, law of If S_n is the number of successes in a sequence of n Bernoulli trials with a probability p of success at each trial, Khintchine's (1924) law of the iterated logarithm states that

$$\limsup_{n \to \infty} \frac{S_n - np}{(2npq \log \log n)^{1/2}} = 1$$

where $q = 1 - p$. There are extensions to more general sequences of variables.

iterative proportional fitting An approach for maximum likelihood estimation of **log-linear models** to a multiway **contingency table**, whereby fitted values are obtained by iterative adjustment of cell entries subject to fixing lower-dimensional margins of the table (Deming and Stephan, 1940).

iterative proportional scaling See **iterative proportional fitting.**

iterative weighted least squares An algorithm whereby maximum likelihood or other estimates are obtained iteratively by writing Newton–Raphson or similar steps in the form of a weighted least squares problem. The weights, responses and design matrix may all change at each iteration. It is used to fit **generalized linear models,** but applies much more widely (Green, 1984).

iteratively reweighted least squares Iterative least squares estimation when the appropriate weight of an observation depends on the predicted value. [See **generalized linear models.**]

J

J-shaped distribution An extremely asymmetrical frequency distribution with the maximum frequency at the initial (or final) frequency group and a declining or increasing frequency elsewhere. The shape of this distribution roughly resembles the letter J or its reverse. Among theoretical curves referred to as J-shaped are the **Pareto** and certain of the **Pearson** system of frequency curves.

jackknife A method of improving estimators whereby the bias, variance and other properties of an estimator T based on a sample of size n can be estimated by judicious combination of the n 'jackknife pseudo-values' $(n-1)(T - T_{-i})$, where $T_{-1}, \ldots, T_{-n}$ denote the values of the estimator obtained when each of the observations is deleted. Deletion corresponds to numerical differentiation of T, and if applied with groups of observations can overcome deficiencies of the original jackknife. The idea was proposed by Quenouille (1949a) and extended by Tukey (1958a).

jackknife variance estimator See **jackknife**.

James–Stein estimator The maximum likelihood estimator of the mean of a multivariate normal distribution of dimension d is inadmissible in terms of overall mean square error when $d \geqslant 3$, in which case James and Stein (1961) gave an improved estimator based on the notion of **shrinkage**.

Jeffreys, Harold (1891–1989) A distinguished British geophysicist, astronomer and applied mathematician. Jeffreys advocated and justified the use of probability to describe rational belief about scientific ideas, and developed powerful methods for interpreting scientific data through probability (Jeffreys, 1939).

Jensen's inequality If X is a random variable defined on an interval I, and g is any convex function on I, then the inequality

$$g\{E(X)\} \leqslant E\{g(X)\}$$

is known as Jensen's inequality.

Jenkins, Gwilym Meirion (1933–1982) A Welsh statistician. Jenkins was co-author with Box of *Time series analysis: forecasting and control* (Box and Jenkins, 1970), which set out what is now referred to as the Box–Jenkins procedure.

Jiřina sequential procedure A procedure for the sequential estimation of tolerance limits due to Jiřina (1952). Under certain conditions it yields a larger probability for a stated coverage than the standard procedure.

jittered sampling A term sometimes used in the analysis of time series (Parzen, 1967) to denote the sampling of a continuous series where the intervals between points of observation are the values of a random variable.

John's cyclic incomplete block designs A form of incomplete block experiment proposed by John (1966) where the t treatments in blocks of k treatments per block can be divided into k cyclic sets of t blocks each. This experiment design was shown (Clatworthy, 1967) to be equivalent to the balanced **incomplete block** design or the partially balanced incomplete block with m **associate classes**.

Johnson–Mehl model A model for crystal growth (compare the **cell model**). Seeds appear according to a Poisson process in space and time, and grow at a constant rate until the space is filled (a seed appearing within an existing cell immediately dies). The resulting cells have quadratic edges or surfaces; although they are not convex, all lines through the seed meet the cell in two points only.

Johnson's system A system of probability distributions based upon transformations of the variable. The original designation by Edgeworth (1898) was 'Method of Translation'. The idea was extended by Johnson (1949) and developed in later papers. The entire family is defined by

$$Y = g\left(\frac{Z - \alpha}{\beta}\right),$$

where α and β are constants, g is a function, Z is the standard normal random variable and Y is a Johnson variable. [See also S_B, S_U **distributions**.]

joint distribution The distribution of two or more variables. The term is equivalent to multivariate distribution and is especially used of two variables. [See **bivariate distribution**.]

joint moment See **product moment**.

joint prediction interval The joint interval (Lieberman, 1961) which is appropriate for some k new predictions ($k > 1$) of the dependent variable at k different values of the independent variables, based upon the original sample of observations.

joint regression The classical regression model assumes that the dependent variable is a function, often linear, of a set of independent variables. If the linear model is inadequate, attempts are sometimes made to allow for cross-product terms in the independent variables, e.g., if there are two independent variables X_1 and X_2 the joint regression equation would be

$$Y = a_1 X_1 + a_2 X_2 + a_3(X_1 X_2)$$

where the third term allows for any interaction between the original variables. Such regression is sometimes said to be 'joint' but the term can be used in other senses.

213

joint sufficiency Estimators $t_1, \ldots, t_k$ are said to be jointly sufficient for parameters $\theta_1, \ldots, \theta_l$ if the likelihood function can be expressed as

$$L(x_1 \ldots x_n; \theta_1 \ldots \theta_l) = L_1(t_1, \ldots, t_k; \theta_1, \ldots, \theta_l)L_2(x_1, \ldots, x_n),$$

where L_2 does not depend on the $\theta_1, \ldots, \theta_l$, although it may depend on other parameters of the system.

Jolly–Seber model A model used to estimate population size and survival rates from **capture–recapture sampling**.

Jonckheere's k-sample test A distribution-free test, suggested independently by Terpstra (1952) and Jonckheere (1954), for trend in the means of k ordered populations when there are available a number of independent realizations from each population.

Jordan's identity A result due to Jordan (1927) which states that the probability $W_{n,r}$ of the event that, among the events $A_1, \ldots, A_n$, precisely r events occur is given by

$$W_{n,r} = \sum_{k=0}^{n-r}(-1)^k \binom{r+k}{k} S_{r+k},$$

where

$$S_k = \sum_{1 \leqslant i_1 < i_2 < \ldots < i_k \leqslant n} P_r(A_{i_1} A_{i_2} \ldots A_{i_k}),$$

for $k = 1, \ldots, n$, and $S_0 = 1$.

judgment sample In the terminology of Deming (1950), in general, any sample which does not arise from a **probability sampling** and in which criteria of balance, fair selection, and so on, are met by the personal judgement of the investigator.

jump matrix For a Markov chain in continuous time, the matrix whose (i, j) element is the conditional probability of a jump from state i to state j, given that a jump from state i occurs. [See also **transition matrix**.]

jump statistic A statistic used to assess the reality of an apparent jump in the mean of a stationary time series. Jowett and Wright (1959) proposed using the mean semi-squared difference (jump) between the means of the last α ($\leqslant \frac{1}{2}n$) terms of one section and the first α terms of the following section.

just identified model A term which is sometimes used to indicate a model which is completely identifiable, but not over-identified.

K

'k'-class estimator In econometric regression analysis, a general class of estimators proposed by Theil (1961) for the parameters in $\mathbf{B}y + \mathbf{\Gamma}z = u$ where y are the endogenous variables, z are exogenous and u are random errors. The method proceeds by least squares in stages. If $k = 0$ we have the ordinary least squares form applied to individual equations, which gives the consistent estimators. The value $k = 1$ gives the two-stage least squares estimates. The limited information estimator is also a member of the k-class, the value of k being the smallest latent root of a determinant dependent on the coefficients in the **reduced form**.

k-means clustering A technique for cluster analysis in which, after k clusters have been formed, each point is examined and reassigned, if necessary, to the cluster whose mean is at the smallest distance from it (Hartigan, 1975). The procedure continues until no further reassignment takes place. As points are moved, the cluster means must be updated, and the distance measure changes if generalized distance is used.

k-ratio t-test See **Duncan's test**.

k-samples problem The problem of determining whether, given k samples, one from each of k populations, the parent populations are different. The usual tests developed in this connection are homogeneity tests for means or variances, but an infinite number of tests are possible; a summary is given in Bradley (1968). [See also **Mosteller's k-sample slippage test**.]

k-statistics A set of symmetric functions of sample values proposed by R.A. Fisher (1928a). The univariate k-statistic of order r is defined as the statistic whose mean value is the rth **cumulant**, κ_r, of the parent population. The statistics have semi-invariant properties and their sampling cumulants can be obtained directly by combinatorial methods. Similarly the multivariate k-statistic, say $k(r, s, \ldots, v)$, is defined as the symmetric function of observations whose mean value is the corresponding cumulant $\kappa_{rs\ldots v}$. It is more usually written $K_{rs\ldots v}$. Another generalization is due to Tukey (1956), who defines a statistic, say $k\{r, s, \ldots, v\}$, as the symmetric function whose mean value is $\kappa_r \kappa_s \ldots \kappa_v$. This statistic is also often written $k_{rs\ldots v}$ and should not be confused with the multivariate k-statistic. [See also **polykays, bipolykays** and **generalized polykays**.]

K-test A distribution-free test for a trend in a series proposed by Mann (1945). If the equally spaced series is $x_1, \ldots, x_n$ and a decreasing trend exists each term will tend to be greater than succeeding terms. The smallest interval K for which $x_i < x_{i+K}$, $i = 1, 2, \ldots, n - K$, is taken as the test statistic, the null hypothesis that no trend exists being rejected if K is small. A similar test, of course, exists for increasing trend.

Kagan–Linnik–Rao theorem This theorem (Kagan, Linnik and Rao, 1965) characterizes the normal distribution by the property of constancy of regression of the sample mean on the deviations from this mean of the sample observations. It states that if $X_1, \ldots, X_n, n \geqslant 3$, are independent and identically distributed random variables with zero expectation $E(X_i) = 0$, and

$$E(\bar{X} | X_1 - \bar{X}, \ldots, X_n - \bar{X}) = 0$$

where $\bar{X} = (1/n) \sum_{i=1}^{n} X_i$, then the X_i are normally distributed. See also Kagan, Linnik and Rao (1975).

Kaiser–Guttman criterion In factor analysis, the number of eigenvalues of the correlation matrix that are greater than unity, taken as a criterion for the number of factors to be fitted.

Kaiser–Meyer–Olkin measure of sampling adequacy The statistic

$$\frac{r_{ij}^2}{r_{ij}^2 + a_{ij}^2},$$

where r_{ij} is the correlation between x_i and x_j, and a_{ij} is the corresponding partial correlation after eliminating all other variables. Small values suggest that the data cannot fit a factor analysis model.

Kalman filter An iterative technique of dynamic linear modelling, estimation, prediction and interpolation used for estimating the parameters of autoregressive moving average time-series models with Gaussian residuals and structural time series models (Kalman, 1960).

Kantorovich's theorem A theorem stating that if X and Y are discrete random variables, X taking on N different values $x_1, \ldots, x_N$ and Y taking on M different values $y_1, \ldots, y_M$, then if the relation

$$E(X^n Y^m) = E(X^n) E(Y^m), \qquad 1 \leqslant n \leqslant N - 1, \qquad 1 \leqslant m \leqslant M - 1,$$

holds, then X and Y are independent.

Kaplan–Meier estimator A non-parametric extension of the **empirical distribution function** used to estimate the **survivor function** based on right-censored data (Kaplan and Meier, 1958). It has many variants such as the **Nelson–Aalen estimator** for use with more complex time-to-event data.

kappa A measure of reliability for nominal classification procedures that 'corrects' the proportion of raw agreement for agreement expected purely by chance given the marginal rates. Many generalizations have been developed based on intuitive, but largely ad hoc, criteria. There has been considerable debate about its utility.

Kapteyn's distribution A generalization of the normal distribution using the method of **Kapteyn's transformation** of the form

$$d\Phi_x = \frac{1}{\sqrt{2\pi}\sigma} \exp\left\{ -\frac{1}{2\sigma^2}(G(x) - \mu)^2 \right\} \left| \frac{dG(x)}{dx} \right| dx.$$

If $G(x) = x$ this reduces to the normal distribution.

Kapteyn's transformation A method proposed by Kapteyn (1903) and Kapteyn and van Uven (1916) for transforming the variable X of a skew frequency function into a variable Z which is normally distributed.

Kärber's method A method for estimating the median effective dose of a stimulus from data on **quantal responses**, advanced by Kärber (1931). Essentially the same method was proposed by Spearman (1908) so that the name **Spearman–Kärber method** is also used.

Kendall, Sir Maurice George (1907–1983) An English statistician. In 1935 Kendall met Yule for the first time. He spent part of his holidays reading statistics books at the St John's College library and he had to seek out Yule who had the key to the library. Their brief conversation would prove significant for when Yule prepared a revision of his text *An introduction to the theory of statistics* (first published in 1911) Kendall became the second author. By 1950 the 14th edition of this book had appeared. Kendall continued a remarkable stream of research papers on topics such as the theory of k-statistics, time series and rank correlation methods, and a monograph *Rank correlation* in 1948. In 1943, despite other heavy commitments, he wrote volume I of *The advanced theory of statistics*, leading to three volumes and many subsequent developments. In 1949 he accepted the second chair of statistics at the London School of Economics and subsequently had distinguished careers also as an industrial consultant and as director of the World Fertility Survey. He masterminded the first edition of this dictionary.

Kendall's S score In rank correlation analysis, Kendall defined a score, for a pair of rankings of n items, as $+1$ if any two are ranked in the same order by the two rankings, -1 if in opposite order and zero if tied in either or both rankings. The total score S is the algebraic sum of the $\frac{1}{2}n(n-1)$ contributions from pairs of items (Kendall and Dickinson, 1990).

Kendall's tau (τ) A coefficient of rank correlation based on the number of **inversions** in one ranking as compared with another. It was proposed by Kendall (1938) as a rank correlation coefficient independent of the nature of underlying variable distributions, but had earlier been considered by Greiner (1909) and by Esscher (1924) as a statistic for the estimation of the correlation parameter in a bivariate normal distribution.

kernel density estimator See **kernel estimator**.

kernel estimator Generally a non-parametric estimator constructed through linear combination of kernels centred at each datum. The kernel is typically a symmetric probability density function, whose extent is determined by a **bandwidth**, which if large will produce a smooth estimate, and if small a rough estimate, thus yielding a **bias–variance trade-off**. Such estimators were first widely used for estimation of the **spectrum** in **time series**, but are now commonly applied for estimation of probability density functions, and in regression contexts (Simonoff, 1996; Wand and Jones, 1995).

kernel smoothing in regression A method of non-parametric regression using a kernel function that is usually a positive symmetric function of the bandwidth parameter and integrates to 1. Choice of bandwidth is a trade-off between bias and variance with a small bandwidth yielding a less smooth curve with smaller bias but larger variance, compared with a larger bandwidth (Wand and Jones, 1995).

Kesten's process A variation of the **Robbins–Munro process** intended to reduce the bias related to the starting point. The step constant is reduced only when there is evidence that doses are straddling the **median lethal dose**.

Khintchine, Aleksandr Yakovlevich (1894–1959) A Russian mathematician. After graduating in 1916, Khintchine remained at Moscow University undertaking research for his dissertation to become a university teacher. Around 1922 he developed his research in the theory of numbers and probability theory. In 1927 he was appointed as a professor at Moscow University and, in the same year, he published *Basic laws of probability theory*. Between 1932 and 1934 he laid the foundations for the theory of stationary random processes culminating in a major paper in *Mathematisches Annalen* (1934).

Khintchine's theorem Let $X_1, X_2, \ldots$ be a sequence of independent and identically distributed variables each with a finite mean μ. Then the variable

$$\bar{X} = \frac{1}{n} \sum_{i=1}^{n} X_i$$

converges in probability to μ as n tends to infinity. This theorem was first proved rigorously by Khintchine (1929). [See **law of large numbers**.]

Kiefer, Jack Carl (1924–1981) Born in Cincinnati, Ohio, Kiefer received his BS in economics and MS in economics and engineering, both at MIT in 1948. He continued his studies at Columbia under Abraham Wald and Jacob Wolfowitz, and received the PhD in mathematical statistics in 1952. After Wald's death in 1951, he followed Wolfowitz to the Cornell Mathematics Department where he remained for the next 18 years, being elected the first Horace White Professor in 1971. His research interests ranged widely over the whole area of mathematical statistics, but nearly half of his over 100 publications dealt with the optimal design of experiments, a field in which he was the internationally recognized leader. Sequential analysis was another area to which he made many fundamental contributions.

Kiefer–Wolfowitz process A **stochastic approximation procedure** proposed by Kiefer and Wolfowitz (1952) for estimating the maximum of a regression function. This concept is related to the **Robbins–Munro process** and has also been developed by other writers.

Klotz's test An inverse **normal scores** analogue of **Mood's W-test**, devised by Klotz (1962). Denoting by $\Phi^{-1}(x)$ the normal (Gaussian) score corresponding to distribution function x, the test statistic is

$$K = \sum_{i=1}^{n} \left[\Phi^{-1} \left(\frac{R_i}{N+1} \right) \right]^2$$

where R_i is the size rank of X_i, the ith among n X's, in the combined sample of $N = n + m$ X's and Y's.

Knox's test A test for **space–time clustering** (Bartlett and Knox, 1964). A contingency table is drawn up showing cases adjacent (within some set distance or interval) and non-adjacent in space and time, and tested by a chi-square test.

Knut Vik square A form of experimental design attributed to the Norwegian Knut Vik. It is also known as a 'knight's-move' square from the association of ideas between the construction of the square design and the move of the chess piece (Larson, 1977). The design may be illustrated by a 5×5 square in which it is desired to have each of five treatments once and once only in each row and column. The rows are formed by cyclic permutations of A, B, C, D, E moving forward two places instead of one:

$$A\ B\ C\ D\ E$$
$$D\ E\ A\ B\ C$$
$$B\ C\ D\ E\ A$$
$$E\ A\ B\ C\ D$$
$$C\ D\ E\ A\ B$$

The design is thus a **Latin square** of a particular type.

kollectiv This word occurs in English with a more specialized meaning than 'aggregate', which is its literal translation from the German. It denotes a population of objects in which each unit bears one of a finite number of identifiable characteristics. In particular, von Mises (1919) used the infinite sequence of such characteristics which bears no systematic properties, the so-called **irregular kollectiv**, as the central concept of his theory of probability. [See also **frequency theory of probability**.]

Kolmogorov, Andrey Nikolaevich (1903–1987) A Russian mathematician, historian and statistician. Kolmogorov graduated from Moscow State University in 1925 and began research under Luzin's supervision in that year. It is remarkable that Kolmogorov published eight papers in 1925, all written while he was still an undergraduate. Another milestone occurred in 1925, when Kolmogorov's first paper on probability appeared. This was published jointly with Khintchine and contains the 'three series' theorem as well as results on inequalities of partial sums of random variables which would become the basis for martingale inequalities and the stochastic calculus. Kolmogorov was appointed a professor at Moscow University in 1931. His monograph on probability theory, *Grundbegriffe der Wahrscheinlichkeitsrechnung*, published in 1933, built up probability theory in a rigorous way from fundamental axioms in a way comparable with Euclid's treatment of geometry. One success of this approach is that it provides a rigorous definition of conditional expectation. Kolmogorov later extended his work to study the motion of the planets and the turbulent flow of air from a jet engine. In 1941 he published two papers on turbulence which are of fundamental importance. In 1954 he developed his work on dynamical systems in relation to planetary motion. He thus demonstrated the vital role of probability theory in physics.

Kolmogorov axioms A set of axioms given by Kolmogorov (1933) for the foundation of probability theory in terms of set and measure theory. They form the starting point of most modern expositions of a mathematical theory of probability.

Kolmogorov equations Two systems of differential equations first derived by Kolmogorov (1931) each of which often uniquely determines a system of transition probabilities for a Markov process. Two different forms are known as the forward and backward equations and correspond respectively to what is called decomposition of the last step and decomposition of the first step.

Kolmogorov inequality A generalization of the **Bienaymé–Tchebychev inequality**. Let X_i ($i = 1, 2, \ldots, n$) be n mutually independent variables with means a_i and variances v_i. Let $A_k = \sum_{i=1}^{k} a_i$ and $V_k = \sum_{i=1}^{k} v_i$, and let $S_k = \sum_{i=1}^{k} X_i$. Then for $t > 0$ the probability that the inequalities

$$|S_k - A_k| < tV_k$$

are simultaneously realized is at least equal to $1 - 1/t^2$.

Kolmogorov representation A representation of the logarithm of the characteristic function $\phi(t)$ of infinitely divisible distributions with finite second moment. The representation has the form

$$\log \phi(t) = i\beta t + \int \frac{e^{itu} - 1 - itu}{u^2} dK(u)$$

where β is a real constant, and the function K coincides, to within a non-negative multiplicative constant, with a distribution function. [See **infinite divisibility**.]

Kolmogorov–Smirnov distributions A name given to a variety of distributions which are exact or approximate distributions of functions of the sample **cumulative distribution function**.

Kolmogorov–Smirnov test A type of **goodness-of-fit** test proposed by Kolmogorov (1933) and developed by Smirnov (1939) and later writers. If $F(x)$ is the population distribution function and $F_n(x)$ the empirical distribution function of the sample, then the test makes use of the statistic $d = \max\{|F(x) - F_n(x)|\}$. This has a distribution which is independent of $F(x)$ provided that the latter is a continuous distribution function. The method can be extended to discontinuous distributions by modifying the associated probability statement from an exact to a minimum level. The test may be used as one of 'goodness of fit' and the d-statistic may also be used to set confidence limits to an unknown probability distribution. It has been extended by Smirnov to test the homogeneity of two distribution functions on the basis of a sample from each. The Smirnov test of homogeneity depends on the greatest difference of the two empirical distribution functions and is distribution-free. [See also **Cramér–von Mises test**.]

Kolmogorov theorem Two theorems stated by Kolmogorov (1928, 1930) which give the conditions under which the **strong law of large numbers** holds: (i) when the random variables are independent; (ii) when the random variables are independent and identically distributed. In (ii), almost sure convergence of arithmetic means occurs if and only if the expected value exists.

Konyus index number The name given to a class of index number rather than to one specific formula. It is an index of prices based on quantities, the 'budget' of the consumer, which are optimal in some field of consumer preference. If the indifference level for consumers is specified in terms of an optimal budget at the base prices, the index is called Laspeyres–Konyus; if by an optimal budget at prices in the period under comparison with the base period, a Paasche–Konyus index. [See **Laspeyres' index**, **Paasche's index**.]

Koroljuk's theorem A theorem due to Koroljuk which states that for a **regular stationary point process** $\{N(t), t \geqslant 0\}$, the mean number of events per unit time, μ, and the intensity λ defined by

$$\lambda = \lim_{t \to 0} \frac{1 - \Pr\{N(t) = 0\}}{t}$$

are identical, including the case where both λ and μ are infinite.

Kounias's inequality A sharpened upper bound to one of the **Bonferroni inequalities**, due to Kounias (1968). It states that if $A_1, \ldots, A_k$ are compatible events, then the probability that at least one of them occurs is less than or equal to

$$\min_i \left\{ \sum_{j=1}^{k} \Pr(A_j) - \sum_{j \neq i} \Pr(A_i A_j) \right\}.$$

kriging Estimation by generalized least squares of a variable defined in space (Krige, 1951), usually two-dimensional. The estimate is usually based on a trend–surface and spatial autocorrelation (using the **semivariogram**).

Kronecker product of design It can be shown that an experiment design D uniquely determines its **incidence matrix** and vice versa. Thus the Kronecker product (in the sense of matrix theory) of two incidence matrices produces a new design matrix and can be used to determine the existence of certain classes of design.

Kronecker product of matrices The product $\mathbf{A} \otimes \mathbf{B}$ of an $m \times m$ matrix $\mathbf{A}$ and an $n \times n$ matrix $\mathbf{B}$ is the $mn \times mn$ matrix whose elements are the products of terms, one from $\mathbf{A}$ and one from $\mathbf{B}$.

Kruskal statistic A rank order statistic for the k-sample problem

$$\frac{12}{N(N+1)} \sum n_i \left(\bar{R}_i - \frac{N+1}{2} \right)^2,$$

where $i = 1, 2, \ldots, k$, n_i is the number in the ith sample, N is the total $\sum n_i$ and $\bar{R}_i$ is the average rank sum in the ith ranking. This statistic, proposed by Kruskal (1952), has been generalized by Basu (1967) for a right-censored sample of r observations.

Kruskal–Wallis test A multi-sample rank randomization test (Kruskal and Wallis, 1952) for identical populations, sensitive to unequal locations. It is a direct generalization of the two-sample **Wilcoxon rank sum test** to the multi-sample case.

Kuder–Richardson formula A formula for estimating the **reliability coefficient** of a test which endeavours to overcome the disadvantages of formulae associated with **split half methods**. There are numerous versions of the formula, which may be written

$$r_{tt} = \left(\frac{n}{n-1} \right) \left(\frac{\sigma_t^2 - \sum pq}{\sigma_t^2} \right),$$

where n is the number of items in the test, p is the proportion passing or satisfactorily responding to an item, $q = 1 - p$ and σ_t^2 is the overall variance of scores (Kuder and Richardson, 1937). [See also **Spearman–Brown formula**.]

Kuiper statistic A goodness-of-fit test statistic similar to the **Kolmogorov–Smirnov test**. For the single-sample problem it is

$$V_n = n^{1/2} \left[\sup_x \left\{ F_n(x) - F(x) \right\} - \inf_x \left\{ F_n(x) - F(x) \right\} \right]$$

and for the two-sample problem

$$V_{m,n} = \left(\frac{mn}{m+n} \right)^{1/2} \left[\sup_x \left\{ F_n(x) - G_m(x) \right\} - \inf_x \left\{ F_n(x) - G_m(x) \right\} \right].$$

The statistic may be used for observations on a circle, as it is independent of the choice of origin.

Kullback–Leibler distance function See **minimum discrimination information statistic**.

Kullback–Leibler information A formulation proposed by Kullback and Leibler (1951) based upon the Shannon–Wiener concept of information which is different from Fisher information. The information number $I(1:2)$ denotes the mean information per observation for discriminating between hypotheses H_1 and H_2 when H_1 is true:

$$I(1:2) = \int_{-\infty}^{\infty} f_1(x) \log \frac{f_1(x)}{f_2(x)} d\psi(x)$$

where $\psi(x)$ is some common measure.

kurtosis A term used to describe the extent to which a unimodal frequency curve is 'peaked'; that is to say, the extent of the relative steepness of ascent in the neighbourhood of the mode. The term was introduced by Karl Pearson (1905), and he proposed as a measure of kurtosis the moment ratio $\beta_2 = \mu_4/\mu_2^2$. It is doubtful, however, whether any single ratio can adequately measure the quality of 'peakedness'. In a bivariate frequency array, if the different arrays corresponding to one variable have different degrees of kurtosis they are said to be heterokurtic or allokurtic, and if they have the same degree, homokurtic or isokurtic. If the moment ratio is adopted as a measure of kurtosis, the value it assumes for a normal distribution, namely 3, is taken as a standard. Curves for which the ratio is less than, equal to or greater than three are known respectively as platykurtic, mesokurtic and leptokurtic. A curve showing the variation in the kurtosis of one variable against values of the other in bivariate variation is called a kurtic curve. [See also **scedasticity**.]

L

Λ-criterion (λ-criterion) An alternative name for a criterion, in hypothesis testing, based on the likelihood ratio, especially the ratio of dispersion determinants given by **Wilks's criterion**.

L-estimator Any estimator that is a linear function of the ordered sample values, with coefficients determined by the order. The trimmed mean and the interquartile range are examples.

l-statistics An alternative designation, proposed by Kendall and Stuart (1958), for the **polykays**.

L-statistics A class of rank order statistics proposed by Dwass (1956) in connection with rank order tests in the two-sample problem; this must not be confused with the Neyman and Pearson L-tests for homogeneity of sample variances. The L-statistic is defined as $\sum a_{N_i} b_{N R_i}$ where the a_{N_i} are $(n/mN)^{1/2}$, $i = 1, \ldots, m$, and $i = m + 1, \ldots, N$ being the two samples under test, the b_N another set of constants and $R_1, \ldots, R_N$ the ranks of the N random variables.

L-tests Tests proposed by Neyman and Pearson (1933) for testing the homogeneity of a set of sample variances. The tests, which are based on likelihood ratios in normal variation, vary according to the precise type of hypothesis under test.

L_1-metric This measures in p dimensions the distance between two points A and B as $\sum_{i=1}^{p} |A_i - B_i|$. It is also called the city-block metric or Manhattan distance.

L_1-norm Another name for **L_1-metric**. [See **L_p-norm**.]

L_2-metric The straight line distance between two points. It is also called the Euclidean distance. In p dimensions the L_2-metric between two points A and B is given by

$$\sqrt{\sum_{i=1}^{p} (A_i - B_i)^2}.$$

L_2 association scheme A partially balanced **incomplete block** design with two **associate classes** is said to have an L_2 association scheme if the number of treatments can be arranged in a square such that any two treatments in the same row or column are first associates; pairs of treatments not in the same row and column are second associates.

L_p-norm This is known also as the Minkowsky metric and is defined as $\left(\sum |A_i|^p\right)^{1/p}$ with $p \geqslant 1$. The minimization of this criterion is called the

L_1-**norm** or **least absolute deviation (LAD) method**. The minimization of the L_p-norm for $p = \infty$ is known as the **minimax** method.

Lachenbruch's method A **cross-validation** technique for estimating misclassification in **discriminant analysis** (Lachenbruch, 1966).

ladder index The time point (epoch) in a **random walk** at which a **ladder variable** begins to occur.

ladder variable If, in a **random walk**, the sum of the mutually independent variables at time point (epoch) n is S_n then, if $S_n > S_j$ ($j = 0, 1, 2, \ldots, n-1$), the variable X_n is an ascending ladder variable. Descending ladder variables are similarly defined where $S_n < S_j$.

lag An event occurring at time $t + k$ ($k > 0$) is said to lag behind an event occurring at time t, the extent of the lag being k. An event occurring k time units before another may be regarded as having a negative lag. By extension, two time series u_t and v_t are said to be lagged in relation to each other or one is said to lag behind the other if the values of one are associated with lagged values of the other; for example, if the production of a commodity at time t, q_t, is regarded as dependent on the price of that commodity at a previous time $t - k, p_{t-k}$, the series q_t is said to be lagged with respect to p_{t-k}. An equation connecting them, such as $q_t = ap_{t-k} + \beta$, is said to contain a lag.

lag correlation The correlation between two series where one of the series has a **lag** with reference to the other. [See **autocorrelation**, **lag covariance**.]

lag covariance The first **product moment** between two series, one of which is lagged in relation to the other. For example, if u_t and v_t are two series defined at $t = 1, \ldots, n$, the lag covariance of u_t and v_t of order $k > 0$ is

$$c_k = \frac{1}{n-k} \sum_{t=1}^{n-k} (u_t - \bar{u}_1)(v_{t+k} - \bar{v}_2),$$

where

$$\bar{u}_1 = \frac{1}{n-k} \sum_{t=1}^{n-k} u_t \quad \text{and} \quad \bar{v}_2 = \frac{1}{n-k} \sum_{t=1}^{n-k} v_{t+k}.$$

In general, $c_k \neq c_{-k}$. Commonly $\bar{u}_1$ and $\bar{v}_2$ are replaced by overall means. See **cross-covariance**.

lag hysteresis The word 'hysteresis' was taken from the theory of electromagnetism and introduced into econometrics by Roos (1925). Later Jones (1937) distinguished between lag hysteresis and skew hysteresis. Lag hysteresis in econometrics is confined to cases of sinusoidal variation in the two variables: skew hysteresis refers to the more realistic case in which the oscillatory movements are asymmetrical. The value of the concept where oscillatory movements are not cyclical is unknown.

lag regression A regression in which the values of the dependent variable and one at least of the independent variables are lagged in relation to each other.

Lagrange multiplier test A **likelihood ratio test** in which the null hypothesis is specified by restrictions using Lagrange multipliers (Silvey, 1959). When this is possible, it ensures that regularity conditions involving the restrictions are obeyed.

lambdagram A graphic device proposed by Yule (1945) in connection with the analysis of time series. For a series of values $x_1, \ldots, x_n$ the lambdagram consists of a coefficient λ_n plotted as ordinate against the sample size n. The coefficient is

$$\lambda_n = \frac{n-1}{n} \sum_{j=1}^{n} r_j = (n-1)\bar{r}_n,$$

where r_j is the jth serial correlation and $\bar{r}_n$ is the mean of the first n serial correlations. λ_n is related to the variance of the mean in sampling from a series whose items are internally correlated:

$$\text{var}(\bar{X}) = \text{var}(X)[1 + (n-1)\bar{r}_n].$$

The coefficient λ_n may be interpreted as an index of the divergence of the samples from samples of n random observations, i.e., of the way in which the n values of the samples are linked together.

Lancaster's partition of chi-squared A total chi-squared value is computed to provide a gross measure of the extent to which the cell frequencies depart from expectation. This value is partitioned into additive components to show how much is attributable to individual classifications and how much is due to both first-order and second-order interaction. This procedure can be extended to cover contingency tables involving more than three ways of classification where higher-order interactions occur. A special merit of this procedure is that it can be used with theoretical parameters or with parameters estimated from the data.

Langevin distributions A family of distributions on the p-dimensional sphere, including the **von Mises** and **Fisher distributions**.

Laplace, Pierre-Simon Marquis de (1749–1827) A French mathematician who was one of the most prominent exponents of nineteenth-century probability theory. Laplace's main probabilistic result was a fairly general central limit theorem which was obtained around 1810. His major probabilistic work, *Théorie analytique des probabilités* (1812), considerably influenced the development of mathematical probability and statistics right through to the beginning of the twentieth century.

Laplace approximation Approximation to an integral based on second-order Taylor expansion of the log integrand and normal approximation about its mode. It is closely related to saddle point approximation and is useful in Bayesian and frequencial inference.

Laplace distribution A frequency distribution of the double exponential type, expressible in the form

$$f(x) = \frac{1}{2\sigma} \exp\left\{-\frac{|x-\mu|}{\sigma}\right\}, \qquad -\infty < x < \infty, \quad -\infty < \mu < \infty, \quad \sigma > 0.$$

It is sometimes known as the first law of Laplace, in contradistinction to the second law which is the same as the **normal distribution**. The median is the maximum likelihood estimate of, and a sufficient statistic for, the location parameter. This feature of the distribution was recognized by Laplace.

Laplace law of succession A rule given by Laplace (1812) concerning the probability of events in further trials when certain trials have been made. If in n previous trials m have yielded an event E, the probability, according to the succession rule, that E happens on the next trial is $(m+1)/(n+2)$. The rule is based on **Bayes' postulate** for unknown probabilities and has been subject to much dispute and undiscriminating application.

Laplace–Lévy theorem A name sometimes given to the **central limit theorem**, which was known to Laplace in its essentials but was not proved rigorously under necessary and sufficient conditions until the beginning of the twentieth century.

Laplace transform If a function $g(t)$ is related to a second function $f(x)$ by the equation

$$g(t) = \int_0^\infty e^{-tx} f(x) dx$$

then $g(t)$ is the Laplace transform of $f(x)$. In statistical theory it is more customary to use the **Fourier transform**, which has certain advantages over the Laplace form; for example, if $f(x)$ is a frequency function the Fourier transform always exists whereas the Laplace transform may not do so for real t.

Laplace's theorem This limit theorem, of which **Bernoulli's theorem** is a corollary, states that if there are n independent trials, in each of which the probability of an event is p, and if this event occurs K times, then

$$\Pr\left(z_1 \leqslant \frac{K-np}{\sqrt{npq}} \leqslant z_2\right) \to \frac{1}{\sqrt{2\pi}} \int_{z_1}^{z_2} e^{-z^2/2} dz$$

as $n \to \infty$ whatever the numbers z_1 and z_2. Generally speaking, the theorem states that the number of successes in n trials is normally distributed for large n.

Laspeyres–Konyus index See **Konyus index number**.

Laspeyres' index A form of index number due to Laspeyres (1871). If the prices of a set of commodities in a base period are $p_0, p_0', \ldots$ and those in a given period $p_n, p'_n, \ldots$, and if $q_0, q'_0, \ldots$ are the quantities sold in the base period, the Laspeyres price index number is written

$$I_{0n} = \frac{\sum(p_n q_0)}{\sum(p_0 q_0)}$$

where the summation takes place over commodities. In short, the prices are weighted by quantities in the base period. Generally, an index number of the above form is said to be of the Laspeyres type even when p and q do not relate to prices or quantities, the characteristic feature being that the weights relate to the base period, as contrasted with **Paasche's index** in which they relate to the given period. [See **Lowe index, Palgrave's index, crossed weight index number.**]

lasso Least squares estimation penalized by a multiple of the sum of absolute values of the linear parameters.

latent class analysis, latent trait analysis The analysis of models in which dependence (usually in contingency tables) is explained by postulating unobservable subdivisions of the population, or unobservable discrete variables. The techniques are closely related to **factor analysis** of **categorical data**.

latent structure In general this phrase refers to a structure expressed in terms of continuous variables or variables which are 'latent' in the sense of not being directly observable. Certain econometric relations, e.g., in terms of 'utility', are of this type, and the models used in **factor analysis** as used in psychology also may be regarded as a kind of latent structure. More recently the term has been applied to studies of attitudes by questionnaire (Lazarsfeld, 1950). The observed replies to questionnaires are expressed in terms of 'latent' distributions of attitude.

latent variable A variable which is unobservable but is supposed to enter into the structure of a system under study, such as demand in economics or the 'general' factor in psychology. Unobservable quantities such as errors are not usually described as latent.

Latin cube An extension of the principle of the **Latin square** so that the layers, and sections in two directions, are effectively Latin squares.

Latin hypercube sampling A form of stratified sampling used for Monte Carlo integration over m variables, the range of each being divided into k and the integrand evaluated at a balanced random subset of the k^m overall strata.

Latin rectangle An experimental design derived from the **Latin square**. It consists of a Latin square with one or more adjacent rows or columns added or omitted. This particular design is one form of the **incomplete Latin square** or **Youden square**.

Latin square One of the basic statistical designs for experiments which aim at removing from the experimental error the variation from two sources, which may be identified with the rows and columns of the square. In such a design the allocation of k experimental treatments in the cells of a k by k (Latin) square is such that each treatment occurs exactly once in each row or column. A specimen design for a 5×5 square with five treatments, A, B, C, D and E, is as follows:

$$A \ B \ C \ D \ E$$
$$B \ A \ E \ C \ D$$
$$C \ D \ A \ E \ B$$
$$D \ E \ B \ A \ C$$
$$E \ C \ D \ B \ A$$

The earliest recorded discussion of the Latin square was given by Euler (1782) but it occurs in puzzles at a much earlier date. Its introduction into experimental design is due to R.A. Fisher.

lattice design See **quasi-factorial design, square lattice**.

lattice distributions A class of distribution, which includes most of the discrete distributions used in statistics, in which the intervals between any one random variable, for which there are nonzero probabilities, are integral multiples of one quantity (which depends on the random variable) and thus form a lattice. By suitable transformations all variables can be made to take values which are integers.

lattice sampling A method of sampling in which substrata are selected, for the sampling of individuals, according to some pattern analogous to the allocation of treatments on a lattice experimental design. For example, if there are two criteria of stratification, each p-fold, so that there are p^2 substrata, it is possible to choose p substrata so that none occurs in more than one 'row' or 'column' of the array representing the p^2 possible substrata; in short, in the manner of a **Latin square**. Similar schemes are possible for three- or more-way classification. Various schemes of the lattice type are known under the name of deep stratification.

lattice square A balanced design related to the **Latin square**, in which each pair of treatments occur together equally often in a row or column.

law of large numbers A general form of this fundamental law relating to random variables may be stated as follows: if X_k is a sequence of mutually independent variables with a common distribution and if the expectation $\mu = E(X_k)$ exists, then for every $\varepsilon > 0$ as $n \to \infty$ the probability

$$\Pr\left\{\left|\frac{X_1 + \ldots + X_n}{n} - \mu\right| > \varepsilon\right\} \to 0.$$

In this form, the law was first proved by Khintchine (1929), but less general forms were known from the time of Jakob Bernoulli onwards. The above form is the so-called 'weak' law. [See also **strong law of large numbers**.]

Le Cam, Lucien (1924–2000) A French statistician. He obtained his BSc from the University of Paris in 1945 and a PhD from the University of California at Berkeley in 1952. Through his life, Le Cam remained very close to Neyman. They co-edited the celebrated Berkeley Symposia volumes. Le Cam's work on asymptotics was published in the book *Asymptotic methods in statistical decision theory* in 1986.

least absolute deviation (LAD) methods Statistical data analysis based on the minimization of the L_1-norm as opposed to the least squares or L_2-norm (Bloomfield and Steiger, 1983).

least absolute deviation (LAD) regression A technique of regression consisting of fitting a model

$$Y = \beta_0 + \beta_1 X_1 + \ldots + \beta_k X_k$$

by minimizing the quantity

$$\sum_{i=1}^{n} |Y_i - \hat{Y}_i|,$$

where the Y_i's are the n observed values and the $\hat{Y}_i$'s are the values predicted by the model. Statistical data analysis using the L_1-norm criterion has been successfully applied for robust estimation, model fitting (constrained and unconstrained), **analysis of variance, hypothesis testing, cluster analysis** and non-parametric method. It has been studied in several contexts under a variety of names: minimum or least sums of absolute errors; least absolute deviations, or errors, or residuals; and L_1-norm method (from minimizing the L_1-norm of the vector of deviations). Historically, the L_1-norm method was suggested and studied in the early work of Boscovich (1757) and successively by Laplace before the least squares work of Gauss and Legendre (Birkes and Dodge, 1993).

least favourable distribution Given the composite hypothesis $H_0 : f_\theta$, $\theta \in \omega$, to be tested against the simple alternative H_1, the general procedure is to reduce the composite hypothesis H_0 to a simple one $H_{0\lambda}$, where λ is a probability distribution function over ω. The maximum power β_2 that can be obtained against H_1 is that of the most powerful test of $H_{0\lambda}$ against H_1. The probability distribution λ is said to be least favourable (for a given level α) if, for all λ', the inequality $\beta_\lambda \leqslant \beta_{\lambda'}$ holds. The notion plays a central role in the asymptotic theory of semi-parametric inference (van der Vaart, 1998). [See **most stringent test.**]

least favourable family A form of empirical exponential family useful for theoretical study of bootstrap and related confidence intervals.

least significant difference test A test for comparing mean values arising in **analysis of variance**. It is an extension of the standard t-test for the difference between two specified mean values. Because the tests between pairs are not independent the error rate is difficult to assess exactly. [See also **multiple comparisons**.]

least squares estimator An estimator obtained by the **least squares method**.

least squares generalized inverse If X is an $n \times p$ matrix, then a $p \times n$ matrix X_l^- is its least squares generalized inverse (g-inverse) if it satisfies $XX_l^- X = X$ and $(XX_l^-)' = XX_l^-$. Several statistical routine packages compute generalized inverses. An algorithm which finds a least squares g-inverse for the design matrix using the **incidence matrix** and the design matrix simultaneously without any rounding off errors for classification models with arbitrary patterns is given by Dodge and Majundar (1979).

least squares method A method of fitting a mathematical form to data by minimizing the sum of squares of deviations between observed and fitted values. It has a number of justifications. It can be regarded as a way of producing a close fit, qualitatively reasonable so long as it is not dominated by the behaviour of a few points. In the branch of numerical analysis called approximation theory it is treated that way to approximate complicated functions by simpler ones such as polynomials or rational functions. If the observations are treated as random variables with expected values linear functions of a vector of unknown parameters and with uncorrelated errors of constant variance, a theorem of Gauss (see **Gauss–Markov theorem**) establishes that the least squares estimates have minimum variance among all linear unbiased estimates (or equivalently among all linear estimates with bounded mean square error). If the errors are independently normally distributed there is a much stronger justification stemming from sufficiency. If the model is nonlinear in the parameters and the errors normally distributed with constant variance the estimates, which in general have to be found iteratively, are maximum likelihood estimates. There are generalizations, e.g., to weighted and to generalized least squares.

least variance difference method A method of generalized linear estimation of structural parameters. The technique is formulated exactly as Aitken's (1935) generalization of Gauss's method of treating independent observations of unequal precision to the case of interdependent observations.

left-censoring This occurs if it is possible that instead of observing the random variable X we observe only that $X < C$. This may happen when X represents a quantitative measurement made with a certain level of detection, as for instance measurement of viral load in HIV-positive patients.

left-truncation If we observe T only if $T > B$, there is left-truncation. This is different from left-censoring in the sense that if $T < B$ the individual does not enter into the sample; this is a selection of the sample. As for

censoring assumptions have to be made about how the truncation mechanism was generated.

Legendre, Adrien-Marie (1752–1833) Legendre's major contribution in statistics was the first description of the **least squares method** as an algebraic fitting procedure (Legendre, 1805). It was subsequently justified on statistical grounds by Gauss and Laplace.

Lehmann alternatives A class of non-parametric alternative hypotheses used by Lehmann (1953) in developing power functions of **rank order statistics**.

Lehmann–Scheffé theorem If T is a sufficient statistic for the parameter τ, then the minimum variance unbiased estimator of τ is given by $E(\hat{\tau}|T)$, where $\hat{\tau}$ is any unbiased estimator of τ (Lehmann and Scheffé, 1950). The theorem is an extension of the **Rao–Blackwell theorem**.

Lehmann's test A two-sample non-parametric test for variances of the **Wilcoxon–Mann–Whitney** type proposed by Lehmann (1951). It is based upon all possible differences between the observations in the two samples.

leptokurtosis See **kurtosis**.

Leslie matrix A square matrix relating the numbers in different age classes in successive generations of a population (Leslie, 1945). In the basic form, the first row consists of fertility rates, the subdiagonal of survival rates, and the remaining elements are zero.

Leslie's test In capture–recapture experiments, the probabilities of capturing each living member of the population must be the same. To test this assumption, Leslie (1958) proposed the statistic

$$L = \frac{\sum_{i=1}^{N}(R_i - \bar{R})^2}{\sum_{j=f}^{g} p_j q},$$

where R_i is the number of recaptures of individual i between the fth and gth sample inclusive; $\bar{R} = \sum_{i=1}^{n} R_i/N$ and p_j is the proportion of recaptures at the jth sample. The statistic is distributed as χ^2 with $N - 1$ degrees of freedom, although it was later shown by Carothers (1971) that $(1 - 1/N)L$ is identical to **Cochran's Q-test**.

level map A graph showing curves in an (x, y) plane corresponding to various values of the constant k in a defining equation $f(x, y) = k$. It is similar to the contours of equal height on a geographical map.

level of a factor See **factorial experiment**.

level of interpenetration See **interpenetrating samples**.

level of significance Many statistical tests of hypotheses depend on the use of the probability distributions of a statistic t chosen for the purpose of the particular test. When the hypothesis is true this distribution has a known form, at least approximately, and the probability $\Pr(t \geqslant t_1)$ or $\Pr(t \leqslant t_0)$ can be determined for assigned t_0 or t_1. The acceptability of the hypothesis is usually discussed, *inter alia*, in terms of the values of t observed; if they have a small probability, in the sense of falling outside the range t_0 to t_1, $\Pr(t \geqslant t_1)$ and $\Pr(t \leqslant t_0)$ small, the hypothesis is rejected. The probabilities $\Pr(t \geqslant t_1)$ and $\Pr(t \leqslant t_0)$ are called levels of significance and are usually expressed as percentages, e.g., 5 per cent. The actual values are, of course, arbitrary, but popular values are 5, 1 and 0.1 per cent. Thus, for example, the expression 't falls above the 5 per cent level of significance' means that the observed value of t is greater than t_1 where the probability of all values greater than t_1 is 0.05; t_1 is called the upper 5 per cent significance point, and similarly for the lower significance point t_0.

leverage point In **regression diagnostics**, a point that has an outlying value of the regressor variables, and therefore a potentially large effect on the regression equation, is said to have high leverage (Chatterji and Hadi, 1988).

Lévy–Cramér theorem The converse of the **first limit theorem**, proved simultaneously by Lévy (1925) and by Cramér (1925). Let $\{\phi_n(t)\}$ be a sequence of characteristic functions corresponding to a sequence of distribution functions $\{F_n(x)\}$. Then if $\phi_n(t)$ tends to $\phi(t)$ uniformly in some finite t-interval, $\{F_n(x)\}$ tends to a distribution function $F(x)$ and $\phi(t)$ is the characteristic function of $F(x)$.

Lévy–Khintchine representation A representation of the logarithm of the characteristic function $\phi(t)$ of an arbitrary **infinite divisible** distribution. The representation has the form

$$\log \phi(t) = i\beta t + \int_{-\infty}^{\infty} \left(e^{itu} - 1 - \frac{itu}{1+u^2} \right) \frac{1+u^2}{u^2} G'(u) du,$$

where $\beta \in \mathbb{R}$ is the centring constant and G coincides with a bounded non-decreasing function. [See also **Kolmogorov representation**, **Lévy representation**.]

Lévy–Pareto distribution The work of Lévy in connection with stable distributions showed that the **Pareto distribution** can be derived from a version of the central limit theorem in which the individual random variables do not have finite variance.

Lévy process A stochastic process with stationary independent increments, the simplest examples being **Gaussian processes** and the **Poisson process**.

Lévy representation A representation of the logarithm of the characteristic function $\phi(t)$ of an arbitrary **infinite divisible** distribution. The representation has the form

$$\log \phi(t) = i\beta t - \sigma^2 t^2 / 2 + \int_0^{\infty} h(t, u) dN(u) + \int_{-\infty}^0 h(t, u) dM(u)$$

where β is real, $\sigma > 0$, $h(t, u) = e^{itu} - 1 - itu/(1+u^2)$ and the functions M and N satisfy a number of conditions. [See also **Kolmogorov representation**.]

Lévy's theorem A synonym for the **first limit theorem** which was first proved rigorously by Lévy and Cramér independently (1925).

Lexian distributions A name given to mixtures of binomial distributions, with common N, studied by Lexis (1875). [See also **Lexis variation**.]

Lexis ratio This ratio provides a measure for distinguishing the three kinds of variation in sampling for attributes: Bernoullian, Lexian and Poissonian. If k samples of $n_1, n_2, \ldots, n_k$ members bear observed proportions of the attribute $p_1, p_2, \ldots, p_k$ the Lexis ratio Q is defined by

$$Q^2 = \frac{\sum_{j=1}^{k} n_j (p_j - p)^2}{(k-1)pq},$$

where p is the proportion of the attribute in all samples together and $q = 1-p$. If the Lexis ratio is equal to unity within sampling limits the sampling is regarded as Bernoullian; if greater than unity, as of Lexian type; if less than unity, as Poissonian. They are also said to possess normal, hypernormal (or supernormal) and subnormal dispersion. [See **dispersion index**, **overdispersion**.]

Lexis theory A general term to describe the theory of sampling for attributes (see **Lexis ratio**) developed by Lexis (1879). In modern terminology it is part of the **analysis of variance** applied to dichotomized material.

Lexis variation A type of sampling variation considered by Lexis (1877). On each of k occasions let n members be drawn at random, and let the probability of success be the same for any member of a set, but vary from one occasion to another, the probabilities being $p_1, p_2, \ldots, p_k$. The mean proportional frequency of occurrence of successes over all occasions is $p = \sum_{i=1}^{k} p_i/k$ and the variance of the number of successes is $npq + n(n-1)\text{var}(p_i)$ where $q = 1 - p$ and var (p_i) is the variance of p_i in the k sets. If all the p_i are equal this reduces to **Bernoulli variation**. In other cases the Lexian is larger than the Bernoullian variance. This effect is encountered in sampling from non-homogeneous strata. The dispersion is said to be supernormal or hypernormal. [See also **Lexis ratio**, **Poisson variation**.]

Liapunov's inequality An inequality due to Liapunov (1901) concerning the relations between the **absolute moments** of a frequency distribution. If $a \geqslant b \geqslant c \geqslant 0$ are three real numbers and ν_a, ν_b, ν_c the absolute moments of orders a, b and c for some arbitrary distribution, then

$$\nu_b^{a-c} \leqslant \nu_c^{a-b} \nu_a^{b-c}.$$

Liapunov's theorem A form of the **central limit theorem** which assumes the existence of absolute third moments. If X_j $(j = 1, 2, \ldots, n)$ is a sequence of independent variables with means m_j, variances σ^2 and absolute third mean-moments ν_j^3, the sum $\sum_{j=1}^n X_j$ is asymptotically normal provided that

$$\lim_{n \to \infty} \frac{\nu}{\sigma} = 0,$$

where

$$\nu^3 = \sum_{j=1}^n \nu_j^3 \quad \text{and} \quad \sigma^2 = \sum_{j=1}^n \sigma_j^2.$$

life expectancy Life expectancy at age x is the average (expected) number of additional years that a survivor to age x ($x = 0$) will live beyond age x. It is calculated by dividing the total number of person-years lived by a cohort beyond age x by the number of survivors to that age. Typically, cross-sectional mortality data are used to calculate life expectancies, thus requiring the assumption that persons living into the future will be subject to the same **age-specific mortality rates** as were observed cross-sectionally.

life table A table showing the number of persons who, of a given number born or living at a specified age, live to attain successive higher ages, together with the numbers who die in the intervals.

lifetime distribution In **renewal theory** and other applications, the distribution of lifetimes from which in particular the **hazard** is derived.

likelihood The density function of continuous variables $X_1, \ldots, X_n$ dependent on parameters $\theta_1, \ldots, \theta_k$ is expressed as $f(X_1, X_2, \ldots, X_n; \theta_1, \theta_2, \ldots, \theta_k)$. If it is considered as a function of the θ's for fixed X's, it is called the likelihood function. Likewise, for discontinuous variation, the likelihood function emanating from the population specified by $f(X_1, \ldots, X_n; \theta_1, \theta_2, \ldots, \theta_k)$ is that frequency function itself taken at $X_1, \ldots, X_n$, considered as a function of the θ's. The likelihood function is usually denoted by L, but for many purposes a more useful function is the logarithm of the likelihood, which is also sometimes denoted by L. If a sample of n independent values $x_1, x_2, \ldots, x_n$ is drawn from a univariate population with frequency function $f(x, \theta_1, \ldots, \theta_k)$, the likelihood of the sample is $\prod_{i=1}^n f(x_i, \theta_1, \ldots, \theta_k)$, with obvious extensions to studies of dependence.

likelihood principle The principle that the information about a model given by a set of data is entirely contained in the likelihood. The strong likelihood principle maintains this position even when the data are collected with a stopping rule dependent on the results, as in sequential trials.

likelihood ratio In its simplest form this is the ratio of the likelihoods of data under two completely specified models. As such it may be used directly, in combination with prior probabilities for the two models and appeal to Bayes' theorem, or as the basis for an optimal test, taking one of the models as the null hypothesis and the other as alternative. In applications of realistic complexity the method is used to test a null hypothesis with nuisance parameters against a parametrically formulated family of alternatives, the null hypothesis being nested within the larger family. Then the ratio of maximized likelihoods, maximizing over all adjustable parameters, first under the null hypothesis and then more generally, is involved. Any optimality properties are asymptotic and the asymptotic distribution under the null hypothesis is, under regularity conditions, related to the chi-square distribution.

likelihood ratio dependence If, in the concept of **regression dependence**, the requirement that the conditioned variable $(Y|X)$ to be stochastically increasing is replaced by one of monotone likelihood ratio in X, then Lehmann (1966) termed this to be likelihood ratio dependence. For example, a bivariate normal distribution $(\rho \geqslant 0)$ has positive likelihood ratio dependence.

likelihood ratio test A test of a hypothesis H_0 against an alternative H_1, based on the ratio of two likelihoods, one derived from each of H_0 and one from H_1. The term is also applied to a test where one or both likelihoods have been maximized over nuisance parameters.

Likert scale A type of composite measure developed by Rensis Likert (1932) in an attempt to improve the levels of measurement in social research through the use of standardized response categories in survey questionnaires. Likert items are those using such response categories as strongly agree, agree, disagree, and strongly disagree. Such items may be used in the construction of true Likert scales as well as other types of composite measures.

Lilliefors' test A test of normality using the **Kolmogorov–Smirnov test** statistic with mean and variance estimated from the data (Lilliefors, 1967).

limited information estimator See 'k'-class estimator.

limited information methods In econometrics, methods of deriving estimates of parameters in a stochastic system which do not use all the information available. The term is usually confined to those methods which give consistent estimates, i.e., are unbiased for large samples. One such method involves the employment of **instrumental variables**. A second is the method of **reduced forms** which applies to systems which are exactly identifiable. Another is the limited information maximum likelihood method which is applied to over-identified systems and incorporates also a reduced form technique. Loosely speaking, this method ignores certain restrictions on the parameters imposed by the structural equations but still produces consistent estimates. Its great advantage is that it can be used without a complete specification of all the equations of the system.

Lincoln index One method of estimating the size of populations of mobile units, e.g., animal populations, relies on the capture, marking, release and recapture of samples from the population under investigation. The technique appears to have been used first by Lincoln (1930) and the statistic formed by dividing the total number of marked units released by the proportion of marked units recaptured – the Lincoln index – can be used to estimate the total population size. The Lincoln index should properly be restricted to populations not subject to birth or immigration. It is slightly biased but the bias can be removed by modification in the estimator.

Lindeberg–Feller theorem A form of the **central limit theorem** which gives a necessary and sufficient condition for the distribution of a sum of independent random variables to be asymptotically normal.

Lindeberg–Lévy theorem A particular case of the **central limit theorem** when all the variables concerned have the same distribution.

Lindley's integral equation A Wiener–Hopf integral equation for the limit distribution of the actual waiting time in a single-server queue with general interarrival and service-time distributions when the **traffic intensity** is less than unity.

Lindley's paradox An apparent difference between Bayesian and non-Bayesian outcomes from hypothesis evaluation. For example, a precise null hypothesis may be strongly rejected by a standard sampling theory test of significance, and yet have high odds by a Bayesian analysis that assumes a small prior probability for the null hypothesis and a diffuse distribution over the alternative hypotheses (Lindley, 1957; Cox and Hinkley, 1974).

line of equal distribution The straight line on the graph of a **Lorenz curve** which passes through the origin and the upper extreme of the curve; when, as is customary, the two variables vary from 0 to 1, this is the line making an angle of 45° with the coordinate axes. It provides a reference line of equal distribution, production, concentration, etc. The gradual approach of a succession of Lorenz curves to this diagonal line indicates that the unequal concentration of distribution is being reduced. This interpretation is similar to that afforded by the change in slope of the **Pareto curve** as indicated by the change in the **Pareto index**.

line sampling A method of sampling in a geographical area. Lines are drawn across the area and all members of the population falling on the line, or intersected by it, are included in the sample. If the lines are straight parallels equally spaced across the area concerned, then the sampling becomes one form of systematic sampling. If, instead of all intercepts on the lines, a series of evenly spaced points are chosen on each line, the sampling is equivalent to choosing the points on a lattice and may also be regarded as two-stage line sampling.

line spectrum A term in spectrum analysis of time series denoting the form of the spectrum diagram when the variance or power is concentrated at discrete frequencies.

linear constraint A condition imposed on certain values of random variable or frequencies which is linear in form. For example, if samples of variables $x_1, x_2, \ldots, x_n$ are drawn their mean, in general, will vary; but if only those samples are considered which have a mean equal to zero, all other samples being ignored, the variables are subject to the linear constraint $\sum_{i=1}^{n} x_i = 0$. The distribution under constraint may be regarded as **conditional**.

linear correlation An obsolete expression once used to denote either (i) the product moment correlation in cases where the corresponding regressions were linear, or (ii) a coefficient of correlation constructed from linear functions of the observations.

linear estimator An estimator which is a linear function of the observations.

linear failure rate distribution A distribution due to Barlow (1968) used to represent distributions of lifetimes in situations where the data show too pronounced a departure from the exponential distribution. Its density function, in standard form, is

$$f(x) = (1 + \theta x) \exp\left\{ -\left(x + \frac{1}{2}\theta x^2 \right) \right\}, \qquad x > 0.$$

linear hypothesis While this term could be used more broadly, it is usually restricted to hypotheses about parameters in a model in which expected values of the observed random variables are linear functions of a vector of unknown parameters and the observations, at least in the simplest case, are independently normally distributed with constant variance. There is a null hypothesis that the vector lies in a specified subspace, often that a set of components are all zero. The test, essentially based on fitting by least squares the full and constrained models and comparing the resulting residual sums of squares, is most simply cast in the form of a variance ratio or F-test. It can be regarded as a special case of a likelihood ratio test although it does not rely on an asymptotic justification. In the special case when the null hypothesis concerns only one parameter (or one parametric function) F is the square of Student's t-statistic for that parameter, a preferable form.

linear mean square regression The line $Y = \alpha + \beta x$, where the values of α and β minimize $E(Y - \alpha - \beta x)^2$. This theoretical line exists even when $E(Y|x)$ is not a linear function of x, and the least squares method gives unbiased estimates of α and β. This justifies the use of linear regression in situations (e.g., in finite populations) when the population regression cannot be assumed linear.

linear model A model in which the equations connecting the variables are in a linear form in parameters.

linear process A stochastic process defined by the formal expression

$$x_t = \sum_{u=-\infty}^{t} g_{t-u} w_u$$

for the discontinuous case or the analogous integral

$$x(t) = \int_{-\infty}^{t} g(t - u)dw(u)$$

for the continuous case, where w_u or $dw(u)$ represent independent and stationary disturbances.

linear programming The procedure used in maximizing, or minimizing, a linear function of several variables when these variables, or some of them, are subject to constraints expressed in linear terms: these may be equations or inequalities. The term 'programming' in this context indicates a schedule of actions.

linear regression See **regression**.

linear structural relation A relationship in linear form among the observed **endogenous variables**.

linear sufficiency A term proposed by Barnard (1963b) for linear parametric functions in the Gauss–Markov estimation model, and extended by Godambe (1966) to the case of sampling from finite populations, where some problems of estimation do not admit a sufficient statistic. It is a weaker form of sufficiency and restricted to linear estimators.

linear systematic statistic A systematic statistic which is a linear function of the observations (see **systematic statistic**). Most 'systematic' statistics in current use are, in fact, linear and the linearity is sometimes taken as understood in describing the linear statistic simply as 'systematic'.

linear trend A trend for which the value is a linear function of the time variable, e.g., $u(t) = a + bt$ where a and b are constants.

linearized maximum likelihood method A method of parameter estimation in multi-equation models in which the Taylor series linear approximations of the partial derivatives of the likelihood function are equated to zero rather than the actual partial derivatives.

lineo-normal distribution See **modified normal distributions**.

link function In **generalized linear models**, the function $f(\mu)$ linking the mean μ of the dependent variable to a linear function of the regressors.

link relative In index number theory, the value of a magnitude in a given period divided by the value in the previous period. [See **chain index**.]

linkage A tendency for particular **alleles** of two genes to occur together (usually indicating that the genes are close together on the same chromosome).

linkage analysis This uses statistical models to infer whether a disease gene is linked to given genetic markers, namely on the same chromosome. Commonly, this is done by testing the null hypothesis that the disease gene and markers are on different chromosomes against the alternative hypothesis that they are on the same chromosome. The location of the disease gene is not known. By finding

239

genetic markers on the same chromosome as the disease gene, its location can be ultimately identified (Ott, 1999).

linked blocks A class of incomplete designs proposed by Youden (1951a) for the purpose of reducing the number of replications normally required in such designs and of restoring the symmetry and simplicity of analysis which results from using lattice designs. For example, a design for 10 treatments involving five blocks with four treatments in a block is as follows:

$$
\begin{array}{llll}
1 & 1 & 2 & 3 & 4 \\
2 & 5 & 5 & 6 & 7 \\
3 & 6 & 8 & 8 & 9 \\
4 & 7 & 9 & 10 & 10
\end{array}
$$

where any two blocks have one linked treatment in common.

linked data Linked data (or matched data) are data from two different sets of microdata that have at least some common members, such that the common members have been identified and cross referenced. As a result, the amount of multivariate data is increased for the members common to both microdata sets. Linkage may be deterministic (based on exact matches of characteristics) or probabilistic (based on approximate as well as exact matches).

linked paired comparison designs An experiment design, proposed by Bose (1956), for comparing n objects with a number of judges (m). Each judge compares r pairs of objects and among these pairs each object appears equally often (α times). Each pair is compared by k (> 1) judges and, given any two judges, there are exactly λ pairs compared by both judges. If a linked paired comparison design exists then there must be a corresponding balanced **incomplete block** design with m treatments and $\frac{1}{2}n(n-1)$ blocks.

linked samples Two samples of the same size in which there is a one–one correspondence between their respective sample units. The link between a pair of corresponding units may be rigid in the sense that one of them uniquely determines the other, or it may be semi-rigid in that one of them restricts the choice of the other, e.g., a pair of linked grids may be separated by a fixed distance. Linking among three or more samples is also possible. [See also **method of overlapping maps**.]

LISREL A model and associated computer program for studying LInear Structural RELationships, including econometric models and models, such as factor analysis, involving latent variables (Jöreskog, 1973).

list sample A sample selected by taking entries from a list of the items constituting the population under review. The usual method of selecting entries is to take them at equal intervals, the starting point being selected at random.

loading See **factor analysis**.

local asymptotic efficiency A development by Konijn (1956) of the concept of **asymptotic relative efficiency** concerned with two-sided tests in which the tails are not necessarily equal.

local likelihood estimation A form of regression smoothing in which log-likelihood contributions are weighted according to the distance of the covariates from the covariate value at which the parameters are to be estimated, often by fitting a low-order polynomial. The Nadaraya–Watson estimator is the simplest example (Fan and Gijbels, 1996).

local statistic A statistic, estimator or test statistic, which is derived from short-term comparisons within a time series. The concept is due to Jowett (1955) and includes the **jump statistic.**

locally asymptotically most powerful test A test of a composite statistical hypothesis defined by Neyman (1959); it is available only for a one-dimensional parameter. [See also **locally asymptotically most stringent test, uniformly most powerful test.**]

locally asymptotically most stringent test A development of the **locally asymptotically most powerful test** to the case of a k-dimensional parameter (Bhat and Nagnur, 1965). [See also **most stringent test.**]

locally most powerful rank order test This test for shift in location parameter with symmetry, as defined by Fraser (1957), is of the form $\sum_{i=1}^{N} z_i E(X_{Ni})$ where X_{Ni} is the ith order statistic from the normal distribution; the $z_i, i = 0, 1$, fall according to y_i being a negative or positive member derived from the y_N ranking of the X_N observations.

location See **measure of location.**

location model A model for **discriminant analysis** with continuous and binary variables introduced by Olkin and Tate (1961) and developed by Krzanowski (1975, 1980, 1982). The model is based on the assumption of a multivariate normal distribution at each location, or combination of binary variables, taking account of the proportions of the different populations at that location.

location parameter A parameter which 'locates' a frequency distribution in the sense of defining a central or typical value such as a mean or mode.

location-scale family A set of random variables $(X_1, \ldots, X_n)$ has a location-scale family distribution with location parameter μ, scale parameter σ, if the joint cumulative distribution function of $(T_1, \ldots, T_n)$ is $F(t_1, \ldots, t_n)$, where $T_i = (X_i - \mu)/\sigma$, and $F(t_1, \ldots, t_n)$ is any n-dimensional cumulative distribution function.

location shift alternative hypothesis If we have a random sample of observations from K $(\geqslant 2)$ populations, in testing the hypothesis that these populations are identical, the alternative hypothesis might be that the only difference between the parent distribution is one of location. This alternative hypothesis is known as the location shift alternative hypothesis. If the alternative hypothesis further specifies that the k population means form an ordered sequence, it is called an ordered alternative hypothesis.

lods A term introduced by Barnard (1949) in connection with certain developments in statistical inference. It is a contraction of the term 'logarithmic odds', the basic probabilities being expressed on a logarithmic scale in terms of odds in favour of or against an event.

log-chi-squared distribution A transformed distribution used in the analyses of heterogeneous variances (Bartlett and Kendall, 1946) and the analyses of Poisson processes (Cox and Lewis, 1966).

log convex tolerance limits A class of tolerance limits proposed by Hanson and Koopmans (1964) to deal with a position intermediate between those for, say, the normal distribution and non-parametric conditions. They use the **Pólya frequency function of order 2** which includes a broad group of distributions.

log F distribution A family of continuous distributions, including the **normal**, **log gamma**, **extreme-value** and **logistic distributions**. If x follows a variance ratio distribution, $\log(x - \alpha)/\beta$ follows a log F distribution. Fisher's $z = (\log F)/2$ is a special case.

log gamma distribution The distribution of $\log X$, when X follows a gamma distribution.

log-linear models A class of generalized linear models for the analysis of categorical data.

log–log transformation The transformation of a probability P to a **response metameter** Y according to the formula

$$Y = \log(-\log P).$$

This was first suggested by Mather (1949) and adapted by Finney (1941) to the estimation of bacterial densities from dilution series.

log–logistic distribution A distribution, analogous to the **lognormal distribution**, noted by Shah and Dave (1963). A variable Y is said to be log–logistic if $\alpha + \beta \log(Y - \gamma)$ has a standard **logistic distribution**.

log-rank test This test, the basis of which was suggested by Mantel (1966), uses the relative death rate in subgroups to form a test for comparing the overall difference between the whole survival curves for different treatments (Cox, 1972). This yields a non-parametric test for the equality of two right-censored survival distributions. It is equivalent to a series of **Mantel–Haenszel tests** applied at the failure instants and gets its name from the relation with the non-parametric test that is optimal when the distributions are exponential (Peto *et al.*, 1977).

log-zero-Poisson distribution A distribution, due to Katti and Rao (1970), obtained by modifying a **compound distribution** formed from a **Poisson** and a **logarithmic series distribution**. The modification consists of 'adding zeros', i.e., assigning extra probability to the event $(x = 0)$.

logarithmic chart A graph whereon one or both axes are scaled in terms of logarithms of the variables. The chart may be called a **semi-** or **double logarithmic chart** according to whether only the ordinate or both the ordinate and abscissa are on a logarithmic scale. In general, the logarithmic method of plotting is used when relative changes are important, since equal linear displacement on a logarithmic scale indicates equal proportional changes in the variable itself.

logarithmic series distribution A frequency distribution developed by R.A. Fisher (1941) in connection with the frequency distribution of species. It is a limiting form of the **negative binomial distribution** with the zero class missing, the frequency of the values $1, 2, 3, \ldots$ being

$$\alpha x, \frac{1}{2}\alpha x^2, \frac{1}{3}\alpha x^3, \ldots,$$

where $1/\alpha = -\log(1-x)$ and x is some parameter; that is to say, the frequency of the value r is the coefficient of x^r in the expansion of $-\alpha \log(1 - x)$.

logarithmic transformation In general, a transformation of a variable x to a new variable y by some such relation as $y = a + b \log(x - c)$. There are a number of contexts in which such transformations are useful in statistics, e.g., to normalize a frequency function, to stabilize a variance, and to reduce a curvilinear to a linear relationship in regression or probit analysis.

logistic curve See **growth curve**.

logistic distribution A frequency distribution of the form $e^t/(1 + e^t)^2$. It closely approximates the normal distribution and may also be shown to be the asymptotic distribution of the mid-range of an exponential-type parent distribution. In its cumulative form it has also been used as a **growth curve** and as a basis for the analysis of binary data.

logistic process A stochastic process associated with the logistic law of growth (see **growth curve**). It is a particular case of a birth and death process in which the rates for these two phenomena are linearly dependent upon population size. If, for a finite constant population, the condition is also made that individuals not possessing some characteristic will eventually do so we have the epidemic model proposed by Bartlett (1946). It may be noted in this connection that 'logistic' in statistical usage has nothing to do with the military use of the word as meaning the provision of material.

logistic regression A model of the dependence of Bernoulli random variables on explanatory variables. The **logit** of the expectation is explained as a linear form of explanatory variables. If we observe $(Y_i, x_i), i = 1, \ldots, n$, where Y_i is a Bernoulli variable and x_i a vector of explanatory variables, the model for $\pi_i = P(Y_i = 1) = E(Y_i)$ is $\mathrm{logit}(\pi_i) = \log\{\pi_i/(1 - \pi_i)\} = \beta_0 + \beta x_i$. The model is especially useful in case–control studies and leads to the effect of risk factors by **odds ratios**.

logit In some problems relating to the proportion of subjects responding to different doses of a stimulus the model based on a **tolerance distribution** is not appropriate. A logistic relation

$$P = \frac{1}{1 + e^{-(\alpha+\beta x)}}$$

may approximately represent the dependence of probability of response P on dose x. The approach has a long history going back essentially to Fisher and Yates (1938), to Wilson and Worcester (1943), to Berkson (1944) in bioassay and to Cornfield (1962) in epidemiology. The method was formalized and related to the theory of exponential families by Cox (1958a).

lognormal distribution If the logarithms of a set of values of random variables are distributed according to the **normal distribution** the variable is said to have a logarithmic normal distribution, or be distributed 'lognormally'.

Lomax distribution A name sometimes given to an alternative form of the **Pareto distribution**, also referred to as the Pareto distribution of the second kind.

longitudinal survey A method of data collection whereby questions are administered to the same participants at two or more successive points in time or during two or more relatively small successive intervals in time. Advantages of this design include the ability to observe outcomes, infer trends from the data, and observe dynamic change. Drawbacks include the fact that locating participants at successive times may be difficult or impossible, and the results are slow to accumulate and are necessarily old. See **causality**, **cross-sectional survey**.

loop plan A name given by Deming (1950) to a method of estimating the variance of an estimator derived from a systematic sample. If a population arranged in a line is sampled by taking units at a fixed interval k apart along the line, starting at random within the initial interval of width k, there is no theoretically valid method of deriving an estimate of the sampling variance from the sample itself. The units are therefore paired (or 'looped together') and each pair is regarded as a sample of two chosen at random within an artificial stratum of length $2k$. On this assumption an estimate of sampling error can be made, although the method gives biased estimators unless the population is arranged at random along the line.

Lorenz curve The curve $L(x) = \int_0^x yf(y)/E(X)$ derived from a density function $f(x), 0 < x < \infty$ (Lorenz, 1905). It is the distribution function of the first moment of the distribution function F, and was introduced for the study of distributions of income.

loss function In the making of decisions on the basis of observations on a variable X, disadvantage may be suffered through ignorance of the true distribution of X. The extent of the disadvantage is often a function of the true distribution and of the decision which is actually made. This function is called the loss function.

loss matrix In the theory of decision functions, a matrix specifying the economic loss or gain incurred according to the various decisions which can be taken and the various situations which can in reality exist.

loss of information This term is used in two entirely different senses: (*i*) to denote the actual loss of information in the ordinary sense, e.g., by the destruction of records; (*ii*) to denote failure to extract all the information which exists in the available data about a particular matter. In the second case the failure may be due to avoidable causes, such as the use of inefficient statistics; or it may, in the technical sense of the word 'information', be due to the fact that no single estimator exists embodying all the 'information' which exists in the sample under scrutiny. [See **ancillary statistic, information, sufficiency.**]

lot A term used in quality control in the sense of aggregate, collection or batch, but usually with a somewhat more specialized meaning. A lot is a group of units of a product produced under similar conditions and therefore, in a sense, of homogeneous origin, e.g., a set of screws produced by a lathe or a set of electric light bulbs produced by a number of similar machines. It is sometimes implicit that the lot is for inspection.

lot quality protection See **average quality protection**.

lot tolerance per cent defective The proportion of defective product allowable, or acceptable, in each lot submitted for inspection under a scheme designed for **lot quality protection**. This is sometimes called 'lot tolerance fraction defective'.

Lotka–Volterra equations Differential equations modelling the interaction between competing species, studied by Lotka (1920) and Volterra (1926). Modifications describe population growth in a limited environment, and prey–predator and host–parasite interactions. Stochastic models of the same type have been studied.

lottery sampling A method of drawing random samples from a population by constructing a miniature of the population, e.g., by inscribing the particulars of each member onto a card and drawing members at random from it, e.g., by shuffling the cards and dealing a set haphazardly. It is the method usually employed at a lottery, hence its name, but suffers from the disadvantage that the preparation of the cards entails considerable labour and strict precautions must be taken in the shuffling process to guard against bias.

245

Lowe index An index number proposed by Lowe (1823) in which average weights are used. If the prices of a set of commodities in either a base or a given period are $p_0, p_0', p_0'', \ldots,$ $(p_n, p_n', p_n'', \ldots)$ and q are the weights the Lowe price index number is written

$$I_{0n} = \frac{\sum p_n q}{\sum p_0 q}$$

where the summation takes place over commodities. The set of periods over which quantities are averaged to obtain the weight q is to some extent at choice. If q relates only to the base period the index number is that of **Laspeyres**; if it relates only to the given period it is that of **Paasche**; if it is the arithmetic mean of the quantity in the base and given period the index number is that of Marshall, Edgeworth and Bowley. [See also **crossed weight index number**, **Marshall–Edgeworth–Bowley index**.]

lower control limit See **control chart**.

lower quartile See **quartile**.

lowess A locally linear scatterplot smoother designed to be highly resistant to outliers, extensions of which have been developed for higher-dimensional situations.

Lugananni–Rice formula An approximate formula for tail probabilities, used in higher-order likelihood asymptotics (Barndorff-Nielsen and Cox, 1989).

lumped variance test A one-sided test of a rectangular hypothesis proposed by Broadbent (1955) in connection with the analysis of multi-modal data involving a **quantum hypothesis**.

M

M-estimate Any estimate T of a parameter θ given by minimizing an expression $\sum \rho(x, T)$, the sum being taken over the sample values of x. This is a generalization of the **maximum likelihood estimator**, and the name is used particularly in connection with robust estimation, many proposed robust estimators being of this type.

M-estimator See **Huber estimator**.

m-rankings, problem of Given m rankings of, say, n objects the problem arises of finding some measure of the general agreement between the rankings and of testing its significance. One such measure is termed the **coefficient of concordance**.

m-statistic See **Mood's W-test**.

Macaulay's formula A moving average formula proposed by Macaulay (1931) to carry out long-period smoothing. It has 43 terms and reduces the variance of a random series to a degree equivalent to a nine-term moving average with equal weights.

Madow–Leipnik distribution An approximation by Madow (1945) and Leipnik (1947) to the distribution of the first serial correlation in samples of n from a Markov normal process with known mean.

magic square design A square array $(n \times n)$ with n^2 integers placed in the cells in such a way that the row, column and two principal diagonal sums are the same, namely $\frac{1}{2}n(n^2 + 1)$. These squares may be used for balancing out linear trend from main effects and lower-order interactions in some factorial designs and in some Latin and Graeco-Latin square designs.

Mahalanobis, Prasanta Chandra (1893–1972) A father figure in Indian statistics, Mahalanobis is well known for his D^2 statistic and pioneering contributions to large-scale sample surveys. He is also remembered for his historic role in modernizing the Indian statistical system, applications including dams and 5-year economic plans and contributions to statistical systems of other countries as chair of the UN Subcommission on Statistics.

Mahalanobis distance A particular development in the general topic of discrimination, in this case concerned with the relationship or 'distance' between two populations. The work of Mahalanobis (1930) resulted in the **D^2 statistic** for which the population parameter (Δ^2) is referred to as the generalized squared distance. It bears a simple relationship to the T^2 statistic developed by Hotelling (1931) and the discriminant function developed by Fisher (1936c) in connection with problems of classification.

Mahalanobis's generalized distance See D^2 statistic.

main effect An estimate of the effect of an experimental variable or treatment averaged over other treatments which may form part of the experiment. Thus, in a balanced experiment involving three factors, A, B, C, each at two levels applied and not applied, the main effect due to A would be the average of the four effects where A was applied, and B and C were applied at each of the two levels, minus the average of the four effects, where A was not applied. One of the reasons for introducing orthogonality into an experimental design is to enable main effects to be separately and efficiently estimated.

Makeham distribution See **Gompertz–Makeham distribution**.

Mallows' C_p statistic If p regressors are selected from a set of k, C_p is defined as

$$\left(\sum (y - y_p)^2 \big/ s^2 \right) - n + 2p,$$

where y_p is the predicted value of y from the p regressors and s^2 is the residual mean square after regression on the complete set of k. The model is then chosen to give a minimum value of the criterion, or a value that is acceptably small. It is essentially a special case of **Akaike's information criterion**.

manifold classification If a population is divided into a number of mutually exclusive classes according to some given characteristic and then each class is divided by reference to some second, third, etc., characteristic, the final grouping is called a manifold classification. While some aspects of experimental design in the factorial form are akin to manifold classification, the term most often occurs with reference to two characteristics, where the manifold classification gives rise to a **contingency table**.

Mann–Kendall test See K-test.

Mann–Whitney test See **Wilcoxon–Mann–Whitney test**.

MANOVA See **multivariate analysis of variance**.

Mantel–Haenszel test Test of the null hypothesis that a set of independent 2×2 contingency tables are separately independent.

Marcinkiewicz's theorem This theorem (Laha and Rohatgi, 1979) proves that if a characteristic function ϕ has the form $\exp\{P(t)\}$ where P is a polynomial, then P is a quadratic polynomial and ϕ is the characteristic function of a normal distribution.

marginal category One of the frequency classes of a **marginal classification**.

marginal classification In a bivariate frequency table it is customary to show, as row and column totals, the univariate frequencies of the two variables separately. This is sometimes called marginal classification. Similarly, for a multivariate frequency array the arrays of one lower dimension formed by summing one of the variables are occasionally said to be marginal in relation to the original array. The frequencies are said to be marginal.

marginal distribution The unconditional distribution of single variables, or combinations of variables, in a multivariate distribution.

marked point process A point process to the points of which are attached random variables, used to model earthquake sizes, insurance losses, etc.

Markov, Andrei Andreevich (1856–1922) Markov, with Liapunov a disciple of Tchebychev, gave rigorous proofs of the central limit theorem. Through his work on Markov chains (which first appeared in his writings in 1906, and was published in the *Izvestiya (Bulletin)* of the Physico-Mathematical Society of Kazan University), the concept of Markovian dependence pervades modern theory and application of random processes. His textbook influenced the development of probability and statistics internationally (Markov, 1913).

Markov chain A stochastic process, with the Markov property, having countable discrete states. Some authors further restrict the term to processes in discrete time.

Markov chain Monte Carlo (MCMC) A simulation-based method of finding probability distributions in which a Markov chain is set up having the target distribution as its stationary distribution. Repeated simulation of the Markov chain then enables the target to be estimated. It was invented in a physical context by Metropolis and Ulam (1949) and refined by Hastings (1970), after both of whom one main version is named. Developments in computing technique have made it an extremely popular technique for the fitting of complex models. The method is widely used for finding Bayesian posterior distributions but is not restricted to that.

Markov estimate An estimate of a parameter derived from an estimator given by the so-called Markov or **Gauss–Markov theorem**.

Markov field See **Markov random field**.

Markov inequality If a variable X is non-negative and has mean equal to a then for $t > 0$ the Markov inequality states that

$$\Pr\{X \geqslant t\} < \frac{a}{t}.$$

Some writers credit this inequality to Tchebychev. [See **Bienaymé–Tchebychev inequality**.]

Markov process A stochastic process such that the conditional probability distribution for the state at any future instant, given the present state, is unaffected by any additional knowledge of the past history of the system.

Markov random field An extension of the Markov condition to stochastic processes in two or more dimensions. Each observation x is associated with a group of neighbours, and the probability distribution of x conditional on all other observations is the same as that conditional on its neighbours. The conditions under which such a scheme is possible are given by the **Hammersley–Clifford theorem**.

Markov renewal process A type of stochastic process closely related to the **semi-Markov process** and first proposed by Pyke (1961). This process $\{N(t); t \geqslant 0\}$ is determined by $(m, \boldsymbol{A}, \mathbf{Q})$ where $\boldsymbol{A}$ is a vector of initial probabilities, $\mathbf{Q}$ a matrix of transition distributions and $N(t)$ a series of counting functions. Where $m = 1$ we have the ordinary renewal process. The intervals between successive transitions are typically not exponentially distributed.

Marshall–Edgeworth–Bowley index An index number formula, proposed by Marshall (1887), Edgeworth (1925) and Bowley (1928) as an alternative to the standard formulae of **Laspeyres** and **Paasche**. It may be written, in terms of prices and quantities, as

$$I_{0n} = \frac{\sum p_n (q_0 + q_n)}{\sum p_0 (q_0 + q_n)}$$

where the suffix 0 refers to the base period and n to the period to which the index relates. The index represents a compromise with no general bias. However, it suffers from the disadvantage of lack of comparability between different years owing to the shifting pattern of weights.

Marshall–Olkin distribution A multivariate exponential distribution introduced by Marshall and Olkin (1967a). The distribution has survivor function

$$\exp\left\{ -\sum_{i=1}^{n} \lambda_i x_i - \sum_{i<j} \lambda_{ij} \max\,(x_i, x_j) - \sum_{i<j<k} \lambda_{ijk} \max\,(x_i, x_j, x_k) - \ldots \right.$$
$$\left. - \lambda_{12\ldots n} \max\,(x_1, x_2, \ldots, x_n) \right\}.$$

martingale Originally, a process known to gamblers under which the loser at a fair game doubled the stakes for the next, and so on at each loss, the paradox being that in the long run the gambler appeared certain to win sooner or later and at that point would have a net gain. More recently the term has been given a precise significance in the theory of stochastic processes. A stochastic process $\{X_t\}$ is called a martingale if $E\,\{|X_t|\}$ is finite for all t, and

$$E\left\{ X_{t_{n+1}} | X_{t_1}, \ldots, X_{t_n} \right\} = X_{t_n}$$

with probability unity for all $n \geqslant 1$ and $t_1 < \ldots < t_{n+1}$. If the equality sign is replaced by $\geqslant$ the process becomes sub-martingale; if it is replaced by $\leqslant$ the process becomes super-martingale, and if replaced by an inequality the process becomes semi-martingale.

martingale residuals The natural residuals in survival data analysis where we are dealing with right-censored (and possibly left-truncated) observations. For instance, with right-censored data we observe for subject i, $\delta_i, \tilde{T}_i$, where δ_i is a censoring indicator taking values 0 or 1 according to the fact that the observation is censored or not; $\tilde{T}_i$ represents the follow-up time. Then the martingale residual is $\delta_i - A(\tilde{T}_i)$, where A is the estimated cumulative **hazard function**.

masking effect (i) The effect of model fitting in making the detection of **outliers** more difficult. (ii) In **kriging**, the effect of intermediate observations in reducing the correlation between estimates.

master sample A sample drawn from a population for use on a number of future occasions, so as to avoid ad hoc sampling on each occasion. Sometimes the master sample is large and subsequent enquiries are based on a subsample from it.

matched samples A pair, or set of, matched samples are those in which each member of a sample is matched with a corresponding member in every other sample by reference to qualities other than the outcome immediately under investigation. The object of matching is to obtain better estimates of differences by 'removing' the possible effects of other variables. For example, if it is desired to investigate acuity of vision for a sample of smokers as compared with a sample of non-smokers, better comparisons can usually be made if, to every member of one sample, there can be associated a member of the other sample of the same sex and about the same age. Difficulties arise in the assessment of significance, however, if the members of a second sample have to be chosen purposively in order to match the first, instead of being chosen at random.

matching If two sequences of a finite number of characteristics $A_1, A_2, \ldots, A_k$ are compared, the jth of one sequence against the jth of the other, a comparison in which both members exhibit the same characteristic is called a match. The number of matches in an observed pair of sequences provides a test of various hypotheses concerning the system which generated them. More generally, p sequences instead of two may be considered and the occurrence of the same characteristic in the jth member of each sequence is also called a match or a multiple match. A distinct interpretation is used in communication theory, where there occurs the problem of matching the message source to the communication channel, in order to secure the maximum efficiency in transmitting messages.

matching coefficient A variant of the **similarity index** proposed by Sokal and Michener (1958) and developed by Goodall (1966).

matching distribution Suppose two sets having n objects each are numbered $1, 2, \ldots, n$ and each is arranged in a random order so as to form n pairs. Then the number of pairs on which the two numbers are the same has the matching distribution. More general distributions are studied, in which there are several qualitative characteristics, e.g., colour or suit instead of numbering in the comparison of two packs of cards or more than two sets are available.

matrix sampling The process of taking a bi-sample from an $R \times C$ population matrix whose elements are $\|x_{IJ}\|$ consists of two samples, r of rows and c of columns, and forming a matrix whose elements are those at the intersections of the selected rows and columns.

maverick A term encountered in the literature of industrial statistics to denote an observation lying so far outside the usual range that it is suspected of not belonging to the population under enquiry.

max-type procedures Class of test procedures for detection of a change(s) related to likelihood ratio approach. Typically, the test statistic is the maximum over weighted test statistics for the related two sample problem.

maximin criterion Any criterion based on maximizing the minimum of a set of statistics. [See also **minimax principle**.]

maximum entropy method A method of estimating spectral window width due to Burg (1975). The entropy, the total integral of the log spectral density, is maximized subject to constraints determined by specified lag covariances.

maximum entropy principle An idea of wide use in information theory and statistics. Subject to constraints, one uses the distribution with maximum entropy. Thus in Bayesian statistics a prior distribution selected in this way may be called least informative.

maximum F-ratio In testing the homogeneity of a set of variances, the ratio of the largest to the smallest, proposed by Hartley (1950) as a simple test alternative to **Bartlett's**.

maximum likelihood estimator (MLE) See **maximum likelihood method**.

maximum likelihood method A method of estimating a parameter or parameters of a population by that value (or values) which maximizes (or maximize) the **likelihood** of a sample. For instance, if the likelihood is $L(x_1, \ldots, x_n, \theta)$ the parameter θ is estimated as the function of the x's, $\hat{\theta}$, for which, under certain regularity conditions,

$$\left(\frac{\partial L}{\partial \theta} \right)_{\theta = \hat{\theta}} = 0, \qquad \left(\frac{\partial^2 L}{\partial \theta^2} \right)_{\theta = \hat{\theta}} < 0.$$

maximum probability estimator A method of estimation proposed by Weiss and Wolfowitz (1967), for which the **generalized maximum likelihood estimator** is a special case, designed to answer problems posed by ordinary maximum likelihood estimation and not covered by the generalized case.

Maxwell distribution The chi-squared distribution with three degrees of freedom, based presumably on Clerk Maxwell's discussion of the energy of particles moving at random in three dimensions.

Maxwell–Boltzmann statistics See **Bose–Einstein statistics**.

McDonald–Kreitman test A test of polymorphism and divergence of two populations of genes based on the differences in populations between synonymous and non-synonymous codons.

McNemar's test A test for binary responses (e.g., death or survival) in paired comparisons. Only those pairs with different responses contribute relevant information (Cox, 1958c).

mean When unqualified, the mean usually refers to the expectation of a variable, or to the arithmetic mean of a sample used as an estimate of the expectation. [See also **geometric mean, harmonic mean**.]

mean absolute deviation See **mean deviation**.

mean absolute error An alternative but much less desirable name for the **mean deviation**.

mean deviation A measure of dispersion derived from the average deviation of observations from some central value, such deviations being taken absolutely, i.e., without reference to algebraic sign. The central value may be the arithmetic mean or the median. Expressed formally the mean deviation is the first **absolute moment**.

mean difference A measure of dispersion due to Gini (1912) and based upon the average of the absolute differences of all possible pairs of variable values. For a continuous variable X with density function $f(x)$ it may be written

$$\Delta_R = \int_{-\infty}^{\infty} \int_{-\infty}^{\infty} |x - y| f(x) f(y) dx dy.$$

For a discontinuous variable X with frequency function $f(x)$

$$\Delta_R = \sum_{j=1}^{N} \sum_{k=1}^{N} |x_j - x_k| f(x_j) f(x_k).$$

If the distribution of X is uniform and a variable value is not regarded as occurring with itself, the preceding reduces to

$$\Delta_R = \frac{1}{N(N-1)} \sum_{j=1}^{N} \sum_{k=1}^{N} |x_j - x_k|.$$

mean direction In directional data, the resultant of a set of unit vectors. It is a sufficient statistic for the preferred direction in the **von Mises distribution** and in Fisher's spherical distribution.

mean excess function The expected value of a random variable, conditional on its exceeding some threshold, $E(X - u \mid X > u)$, empirical versions of which are useful in tail estimation (Coles, 2001).

mean likelihood estimator An estimator proposed by Barnard (1959) which minimizes the mean square error when the prior distribution is uniform.

mean linear successive difference The same as **mean successive difference**.

mean range The arithmetic mean of the ranges of a set of samples of the same size. A multiple of the mean range may be used as an estimator for the population standard deviation.

mean residual life The mean residual life at time t is the expected remaining lifetime or survival time, i.e., $E(T - t \mid T > t)$. [See also **biometric functions**.]

mean semi-squared difference The statistic used in the study of **serial variation**.

mean square In general, the mean square of a set of values is the arithmetic mean of the squares of their differences from some given value, namely their second moment about that value. More generally the mean square may be from a value fitted under some model. When the mean square is regarded as an estimator of certain parental variance components the sum of squares about the observed mean is usually divided by the number of degrees of freedom, not the number of observations.

mean square consecutive fluctuation estimator A generalized mean square successive difference proposed by Ruben (1963) as an estimator for the interaction or migration parameter in an emigration/immigration process.

mean square contingency See **contingency**.

mean square deviation The second moment of a set of observations about some arbitrary origin. If that origin is the mean of the observations the mean square deviation is the equivalent of the **variance**. An equivalent expression, especially when the observations are variable values, is mean square error. This latter term also occurs in older writings in the sense of variance but should not be employed in that sense.

mean square error The mean square deviation of an estimator from the true value, equal to the variance plus the squared bias.

mean square successive difference An estimate of the population variance may be based upon the first difference of a series of independent observations, $x_1, x_2, \ldots, x_n$, by the formula

$$\delta^2 = \frac{1}{n-1} \sum_{i=1}^{n-1} \{x_{i+1} - x_i\}^2.$$

δ^2 is called the mean square successive difference. Its mean value in random samples from a normally distributed population is $2\sigma^2$, whence $\frac{1}{2}\delta^2$ affords an unbiased estimator of σ^2 in the absence of serial correlation in the x's. In this connection it is related to **von Neumann's ratio**: alternative forms using second-order differences developed by Kamat and Sathe (1962). [See also **modified mean square successive difference**.]

mean successive difference In a time series, the arithmetic mean of the difference of successive values, i.e., for a series $x_1, x_2, \ldots, x_n$, is

$$d = \frac{1}{n-1} \sum_{i=1}^{n-1} |x_i - x_{i+1}|.$$

The intractability of absolute quantities leads to a preferred use of the mean square of differences.

mean values A general class of functions of distributions of which **moments** constitute a special case. If a variable has a density function $f(x)$ and t is some function defined within the range of the distribution, the mean value, or mathematical expectation, is defined as

$$E[t(X)] = \int_{-\infty}^{\infty} t(x) f(x) dx,$$

subject, of course, to existence. The definition may readily be generalized to n-dimensional variation. The various moments of a distribution are derived by substituting the appropriate function for $t(x)$. [See **expected value**.]

measure of location A quantity which purports to locate a distribution, or a set of sample values derived therefrom, by means of a value which is, in some sense, central or typical, e.g., the arithmetic mean, the median or the mode.

medial test A graphical test of association between two variables. The **scatterplot** for the pairs of observations is divided into four quadrants by lines, parallel to abscissa and ordinate, passing through the medians of the variables. Association is judged by the number of points falling into the positive quadrant as compared with the number expected, namely one-quarter of the observations, on the hypothesis of no association. If the total number of points is n and the total number in the positive quadrant and its opposite quadrant is d, the coefficient $2d/n - 1$ has been termed (Quenouille, 1952) the medial correlation coefficient. A test by Olmsted and Tukey (1947), based on the outlying members in each quadrant, is known as the corner test.

median The value of the variable which divides the total frequency into two halves. As a partition value the median may be defined for a continuous frequency distribution by the equation

$$\int_{-\infty}^{M} f(x) dx = \int_{M}^{\infty} f(x) dx = \frac{1}{2},$$

M being the median value. For a discontinuous variable ambiguity may arise which can only be removed by some convention. For a total frequency of $2N + 1$ items the median is the variable value of the $(N + 1)$th item: for $2N$ items it is customary to take the average of the Nth and $(N + 1)$th item.

median absolute deviation (MAD) The absolute deviation of a variable from its median, empirical versions of which are used for resistant estimation of scale. See also **mean deviation**.

median direction A measure of location for angular data. Consider a sample of angular observations as a sample of points on the unit circle. Then any one point P of these points which has the properties that (i) half of the sample points are on each side of the diameter PQ through P, and (ii) the majority of the sample points are nearer to P than Q, is called a median of the sample. The vector OP is called a median direction of the sample.

median effective dose A term proposed by Trevan (1927) to characterize the potency of a stimulus by reference to the amount which produces a response in 50 per cent of the cases where it is applied. The median effective dose is sometimes written ED_{50}; so, by a natural extension, many other effective doses for different quantile values may be styled, e.g., ED_{75} or ED_{90}.

median F-statistic If a sample of $2m + 1$ items is ranked and transformed to a **uniform distribution** the normalized random variable

$$M_{2m+1} = 2\sqrt{2m + 3} \left\{ F(x'_{m+1}) - \frac{1}{2} \right\}$$

converges in distribution to the standardized normal variable. The statistic $F(x'_{m+1})$ proposed by Birnbaum and Tang (1964) is the sample median and M_{2m+1} is distribution free with respect to the distribution of the original observations.

median lethal dose A particular name for the **median effective dose** when the response is death. It is often written LD_{50}.

median regression curve The type of regression line or curve derived from the **Brown–Mood procedure**.

median survival time The 50th percentile of the distribution of the survival time (or lifetime) T, i.e., $S(t_{0.5}) = 0.5$, where S is the **survivor function**.

median test A rank order test proposed by Westenberg (1948) and Mood (1950) which rejects the hypothesis of identity of two populations, in the one-sided case, when there are too few observations from one sample larger or smaller than the median of the combined sample.

median unbiased confidence interval A $(1 - \alpha)$ confidence interval bounded by two confidence limits θ' and θ'' at $(1 - \alpha)/2$ (Birnbaum, 1961b).

median unbiasedness A concept, one of a group proposed by Brown (1947) and subsequently developed, for example, by van der Vaart (1961), whereby the expected value of an estimator being equal to the parameter is replaced by the median value as a demonstration of unbiasedness.

Merrington–Pearson approximation An approximation to the non-central t-distribution proposed by Merrington and Pearson (1958) based upon the **type IV distribution**.

mesokurtosis See **kurtosis**.

meta-analysis See **overviews**.

metameter A transformed value of a dose or a response, e.g., logarithm or probit, obtained by using a transformation equation that is independent of all parameters. It is adopted mainly to simplify the analysis or the expression of the dose–response relationship (Finney, 1949).

method of overlapping maps A device for the selection of **linked samples**. Thus to reduce travel costs in a multi-purpose survey where it is desired to select a sample of villages with probability proportional to (i) population, for population enquiry, and (ii) to area, for direct physical observation of fields for land utilization enquiry, the villages may be ordered in a serpentine manner, and represented twice on a straight line of fixed length which is completely covered twice, once by segments proportional to population and a second time to area. A point thrown at random on this overlapping map will select a linked pair of villages, each with the desired probability, and the two are likely to be in the same neighbourhood, if not identical.

Metropolis–Hastings algorithm One of the earliest rejection sampling algorithms widely used in **Markov chain Monte Carlo** simulation.

mid-mean The arithmetic mean of the sample observations remaining after trimming the smallest and the largest 25 per cent of the original sample. If the original sample size is not a multiple of 4, the two extreme remaining observations after trimming are appropriately weighted.

mid-range For a set of values $x_1, x_2, \ldots, x_n$ arranged in order of magnitude the mid-range is defined as $\frac{1}{2}(x_n + x_1)$. It is synonymous with centre, but 'mid-range' usually refers to a sample and 'centre' to the parent distribution.

mid-rank method See **tied ranks**.

Miller distribution A name sometimes given to a special case of a distribution obtained from the **Poisson–Pascal distribution**.

Miller's jackknife test An asymptotically distribution-free test, based on the **jackknife**, of the equality of the scale parameters of two distributions known to be identical in all respects except possibly location and scale.

Mills' ratio The ratio of the area of the 'tail' of a distribution to the bounding ordinate. For a normal deviate x, the function

$$e^{x^2/2} \int_x^\infty e^{-u^2/2} du.$$

The ratio occurs naturally in the computation of values of the normal integral and was considered by Laplace, who gave a continued fraction for it. It was tabulated by Mills (1926).

minimal essential completeness See **completeness (of a class of decision functions)**.

minimal sufficient statistics A vector of statistics is minimal sufficient if it has a minimal number of components, in which case it is a function of all other sufficient vectors for the parameters in question.

minimally connected design A connected $a \times b$ design with $a + b - 1$ observations is said to be minimally connected. The number of minimally connected $a \times b$ designs is $a^{b-1} b^{a-1}$ (Birkes and Dodge, 1986).

minimax estimation The estimation of parameters by the application of the **minimax principle** to a **risk function**. In particular it may be shown that a Bayes' estimator which has a constant risk function is also a minimax estimator.

minimax principle A principle introduced into decision function theory by Wald (1950). It supposes that decisions are taken subject to the condition that the maximum risk in taking a wrong decision is minimized. In the theory of games, from which the concept is taken, it is not open to the same objection, prudent players being entitled to assume that their adversaries will do their best.

minimax robust estimator A term introduced by Huber (1964) to describe that estimator (perhaps restricted to be of a particular type) whose maximum variance is the smallest over the family of possible underlying probability distributions.

minimax strategy If a strategy is selected from a group of **admissible strategies** as being the one which, on a basis of the expected loss, has the smallest maximum loss this strategy will be a minimax strategy.

minimum χ^2 A method of estimation based upon the χ^2 **goodness-of-fit** statistic. The method determines values of the parameters so as to minimize χ^2 calculated from observed frequencies and 'expected' frequencies expressed in terms of the parameters. The method is troublesome to apply in general because of the difficulty of expressing the observed frequencies explicitly in terms of the parameters under estimate. A modified minimum χ^2 method (Jeffreys, 1938) simplifies the method to some extent by minimizing the statistic $\chi^2 = \sum (\lambda_j - l_j)^2 / l_j$ where λ_j is the theoretical and l_j the observed frequency in the jth group. For large samples the estimators from the two methods are asymptotically equivalent and they also tend to the values of **maximum likelihood estimators**.

minimum completeness See **completeness (of a class of decision functions)**.

minimum discrimination information statistic This statistic is based upon the principle of **information** in a sample and can be considered as the 'divergence' of the alternative hypothesis from the null hypothesis. The distribution of this statistic involves central and non-central χ^2 and it has additive properties. For a wide class of problems concerning contingency tables, this statistic takes the form of $-2 \log \lambda$, where λ is a likelihood ratio.

minimum logit χ^2 A method of estimation in bioassay proposed by Berkson (1944) and subsequently developed by him and other authors. The **logistic distribution** is used as the **tolerance distribution** and the principle of minimum χ^2 rather than maximum likelihood. It has also been shown by Taylor (1953) that such estimators are **regular best asymptotically normal**. It is best regarded as an example of weighted least squares estimation with empirically estimated weights and applied to transformed data.

minimum normit chi-squared estimator A method of estimating the cumulative normal distribution proposed by Berkson (1955b) which involves minimizing the following quantity, distributed asymptotically as χ^2:

$$\chi^2(\text{normit}) = \sum n_i \frac{z_i^2}{p_i q_i} (\nu_i - \hat{\nu}_i)^2$$

where n_i is the number exposed to stimulus x_i, $p_i = 1 - q_i$ is the proportion affected, and ν_i and $\hat{\nu}_i$ are the observed and estimated **normit** with z_i the normal curve ordinate at the point where the area divides p_i and q_i. [See also **minimum** χ^2.]

minimum spanning tree A tree, in the sense of graph theory, including all points and of minimum length. Statistical applications are in cluster analysis; the shortest dendrogram, when cluster distances are nearest-neighbour distances, is a minimum spanning tree.

minimum variance As applied to estimators, this term denotes the property of possessing the least variance among the members of a defined class. A minimum variance estimator exists only where there exists a sufficient estimator. [See also **Cramér–Rao inequality**.]

minimum variance linear unbiased estimator In the case of the **linear hypothesis** the **least squares method** provides estimators that are unbiased, linear in the observations, and with minimum variance. [See also **Gauss–Markov theorem**.]

Minitab A statistical and graphics package accessible to both statistical and non-statistical users. It is particularly valuable as an undergraduate teaching tool and serves as a precursor to the use of more advanced systems such as **SAS**. Minitab provides a wide range of data exploration and general statistical methods, and its statistical quality control capabilities are popular in industry. Product details and availability can be found at http://www.minitab.com.

Minkowski's inequality A result due to Minkowski which states that for any random variables X and Y

$$(E|X + Y|^r)^{1/r} \leqslant (E|X|^r)^{1/r} + (E|Y|^r)^{1/r},$$

provided $1 \leqslant r < \infty$.

MINQUE MInimum Norm Quadratic Unbiased Estimation, introduced by Rao (1971), for estimating the variances σ_i^2 for the Gauss–Markov linear model

$$\boldsymbol{Y} = \mathbf{X}\boldsymbol{\beta} + \boldsymbol{e}$$

where $\boldsymbol{Y}$ is a vector of n observations, $\mathbf{X}$ is an $n \times m$ known matrix, $\boldsymbol{\beta}$ is a vector of m unknown parameters and $\boldsymbol{e}$ is an error vector with a dispersion matrix of diagonal form.

misclassification When a subject is falsely classified into a category in which the subject does not belong. It may result from misreporting by study subjects, from the use of less than optimal measurement devices, or from random error. In particular, the proportion of observations misclassified in discriminant analysis.

missing at random (MAR) Used to describe situations where response and explanatory variables are recorded but the response may be missing with a probability independent of its unobserved value. It is missing completely at random if the probability is also independent of the explanatory variable and observed at random if the probability is independent of the explanatory variable but not the response. Non-ignorable non-response occurs when the probability depends on the unseen response value (Rubin, 1976).

missing completely at random (MCAR) See **missing at random**.

missing data It often happens in studies that observations which were planned are missing (Dodge, 1985). Missing data in a survey may occur when there are no data whatsoever for a respondent (non-response) or when some variables for a respondent are unknown (item non-response) because of refusal to provide or failure to collect the response. The concept is particularly relevant in longitudinal studies where a measurement was planned at a particular time and has not been done. It is essential then to make assumptions about the mechanism leading to missing data, for instance the **missing at random** assumption. Missing data may also occur in regression models at the level of explanatory variables.

missing plot technique The name given by Allan and Wishart (1930) and Yates (1933) to the process of analysing material which was designed to conform to an experimental pattern but from which certain values are missing through circumstances beyond the control of the experimenter. The use of the word 'plot' arose from the agricultural background of the original investigations noted above but methods are applicable to missing values generally. [See also **R-process, connectedness, EM algorithm**.]

Mitscherlich equation Amounts of an element in the soil can sometimes be estimated using Mitscherlich's equation

$$Y = A(1 - bR^x)$$

where Y represents yield and x the amount of fertilizer applied.

mixed autoregressive–moving average (ARMA) process A stochastic process $\{u_i\}$ in discrete time that is driven, by independent and identically distributed random disturbances ε_t with expectation zero, in accordance with the model

$$(1 - \phi_1 B - \phi_2 B^2 - \ldots - \phi_p B^p)(u_t - \mu) = (1 - \theta_1 B - \theta_2 B^2 - \ldots - \theta_q B^q)\varepsilon_t,$$

i.e., $\phi(B)\bar{u}_t = \theta(B)\varepsilon_t$ where B is the unit time backward shift operator. The non-negative integers (p, q) are jointly referred to as the order of the ARMA process. The ARMA model (with the polynomial equation $\phi(B) = 0$ having all its roots outside the unit circle) is a useful method for parsimoniously representing stationary time series. Special cases of the ARMA process are the **autoregressive process** of order p obtained by setting $q = 0$, and the **moving average** process of order q obtained by setting $p = 0$.

mixed autoregressive–regressive system A system of equations which could represent a particular econometric model might be an autoregressive set of Y's regressed upon fixed X's:

$$\sum_{j=0}^{k} \alpha_j Y_{t-j} = \sum_{l=1}^{q} \beta_{lt} X_t + \varepsilon_t.$$

This may be rewritten in the form

$$Y_t = \sum_{l=1}^{q} \beta_{lt} X_t - \sum_{j=1}^{k} \alpha_j Y_{t-j} + \varepsilon_t,$$

giving the first term on the right-hand side as an ordinary regressive component and the second term a component in autoregressive form.

mixed distribution A term for the distribution which results from the parameter of a given distribution being itself a random variable. [See also **compound distribution, generalized distribution**.]

mixed exponential response law There are a number of situations where the distribution of the response exhibited by members of a population is negative exponential but the presence of definite strata creates a mixture of such distributions. The form of a 'mixed' law depends critically on the weight function used to combine the separate distributions.

mixed factorial experiments An experiment in factorial form where the number of levels for the factors varies from one factor to another. For example, an experiment involving one factor at two levels, one at three levels and one at four levels would be a three-factor experiment of the 'mixed' type.

mixed model This term is used in at least four slightly different senses: a model is termed 'mixed' if (i) its equations contain both determinate and stochastic elements; (ii) its equations contain both difference and differential terms; (iii) it contains **endogenous** as well as **exogenous** elements; or (iv) in an analysis of variance context for a two-way layout, the rows correspond to a **model I** factor, i.e., fixed effects, and the columns correspond to a **model II** factor, i.e., random effects. This version is also sometimes known as model III (Plackett, 1960).

mixed sampling Where a sampling plan envisages the use of two or more basic methods of sampling it is termed mixed sampling. For example, in a multi-stage sample, if the sampling units at one stage are drawn at random and those at another by a systematic method, the whole process is 'mixed'. Usage is not uniform, but where samples at one stage were drawn at random with replacement and at another stage were drawn at random without replacement, it would seem better not to describe the whole process as 'mixed', the essential basic method of random selection being employed throughout.

mixed spectrum A **spectral density function** of a (general stationary) stochastic process which contains both discrete and continuous components.

mixed strategy See **strategy**.

mixed-up observations A phrase which has been used to describe a situation where the identity of some observations may be lost but their total value is known. It is too imprecise for general use in view of the many ways in which observations can be mixed up.

mixture distribution A distribution specified as a mixture, in fixed proportions, of distributions with different parameters or functional forms. The term is often reserved for mixtures with a finite number of components (Johnson, Kotz and Balakrishnan, 1995). [See also **compound distributions**.]

mode The mode was originally conceived of as that value of the variable which is possessed by the greatest number of members of the population. Although the idea of the most frequently encountered or fashionable value of the variable is probably very old, it was not generally used in statistics until popularized by Pearson (1894a). The concept is essentially of use only for continuous distributions, although it can be extended to the discontinuous case. More formally, if $f(x)$ is a frequency function with continuous derivatives, a mode is a value of

x for which

$$\frac{df(x)}{dx} = 0, \quad \frac{d^2f(x)}{dx^2} < 0.$$

There may thus be more than one mode of a distribution, though the practical occurrence of multi-modality is comparatively rare.

model A formalized expression of a theory or the causal situation which is regarded as having generated observed data. In statistical analysis the model is generally expressed in symbols, that is to say in a mathematical form, but diagrammatic models are also found. The word has recently become very popular and possibly somewhat overworked.

model I (or first kind) A term introduced by Eisenhart (1947) to denote analysis of variance based upon the least squares analysis of the general linear model. Essentially it is an analysis of mean values and is frequently termed the fixed effects or constants model. [See also **variance component**.]

model II (or second kind) A term in the general scheme proposed by Eisenhart (1947) to denote analysis of variance based upon a vector of random variables instead of a vector of parameters (means). It is also termed the components of variance model. [See also **variance component**.]

model averaging The process of explicitly fitting a number of different models, usually from within some broad family, to the same data and averaging the answers with appropriate weights, often calculated as the posterior probabilities of the models under some Bayesian formulation. The interpretation is reasonably clear when the objective is the prediction of a future observation, but when the objective is estimation of an unknown parameter it may be difficult to ensure that the interpretation of the parameter is the same for all models.

modified binomial distribution Consider a sequence of n trials with constant probability p of success in an individual trial with the restriction that as soon as a success is experienced, the subsequent $m - 1$ trials result in failures. The number of successes in the above sequence has a modified form of the binomial distribution. A further modification assumes that the probability of a success in the first trial was p and regards the first observation in the experiment as a random start in an infinite sequence of similar trials.

modified control limits In the case of statistical quality control using the Shewhart **control chart**, the use of reject limits as the control limits has been termed 'modified control limits'.

modified exponential curve See **growth curve**.

modified Latin square A generalization of the **semi-Latin square** described by Rojar and White (1957).

modified mean This term is found in two different senses. In the first, it refers to the mean of the highest and lowest values of a set of values, or what is more generally known as the **mid-range**. In the second, it refers to the mean of a set of observations from which certain values have been rejected as atypical.

modified mean square successive difference For a series of $2m$ independent observations this quantity is

$$\delta_0^2 = \frac{1}{4(m-1)} \sum_{i=1}^{2m-1} (x_{i+1} - x_i)^2, \qquad i \neq m.$$

modified normal distributions A class of distributions, introduced by Romanowski (1964), obtained by ascribing a truncated (from above) **Pareto distribution** to the variance of a **normal distribution** with zero mean. The cumulative distribution function is

$$\Pr[X < x] = \frac{a+1}{\sigma\sqrt{2\pi}} \int_{-\infty}^{x} \int_0^1 t^{a-1/2} e^{-y^2/2t^2\sigma^2} \, dt dy.$$

Romanowski distinguishes a number of special cases, depending on the value of a in the above expression:

$$
\begin{array}{ll}
a = 0 & \text{equi-normal} \\
a = \tfrac{1}{2} & \text{radico-normal} \\
a = 1 & \text{lineo-normal} \\
a = 2 & \text{quadri-normal} \\
a = \infty & \text{standard normal distribution.}
\end{array}
$$

modified profile likelihood A modification to the profile likelihood function based on higher-order asymptotic theory in which inference based on the modified function has improved small-sample performance.

modified von Neumann ratio A minor modification of **von Neumann's ratio** due to Geisser (1957) which is twice the ratio of the **modified mean square successive difference** to the pooled variance of observations 1 to m and $m + 1$ to $2m$.

moment In general, the **mean value** of a power of a variable. For a univariate value x with density function $f(x)$ the rth moment of the variable $g(x)$ is

$$\int_{-\infty}^{\infty} (g(x))^r f(x) dx.$$

More generally, for a multivariate distribution $f(x_1, x_2, \ldots, x_p)$, the moment of order $(r_1, r_2, \ldots, r_k)$ of the functions $g_1, g_2, \ldots, g_k$ is the expectation

$$\int_{-\infty}^{\infty} \cdots \int_{-\infty}^{\infty} g_1^{r_1} \ldots g_k^{r_k} f(x_1, \ldots, x_p) dx_1 \ldots dx_p.$$

In particular the rth moment of a variable X is given by

$$\mu_r = \int_{-\infty}^{\infty} x^r f(x) dx$$

and the rth moment about a particular fixed value a by

$$\int_{-\infty}^{\infty} (x - a)^r f(x) dx,$$

again with obvious generalizations to the multivariate case.

moment estimator Estimator(s) of the unknown parameter(s) of a distribution by fitting **moments**.

moment generating function A function of a variable t which, when expanded as a power series in t, yields the moments of a frequency distribution as coefficients of the powers. For example, the characteristic function is a moment generating function by virtue of the formal expansion

$$\phi(t) = \int_{-\infty}^{\infty} e^{itx} f(x) dx = \sum_{r=0}^{\infty} \frac{(it)^r}{r!} \mu_r'.$$

By an easy extension from the univariate case there may be derived moment generating functions for multivariate distributions, the characteristic function being one such.

moment matrix If p variables $X_1, X_2, \ldots, X_p$ have second-order moments typified by μ_{ij} as the product moment of X_i and X_j, the matrix whose ith row and jth column is μ_{ij} is called the moment matrix. It is, in fact, not a display of all the joint moments of the variables, but only those of the second order. If the moments are taken about the respective variable means the matrix becomes the covariance or dispersion matrix.

moment ratio A ratio in which the numerator and the denominator are moments or simple functions of moments. In certain cases the moment ratios may be interpreted as characteristics of the probability distribution (Pearson, 1895). For example, the most common of the moment ratios are those referring to the shape of the distribution known also as beta coefficients:

a measure of skewness: $\beta_1 = \mu_3^2 / \mu_2^3$
a measure of kurtosis: $\beta_2 = \mu_4 / \mu_2^2.$

More general ratios of this type, due to Pearson, are

$$\beta_{2n+1} = \frac{\mu_3 \mu_{2n+3}}{\mu_2^{n+3}}$$

$$\beta_{2n} = \frac{\mu_{2n+2}}{\mu_2^{n+2}}.$$

[See also **g-statistics**, **gamma coefficients**.]

moments, method of A method of curve fitting which proceeds by identifying the lower moments of the observed data with those of the particular curve form being fitted. This method has been generally associated with the fitting of frequency distributions of the Pearson type. Where sampling questions are involved the method is generally not the most efficient.

monotone likelihood ratio If there is a family of density functions $p_\theta(x)$ and if, for $\theta < \theta'$, the distributions P_θ and $P_{\theta'}$ are distinct, then if $p_{\theta'}(x)/p_\theta(x)$ is monotone increasing we have a family which is monotone likelihood ratio.

monotone regression A term introduced by Lombard and Brunk (1963) which embraces both the ideas of 'order preserving' as well as 'order reversing' in regression analysis. [See also **isotonic regression function**.]

monotonic structure An alternative term proposed by Barlow and Proschan (1965) for **coherent structure**.

Monte Carlo EM algorithm An **EM algorithm** in which the conditional expectation is computed by Monte Carlo simulation.

Monte Carlo method A term which has been used with several different meanings: (i) to denote the approximate solution of distributional problems by sampling experiments; this usage is not to be recommended; (ii) to denote the solution of mathematical problems arising in a stochastic context by sampling experiments; for example, the **Fokker–Planck equation** arises in several physical problems, but it also arises in a probability problem, and hence sampling can be used to obtain approximate solutions applicable to the physical case; (iii) by extension of (ii), the solution of any mathematical problem by sampling methods; the procedure is to construct an artificial stochastic model of the mathematical process and then to perform sampling experiments upon it.

Monte Carlo test A significance test of size r/n to determine whether a set of data can be regarded as a random sample from a given distribution or a realization of a specified stochastic process. A total of $n - 1$ simulations are carried out, and a suitable test statistic calculated from the data and from each simulation. If the observed value occurs among the extreme r values, it is judged significant (Barnard, 1963a).

monthly average By analogy with annual averages and **moving averages** generally this term ought to refer to the average of values of a time series occurring within a month, the resulting figure being representative of that particular month. In practice the phrase is sometimes used to denote the averaging of monthly values occurring in the same month, e.g., January from year to year, the object being to provide a pattern of seasonal fluctuation. This is objectionable and a better expression would be 'seasonal average by months'.

Mood–Brown estimation See **Brown–Mood procedure**.

Mood–Brown median test A distribution-free test of the difference between k-populations proposed by Brown and Mood (1951) based upon the overall median of the k samples. If A and B are the numbers of observations above and below the grand median $\tilde{M}$ then the Mood–Brown statistic is

$$\frac{1}{AB} \left\{ \sum_{i=1}^{k} (n_i^+ N' - n_i' A)^2 / n_i' \right\}$$

where n_i^+ are the sample observations above $\tilde{M}$ and $N' = A + B = \sum n_i'$.

Mood–Brown procedure See **Brown–Mood procedure**.

Mood's W-test A distribution-free procedure for dispersion proposed by Mood (1954) for the two-sample problem. If there are two samples m, n from distributions $F(x)$ and $G(y)$, the observations in the combined sample $m + n$ can be ranked and the statistic W formed as follows:

$$W = \sum_{i=1}^{n} \left(r_i - \frac{m + n + 1}{2} \right)^2$$

where r_i is the rank of the ith observation from the sample of n from $G(y)$. It should be noted that subsequent writers tend to refer to this as the M-test or statistic.

Moore–Penrose inverse If an $n \times m$ matrix $\mathbf{A}$ and an $m \times n$ matrix $\mathbf{X}$ satisfy (i) $\mathbf{AXA} = \mathbf{A}$, $\mathbf{X}$ is said to be a **generalized inverse** or g-inverse of $\mathbf{A}$. The further conditions (ii) $\mathbf{XAX} = \mathbf{X}$, ($iii$) $(\mathbf{AX})' = \mathbf{AX}$, ($iv$) $(\mathbf{XA})' = \mathbf{XA}$, define a unique matrix $\mathbf{X}$ known as the Moore–Penrose inverse.

moral graph A graph formed from a directed acyclic graph by placing edges between 'parent' nodes and replacing directed by undirected edges.

Moran's test statistic Proposed by Moran (1951), a test of the null hypothesis that data from an observed series of events arise from a **Poisson process** against the alternative hypothesis that they arise from a **renewal process**. [See also **Sherman's test statistic**.]

Morgenstern distributions The family of bivariate distributions (Morgenstern, 1956) that are constructed from given univariate distributions by means of the relation

$$H(x, y) = F(x)G(y) \left\{ 1 + \alpha(1 - F(x))(1 - G(y)) \right\}, \qquad -1 \leqslant \alpha \leqslant 1$$

between their distribution functions. When (X, Y) are continuous random variables their joint probability density function is given by

$$h(x, y) = f(x)g(y) \left\{ 1 + \alpha(2F(x) - 1)(2G(y) - 1) \right\}, \qquad -1 \leqslant \alpha \leqslant 1.$$

On putting $F(x) = x, G(y) = y$, the probability density of Morgenstern's uniform distribution is found to be

$$h(x,y) = 1 + \alpha(2x - 1)(2y - 1), \qquad 0 \leqslant x, y \leqslant 1, \qquad -1 \leqslant \alpha \leqslant 1.$$

morphometrics The numerical study of shape.

Mortara formula A method proposed by Mortara (1949) to enable age-specific fertility rates to be calculated from census results

$$f(y) = \frac{K_{y+1}}{L_{y+1}} - \frac{K_y}{L_y} = F(y+1) - F(y)$$

where L_y denotes the number of women ($y \pm 0.5$) years old at the census date and K_y the total number of live children born to these women at that date.

Moses' ranklike dispersion test A distribution-free rank test (Moses, 1963) of the equality of the scale parameters of two identically shaped populations, applicable when nothing is known about the population medians. [See **Ansari–Bradley dispersion test**.]

Moses' test A two-sample distribution-free test based upon the hypergeometric distribution proposed by Moses (1952) for similarity of proportions in two populations.

most efficient estimator An unbiased estimator whose sampling variance is not greater than that of any other unbiased estimator is called a most efficient estimator, the qualification 'unbiased' being understood. For biased estimators the same expression is sometimes used to denote an estimator for which the mean square error is minimal. If an estimator is consistent, that is to say asymptotically unbiased, the mean square error is asymptotically equal to the variance and the two usages coincide. Such an estimator, if its variance is minimal, is called asymptotically most efficient.

most powerful critical region The **critical region** which has the highest **power** in testing a hypothesis.

most powerful rank test A two-sample rank order test which is most powerful under some alternative hypothesis; for example, normal distributions differing only in the mean.

most powerful test A test of a hypothesis which is most powerful against an alternative hypothesis. [See **power, uniformly most powerful test**.]

most selective confidence intervals An alternative name proposed by Kendall (1946) for what Neyman (1937a) designated as 'shortest' confidence intervals, the objection to Neyman's term being that such intervals were not necessarily shortest in terms of length. Whereas 'shortest' confidence intervals should be concerned only with the narrowness of the intervals the concept of 'most selective' confidence intervals gives due weight to the frequency with which alternative values of the parameter are covered. The most selective set of confidence intervals covers false values of the parameters with minimum frequency.

most stringent test A test of a statistical hypothesis (H_0) is said to be most stringent if it minimizes the maximum difference by which the test falls short, with respect to a particular class of alternative hypotheses (H_1), of the power that could be attained with respect to these alternatives. A **uniformly most powerful test** is necessarily most stringent.

Mosteller's k-sample slippage test A statistic proposed by Mosteller (1948) to detect the existence of an extreme population. It is based upon the number of observations in one sample greater than all observations in the remaining $k - 1$ samples. It was extended to the case of unequal sample sizes by Mosteller and Tukey (1950).

MOSUM procedures Statistical procedures for detection of changes based on moving sums of residuals; recursive version is based on partial sums recursive residuals, non-recursive version is based on partial sums non-recursive residuals.

mover–stayer model This generalization of the Markov chain model assumes two types of individuals in the population under consideration. First, the 'stayer' who, with probability one, remains in the same category during the entire period of study, and secondly, the 'mover' whose changes in category over time can be described by a Markov chain with constant transition probability matrix.

moving annual total A series derived from an observed time series in which each term consists of the current observation and those immediately preceding it for the period of a year, e.g., the moving annual total of a monthly series would consist of the sums of 12 consecutive monthly values. It may be regarded as the first stage in the computation of a simple **moving average** for which the span or extent is one year. This derived series, however, has an existence in its own right since the upper curve of the three curves on a **Z-chart** is a moving annual total.

moving average If a time series is $x_1, x_2, \ldots, x_n$ and there are chosen a set of weights $w_0, w_1, \ldots, w_k$, $\sum w_i = 1$, the series of values

$$u_t = \sum_{j=0}^{k} w_j x_{t+j}, \qquad t = 1, 2, \ldots, n - k,$$

are the moving averages of the series. In practice it is usual to choose k to be odd, say $2p + 1$, and to locate the corresponding u_t at the middle of the span of $2p + 1$ values which contribute to it. By a suitable choice of weights the series can be represented locally by the values of a polynomial and hence 'smoothed'. If all the weights are equal to $1/k$ the moving average is said to be simple and can be constructed by dividing the **moving total** by k. [See also **trend, smoothing**.]

moving average disturbance In an equation expressing a relationship between variables it is sometimes convenient to include a final term which serves to summarize the effect of factors not separately specified. If such a term z_t takes the form of a **moving average** process, say

$$z_t = \alpha_0 \varepsilon_t + \alpha_1 \varepsilon_{t-1},$$

then the equation is said to possess a moving average disturbance.

moving average method A method for estimating the **median effective dose** of a stimulus from data on quantal responses suggested by Thompson (1947). A moving average of span k formed from the proportions of subjects responding to the various doses of the stimulus is associated with the corresponding average doses. The method then proceeds by linear interpolation between the successive values of the first moving average to estimate the value for which the smoothed response would be 0.50. This method is valid only when the tolerance distribution is symmetrical.

moving average model The representation of a stationary stochastic process in terms of a moving average of possibly infinite length. The observation x_t is a linear combination of uncorrelated terms ε_t of the form

$$x_t = \sum_{j=0}^{\infty} b_j \varepsilon_{t-j}.$$

moving observer technique A method of enumerating a moving population in which the observer moves among the population. If, for example, it is required to estimate the number of people in a street, the observer walks in one direction making a net count of people he or she passes in whatever direction they are moving, deducting those who overtake him or her. This process is repeated in the reverse direction and the average of these two counts gives an estimate of the average number of people in the street during the time of the count.

moving range A concept similar to that of moving average. If the total number of measurements on a variable x is N, by taking absolute differences of successive pairs $|x_i - x_{i-1}|$, we have $N - 1$ values which can be recorded as (moving) ranges of successive samples, each of size 2. This can be generalized to $|x_{i+k} - x_i|, i = 1, 2, \ldots, N - k$, in order to afford a greater degree of smoothing.

moving seasonal variation A pattern of seasonal variation which changes with time. It is usually obtained by determining a seasonal pattern for a certain number k of consecutive years and 'moving' the set of k years along the series as for a **moving average**, so obtaining a seasonal pattern for each year based on the previous k years. The method has the advantage that it avoids the under- and over-correction that might be induced by a fixed seasonal pattern. However, it may well become too flexible and remove more than the true seasonal movement. It is also difficult to project into the future except under assumptions which render it little different from the fixed seasonal pattern.

moving summation process If a random stochastic process is written as $\{\varepsilon_t\}$ then the sums formed by

$$\xi_t = \alpha_0\varepsilon_t + \alpha_1\varepsilon_{t-1} + \alpha_2\varepsilon_{t-2} + \ldots, \qquad -\infty < t < \infty$$

form a stationary stochastic process $\{\xi_t\}$ subject to certain convergence conditions on the coefficients α. Such a process is defined by Kolmogorov (1941) as a moving summation process.

moving total For a series of ordered terms $x_1, x_2, \ldots, x_n$ the sums

$$\sum_{i=1}^{k} x_i, \qquad \sum_{i=2}^{k+1} x_i, \qquad \sum_{i=3}^{k+2} x_i, \qquad \ldots$$

are called moving totals. When divided by k they provide a **moving average** with equal weights.

moving weights In most cases where a moving average is taken of a time series the weights composing the average are constants independent of time. For certain purposes, however, it is desirable to have weights which themselves reflect changing circumstances, as for instance in index numbers of prices where the quantities purchased may alter as time goes on. In such cases the weights themselves may be moving averages of time series and are said to be moving weights. More generally the term can be used to describe any set of weights which change with time.

multi-binomial test A term introduced by Bradley (1953) to describe a test of the hypothesis that, by the method of **paired comparisons**, a set of t objects $A_i, i = 1, 2, \ldots, t$, are of equal merit. The test consists of pooling the $\frac{1}{2}t(t-1)$ independent binomial tests which can be made into a **preference table** by means of the statistic

$$4n \sum_{i<j} \left(p_{ij} - \frac{1}{2} \right)^2$$

which, under Bradley's alternative hypothesis, has asymptotically a non-central χ^2 distribution with $\frac{1}{2}t(t-1)$ degrees of freedom. The p_{ij} are the binomial probabilities in the right-hand upper triangle in the preference table.

multicollinearity If in linear regression analysis the explanatory variables are exactly linearly related, only certain combinations of regression coefficients can be estimated. If the explanatory variables are nearly linearly related, the full vector of regression coefficients will be very poorly estimated. Both situations are referred to as instances of collinearity.

multi-dimensional scaling A set of models and associated methods for constructing a geometrical representation of (i) the proximity (similarity or dissimilarity) relations or (ii) the dominance relations between elements in one or more sets of entities. Each dimension of the data can be the same or a different mode (Carroll and Arabie, 1996).

multi-equational model A model of a system where the variables are interconnected by more than one equation. An alternative name is **simultaneous equations model**.

multi-factorial design See **factorial experiment**.

multi-level continuous sampling plans A type of sampling for control of a continuous process which permits sampling at more than one partial level and complete inspection. In fact, the plan proposed by Lieberman and Solomon (1955) was a random walk with reflecting barriers.

multilinear process A type of stochastic process, introduced by Parzen (1957), which may be represented as

$$X(t) = \sum a(\nu_1 \ldots \nu_k) W_1(t - \nu_1) \ldots W_k(t - \nu_k)$$

where k is a positive integer, $a(\nu_1 \ldots \nu_k)$ constants subject to certain conditions.

multi-modal distribution A frequency distribution with more than one modal value. Such distributions are comparatively rare when they are derived from homogeneous material and, in fact, multi-modality is often accepted as presumptive evidence that the underlying variation is a mixture of different distributions.

multinomial distribution The discrete distribution associated with events which can have more than two outcomes: it is a generalization of the **binomial distribution**. If there are k possible incompatible and exhaustive results of some chance event for which the separate probabilities are p_i ($i = 1, 2, \ldots, k$) then in n trials the distribution of x_1 events of the first kind, x_2 of the second kind, $\ldots$, x_k of the kth kind is

$$f(x_1, x_2, \ldots, x_k) = n! \prod_{i=1}^{k} \frac{p_i^{x_i}}{x_i!}, \qquad \sum_{i=1}^{k} x_i = n,$$

that is to say, the term involving $\prod p_i^{x_i}$ in the multinomial expansion of $(p_1 + p_2 + \ldots + p_k)^n$.

multi-phase sampling It is sometimes convenient and economical to collect certain items of information from the whole of the units of a sample and other items of usually more detailed information from a subsample of the units constituting the original sample. This may be termed two-phase sampling, e.g., if the collection of information concerning variable Y is relatively expensive, and there exists some other variable, X, correlated with it, which is relatively cheap to investigate, it may be profitable to carry out sampling in two phases. At the first phase, X is investigated, and the information thus obtained is used either

(i) to stratify the population at the second phase, when Y is investigated, or (ii) as **supplementary information** at the second phrase, a ratio or regression estimate being used. Two-phase sampling is sometimes called 'double sampling'. Further phases may be added if desired. It may be noted, however, that multiphase sampling does not necessarily imply the use of any relationships between the variables X and Y.

multiple bar chart A chart depicting two or more characteristics in the form of bars of length proportional to the magnitude of the characteristics. For example, a chart comparing the age and sex distribution of two populations may be drawn with sets of pairs of bars, one bar of each pair for each population, and one pair for each age group. [See also **component bar chart**.]

multiple changes Change point problem when more than one change is expected.

multiple classification An alternative name for **manifold classification**. Sometimes the expression is restricted to the case of quantitative variables.

multiple comparisons Issues arising when a number of different comparisons can be made from a single set of data. There are various formulations of the problem especially in the context of **analysis of variance** (Dodge and Thomas, 1980). [See **Duncan's test, Gabriel's test, Newman–Keuls test, Scheffé test, Tukey's test**.]

multiple correlation, coefficient of The product moment correlation between the actual values of the 'dependent' variable in multiple regression and the values as given by the regression equation. It measures the closeness of representation by the regression line and may also be regarded as the maximum of the correlation coefficient between the 'dependent' variable and all linear functions of a set of two or more of the 'independent' variables. The coefficient is usually denoted by R but is regarded as essentially non-negative, the quantity R^2 being the one which occurs in practice; R^2 is also known as the **coefficient of determination**, and represents the proportion of the sum of squares accounted for by the regression.

multiple decision methods A general term to describe statistical techniques for dealing with problems where the possible outcomes or decisions are more than two.

multiple decision problem The problem of choosing one hypothesis or decision from a set of k mutually exclusive and exhaustive hypotheses or decisions on the basis of some observations on a random variable. The simple problem of testing a statistical hypothesis against a single alternative can be formulated as a special case of this, when $k = 2$.

multiple factor analysis In current usage this is equivalent to **factor analysis**. Historically, the analysis of psychological material into factors grew from one-factor to two-factor and then to m-factor complexes and the last was called 'multiple' to distinguish it from the simpler forms. This no longer seems necessary.

multiple imputation Replacement of missing data by multiple values leading to many sets of complete data, variability among which is intended to mimic uncertainty due to the missing data (Rubin, 1987).

multiple partial correlation, coefficient of An extension of the classical correlation coefficient by Cowden (1952), based on a suggestion of Hotelling (1926), to the case of multiple correlation between the 'dependent' variable and two or more 'independent variables' when all these have been adjusted for the effect of one or more other variables.

multiple phase process A stochastic birth process (Kendall, 1948) in which an individual, after being 'born', passes through k successive phases and can only subdivide or give birth itself after the kth phase. The lifetimes in each phase are usually taken to be independently distributed. If $k = 1$ the process becomes the simple **birth process**. Also known as Erlang's method.

multiple Poisson distribution The joint distribution of independent Poisson variables. It may also be regarded as an approximation to the **multinomial distribution** when m is large and $np_i = \lambda_i$ in the form

$$e^{(\lambda_1 + ... + \lambda_{r-1})} \frac{\lambda_1^{k_1} \lambda_2^{k_2} \ldots \lambda_{r-1}^{k_{r-1}}}{k_1! k_2! \ldots k_{r-1}!} .$$

multiple Poisson process A term sometimes used for a generalization of a **Poisson process** where there are simultaneous occurrences at the event points.

multiple random starts An idea for selection procedure under systematic sampling originally due to Tukey and developed among others by Gautschi (1957). The method is to choose a random sample of s without replacement from the first k elements and subsequently every kth multiple element of those selected.

multiple range test A method of comparing mean values arising in analyses of variance by using the range of subsets of means, or of ranges of sets of observations contributing to mean values. [See also **Duncan's**, **Gabriel's**, **Newman–Keuls**, **Scheffé** and **Tukey's tests**.]

multiple recapture census A sequential method of **capture–recapture sampling** where the method of capture does not kill or affect the behaviour of the sample unit and where the experimenter has full control over the sampling and marking.

multiple record system See **capture–recapture sampling**.

multiple regression The regression of a dependent variable on more than one 'independent' or 'explanatory' variable.

multiple sampling See **double sampling, sequential analysis**.

multiple smoothing method A generalization by Brown and Meyer (1961) of the simple model for **exponential smoothing** to take account of polynomial trends. This approach should not be confused with the iterated application of a moving average to a time series.

multiple stratification If a sample is stratified according to two or more factors it is said to be multiply stratified. In practice multiple stratification is difficult to carry out because the information with which to divide the population into substrata is often not available. [See **control of substrata**.]

multiple time series A time series in which a vector of observations is available at each time point. Many of the simpler forms of analysis hinge on generalizing the notion of one-dimensional autocorrelation functions and spectra to several dimensions.

multiplicative process A synonym of **branching process**.

multiplicity If several tests of size α are performed, the global type I error is much higher than α. In epidemiological studies this is likely to lead to many spurious results because often many variables are tested and several models are tried. The **Bonferroni rule** is an easy way to correct the type I error.

multi-stage estimation A generalization to n stages (Blum and Rosenblatt, 1963) of the principle of estimation first proposed for two stages by Stein (1945).

multi-stage sampling A sample which is selected by stages, the sampling units at each stage being subsampled from the (larger) units chosen at the previous stage. The sampling units pertaining to the first stage are called primary or first-stage units, and similarly for second-stage units, etc. Where the sampling frame has to be constructed in the course of the sampling operation, multi-stage sampling has the additional advantage that only the parts of the population selected at any stage need to be listed for sampling at the next stage.

multistate model Used in epidemiology to represent the evolution of subjects through different health states. A process with a finite number of states is associated to each individual. Generally the structure of the model is given, i.e., the number of states, and which transitions are possible: it is then sought to estimate the **transition intensities** between states.

multi-temporal model See **dynamic model**.

multi-valued decision The ordinary test of a statistical hypothesis involves, according to some, a two-valued decision: accept or reject, though other writers contend that failure to reject is not the same thing as acceptance. Where there are more than two possible decisions, for instance, to sell/to stock/to destroy, the decision is said to be multi-valued. The simple plan for acceptance sampling by sequential methods is a case of a multi-valued decision problem. At each successive stage in the sampling a decision has to be taken: accept/reject/continue sampling.

multivariate analysis This expression is used rather loosely to denote the analysis of data which are multivariate in the sense that each member bears the values of p response or dependent variables. The principal techniques of multivariate analysis, beyond those admitting of a straightforward generalization, e.g., regression, correlation and variance analysis, are **factor analysis, component analysis, classification, discriminant analysis, canonical correlation** and various generalizations of homogeneity tests, as illustrated by the D^2 statistic. [See also **Hotelling's T^2**, **Wishart distribution**, **Wilks's criterion.**]

multivariate analysis of variance (MANOVA) A generalization of the **analysis of variance** for the test of equality of means of l univariate normal populations having a common unknown variance (Flury, 1997).

multivariate analysis of variance (MANOVA) table Formal analogue of analysis of variance table in which sums of squares and products of component observations are divided into components. Various procedures of analysis based on the multivariate normal distribution can be derived from the table.

multivariate beta distribution If there are two symmetric matrices $\mathbf{A}$ and $\mathbf{B}$ of order p with independent Wishart distribution (f_1, f_2 degrees of freedom), there exists a lower triangular matrix $\mathbf{C}$ so that $\mathbf{A} + \mathbf{B} = \mathbf{CC}'$. Let $\mathbf{A} = \mathbf{CLC}'$; then the distribution of $\mathbf{L}$ is the multivariate beta distribution (Kshirsagar, 1961). It has the form

$$f \propto |\mathbf{L}|^{(f_1 - p - 1)/2} |\mathbf{I} - \mathbf{L}|^{(f_2 - p - 1)/2}.$$

multivariate binomial distribution An s-variate generalization of the binomial distribution. For example, if a member can bear the value of two dichotomized variables with probabilities

		Variable 1	
		Present	Absent
Variable 2	Present	p_{11}	p_{10}
	Absent	p_{01}	p_{00}

the bivariate binomial in samples of n is given by

$$(p_{11} + p_{10} + p_{01} + p_{00})^n.$$

multivariate Burr's distribution A multivariate form of **Burr's distribution** developed by Takahasi (1965).

multivariate distribution The simultaneous distribution of a number p of variables ($p > 1$) or, equivalently, the probability distribution of p variables.

multivariate exponential distribution A distribution in life assessment of systems. The complement of the distribution function is given by $\Pr\{X_1 > x_1, X_2 > x_2, \ldots, X_n > x_n\}$ which is equal to

$$\exp\left\{-\sum_{s\in S}\lambda_s\max(x_is_i)\right\},$$

where S is a set of vectors; marginal distributions are ordinary negative exponential (Marshall and Olkin, 1967b).

multivariate F-distribution If there are two non-symmetric matrices $m \times p$ and $m \times n$ ($p \leqslant m \leqslant n$) denoting matrix variables $\mathbf{X}$ and $\mathbf{Y}$ and all columns are normally and independently distributed with covariance Σ, the multivariate F-distribution is that of $\mathbf{X}'(\mathbf{YY}')^{-1}\mathbf{X}$ provided $E(\mathbf{X}) = E(\mathbf{Y}) = 0$.

multivariate hypergeometric distribution Suppose n balls are drawn at random without replacement from an urn which contains N balls of $(s+1)$ different colours, M_i being of the ith colour, $i = 0, 1, 2, \ldots, s$. If X_i is the number of balls of the ith colour, $i = 1, 2, \ldots, s$, drawn in the n draws then $(X_1, X_2, \ldots, X_s)$ has the s-dimensional hypergeometric distribution. This distribution is also called the factorial multinomial distribution.

multivariate inverse hypergeometric distribution The limiting form of the negative **multinomial distribution** when $N \to \infty$ and the $M_i \to \infty$ so that $M_i/N \to p_i$ of the negative multinomial. [See also **inversion**.]

multivariate L_1-mean An L_1-mean $\boldsymbol{\mu} = (\mu_1, \ldots, \mu_p)$ of multivariate data $\mathbf{X} = (\boldsymbol{x}_1, \ldots, \boldsymbol{x}_n)'$ as a matrix of $\mathbb{R}^{n\times p}$, in which each data point $\boldsymbol{x}_i = (x_{i1}, \ldots, x_{ip})$ is an element of $\mathbb{R}^p$, is a point that minimizes

$$\sum_{i=1}^{n}\left(\sum_{j=1}^{p}|x_{ij}-\mu_j|\right)^2.$$

In the case $p = 1$, the L_1-mean is equivalent to the usual mean. The multivariate L_1-mean was introduced by Dodge and Rousson (1999).

multivariate moment See **product moment**.

multivariate multinomial distribution A multivariate distribution of the multinomial type. It is difficult to write down in generality but an example will be found under **bivariate binomial distribution**.

multivariate negative binomial distribution An alternative name for the **negative multinomial distribution**.

multivariate negative hypergeometric distribution An alternative name for the **multivariate inverse hypergeometric distribution**.

multivariate normal distribution A generalization of the univariate normal distribution to the case of p variables ($p \geqslant 2$). If the ith variable x_i has mean m_i and the covariance (dispersion) matrix of the variables is $(v_{ij}), i, j = 1, 2, \ldots, p$, with an inverse (v^{ij}), the multivariate normal distribution has the frequency function

$$\frac{|v^{ij}|^{1/2}}{(2\pi)^{p/2}} \exp\left\{-\frac{1}{2}\sum_{i,j=1}^{p} v^{ij}(x_i - m_i)(x_j - m_j)\right\}.$$

multivariate Pareto distribution An extension by Mardia (1962) of the two forms of **bivariate Pareto distribution**. The conditional distributions are of Pareto form with displaced origins.

multivariate Pascal distribution If a population of individuals can be described by one or more of characters $A_1, \ldots, A_s$ with probabilities of individuals as $p_1, p_2, \ldots, p_s$, then random observations are taken until k individuals possessing none of the characteristics are recorded, the x_i relating to occurrences of A_i having an s-variate Pascal distribution. It is a special case of the multivariate negative binomial distribution.

multivariate Poisson distribution An s-variate generalization of the Poisson distribution and the limiting distribution as n tends to infinity and p_i tends to zero of the multivariate binomial distribution.

multivariate Pólya distribution Suppose an urn contains N balls of $(s+1)$ different colours, a_i being the ith colour, $i = 1, 2, \ldots, s$, and b of the $(s+1)$th colour. Suppose n balls are drawn one after another, with replacement, such that at each replacement c new balls of the same colour are added to the urn. If X_i denotes the number of balls of the ith colour in the sample, $i = 1, 2, \ldots, s$, then $(X_1, X_2, \ldots, X_s)$ has the s-variate Pólya distribution.

multivariate power series distribution Let

$$f(\theta_1, \theta_2, \ldots, \theta_s) = \sum_{x_1, x_2, \ldots, x_s} a_{x_1 x_2 \ldots x_s} \theta_1^{x_1} \theta_2^{x_2} \ldots \theta_s^{x_s}$$

be a convergent power series in $\theta_1, \theta_2, \ldots, \theta_s$ such that $a_{x_1 x_2 \ldots x_s} \theta_1^{x_1} \theta_2^{x_2} \ldots \theta_s^{x_s} \geqslant 0$, $x_i = 0, 1, 2, \ldots$, $i = 1, 2, \ldots, s$, for all $(\theta_1, \theta_2, \ldots, \theta_s) \in \Theta$, the s-dimensional parameter space. Then the s-variate power series distribution with the series function $f(\theta_1, \theta_2, \ldots, \theta_s)$ is defined by the probability function

$$\Pr(x_1, x_2, \ldots, x_s) = \frac{a_{x_1 x_2 \ldots x_s} \theta_1^{x_1} \theta_2^{x_2} \ldots \theta_s^{x_s}}{f(\theta_1, \theta_2, \ldots, \theta_s)}.$$

multivariate processes A class of **stochastic processes** involving more than one random variable. Also known as simultaneous or vector processes and not to be confused with multi-dimensional processes which are concerned with more than one parameter.

multivariate quality control Control of quality in which each item for inspection must conform to standards in respect of more than one variable, as, for example, the length, breadth and height of a metal block.

multivariate signed rank test Generalization of the **Wilcoxon signed rank test** for the case of matched bivariate and trivariate properties (Bennett, 1965).

multivariate Tchebychev inequalities A development by Olkin and Pratt (1958) based upon the bivariate form (Berge, 1938). For example, for p uncorrelated random variables $(\rho = 0)$

$$\Pr(|Y_i| \geqslant k_i\sigma_i \text{ for some } i) \leqslant \sum k_i^{-2}$$

and $\rho = -1/(p-1)$

$$\Pr(|Y_i| \geqslant k_i\sigma_i \text{ for some } i) \leqslant (p-1)/k^2.$$

Murthy's estimator In sampling without replacement, an estimator of the population mean and variance proposed by Murthy (1957) and based upon order statistics. It is of limited use for sample size greater than three owing to the heavy computation involved.

N

natural conjugate prior If t is a vector of k statistics based upon a sample of size n, and t is sufficient for the p parameters θ, then any prior probability density function for θ that is proportional to the conditional probability density of t given θ and n, with the constant of proportionality independent of θ, is defined as a natural conjugate prior probability density function. To represent prior information, an investigator assigns suitable values to t and n to obtain the **informative prior**.

nearest-neighbour clustering See **single-linkage clustering**.

nearest-neighbour methods Techniques for analysing point processes that depend on distances between nearest neighbours (or kth nearest neighbours) as opposed to techniques based on point counts in subsets.

nearly best linear estimator A form of estimator proposed by Blom (1958) making use of approximations for the problem of minimum variance which is central to the 'best' linear estimator.

negative binomial distribution A distribution in which the relative frequencies (probabilities) are arrayed by a binomial with a negative index. For example, if the variable values are 0, 1, 2, etc., the frequency at $x = j$ is the coefficient of t^j in the expansion of $(1 - pt)^{-n}(1 - p)^n$ in powers of t. In general, n lies between zero and infinity (not necessarily integral). As n increases, the distribution approaches the **Poisson** form, the variance always being greater than the mean. When n is integral, it is sometimes called the **Pascal distribution**. A number of distinct probability models generate the distribution.

negative exponential distribution See **exponential distribution**.

negative factorial multinomial distribution An alternative name for the **multivariate inverse hypergeometric distribution**.

negative hypergeometric distribution The analogue of the **negative binomial** when sampling is from a finite population of attributes, without replacement, i.e., it is the distribution of the sample number required in such circumstances to reach a preassigned number of 'successes'.

negative moments The negative moments of a frequency distribution are the moments of reciprocal powers of the variable; for example, $E(1/X^k)$ where k is the order of the moment. Their existence, of course, depends upon convergence and they are little used. In fact, owing to possible confusion with ordinary moments, which have negative sign, the preferable expression would be 'moments of negative order'.

negative multinomial distribution If there is a sequence of independent trials in each of which there can be $s + 1$ mutually exclusive outcomes $A_0, \ldots, A_s$ with probabilities $p_0, \ldots, p_s$ ($\sum p_i = 1$) and $x_1, \ldots, x_s$ are the occurrences of $A_1, \ldots, A_s$ before A_0 occurs k times, then $(x_1, \ldots, x_s)$ has the negative multinomial distribution. This distribution can also arise as a compound of Poisson variables where the parameter has a gamma distribution: in this form it is also known as the multivariate negative binomial distribution.

Nelson–Aalen estimator This estimates the cumulative hazard function Λ from right-censored and possibly **left-truncated** data:

$$\hat{\Lambda}(t) = \sum_{i, \ t \leqslant t_{(i)} \leqslant t} \frac{1}{R(t_{(i)})},$$

where $t_{(i)}$ are the ordered observation times of failure and $R(t)$ is the number of subjects at risk at time t. The **Breslow estimator** generalizes it to the case of the **Cox proportional hazards model**. It gives rise to a discrete measure which is identical to that obtained by the **Kaplan–Meier estimator**.

nested balanced incomplete block design An experiment design with two systems of blocks, the second within the first such that ignoring either system leaves a balanced incomplete design whose blocks are those of the other system. In the nesting system each block from the first part contains m blocks from the second.

nested case–control studies A case–control study in which the cases and the controls are taken from a cohort study. Relatively to the analysis of the full cohort it spares resources because the number of possible controls is generally much higher than the number of cases; the nested case–control study allows the number of controls to be fixed.

nested design A class of experimental design in which every level of a given factor appears with only a single level of any other factor. Factors which are not nested are said to be crossed. If every level of one appears with every level of the others, the factors are said to be completely crossed; if not, they are partly crossed.

nested hypotheses A sequence of hypotheses in which the hypothesis at any stage is contained in all hypotheses later in the sequence.

nested methods Two parametric models are said to be nested if one reduces to the other when restrictions are placed on its parameters.

nested sampling A term used in two somewhat different senses: (i) as equivalent to **multi-stage sampling** because the higher-stage units are 'nested' in the lower-stage units; (ii) where the sampling is such that certain units are imbedded in larger units which form part of the whole sample, e.g., the **entry plots** of clusters are 'nested' in this sense.

network information criterion (NIC) An information criterion intended to allow for possible model misspecification, differing from Akaike's criterion by replacement of p by the trace of the product of the inverse observed information matrix and sum of squared contributions to the score vector.

network of samples An alternative name for a set of **interpenetrating samples**.

network sampling A procedure developed by Sirken (1970) for the measurement of characteristics in rare populations. For any counting rule for linking enumeration units with elements, parameters can be defined which characterize the network linking the enumeration units to the members of the rare population group.

neural network A regression model in which the responses are nonlinear functions of inputs through layers of connected hidden variables, originally by treating biological neurons as binary thresholding devices. They are flexible models useful for discrimination and classification (Ripley, 1996) and are implemented by a computerized 'black-box' trained by a training data set.

new better than used distribution A non-negative random variable X is said to have a new better than used (NBU) distribution if, for all $x, y \geqslant 0$, $\Pr(X \geqslant x + y | X \geqslant x) \leqslant \Pr(X \geqslant y)$. If x is the life-length of an item, it has an NBU distribution if, for all $x, y \geqslant 0$, the probability that it survives at least an additional period y, given that it has survived to time x, is less than or equal to the probability that a new item will survive an initial period y. NBU distributions are a wider class than **increasing failure rate** (IFR) distributions, which they encompass.

Newcomb, Simon (1835–1909) A Canadian astronomer who contributed especially to the treatment of outliers in statistics and to the application of probability theory to data (Newcomb, 1859–1861).

Newman–Keuls test A version of a multiple range comparison procedure by Newman (1939) and Keuls (1952) where the sample ranges are tested against the studentized range of the subsets rather than the ranges of mean values. It is a step-by-step procedure subsequently modified by Duncan (1955).

Newton–Raphson method An iterative method of solving equations numerically in which the successive approximations are determined using first derivatives, or numerical estimates of them. Also known as Newton's method.

Neyman allocation A method of allocating sample numbers to different strata in order to secure unbiased estimators of parent mean values with minimal variance. The numbers allocated, for large samples, are proportional to the standard deviations of the variable under examination in the respective strata (as well as to the stratum number). This method was advanced by Neyman (1934). [See also **optimum allocation, proportional sampling**.]

Neyman model A term proposed by Ogawa (1963) to denote experimental designs which involve situations containing technical errors in the sense that variations in replicate observations are solely due to experimental techniques. [See also **Fisher model.**]

Neyman–Pearson lemma An important result due to Neyman and Pearson (1928). It states that the best **critical region** of size α for testing a simple null hypothesis H_0 against a simple alternative hypothesis H_1 is the region W_b in sample space for which

$$\frac{L_1}{L_0} \geqslant b$$

where L_i is the likelihood function given the hypothesis H_i ($i = 0, 1$) is true, and b is such that the integral of L_0 over the critical region is equal to α.

Neyman–Pearson theory A general theory of testing hypotheses, due to Neyman and Pearson (1933). It is based upon the consideration of two types of errors which may be incurred in judging a statistical hypothesis. Since the 1930s the theory has been developed by other writers and led to the more general theories, e.g., the statistical decision functions of Wald. The probabilities of errors incurred by rejecting a hypothesis H_0 when it is true and accepting it when it is false – the **errors of the first and second kind** – are usually written α and β. The function $1 - \beta(H_1)$, namely the probability with which the hypothesis H_0 is rejected when some alternative hypothesis H_1 is true, is called the **power function** of the particular test. It is the concept of the power function coupled with that of the two kinds of errors that form the principal features of the Neyman–Pearson theory.

Neyman–Scott model A clustering model proposed by Neyman and Scott (1958) for the distribution of galaxies and also in ecological applications. The process of clusters is supposed stationary and Poisson, while conditional on a given cluster size; the cluster members are supposed independently and identically distributed about the cluster centre with a common distribution function.

Neyman-shortest unbiased confidence intervals An optimum set of confidence intervals which, among all unbiased α-confidence intervals, uniformly minimizes the probability of covering false values. The word 'shortest' may be misleading in that the intervals do not necessarily have minimal length. The alternative 'most selective' has been proposed.

Neyman's factorization theorem A theorem first proved rigorously by Neyman (1935). It states that if X is a vector random variable with frequency function $f(x|\theta)$ then, subject to certain regularity conditions, a vector function $T = t(x)$ is sufficient for θ if and only if $f(x|\theta)$ factorizes into a product of a function of $t(x)$ and θ and a function of x alone, i.e.,

$$f(x|\theta) = g(t(x), \theta) \cdot h(x).$$

Neyman's ψ^2 test The first of the **smooth tests** of goodness of fit, developed as alternatives to the χ^2 test initially by Neyman (1937b). The test statistic is

$$\psi_k^2 = \sum_{r=1}^k u_r^2 = \frac{1}{n} \sum_{r=1}^k \left\{ \sum_{i=1}^n \pi_r(\gamma_i) \right\}^2$$

where the γ_i are transformed observations from x_i, by probability integral transforms, and $\pi_r(\gamma)$ linearly transformed Legendre polynomials.

Nile, problem of the A problem posed by Fisher (1936b). The fertility of an area depends on an unknown parameter θ (the height of the Nile) and it is required to define contours dividing the area into sections with total fertility in a constant ratio for all θ. The possibility of a solution depends on the existence of an ancillary statistic.

Noether's test for cyclical trend A test for trend (Noether, 1956) in a sequence of measurements on a continuously distributed variable, sensitive both to monotonic trend and to cyclical trend, i.e., to a trend which periodically reverses direction. It is based on the fact that if there is no trend of any kind, then any three consecutive measurements will with probability $\frac{1}{3}$ form a monotonic sequence, whilst for almost any form of trend the probability of this event will exceed $\frac{1}{3}$.

noise A convenient term for a series of random disturbances borrowed, through communications engineering, from the theory of sound. In communication theory noise results in the possibility of a signal sent, x, being different from the signal received, y, and the latter has a probability distribution conditional upon x. If the disturbances consist of impulses at random intervals it is sometimes known as 'shot noise'. [See also **white noise**.]

nomogram A form of line chart upon which appears scales for the variables involved in a particular formula in such a way that corresponding values for each variable lie on a straight line which intersects all the scales.

non-central beta distribution The distribution of the ratio

$$\frac{X_1}{X_1 + X_2}$$

where X_1 follows a non-central $\chi_{\nu_1}^2$ distribution and X_2 follows a $\chi_{\nu_2}^2$ distribution. If both X_1 and X_2 are χ^2 non-central the resultant distribution is the doubly non-central beta distribution.

non-central confidence interval A confidence interval which is not **central**.

non-central χ^2 distribution The distribution of the sum of squares of independent normal variables with unit variance but not with zero mean. The distribution was obtained by Fisher (1928b) as a special case of the distribution of the multiple correlation coefficient. It has several applications and in particular is required in order to determine the power function of the chi-squared test. The difficult problem of tabulation was investigated by Patnaik (1949) who derived certain approximations which used existing tabled functions.

non-central F-distribution The distribution of the ratio of a non-central χ^2 to a central χ^2. Fisher gave the distribution (1928b). Wishart (1932) considered in it the form of the distribution of the correlation ratio. In a particular case the form reduces to that of the **non-central t-distribution**. The use of the distribution in evaluating the power of analysis of variance tests was extended by the approximations investigated by Patnaik (1949), a development of work by Tang (1938) and Hsu (1941).

non-central multivariate beta distribution If, in a **multivariate beta distribution**, the matrix $\mathbf{B}$ has a **non-central Wishart distribution** then the distribution of $\mathbf{L}$ is of the non-central multivariate beta form.

non-central multivariate F-distribution The distribution of $\mathbf{X}'(\mathbf{Y}\mathbf{Y}')^{-1}\mathbf{X}$, which depends upon $\mathbf{M}'\boldsymbol{\Sigma}^{-1}\mathbf{M}$, where the matrix variables $\mathbf{X}$ and $\mathbf{Y}$ are $m \times p$ and $m \times n$ respectively $(p \leqslant m \leqslant n)$ with columns all independently normally distributed with covariance $\boldsymbol{\Sigma}$ and $E(\mathbf{X}) = \mathbf{M}$, $E(\mathbf{Y}) = 0$.

non-central t-distribution The distribution of the ratio $(x - \alpha)/s$ where x is a normal variable with zero mean and variance σ^2, α is a nonzero constant and s is distributed as $x\sigma/\sqrt{\nu}$ with ν degrees of freedom independently of x. The distribution is a simple transform of a particular case of the **non-central F-distribution** and is used, among other things, to determine the power function of 'Student's' t-test of the significance of the mean.

non-central Wishart distribution The distribution of the second-order dispersions, about arbitrary origins, of a multivariate normal distribution, reducing to a Wishart distribution when the origins are the sample means. [See **Wishart distribution**.]

non-centrality parameter A parameter of the distribution of a non-central test statistic that vanishes in the central case.

non-circular statistic See **circular formula**.

non-compliance In a clinical trial, there are some patients who do not follow the assigned treatments (Robins, 1998). Also called non-adherence.

non-determination, coefficient of The square of the **coefficient of alienation**; that is to say, if r is the correlation between two variables, the coefficient of non-determination is $1 - r^2$. [See also **total determination, coefficient of**.]

non-ignorable non-response See **missing at random**.

nonlinear correlation This term is meant to relate in broad terms to the correlation between variables where the regression is not linear. [See also **correlation ratio**.]

nonlinear regression A regression model that is nonlinear in the parameters. The term has also been used for regression equations nonlinear in the regressors, but this should be regarded as obsolete.

non-normal population A population for which the frequency distribution is not the **normal distribution**. The term does not mean abnormal in the sense of unusual.

non-null hypothesis In general, a hypothesis alternative to the one under test; the **null hypothesis**. In some contexts, however, it is given the meaning of a hypothesis under test where the effect is not equal to zero.

non-orthogonal data This expression originates in the use of 'orthogonal' to denote independence. Data are said to be non-orthogonal if they lead to estimates of various effects which are not independent of one another or, perhaps, of other features such as block differences which are a nuisance in the analysis. The disadvantage of such material is that effects thus mixed up may be genuinely inextricable or may require a more complicated technique for their disentanglement, leading to enhanced variance and model dependence.

non-parametric See **distribution-free method**.

non-parametric confidence intervals Confidence intervals based on non-parametric tests; for example, a confidence interval for the difference of two medians may be based on the Wilcoxon test, by finding two values which, added to all the observations in one sample, make the difference just significant in either direction. Such intervals are not invariant under monotonic transformations, as is the test.

non-parametric delta method Computation of variance in non-parametric problems by empirical von Mises expansion (Davison and Hinkley, 1997).

non-parametric maximum likelihood In a non-parametric approach some functions are not restricted to belong to a finite-dimensional space. One can still write the likelihood and maximize it over the space of functions considered. For instance, the **Nelson–Aalen estimator** of the cumulative hazard function and the **Kaplan–Meier estimator** of the survival function are estimators of the non-parametric maximum likelihood.

non-parametric tolerance limits Tolerance limits which do not depend on the parameters of the parent population from which a sample is drawn. It seems possible, but is by no means universal practice, to draw a distinction between non-parametric limits, in which the parent distribution is known in form, and distribution-free limits in which the form of the parent is unknown. [See **tolerance limits, distribution-free method**.]

non-random sample A sample selected by a non-random method. For example, a scheme whereby units are selected purposively would yield a non-random sample. Again, a sample obtained by taking members at fixed intervals on a list is a non-random sample unless the list was arranged in a random order. [See also **quasi-random sampling**.]

non-recurrent state See **recurrent state**.

non-recursive residual Residuals used in non-recursive change point problems, the kth non-recursive residual is usually the difference of the kth observation and its prediction for this valued based on all observations; more generally, the kth non-recursive residual is a functional of the kth observation and its prediction based on all observations.

non-regular estimator See **regular estimator**.

non-response In sample surveys, the failure to obtain information from a designated individual for any reason (death, absence, refusal to reply) is often called a non-response and the proportion of such individuals of the sample aimed at is called the non-response rate. It would be better, however, to call this a 'failure' rate or a 'non-achievement' rate and to confine 'non-response' to those cases where the individual concerned is contacted but refuses to reply or is unable to do so for reasons such as deafness or illness. Non-availability of information in other situations, e.g., arrival of the investigator for crop cutting experiments after harvesting, may also be termed non-response, or better, non-achievement. When several items of information are to be collected for the same sample unit, it may so happen that information is not available for some of the items but available for others. The term non-response is usually not applied in such a situation, but incomplete response or incomplete achievement may be used.

non-sampling error An error in sample estimates which cannot be attributed to sampling fluctuations. Such errors may arise from many different sources such as defects in the frame, faulty demarcation of sample units, defects in the selection of sample units, mistakes in the collection of data due to personal variations or misunderstandings or bias or negligence or dishonesty on the part of the investigator or of the interviewee, mistakes at the stage of the processing of the data, etc. The term 'response error' is sometimes used for mistakes in the collection of data and would not, strictly speaking, cover errors due to non-response. The use of the word 'bias' in the place of error, e.g., 'response bias', is not uncommon. The term 'ascertainment error' is preferable as it would include errors due to non-response and also cases of collection of data by methods other than interviewing, e.g., direct physical observation of fields for crop estimates.

nonsense correlation See **illusory association or correlation**.

non-singular distribution A multivariate distribution in, say, p variables which cannot, by a transformation of the variables, be converted into a distribution with fewer than p variables. A normal distribution is non-singular if and only if the dispersion matrix or the **correlation matrix** is of rank p.

normal deviate The value of a deviate of the normal distribution. [See also **normal equivalent deviate**.]

normal dispersion See **Lexis ratio**.

normal distribution The continuous probability distribution of infinite range represented by the equation

$$f(x) = \frac{1}{\sqrt{2\pi}\sigma} \exp\left\{ -\frac{(x-\mu)^2}{2\sigma^2} \right\}, \qquad -\infty < x < \infty,$$

where μ is the mean and σ the standard deviation. In Continental writings the distribution is often known as the Gaussian, the Laplacian, the Gauss–Laplace or the Laplace–Gauss distribution, or the Second Law of Laplace. It was apparently first discovered by de Moivre (1733) as the limiting form of the binomial distribution. When $\mu = 0$ and $\sigma = 1$, the probability distribution is called the standard normal distribution.

normal equations The set of simultaneous equations arrived at in estimation by the **least squares method**.

normal equivalent deviate (NED) If P is a proportion or a probability and Y is defined by

$$P = \int_{-\infty}^{Y} \frac{1}{\sqrt{2\pi}} e^{-\frac{1}{2}x^2} dx$$

then Y is termed the NED of P. This quantity or the **probit** is often used in the analysis of a stimulus–response relationship.

normal inspection The amount of inspection required by the initial application of a sampling inspection plan. It is undertaken so long as the quality of the product is close to the acceptable quality level laid down by the plan. If the actual quality improves consistently reduced inspection may be introduced, and conversely if the quality deteriorates tightened inspection is required.

normal probability paper A specially ruled graph paper with a variable x as abscissa and an ordinate y scaled in such a way that the graph of the distribution function, y, of the normal distribution, is a straight line.

normal scores test Randomization test in which, before computation of the test statistic, observations are transformed twice, first to size ranks, and then by a transformation which involves the unit normal distribution. The original idea proposed by Fisher and Yates (1938) was that the second transformation should be the replacement of ranks by the expected values of the corresponding unit normal order statistics, to give an expected normal scores test. If Φ denotes the unit normal distribution function, and the transformation of the ranks R_i, $1 \leqslant R_i \leqslant N$, are of the form $\Phi^{-1}(R_i/(N+1))$, then the test is called an inverse normal scores test. [See **Klotz's test** and **van der Waerden's test**.] If the ranks are replaced by the corresponding members of a random sample of the same size from a unit normal distribution, the test is a random normal scores test. Such tests have the advantage that the null distribution of the test statistic is usually trivially related to one of the well-tabulated classical distributions, often the **normal** or the **chi-square**. [See **Terry–Hoeffding test**.]

normalization of frequency function The transformation of a random variable so that its probability function becomes normal, or approximately so.

normalization of scores In the analysis of data resulting from educational or psychological tests it is often desirable to convert each set of original scores to some standard scale. The process of doing so is called a normalization of the scores, but this may mean the reduction to a norm or common standard, not necessarily to the scale of a normal (Gaussian) distribution. Nevertheless, one method in common use is to determine the percentiles of the scores and then express these as corresponding deviations from the mean of a normal distribution. This particular device is assisted by the use of **normal probability (graph) paper**. [See also *T*-score, *z*-score.]

normalizing transform Transformations of variables with the object of reducing the distribution to the normal or very closely so. It may be noted that a transformation to achieve, say, stabilization of variance may also yield a large measure of normalization. Usually, this expression denotes a transformation of a random variable so that its distribution approximates, or is exactly equal to, the normal (Gaussian) form. It also occurs in the sense of a transformation which reduces the distribution to some other standard form which is regarded as the norm.

normit A contraction, or diminutive term, for **normal deviate** proposed by Berkson (1955a) for use in connection with a method of analysing quantal response data due to Urban (1910). [See also **probit**.]

nucleotide diversity The average pairwise difference between nucleotide sites in a sample of DNA sequences.

nugget effect In **geostatistics**, the **semivariogram** often seems to drop to a nonzero limiting value as h approaches zero. This is associated with variation on a much smaller scale than the general trend represented, such as might be associated with the presence of individual 'nuggets' or pure measurement errors.

nuisance parameters In the theories both of estimation and of tests of significance there arises the problem of finding a sampling distribution which is independent of certain unknown parameters of the population. Although these parameters are essential to the specification of the population, they are a 'nuisance' in the formulation of exact statements about certain other parameters. The classical case of a nuisance parameter arises in the setting of confidence intervals for the mean, which depend on the unknown parent variance when the distribution of the mean is itself used; the difficulty in this case is overcome by the use of Student's distribution which does not depend on the parent variance.

null hypothesis In general, this term relates to a particular hypothesis under test, as distinct from the alternative hypotheses which are under consideration. It is therefore the hypothesis which determines the probability of the **type I error**. In some contexts, however, the term is restricted to a hypothesis under test of 'no difference'.

null recurrent state In the theory of stochastic processes, a **recurrent state** for which the average return time is infinite.

numerical taxonomy Methods for grouping objects based on attributes and/or measurements. It is usually used to refer to the identification of species or sub-species in biological material, but sometimes more generally as synonymous with cluster analysis.

Nyquist frequency In connection with data consisting of observations recorded at equal distances of time, the frequency of a sinusoidal term whose period is twice the time interval between successive observations (Nyquist, 1928).

Nyquist interval For time series consisting of a series of harmonic components with frequencies in a limited range, the Nyquist interval is such that the highest admissible frequency has a period of two intervals.

Nyquist–Shannon theorem A theorem with applications in communications engineering, where it is known as 'the sampling theorem' and variously ascribed to Nyquist or Shannon (1949). The theorem states that, if the probability density f has a characteristic function vanishing outside the interval $(-a, a)$, then f is uniquely determined by the values

$$\frac{1}{2}\pi f(n\pi/\lambda)$$

for any fixed $\lambda \geqslant a$, and these values induce an arithmetic probability distribution.

O

o_P and O_P notation Denote by ξ_n a sequence of real-valued random variables, by $r(n)$ a sequence of strictly positive numbers ($n \in \mathbb{N}$). The notation $\xi_n = o_P(r(n))$ means that $\xi_n/r(n)$ converges to zero in probability, as $n \to \infty$; hence, $\xi_n = o_P(1)$ simply means that ξ_n converges to zero in probability. The notation $\xi_n = O_P(r(n))$ means that $\xi_n/r(n)$ is bounded in probability, or uniformly tight, as $n \to \infty$; that is, for all $\varepsilon > 0$ there exist $B > 0$ and $N \in \mathbb{N}$ such that $\Pr\left[r^{-1}(n)|\xi_n| > B\right] < \varepsilon$ for all $n \geqslant N$.

ω^2-test An alternative name for the **Cramér–von Mises test**.

oblimax See **factor rotation**.

oblique factor In educational or psychological testing a factor which is correlated with one, or more, other factors is said to be oblique. In one geometrical representation of the situation the vectors which represent them are no longer orthogonal one to another. The use of oblique factors is confined largely to psychology. [See also **factor analysis, factor pattern**.]

observable variable A mathematical or stochastic variable, the values of which can be directly observed, as distinct from unobservable variables which enter into structural equations but are not directly observable.

observational error This term ought to mean an error of observation but sometimes occurs as meaning a response error. [See **non-sampling error**.]

observational study An observational study differs from an experiment, i.e., it is an empirical investigation without using controlled experimentation. The objective of an observational study may be to establish empirical predictors or to elucidate cause-and-effect relationships (Cochran, 1965).

observed at random (OAR) See **missing at random**.

observed information matrix The matrix of negative second derivatives with respect to the parameter of a log-likelihood function, typically evaluated at the maximum likelihood estimator, for which in many cases its inverse gives an approximate covariance matrix of the **maximum likelihood estimators**.

Occam's razor The principle that explanations, or often in the statistical context models, should not be complicated beyond necessity. Also called the principle of parsimony, and attributed to the fourteenth-century theologian William of Occam.

occupancy problems The general class of problems in probability which deal with the random distribution of r objects over n cells with particular reference to the numbers of objects falling in particular cells. [See also **Bose–Einstein statistics**.]

odds The ratio of a probability to its complement: $\pi/(1-\pi)$.

odds ratio A measure of the effect of risk factor in epidemiology. For a binary factor, the ratio of the odds

$$\frac{\Pr(Y=1|X=1)/\Pr(Y=0|X=1)}{\Pr(Y=1|X=0)/\Pr(Y=0|X=0)}.$$

The odds ratio (OR) can be estimated in **case–control studies** and is easily computed with a **logistic regression** model. For rare events its value is close to that of the **relative risk**.

official statistics Most countries have survey programmes on, for instance, the measurement of unemployment, population counts, retail trade, livestock, crop yields and transportation. Almost every country in the world has one or more government agencies (usually national statistical institutes) that supply decision-makers and other users including the general public and the research community with a continuing flow of information on these and other topics. This bulk of data is usually called official statistics. Official statistics should be objective and easily accessible and produced on a continuing basis so that measurement of change is possible (Biemer and Lyberg, 2003).

ogive A general name for the Galton ogive. [See also **distribution function**.]

one-sided test A test of a hypothesis for which the region of rejection is wholly located at one end of the distribution of the test statistic; that is to say, if the statistic is t the region is based on values for which t exceeds some t_1 or for which t does not exceed some t_0, but not both.

one-way classification When a set of variable values can be classified according to the k classes of a single factor such a classification is termed a 'one-way' classification and forms the basis for the simplest case of **analysis of variance**.

open sequential scheme A sequential sampling scheme which does not impose an ultimate limit to the size of the sample. Many sequential sampling schemes will terminate with a high degree of probability even when no limit to the sample size is imposed; and such probability may be so high as to be 'almost certain'. Such schemes are nevertheless called open. [See **closed sequential scheme**.]

open-ended classes If, in a frequency distribution, the initial class interval is indeterminate at its beginning and/or the final class interval is indeterminate at its end, the distribution is said to possess 'open-ended' classes. This feature is undesirable owing to its effect upon certain calculations which require the central value of the class interval, e.g., the power moments. On the other hand, it usually has no effect upon the calculation of the quantiles and in particular the median value of the variable.

open-ended question A question which does not admit of a limited number of definite answers, as for example, 'What do you think of the present government?', as opposed to a closed-ended question such as 'Are you in favour of the present government?', the replies to which can be classified as 'Yes', 'No', 'Don't know'. The distinction is important and useful in practice, although it may be criticized from certain logical and psychological standpoints.

operating characteristic In the theory of decisions, and especially in quality control and sequential analysis, a description of the behaviour of a decision rule which provides the probability of accepting alternative hypotheses when some null hypothesis is true. For example, if the hypothesis is specified as $H(\theta)$, dependent on a parameter θ, an operating characteristic function might show, with θ as variable, the probability of accepting $H(\theta)$ when the true value is θ_0. The graph of this probability as ordinate against θ as abscissa is called the operating characteristic curve or OC curve. The OC function may be regarded as the complement of the **power function** in the theory of testing hypotheses. The expression 'performance characteristic' also occurs, sometimes synonymously and sometimes in a more general sense as describing the consequences of the decision rule for different null hypotheses.

operational research The study of systems as they operate in reality. The more mathematical aspects are optimization (mathematical programming) and stochastic modelling, e.g., of queueing and inventory systems (Winston, 1994).

opinion survey A sample survey which aims at ascertaining or elucidating opinions possessed by the members of a given human population with regard to certain topics.

optimal asymptotic test An alternative name for a **locally asymptotically most powerful test**.

optimal stopping rule Assume that we can observe sequentially random variables $Y_1, Y_2, \ldots$ having a known joint distribution. Suppose that we must stop the observation process at some point, and that if we stop at the nth stage, we receive a 'reward' x_n, a known function of $Y_1, Y_2, \ldots, Y_n$. An optimum stopping rule is one which maximizes the expected reward.

optimality of design See **design optimality**.

optimum allocation In general, the allocation of numbers of sample units to various strata so as to maximize some desirable quantity such as precision for fixed cost. Secondarily, allocation of numbers of sample units to individual strata is an optimum allocation for a given size of sample if it affords the smallest value of the variance of the mean value of the characteristic under consideration. Optimum allocation in this sense for unbiased estimators requires that the number of observations from every stratum should be proportional to the standard deviation in the stratum as well as to the stratum number.

optimum linear predictor A predictor for future observations of a stochastic process which is constrained to be a linear combination of past observations and which minimizes the mean square error of prediction.

optimum statistic An expression which is usually synonymous with **best estimator**. If it refers to a statistic used for testing hypotheses it is usually known as an optimum test statistic.

optimum stratification In a general sense a system of stratification in sampling which optimizes some given criteria. Usually the criteria relate to a set of estimated means and require the minimization of their generalized variance.

optimum test A test which can be shown to possess a certain desirable characteristic or group of characteristics to a greater degree than any other test of the same class.

Ord–Carver system A system of discrete distributions analogous to the Pearson system for continuous variables first proposed by Carver (1919) and developed by Ord (1967).

order of interaction See **interaction**.

order of stationarity A stochastic process $\{x_t\}$ is said to be stationary to the rth order if the expectation $E(x_{t_1}^{\alpha_1} x_{t_2}^{\alpha_2} \ldots x_{t_n}^{\alpha_n})$ exists for any sequence $\{t_i\}$ and all $\alpha_1 + \alpha_2 + \ldots + \alpha_n \leqslant r$, and if such expectation depends only on the $n-1$ differences $t_2 - t_1, \ldots, t_n - t_1$.

order statistics When a sample of variable values is arrayed in ascending order of magnitude these ordered values are known as order statistics. Examples are the smallest value of a sample and the median. More generally, any statistic based on order statistics in this narrower sense is called an order statistic, e.g., the range and the interquartile distance.

ordered alternative hypothesis See **location shift alternative hypothesis**.

ordered categorization A categorization in which, although the variable is not expressible in terms of an underlying measurable variable, it may nevertheless be arranged in order. Where even this is not possible the categorization is said to be unordered. Thus, the classification of persons according to social classes is in some scheme ordered; that according to the type of crime they commit is not.

ordered series A set of variable values which possess a natural sequence in time or space. In another sense, an ordered series is a set of variable values which have been arrayed in some specific manner related to their values, e.g., from the lowest value to the highest value, or from the earliest available value to the latest available value.

orderly stationary point process See **regular stationary point process**.

ordination Reduction of dimensionality of multivariate data, originally to 'ordering' along a single axis, but by extension also to two- or three-dimensional representations. [See **principal components, principal coordinates analysis, multi-dimensional scaling.**]

Ornstein–Uhlenbeck process A model for the velocity of a particle in Brownian motion. The defining equation may be written

$$dU(t) = -\alpha U(t)dt + dY(t),$$

where $Y(t)$ is a **Wiener process**. It was introduced by Ornstein and Uhlenbeck (1930).

orthant probabilities Generally, for an n-dimensional distribution the probabilities of individuals falling into the 2^n orthants into which the sample space is divided by the coordinate planes. In particular, the probability that the components of a multivariate normal distribution with zero means are all positive or all negative.

orthogonal This word occurs in several distinct senses: (i) in its mathematical sense as meaning perpendicular, e.g., in relation to a pair of coordinate axes; (ii) in relation to a set of mathematical functions (see **orthogonal polynomials**); (iii) in relation to two variables or functions of variables, which are said to be orthogonal if they are statistically independent, or for linear functions uncorrelated; (iv) in relation to an experimental design, which is called orthogonal if certain contrasts can be regarded as statistically independent; (v) for parameters, when their maximum likelihood estimates are asymptotically independent.

orthogonal arrays A concept introduced by Rao (1946a) in connection with the design of factorial experiments. Orthogonal arrays are a generalization of **orthogonal squares**. For example, an $(s^2, k, s, 2)$ orthogonal array is equivalent to $k - 2$ mutually orthogonal $s \times s$ **Latin squares**. See also Hedayat, Sloane and Stufken (1999).

orthogonal design See **orthogonal**.

orthogonal polynomials If $P_i \equiv P_i(x)$ is a polynomial with the coefficient of x^i not zero and $F(x)$ is a distribution function, then $P_0, P_1, \ldots, P_n$ form a set of orthogonal polynomials if $\int P_i P_j dF(x) = 0$ $(i \neq j)$, where the integral may include summation of discrete values. If, also, $\int P_i^2 dF(x) = 1$ the set is said to be orthonormal.

orthogonal process A stochastic process $\{X_t\}$ such that $E\{|X_t|^2\} < \infty$ and for which $E\{X_s \bar{X}_t\} = 0$ $(s \neq t)$, $\bar{X}$ being the complex conjugate of X. A stochastic process with orthogonal increments is one for which $E\{|X_t - X_s|^2\} < \infty$ and for which $E\{(X_{t_2} - X_{s_2})(\bar{X}_{t_1} - \bar{X}_{s_1})\} = 0$, where s_1, t_1 and s_2, t_2 do not overlap.

orthogonal projection A generalization of the everyday-life procedure of 'dropping a perpendicular' in which, for example, the observed responses are represented by a vector in a linear vector space and the least squares estimates are obtained by orthogonal projection onto a linear subspace representing the model. The idea was first formalized in a statistical context by Bartlett (1933).

orthogonal regression Given a set of bivariate values (x, y) represented as a set of points with Cartesian coordinates (x, y), the so-called 'orthogonal regression' is the straight line such that the sum of squares of perpendiculars from the points onto the line is a minimum. The term 'regression' is open to objection in this context. It is true that the 'orthogonal regression' minimizes the sum of squares in the direction perpendicular to itself, whereas the ordinary regressions minimize those sums in the direction of the coordinate axes. But **regression** is a property of conditional variables and no such interpretation can be given to the 'orthogonal regression'. In particular it is not invariant under a change of scale.

orthogonal squares If two **Latin squares** can be superimposed so that every letter of the first and every letter of the second occupy somewhere the same position the original squares are said to be orthogonal squares. For example, the two squares

$$
\begin{array}{ccc}
\text{ABCD} & & \text{ABCD} \\
\text{BADC} & \text{and} & \text{CDAB} \\
\text{CDAB} & & \text{DCBA} \\
\text{DCBA} & & \text{BADC}
\end{array}
$$

combine to give

$$
\begin{array}{cccc}
\text{AA} & \text{BB} & \text{CC} & \text{DD} \\
\text{BC} & \text{AD} & \text{DA} & \text{CB} \\
\text{CD} & \text{DC} & \text{AB} & \text{BA} \\
\text{DB} & \text{CA} & \text{BD} & \text{AC}
\end{array}
$$

If this form is written with the second letter in Greek the design is said to be a **Graeco-Latin square**. The second letter may be identified with a further source of variation to be considered in an experiment. More generally, if a number of squares are orthogonal in pairs they are said to be mutually orthogonal. [See **hyper-Graeco-Latin square**.]

orthogonal tests Another name for tests which are independent.

orthogonal variable transformation A linear transformation of variables $X_1, X_2, \ldots, X_n$ to variables $Y_1, Y_2, \ldots, Y_n$ of the form

$$
Y_i = \sum_{j=1}^{n} d_{ij} X_j
$$

such that

$$\sum_{j=1}^{n} d_{ij}d_{kj} = 0, \qquad i \neq k.$$

That is to say, a linear transformation with an orthogonal matrix. The most usual type also has

$$\sum_{j=1}^{n} d_{ij}^2 = 1, \qquad \text{for all } i,$$

in which case the transformation is a particular case of isometry of the n-dimensional Euclidean space.

orthonormal system See **orthogonal polynomials**.

oscillation An oscillation in a time series or, more generally, in a series ordered in time or space is a more or less regular fluctuation about the mean value of the series. In this sense it is to be sharply distinguished from a cycle, which is strictly periodic; thus, while a cyclical series is oscillatory an oscillatory series is not necessarily cyclical.

oscillatory process A class of stochastic processes, usually non-stationary but which includes all second-order stationary processes, where the covariance function may be represented as

$$R_{s,t} = \int_{-\infty}^{\infty} \phi_s(\omega)\phi_t^*(\omega)d\mu(\omega)$$

and the family of oscillatory functions $\phi_t(\omega)$ exists. The $*$ denotes a complex conjugate and $\mu(\omega)$ a measure on the real line involving ω, the angular frequency.

outcome 'Outcomes' are typically referenced in terms of mortality, disease, or disability rates for a given portion of the population. This is a term used in evaluation and analysis to describe the effects of a particular intervention or experiment, and thus as a response variable or object of study. For example, the outcome of a medical treatment could be recovery, continuation of the disease, or death. [See **causality**, **longitudinal survey**.]

outlier prone distributions Neyman and Scott (1971) considered that certain families of distributions, e.g., **gamma distributions** and **lognormal distributions**, are likely to produce 'outliers' as they are implicitly defined by the tests to detect them. Extension of the concept to individual distributions has been made by Green (1976).

outliers In a sample of n observations it is possible for a limited number to be so far separated in value from the remainder that they give rise to the question whether they are not from a different population, or that the sampling technique is at fault. Such values are called outliers. Tests are available to ascertain whether, under certain assumptions, they can be accepted as homogeneous with the rest of the sample.

overall sampling fraction It is sometimes necessary to qualify the term 'sampling fraction' when more than one act of sample selection is involved. Thus in a three-stage sampling scheme if the (sub)sample within a given first-stage sample unit is **self-weighting** for the estimation of the first-stage unit total then the reciprocal of the corresponding raising factor is called the overall sampling fraction for the specified first-stage unit. Overall sampling ratio (or rate) is also used. [See also **sampling fraction**.]

overdispersion Data are said to be overdispersed if the variance is greater than the theoretical value under some model. This applies particularly to counts and proportions that would follow the Poisson or binomial distribution if independent; the index of dispersion is greater than unity. Non-independence can also lead to underdispersion. Confusion with dispersal must be avoided. If seed dispersal is inefficient, the resulting plants will be clustered, and the distribution of counts will show overdispersion.

over-identification See **identifiability**.

overlap design An alternative name proposed by Thompson and Seal (1964) for **serial designs** in the context of routine quality control and experiments.

overlapping sampling units Usually the population of elementary units or basic cells is broken up for purposes of sampling into clusters or grids of units or cells which are mutually exclusive; that is, every elementary unit or basic cell belongs to one and only one sampling unit. It is, however, possible to have a system of sampling units in which the same elementary unit or cell may occur in more than one sampling unit, in which case we have an overlapping system. If properly used such a system provides unbiased estimates.

overviews The summarization of evidence on an issue by combined analysis of distinct studies. Previously discussed under the heading 'analysis of series of experiments' and more recently as meta-analysis. The first systematic discussion of principles is by Yates and Cochran (1938).

P

P–P plot Probability–probability plot; a similar graphical display to a **Q–Q plot**. See also **probability plot**.

p-statistics A set of statistics introduced by Roy (1939) in multivariate analysis. They are closely allied to the sample values of the characteristic roots of dispersion matrices.

p-value The exact significance probability of obtaining a value of a statistic at least as extreme, in relation to the null hypothesis, as that observed. See Gibbons and Pratt (1975).

Paasche–Konyus index See **Konyus index number**.

Paasche's index A form of index number due to Paasche (1874). If the prices (quantities) of a set of commodities in a base period are $p_0, p'_0, p''_0, \ldots$ ($q_0, q'_0, q''_0, \ldots$) and those in the given period are $p_n, p'_n, p''_n, \ldots$ ($q_n, q'_n, q''_n, \ldots$), the Paasche price index number is written

$$I_{0n} = \frac{\sum p_n q_n}{\sum p_0 q_0}$$

where the summation takes place over commodities. In short, the prices are weighted by the quantities of the given period, as distinct from the **Laspeyres index** where they relate to the base period. Generally, an index number of the above form is said to be of the Paasche type even when p and q do not relate to prices and quantities, the characteristic feature being that the weights relate to the given period. [See also **Lowe index, Palgrave index, crossed weights**.]

paired comparison The comparison of a set of objects in pairs, each pair AB being placed in a preference relationship: A preferred to B or B preferred to A or, in more general conditions, neither preferred to the other. The method is used where order relations are more easily determined than measurements, e.g., in investigating taste preferences. More generally, the expression is used to denote the comparison of two samples of equal size where members of one can be paired off against members of the other with possibly a randomized allocation of two treatments, one to each member of the pair.

paired *t*-test A single-sample *t*-test appropriate for comparing two samples in which individuals are paired and observed values within pairs are positively correlated. For example, a pair might consist of 'before' and 'after' measurements on the same individual, or of two log-transformed fluorescence intensities from a DNA sequence spotted on a microarray slide. In each case, the *t*-test is performed on the differences between the pairs of measurements within each individual or spot. [See **paired comparison**.]

pairwise independence When each pair of a set of variables is independent, they are said to be pairwise independent. It does not imply mutual independence.

Palgrave's index An index number recommended in Palgrave's *Dictionary of political economy* (1925). If the prices of a set of commodities in the base period (or the given period) are represented by $p_0, p_0', p_0'', \ldots$ (or $p_n, p_n', p_n'', \ldots$) and the corresponding quantities by $q_0, q_0', q_0'', \ldots$ (or $q_n, q_n', q_n'', \ldots$) Palgrave's index is given by

$$I_{0n} = \frac{\sum p_n q_n \, (p_n/p_0)}{\sum p_n q_n}$$

where the summation takes place over the commodities. It is thus an index of **price relatives** weighted by the total value of commodities in the given period. [See also **Laspeyres' index**, **Paasche's index**.]

Palm function A set of functions introduced by Palm (1943) in connection with point processes. The functions $\phi_k(t)$ may be interpreted as the conditional probability that k points occur in a period of length t assuming a single point has occurred during the first moment, or small finite time period τ, of the period (Cox and Isham, 1980).

panel data Data, especially in a social science setting, where a group of individuals (the panel) are interviewed at intervals to follow development of key features.

panel study A longitudinal study of the same group of people over time.

Papadakis's method The analysis of field experiments based on the correlations among neighbouring plots, suggested by Papadakis (1937).

parallel line assay An important method for the bioassay of a test preparation against a standard. If the expected response is linearly related to the logarithm of dose the regressions of response on log dose for the two preparations will often be parallel and the horizontal distance between them will estimate the logarithm of the relative potency.

parameter This word occurs in its customary mathematical meaning of an unknown quantity which may vary over a certain set of values. In statistics it most usually occurs in expressions defining frequency distributions (population parameters) or in models describing stochastic situations (e.g., regression parameters). The domain of permissible variation of the parameters defines the class of population or model under consideration and is called the parameter space.

parameter of location (or scale) A parameter of a frequency function which can be identified with some measure of location (or scale).

parameter orthogonality Component parameters such that for a particular model the corresponding off-diagonal element of the Fisher information matrix is zero. The idea was first studied by Huzurbazar (1950) and in a more recent context by Cox and Reid (1989). There is a relation with the apparently different notion of orthogonality in experimental design.

parameter point If a class of frequency functions depends on certain parameters, e.g., if the univariate function of x is $f(x, \theta_1, \theta_2, \ldots, \theta_k)$, the domain of variations of the θ's is called the parameter space and any particular set of θ's determines a point in that space.

parametric hypothesis A **hypothesis** concerning the parameter(s) of a distribution.

parametric programming A development of **linear programming** in which the parameters of the objective function, formed by the constraints, are allowed to vary in a determinate fashion.

Pareto, Wilfredo (1848–1923) An Italian economist famous for his studies on income distribution. After his studies in Turin, he became a lecturer in mathematics at the University of Florence. Later he became professor of economics at the University of Lausanne.

Pareto curve An empirical relationship describing the number of persons y whose income is x, first advanced by Pareto (1897) in the form

$$y = \alpha k^{\alpha} x^{-(\alpha+1)}, \qquad 0 \leqslant x \geqslant k,$$

where k and α are positive constants. The expression is now used to denote any distribution of this form, whether related to incomes or not. The variable X may be measured from some arbitrary value, not necessarily zero.

Pareto distribution See **Pareto curve**.

Pareto index The coefficient α in the expression for the Pareto curve is generally referred to as the Pareto index. It affords evidence of the concentration of incomes, or, more generally, of the concentration of variable values in distributions of the Pareto type. [See also **concentration, coefficient of**.]

Pareto-type distribution A loosely used expression to denote any distribution shaped similarly to that of Pareto. It is probably better avoided unless the distribution is actually Paretian apart from origin and scale.

parsimony, principle of The general principle that among satisfactory models those with fewest parameters should be preferred. It is, in particular, cited in selecting **Box–Jenkins models** for time series.

partial confounding If, in a **factorial experiment** with several replicates, there are interactions which are **confounding** in some replicates but not in others, these interactions are said to be partially confounding.

partial correlation The correlation between two variables after allowing for the effect of other variables. For variables X, Y and Z, if r_{XY} denotes the correlation between X and Y, the partial correlation between X and Y allowing for Z is given by

$$r_{XY|Z} = \frac{r_{XY} - r_{XZ} \cdot r_{YZ}}{\sqrt{1 - r_{XZ}^2}\sqrt{1 - r_{YZ}^2}}.$$

partial correlogram The empirical partial autocorrelation function of a time series.

partial least squares (PLS) A method for constructing predictive models when the independent variables are too many and highly correlated. Wold (1966) developed it in the 1960s as an econometric technique. PLS has been applied to monitoring and controlling industrial processes.

partially balanced arrays Generalizations of orthogonal arrays introduced by Chakravarti (1956). They permit a multi-factorial design to deal with a given number of factors by requiring a reduced number of assemblies, i.e., columns in the matrix $\mathbf{A} = \{a_{ij}\}$ where the rows are the factors and the elements a_{ij} the levels of those factors.

partially balanced incomplete block design An experimental design in incomplete blocks for which the layout, though not completely balanced, is partially balanced in the sense that each treatment is tested the same number of times and certain other symmetries exist. This class of design, introduced by Bose and Nair (1939), avoids the large number of replicates which may be required by a completely balanced design. [See also **block, incomplete block**.]

partially balanced lattice In certain cases it is possible to arrange a lattice design so that each effect or interaction is confounded in only some of the replicates constituting the whole design. The lattice is then sometimes described as semi-balanced or partially balanced. [See also **square lattice**.]

partially balanced linked block design The condition for a **partially balanced incomplete block design** also to be a **linked block** design is that its dual, i.e., the design created by interchanging blocks and treatments, is also balanced.

partially consistent observations A term proposed by Neyman and Scott (1948) in the problem of deriving consistent estimators. If a set of observations depends partly on parameters which are common to all and partly on parameters which are specific to the individual observation they are said to be partially consistent. More precisely, if the probability laws of the variables X_i depend (i) on a finite number of parameters which appear in an infinity of variables of the sequence X_i and (ii) on an infinity of parameters each of which appears in the probability law of only a finite number of the variables, the situation is described as one of partially consistent variables. The parameters in the first class are called structural; those in the second class are called incidental.

partially linked block design An extension by Nair (1966) of the **linked block** design to yield new **partially balanced incomplete block designs**

in a dual relationship, i.e., the treatments in the one design become the blocks in the other.

partition of chi-squared (χ^2) In certain circumstances the sum of squares of standardized normal variables about their mean, which is distributed as χ^2, can be meaningfully divided into two or more parts each of which is also distributed as χ^2 independently of the others. This is known as a partition of χ^2.

Pascal distribution An unnecessary alternative name for the **negative binomial distribution** with integral index, presumably because some untraced individual thought that Pascal discovered it.

patch In the terminology of **Mahalanobis**, a compact cluster of units whose variable values all fall in a specified class interval, or if quantitative in a specified category, is called a patch. A further condition is that the cluster should be complete and inextensible. The term 'contour level' is also used.

path analysis A method introduced by the geneticist Sewell Wright (1921) to describe by a graphical representation the directed dependencies among a set of variables. The quantitative representation involved partial and total correlations of multivariate normal distributions. It was later introduced into sociology, econometrics and graphical models.

path coefficients, method of A method of analysis proposed by Wright (1921) for the purpose of relating the matrix of zero-order correlations between the variables in a multiple system to various functional relations which are supposed to connect the variables of that system. Each path coefficient, a function of the standardized variables, measures the fraction of the standard deviation of the dependent variable for which a designated factor is deemed responsible and the term derives from a particular diagrammatic approach used in the exposition. The method of path coefficients is related to ordinary **multiple regression** analysis.

pattern function A function of the sample number n used in connection with the evaluations of sampling cumulants of **k-statistics**. The name derives from the fact that the function depends on the configuration of zeros in an array representing a bipartition.

pattern recognition A branch of computer science concerned with identification of objects of known classes, or grouping of objects. The first stage gives the pattern a digital code, the second stage is closely analogous to discriminant analysis or cluster analysis. The techniques, however, are instrumental and do not, as a rule, involve any consideration of underlying distributions.

patterned sampling An alternative name for **systematic sampling**.

pay-off matrix In the theory of games, a matrix specifying how money or its equivalent is to pass from one player to the other for all the possible outcomes of a two-person game. [See also **loss matrix**.]

peak An observation in an ordered series is said to be a 'peak' if its value is greater than the value of its two neighbouring observations.

peak over threshold An approach to extreme-value statistics originating in hydrology in which exceedances over or under a threshold are modelled by the generalized Pareto distribution, their times being taken to be a Poisson process (Davison and Smith, 1990).

Pearl–Read curve Another name for the logistic curve, a general form of **growth curve**.

Pearson, Egon Sharpe (1885–1980) Egon Pearson continued and developed the pioneering work of his father, Karl, in teaching and research in statistics at University College London and, with Neyman, made fundamental contributions to the theory of hypothesis testing. He was active also in industrial statistics and the compilation of statistical tables.

Pearson, Karl (1857–1936) The English applied mathematician Karl Pearson was founder of the Biometric School and the co-founder of the journal *Biometrika*. He made prolific contributions to statistics, eugenics and the scientific method. Stimulated by the applications of W.F.R. Weldon and F. Galton, he laid the foundations of much of modern mathematical statistics.

Pearson chi-squared test A chi-squared test based on

$$\sum (O - E)^2 / E,$$

where O and E stand for observed and expected frequencies (in contradistinction to other tests based on the **chi-squared distribution**).

Pearson coefficient of correlation The **product moment** coefficient of correlation is sometimes referred to as the Pearson coefficient of correlation because of K. Pearson's part in introducing it into general use.

Pearson criterion See **criterion**.

Pearson curve A distribution from the family of frequency distributions developed by K. Pearson (1894*a*, 1895). The basic equation of the family is

$$\frac{df}{dx} = \frac{(x - a)f}{b_0 + b_1 x + b_2 x^2}$$

where f is the density function. The constants of this equation may be expressed in terms of the first four moments, if these exist. The explicit solutions are classified into types according to the nature of the roots of the equation $b_0 + b_1 x + b_2 x^2 = 0$. By appropriate transformations, many of the important distributions of statistics can be derived from this basic equation. [See also **type I** to **type XII distributions**.]

Pearson–Durbin ratio The ratio, based on Durbin's (1961) modification of Pearson's probability test, is a procedure for discriminating between models from separate families of hypotheses. Another example is the **Kolmogorov–Smirnov test**.

Pearson measure of skewness A measure of skewness proposed by K. Pearson in the form

$$\text{skewness} = \frac{\text{mean} - \text{mode}}{\text{standard deviation}}$$

which, however, suffers from the general indeterminate nature of the mode. For distributions of the Pearson system it may be expressed as

$$\text{skewness} = \frac{\sqrt{\beta_1(\beta_2 + 3)}}{2(5\beta_2 - 6\beta_1 - 9)}$$

where β_1 and β_2 are the first two **moment ratios**.

Pearson residual In regression modelling, a residual formed as the standardized difference between a response value and its fitted mean. In some cases the Pearson goodness-of-fit statistic is the sum of squared Pearson residuals (McCullagh and Nelder, 1989).

Peek's inequality An improvement (Peek, 1933) of the **Camp–Meidell inequality** making use of the mean and the ratio of mean deviation to standard deviation (γ):

$$\Pr\left(\left|X - \bar{X}\right| \geqslant t\sigma\right) \leqslant \frac{4}{9}\frac{1 - \gamma^2}{(t - \gamma)^2}.$$

peeling algorithm An algorithm for computing the likelihood of a stochastic graph in a **Markov process** by taking into account parents and children of each node.

penalized estimation Any form of estimation procedure in which the objective function incorporates a penalty, usually intended to trade off quality of fit and complexity. Penalized least squares and penalized likelihood estimation are common examples.

penalized least squares A way to balance fitting the data closely and avoiding expensive roughness or rapid variation. A penalized least squares estimate is a surface that minimizes the penalized least squares over the class of all surfaces satisfying sufficient regularity conditions.

penalized likelihood To obtain smooth estimates of density functions or hazard functions it is necessary to penalize the likelihood. The penalization is a norm of the function to be estimated (for instance, the L^2 norm of its second derivative) chosen so that unsmooth functions have a large penalization.

penalized quasi-likelihood A modification of the quasi-likelihood to prevent estimates of a block of similar parameters being too dispersed. It is essentially equivalent to assigning the parameters a proper prior distribution.

pentad criterion In factor analysis, an extension of the tetrad criterion developed by Kelley (1925) and Holzinger (1937) and based on sets of five correlations from the correlation matrix. [See **hierarchy**.]

percentage diagram A diagram which exhibits a simple analysis of statistical data in terms of percentages. The actual form of the diagram can vary; examples are the **bar chart** and the **pie chart**.

percentage distribution A **frequency distribution** with the total frequency equated to 100 and the individual class frequencies expressed in proportion to that figure.

percentage point A value, especially as used in a significance test, exceeded only with specified (percentage) probability.

percentage standard deviation See **variation, coefficient of**.

percentile See **centile, quantile**.

percentile confidence interval A bootstrap confidence interval whose limits are empirical quantiles of the simulated statistic and hence are transformation-invariant. The basic percentile interval with level $(1-2\alpha)$ uses the α and $(1-\alpha)$ quantiles of the simulated statistic, while the more sophisticated **BCa confidence intervals** uses quantiles adjusted for skewness and bias of the simulated distribution.

percolation process A stochastic process where the physical interpretation is the dispersion of a fluid through a medium influenced by a random mechanism associated with the medium. This is opposed to a **diffusion process** where the random mechanism is associated with the fluid. For example, a regular mosaic of elements that are conducting or non-conducting, with probabilities p and $1 - p$, arranged at random, has a certain probability of being a conductor; that is, of containing a path of conducting elements in contact with each other.

performance characteristic See **operating characteristic**.

period A term used to describe regularities of recurrence in ordered series, sometimes rather vaguely. Strictly, the word should relate to a period in the mathematical sense; that is to say, a term $u(t)$ has period ω if $u(t + \omega) = u(t)$ for all t, and if a series can be analysed into a sum of such functions, the corresponding set of ω's are the periods of the series. More loosely, the expression is used to denote the interval or average interval between identifiable points of recurrence, e.g., between peaks or troughs of the series. It is better to avoid this usage in general, since the intervals between successive peaks etc. in most time series are not equal and the underlying model may not generate a periodic sequence.

period of a state The period of a **return state** k of a **Markov chain** is the greatest common divisor λ_k of the set of integers n for which $f_{kk}(n) > 0$, where $f_{kk}(n)$ is the probability that the (first) **return time** of state k is n. If

λ_k is greater than unity, state k is called 'periodic'; if $\lambda_k = 1$ the state is called aperiodic.

periodic process If any realization of a **stationary** or **stochastic process** yields a series which is strictly periodic then the process is a periodic process.

periodogram A diagram used in the harmonic analysis of an oscillatory series. If the value of the series at time t is u_t the procedure is to calculate, for $\lambda = 2\pi p/n$ $(p = 1, \ldots, [\frac{1}{2}n])$,

$$A = \frac{2}{n} \sum_{t=1}^{n} u_t \cos \lambda t, \qquad B = \frac{2}{n} \sum_{t=1}^{n} u_t \sin \lambda t.$$

The function $S^2 = A^2 + B^2$ is known as the intensity of the frequency $\lambda/2\pi$ or of the period $2\pi/\lambda$. Graphed against the period as abscissa it gives the periodogram. When multiplied by a constant involving n and graphed against λ it gives the estimated **power spectrum**. Originally designed to isolate strictly periodic components in a noisy signal, it needs smoothing if the underlying process has a continuous spectrum.

Perks's distribution The distribution of the mean in samples from a population of the form

$$\frac{2\lambda}{\pi \left(e^{\lambda x} + e^{-\lambda x}\right)}$$

is sometimes referred to as the Perks (1932) distribution.

permissible estimator See **strictly dominated estimator**.

permutation tests A class of distribution-free tests based upon the fact that any ordering of a random sample of n items has the same probability $1/n!$. In particular, all tests based on ranks are of this type. [See **Pitman's tests**].

persistency A term applied, mainly by meteorologists, to a time series to denote regularity of recurrence. Bartels (1935a) endeavoured to distinguish between 'true' persistency in the sense of periodicity and 'quasi' persistency to denote oscillatory behaviour of a less durable and regular kind.

persistent state A **recurrent state** for which the average **return time** is finite.

person-years Demographic rates commonly contain in the numerator a count of the number of events occurring within some defined time period, and in the denominator an estimate of the number of years lived by persons in the population during that time period. A given person can thus be counted more than once in the denominator. The number of person-years functions in part as an indicator of the amount of the population's exposure to the risk of the event.

perturbation techniques Methods for investigating the robustness or stability of statistical models when the observations suffer small random increments.

Peters' method A method of estimating the standard deviation of a distribution which is approximately normal by multiplying the mean deviation by 1.253, this being the ratio $\sqrt{\pi/2}$ appropriate to the normal distribution (Peters, 1856).

p-function A function introduced by Kingman (1964). If E_t is the event that a phenomenon θ occurs at a time $t \geqslant 0$, and

$$\Pr\{E_{t_i} \cap E_{t_2} \cap \ldots \cap E_{t_n}\} = p(t_1)\Pr\{E_{t_2-t_1} \cap E_{t_3-t_1} \cap \ldots \cap E_{t_n-t_1}\}$$

for all positive integers n and $0 \leqslant t_1 \leqslant t_2 \leqslant \ldots \leqslant t_n$, then $p(\cdot)$, with $p(0) = 1$, is called a p-function; θ is described as a regenerative phenomenon that surely happens at $t = 0$, and $p(t) = \Pr\{E_t\}$.

phase The interval between the **turning points** of a series which is ordered in time or space is termed a phase. The distribution of phase lengths provides one test of random order. The expression is also used in its customary mathematical sense relating to the angle α in sine or cosine terms such as $\sin(\theta t + \alpha)$.

phase confounded designs A method of reducing the block size required for a full design, e.g., in cyclic rotation experiments when **reduced designs** are not available: Patterson (1964) used a method of partially confounding some of the test crop comparisons.

phase diagram A name proposed by Frisch (1937) for a diagram showing two time series x_1 and x_2 plotted as ordinate and abscissa. If the fluctuations of these two variables keep in step then the line joining the plotted points will trace a definite pattern, e.g., similar to an ellipse for oscillatory series.

phase function See **gain function**.

phase spectrum The sample phase spectrum indicates whether the frequency components in one of two series lead or lag the components at the same frequency in the other series; it is a description of the covariance between the two series.

Phi-coefficient A term equivalent to the coefficient V defined under **coefficient of association**. It can be defined as the correlation between binary variables representing the rows and columns of a 2×2 contingency table. [See also **contingency**.]

pictogram A method of visual presentation of statistical quantities by means of drawings or pictures of the subject matter under discussion. The method is restricted to the presentation of simple relationships and in order to overcome the unsatisfactory nature of crude comparisons by the eye of objects of different size it is now customary to represent a unit value of the data by a standard symbol and present the appropriate number of repetitions of this standard symbol to depict the magnitude under discussion. This virtually changes the style of the diagram to a pictorial bar chart. The system has become known as the isotype method.

pie chart A method of diagrammatic representation whereby the components of a single total can be shown as sectors of a circle. The angles of the sectors are proportional to the components of the total. Additional visual aid can be obtained with coloured shading or cross-hatching. Also known as a pie diagram.

pie diagram A more picturesque term for the **pie chart**.

Pillai's trace test A method for testing the following: (i) the equality of mean vectors of n p-variate normal distributions with a common but unknown covariance matrix; (ii) independence between two sets of variables distributed as a normal distribution with unknown means vector; (iii) equality of covariance matrices of two p-variate normal distributions with unknown mean vectors.

pilot survey A survey, usually on a small scale, carried out prior to the main survey, primarily to gain information to improve the efficiency of the main survey. For example, it may be used to test a questionnaire, to ascertain the time taken by field procedure or to determine the most effective size of sampling unit. The term 'exploratory survey' is also used, but in the rather more special circumstance when little is known about the material or domain under enquiry.

pistimetric probability A probability measure, analogous to **fiducial probability**, proposed by Roy (1960) and based upon an etymology relating to 'trust, faith, belief'. Its motivation refers to decision taking in the light of scarce information rather than experimental science.

Pitman, Edwin James George (1897–1993) An Australian mathematician and statistician. Pitman made rigorous yet applicable contributions to theory in areas as diverse as non-parametric inference and the properties of characteristic functions, and developed concepts such as 'closeness', 'asymptotic relative efficiency' and 'sufficient statistics' (Pitman, 1936; Pitman, 1938).

Pitman efficiency The concept of **asymptotic relative efficiency** introduced by Pitman (1949) for comparing the large-sample efficiency of two different tests of a null hypothesis.

Pitman estimator An estimator of the location parameter (ξ) of a distribution proposed by Pitman (1939a). It is an unbiased estimator and optimal in the class of all estimators of θ as the general form is

$$t_n = \frac{\int \xi \prod^n f(x_i - \xi) d\xi}{\int \prod^n f(x_i - \xi) d\xi} \ .$$

It was one of the earliest examples of the principle of **invariance**.

Pitman–Morgan test Morgan (1939) and Pitman (1939b) independently derived the test statistic

$$t = r_{uv} \sqrt{\frac{n-2}{1-r_{uv}^2}}$$

to compare the marginal variances in a bivariate normal distribution of two random variables.

Pitman's tests Distribution-free randomization tests developed by Pitman (1937*a*, 1937*b*) for testing differences of means in two samples, for homogeneity of means in several samples, and for correlation in a bivariate sample. [See **concordant sample**.]

pivotal quantity A function of the sample values and one or more parameters with a distribution that is independent of the parameters. The concept is mainly used in the construction of confidence intervals. For example, if x is normally distributed with estimated variance s^2, $(\bar{x} - \mu)\sqrt{n}/s$ follows a t-distribution and may be used to define confidence limits for μ.

placebo A pharmaceutical treatment having no active ingredient. The purpose of a placebo is to provide a control group in the assessment of a new drug during a clinical trial or to provide a psychological benefit to patients, who believe that they are receiving definitive treatment. In many contexts it is reasonably well established that a placebo gives 'better' responses than no treatment at all, the so-called placebo effect.

Plackett's uniform distribution See **contingency-type distributions**.

plaid square The use of a **quasi-Latin square** in the form of a **split plot design** in such a way that different treatments are applied to whole rows and columns of the square (Yates, 1937*a*). Thus the main effects of these treatments are confounded with rows and columns and are estimated with low precision.

platykurtosis See **kurtosis**.

plot In experimental design this term usually refers to the basic unit of the experimental material. Although it derives from the physical unit of a plot of land in an agricultural trial, its interpretation is very much more general according to the subject matter of the particular design. [See also **split plot design**.]

point biserial correlation A modification of the **biserial correlation** to the case where one variable, instead of being based on a dichotomy of an underlying continuous variable, is discontinuous and two-valued.

point bivariate distribution An alternative name for a bivariate distribution of two discrete variables.

point estimation One of the two principal bases of estimation in statistical analysis. Point estimation endeavours to give the best single estimated value of a parameter, as compared with **interval estimation**, which proceeds by specifying a range of values. Criteria for comparing point estimates include unbiasedness, variance and mean square error, but are usually somewhat arbitrary except in a decision-making context.

point of indifference The central point of the operating characteristic curve: the percentage defective in the bulk that will be accepted or rejected equally often. Alternatively 'point of control', it occurs in attribute sampling schemes.

point process Stochastic process consisting of point events occurring irregularly in time and/or space (Cox and Isham, 1980).

point sampling A method of sampling a geographical area by selecting points in it, especially by choosing points at random on a map or aerial photograph.

Poisson, Siméon-Denis (1781–1840) French mathematician and mathematical physicist. His contribution to probability theory is not confined to the distribution which bears his name or to the expression '**Law of Large Numbers**' (Poisson, 1835), but bears on various areas, ranging from pure mathematics to the mathematics of artillery.

Poisson approximation Any approximation based on the Poisson distribution, the simplest use being for binomial probabilities.

Poisson beta distribution A compound distribution proposed by Holla and Bhattacharya (1965) where the parameter λ of a Poisson distribution is itself distributed as a beta of the first or second kind.

Poisson binomial distribution A discrete distribution of the number of successes in n independent trials with probability p_j of success in the jth trial. The binomial distribution is a special case where the parameters are $n, p_1, p_2, \ldots, p_n$ and $p_1 = p_2 = \ldots = p_n = p$. If p_j is specified as a function of a random variable p and a constant c_j, the resulting distribution obtained by integrating over the frequency of p is sometimes called a Poisson–Lexis distribution.

Poisson clustering process A term proposed by Bartlett (1963) to cover a complex Poisson process where each event in the basic Poisson process is followed by a sequence of associated events, themselves forming a subsidiary process not necessarily Poisson, before the succeeding event in the main process. Where the subsidiary processes are also Poisson we have a **doubly stochastic Poisson process**.

Poisson distribution The Poisson distribution is used as a model when counts are made of events or entities that are distributed at random in space or time. The Poisson distribution developed by Siméon-Denis Poisson has probability function $f(k)$ with mean λ and is

$$f(k) = (\lambda^k e^{-\lambda})/k!, \quad \text{where } k = 0, 1, 2, \ldots, \quad \text{and } \lambda > 0.$$

Here $f(k)$ is the probability of k occurrences of the event in one unit of space or time.

Poisson forest A descriptive term for a two-dimensional Poisson process.

Poisson index of dispersion An index appropriate to events obeying a Poisson distribution. If k samples of the same size have frequencies of occurrence

$x_1, x_2, \ldots, x_k$, with mean $\bar{x}$, the index is

$$\frac{1}{\bar{x}} \sum_{i=1}^{k} (x_i - \bar{x})^2 .$$

If the samples emanate from the same Poisson population this is distributed approximately as χ^2 with $k - 1$ degrees of freedom, a fact which provides a test for consistency with a Poisson distribution. [See also **binomial index of dispersion, Lexis ratio.**]

Poisson probability paper Graph paper showing curves of the relationship between P and λ where

$$P = e^{-k} \left\{ \frac{\lambda^c}{c!} + \frac{\lambda^{c+1}}{(c+1)!} + \ldots \right\}.$$

The axes may be calibrated linearly in P and λ but other scales are in use.

Poisson process A point process with independent increments at constant intensity, say λ. The count after time t thus has a Poisson distribution with mean λt and the distribution of intervals between successive points has an exponential distribution of mean $1/\lambda$.

Poisson truncated normal distribution A distribution compounded of half the normal distribution in λ ($\lambda \geqslant 0$) and a Poisson distribution with parameter λ.

Poisson variation A type of sampling variation considered by Poisson (1837). On each of k occasions let n members be chosen at random, and let the probabilities be the same for all occasions, but such that the probability of success at the drawing of the ith member is p_i ($i = 1, 2, \ldots, n$). The mean number of successes on any occasion is

$$\sum_{i=1}^{n} p_i$$

and the variance is $npq - n\,\mathrm{var}(p_i)$ where $p = \sum p_i/n$, $q = 1-p$ and $\mathrm{var}(p_i)$ is the variance of p_i among the possible values. If all the p_i are equal this reduces to **Bernoulli variation**. In other cases the Poissonian variance is smaller than the Bernoullian variance. This effect is encountered in sampling where the numbers are systematically spread over different strata. The dispersion is said to be subnormal. [See also **Lexis ratio, Lexis variation.**]

Poisson–Dirichlet process The distribution of **allele** gene frequencies in a population, which is related to a non-homogeneous Poisson process with a finite number of points which are scaled to add to unity.

Poisson–Lexis distribution See **Poisson binomial distribution.**

Poisson–Markov process A stochastic process, discussed by Patil (1957), whose probability transition matrix is that of a Markov process and at any point in time the space distribution is that of k independent Poisson variables.

Poisson–normal distribution A distribution resulting from the expected value of a **Poisson distribution** itself taking a truncated **normal distribution**. It is also sometimes known as the Gauss–Poisson distribution.

Poisson–Pascal distribution A name sometimes given to the distribution obtained when a **negative binomial distribution** and a **Poisson distribution** are convoluted.

Poisson's law of large numbers An extension by Poisson of **Bernoulli's theorem**. Both are special cases of results deducible from the **Bienaymé–Tchebychev inequality**. If the probability of an event varies from one trial to another and a set of n trials is $p_1, p_2, \ldots, p_n$, and if there are K successes in n trials, then in repeated sampling

$$\Pr\left\{\left[\frac{K}{n} - E\left(\frac{K}{n}\right)\right] > t\sqrt{\frac{1}{n}\sum_{i=1}^{n}p_i q_i}\right\} \leqslant \frac{1}{t^2},$$

where $E\left(K/n\right)$ is the mean proportion of successes

$$\sum_{i=1}^{n}p_i/n.$$

polar-wedge diagram An alternative name for a **circular histogram** or a **rose diagram**.

Politz and Simmons technique A sample survey procedure developed by Politz and Simmons (1949, 1950) that under appropriate conditions approximately eliminates bias in the estimation of population parameters when the information is collected by means of house calls and **call-backs** are not possible owing to time or budget constraints.

Pollaczek–Khintchine formula Similar to **Pollaczek's formula** (1930), but obtained directly by Khintchine (1932) and acknowledged as such.

Pollaczek–Spitzer identity An important theorem on the distribution of the maxima of successive sums of random variables, due to Pollaczek (1952) and Spitzer (1956b). It states that if $X_1, X_2, \ldots$ is a sequence of independent and identically distributed random variables,

$$S_n = X_1 + X_2 + \ldots + X_n, \qquad S_0 = 0, \qquad T_n = \max_{0 \leqslant k \leqslant n} S_k, \qquad S_n^+ = \max(0, S_n),$$

and if $\phi_n(t)$, $\psi_n(t)$ are the characteristic functions of T_n and S_n^+ respectively, then

313

$$\sum_{i=0}^{\infty} \phi_i(t)Z^i = \exp\left\{\sum_{j=1}^{\infty} j^{-1}\psi_j(t)Z^j\right\},$$

the two series being convergent for $|Z| < 1$.

Pollaczek's formula A formulation of the equilibrium situation for a single-server queue with Poisson input and a general service time distribution.

Pólya–Aeppli distribution The compound of a **geometric distribution** of a parameter λ and a **Poisson distribution** with that value of λ (Pólya, 1930).

Pólya–Eggenberger distribution This distribution may be derived from **Pólya's distribution** with parameters $p = b/(b+r), \gamma = c/(b+r)$ and n the sample size, as the limiting form $p, \gamma \to \theta$ and $n \to \infty$ such that $np \to h$ and $n\gamma \to \theta$ which are the parameters of the limit distribution. It may also be derived from the **negative binomial distribution** by putting the parameters k and p of that distribution in the form $k = h/\theta$ and $p = 1/(1+\theta)$ (Pólya and Eggenberger, 1923).

Pólya frequency function of order 2 If there are two sets of increasing numbers $x_1 < x_2$ and $t_1 < t_2$ and the determinant of the matrix $|f(x_i - t_j)|_{1,2} \geqslant 0$, then f will be a Pólya frequency function of order 2. This group includes the normal, exponential, gamma, beta, logistic and uniform distributions.

Pólya process A particular case of a **birth process** in which the parameter λ_n is given by

$$\lambda_n(t) = \frac{1 + an}{1 + at},$$

a being a constant.

Pólya's distribution A discontinuous frequency distribution considered by Pólya (1930) in connection with **contagious distributions**. It may be generated by drawing with replacement from an urn containing b black and r red balls under the condition that as every ball drawn is returned an additional c balls of the same colour are added to the urn. It is a particular case of the **negative binomial distribution**.

Pólya's theorem There are two meanings. (i) It states that every convex function f on the interval $[0, \infty)$, satisfying the requirement that $f(0) = 1, f(t) \to 0$ as $t \to \infty$, and defined for negative t by symmetry, is a characteristic function (Pólya, 1949). (ii) A result due to Pólya which states that if a sequence of distribution functions $F_1, F_2, \ldots$ tends to F and F is continuous then the convergence is uniform, i.e.,

$$\lim_{n \to \infty} \sup_x |F_n(x) - F(x)| = 0.$$

polychoric correlation An extension of **tetrachoric correlation** to the case of an $m \times n$ table where it may be assumed that the two underlying variables are jointly normally distributed.

polycross designs Experimental designs in plant breeding ensuring that each pair of genotypes occur as neighbours the same (or approximately the same) number of times.

polykay A generalization by Tukey (1950, 1956) of the Fisher k-**statistics**.

polynomial trend A trend line of the general form

$$y = \alpha_0 + \alpha_1 t + \alpha_2 t^2 + \alpha_3 t^3 + \ldots + \alpha_n t^n$$

fitted to a series which is ordered in time or space. The coefficients $\alpha_i, i = 0, 1, 2, \ldots, n$, may be estimated by the method of least squares.

polynomial regression Regression modelling in which the mean response is a linear function of a system of polynomials, often chosen to be orthogonal.

polyspectra The spectrum of a time series may be regarded as the Fourier transform of its autocorrelations, which depend on the product of terms $u(t)u(t + k)$. The transform of the product of more than two terms is called a polyspectrum, e.g., for three terms we have the **bispectrum**.

polytomic table A contingency table with more than two categories in the row and column classifications.

pooling of classes The amalgamation of frequencies in a group of classes to form one frequency in a more comprehensive class. This procedure often serves to eliminate blanks or small subclass numbers in a complex analysis.

pooling of error In some situations where several sets of data are regarded as generated under the same model it is possible to construct several independent 'residual' sums of squares which, under suitable assumptions, all provide estimators of the error variance. These sums of squares may be 'pooled' by adding them together, the resulting estimator of the error variance then being based on more degrees of freedom. This is described as 'pooling the error' or, preferably, as 'pooling the residual sums of squares'.

population In statistical usage the term population is applied to any finite or infinite collection of individuals. It has displaced the older term 'universe', which itself derived from the 'universe of discourse' of logic. It is practically synonymous with 'aggregate' and does not necessarily refer to a collection of living organisms.

population genetics The quantitative study of changes in the genetic structure of a population.

positive binomial distribution The distribution formed from the **binomial distribution** by omitting the value 0. The name is also sometimes given to the usual form of the binomial distribution in order to distinguish it from the **negative binomial distribution**.

positive hypergeometric distribution The distribution formed from the **hypergeometric distribution** by omitting the zero value. [See also **decapitated distribution**.]

positive recurrent state An alternative name for a **persistent state**.

positive skewness See **skewness**.

post cluster sampling A term proposed by Dalenius (1957) to cover the situation where lack of information on the composition of clusters indicates selecting an initial random sample from which the clusters are then formed.

posterior probability The probability of a hypothesis or set of values of a parameter or of a future observation calculated after (i.e., conditionally on) some data as contrasted with the prior probability of the same event before the data are available. The two are related by **Bayes' theorem**. This relation is the basis of the Bayesian approach to statistical inference.

Potthoff's test Potthoff (1963) suggested a test based on the Mann–Whitney statistic for the generalized Behrens–Fisher problem when the underlying distribution is symmetric.

power In general, the power of a statistical test of some hypothesis is the probability that it rejects the null hypothesis when that hypothesis is false. The power is greatest when the probability of an **error of the second kind** is least.

power efficiency Alternative name for **relative efficiency** of a test.

power function When the alternatives to a null hypothesis form a class which may be specified by a parameter θ the power of a test of the null hypothesis considered as a function of θ is called the power function. Exhibited graphically with the power as ordinate against θ as abscissa it provides a clear picture of the 'performance' of the test. Comparisons among a number of different tests are made by comparing their power functions.

power spectrum An alternative name for the **spectral function**.

power sum The sum of a series of observations on a variable each of which has been raised to the same power. Such quantities occur most frequently in the calculation of moments or similar symmetric functions of the observations.

power transformation A data transformation of the form $y = x^p$, sometimes including $y = \log x$, corresponding to $p \to 0$.

precedence test This test, proposed by Nelson (1963), is equivalent to the **exceedances test** and concerns the hypothesis that two samples come from the same population. It consists of counting the number of observations in the sample yielding the smallest observation which precede the observation of the rth rank in the other sample.

precision The property of the set of measurements of being very reproducible or of an estimate of having small random error of estimation. It is to be contrasted with accuracy, which is the property of being close to some target or true value.

precision matrix Inverse of covariance matrix. See **concentration matrix**.

precision, modulus of In the theory of errors of observation, the reciprocal of the standard deviation multiplied by $\sqrt{2}$. It may be interpreted as the parameter h in the general equation for the normal distribution, or error function:

$$y = h\sqrt{\pi}e^{-h^2x^2}$$

where $h = 1/(\sigma\sqrt{2})$, σ being the standard deviation. As h increases, the normal curve becomes relatively narrower, i.e., the variability is reduced and, hence, the modulus of precision measures the closeness with which the observations cluster. [See also **probable error**.]

predetermined variable In the statistical analysis of models, particularly of the economic kind, a variable may be classified as an **endogenous variable** or an **exogenous variable** according to whether it represents an integral part of the system or influences impinging on it from without. Some of the variables may also appear as 'lagged': that is to say, as values occurring at some prior point of time. A predetermined variable is one whose values at any point of time may be regarded as known, and therefore includes either an exogenous or a lagged endogenous variable. The remaining variables are sometimes known as 'jointly determined' or 'currently exogenous' variables.

prediction In general, prediction is the process of determining the magnitude of statistical variables at some future point of time. In statistical contexts the word may also occur in slightly different meanings, e.g., in a regression equation expressing a dependent variable Y in terms of explanatory X's, the value given for Y by specified values of X's is called the 'predicted' value even when no temporal element is involved.

prediction interval The interval between the upper and lower limits attached to a predicted value to show, on a probability basis, its range of error.

predictive decomposition See **decomposition**.

predictive inference Statistical inference in which the objective is not the estimation of parameters but the prediction of future observations from the same, or related, random system as generated the data. Solution is formally easy within a Bayesian formulation and possible but rather less simple in a non-Bayesian setting (Geisser, 1993).

predictive likelihood A modified likelihood proposed by Lauritzen (1974) and Hinkley (1979) for prediction of further observations z from a sequence of observations x dependent on a common parameter θ. The predictive likelihood is defined as

$$f\left(\frac{z}{s}, t\right) L^*\left(\frac{t}{s}\right),$$

where

$$L^*\left(\frac{t}{s}\right) = f\left(\frac{s}{r}\right).$$

Here s, t and r are minimal sufficient statistics given x, z and (x, z) respectively.

predictive regression analysis Methods of regression analysis aimed at minimizing errors in predicting the dependent variable, rather than in the parameters. A similar distinction is made in discriminant analysis.

predictive sample reuse method See **cross-validation**.

predictor See **fixed variable, independent variable**.

pre-emptive discipline A form of **priority queueing** whereby the arrival of an element of higher priority can actually displace the lower-priority element actually in the service channel. When this displaced item returns to service the system must distinguish between 'resumption' at point of break or 'repeat' which ignores the earlier partial service.

preference-field index number A synonym for **Konyus index number**.

preference table The $\binom{t}{2}$ **paired comparisons** of t objects, where ties are not permitted, may be displayed by means of 0, 1 variables in a two-way table known as a preference table.

prevalence, prevalence rate The prevalence of a characteristic is the number of existing cases of the characteristic (typically a disease or health condition) in a population at a designated time. Prevalence data provide an indication of the extent and burden of a health problem and thus may have implications for the scope of health services needed in the community. Prevalence is measured either at a point in time or during a period of time. The prevalence rate is the number of existing cases of the characteristic divided by the size of the population in which the characteristic was identified and counted.

prevision A comprehensive term introduced by de Finetti (1974) that for random variables in general is synonymous with mathematical expectation, and in the particular case of random events means probability.

pre-whitening A transformation used in the measurement of spectra. Degradation of results by computational noise and distortion may often be reduced by analysing the spectrum of a transformed input, whose spectrum has been altered in a specific manner, in order to make its spectrum more nearly white, i.e., to be close to complete randomness. Such a transformation aids accurate computation and is termed pre-whitening.

price compensation index An index number of consumers' prices, constructed as a chain index on the basis of a consumer income, which varies so as to maintain a constant standard of living. The Laspeyres–Konyus index is of this type. [See **Konyus index number.**]

price index An index number which purports to combine several series of price data into a single series expressing an average level of prices, e.g., of retail prices or of prices of manufactured products. [See **Laspeyres, Paasche, Marshall–Edgeworth–Bowley** and **'ideal' index numbers.**]

price relative The ratio of the price of a commodity in the given period to the price of the same commodity in the base period; such ratios enter into price index numbers of the Laspeyres or Paasche form.

Priestley's $P(\lambda)$-test A test for a discrete component in the spectrum of a time series, corresponding to a strictly periodic component in the original series (Priestley, 1962).

primary unit This term is used in at least two senses. The first concerns a statistical unit of record which is basic in the sense that it does not depend upon any derived calculations, for example: persons, kilometres, tonnes, litres, thousands of an article. The second usage of this term arises in sample surveys. Where a population consists of a number of units which may be grouped into larger aggregates but are not subdivided the units are called primary. For example, if a town is divided into districts, each of which is divided into blocks, each block comprising a number of houses; and if a sample of houses is desired, the house would be the primary unit. [See **multi-stage sampling.**]

principal components If each member of an aggregate bears the values of p variables $1, \ldots, p$ it is, in general, possible to find a linear transformation to p new variables $\xi_1, \ldots, \xi_p$, which (i) are uncorrelated and (ii) account in turn for as much of the variation as possible in the sense that the variance of ξ_1 is a maximum among all linearly transformed variables; the variance of ξ_2 is a maximum among all linearly transformed variables orthogonal to ξ_1, and so on. Such variables are called principal components.

principal components regression Regression analysis using principal components to circumvent difficulties due to collinearity in the explanatory variables.

principal coordinates analysis A special case of multi-dimensional scaling where one has information on the pairwise relationships between a set of n objects, represented as a matrix. Commonly a plot of the objects in two dimensions is output to give a visualization of the distances in the matrix.

principle of equipartition The division of the range of a frequency function into a number of parts such that the frequencies corresponding to each part are equal. This method is sometimes used in tests of goodness of fit; it is also used in sample surveys for the construction of strata (Kitagawa, 1956). In physics it has an entirely different meaning concerned with the distribution of energy between different degrees of freedom.

prior probability See **posterior probability, Bayes' theorem.**

priority queueing A queueing system where the actual order of service for the arrivals is determined by some scheme of relative priority, i.e., not a first-come first-served discipline.

probability A basic concept which may be taken either as undefinable, expressing in some way a 'degree of belief', or as the limiting frequency in an infinite random series. Both approaches have their difficulties and the most convenient interpretation of probability theory is a matter of objectives. Fortunately both lead to much the same calculus of probabilities.

probability density function See **probability distribution.**

probability distribution A distribution giving the probability of a value x as a function of x; or more generally, the probability of joint occurrence of a set of values of random variables $X_1, \ldots, X_p$ as a function of their values. It is customary, but not the universal practice, to use 'probability distribution' to denote the probability mass or probability density of either a discontinuous or continuous variable and some such expression as 'cumulative probability distribution' to denote the probability of values up to and including the argument x.

probability element The probability associated with a small interval of a continuous variable, written in some such form as $f(x)dx$, or in general

$$f(x_1, x_2, \ldots, x_n)dx_1 dx_2 \ldots dx_n,$$

where f denotes a **probability density function.**

probability integral An alternative name for the **distribution function** or the **cumulative distribution function** for continuous variables. For example, the probability integral of a continuous variable X is a function $F(x)$ having the property that

$$F(a) = \Pr\{X \leqslant a\} = \int_{-\infty}^{a} f(x)dx$$

where $f(x)$ is the probability (density) function.

probability integral transformation If X is a continuous random variable with density function $f(x)$ and distribution function $F(x)$, a transformation to a new variable Y given by

$$Y = \int_{-\infty}^{X} f(u)du = F(X)$$

is called the probability integral transformation. Y is uniformly distributed in the range $0 \leqslant y \leqslant 1$.

probability limits Upper and lower limits assigned to an estimated value for the purpose of indicating the range within which the true value is supposed to lie according to some statement of a probabilistic character. For example, confidence limits, control chart limits and fiducial limits are probability limits. They may be contrasted with the numerical limitations sometimes placed upon aggregates in descriptive statistics which are indicative of possible errors of collection or compilation rather than probability statements.

probability mass A term which is sometimes used to describe the magnitude of a probability or the relative frequency of observations located at a particular variable value, as distinct from being spread over a continuous range.

probability moment If a probability distribution is given by $dF = y\,dx$ the rth probability moment is defined by

$$\Omega_r = \int_{-\infty}^{\infty} y^r\,dx.$$

probability paper A graph paper with the grid along one axis specially ruled so that the distribution function of a specified distribution can be plotted as a straight line against the variable as abscissa. These specially ruled grids are available for the normal, binomial, Poisson, lognormal, extreme-value and Weibull distributions.

probability plot A plot of sample order statistics against quantiles of a theoretical distribution, useful for assessing adequacy of distributional assumptions.

probability ratio test This term is often used (in preference to **likelihood ratio test**) in connection with sequential tests.

probability sampling Any method of selection of a sample based on the theory of probability; at any stage of the operation of selection the probability of any set of units being selected must be known. It is the only general method known which can provide a measure of precision of the estimate. Sometimes the term random sampling is used in the sense of probability sampling. [See **non-random sample**.]

probability surface A bivariate frequency function; that is to say, the three-dimensional representation of a bivariate frequency (probability) distribution with the frequency (probability density) along one axis and the variables along the other two axes.

probability weighted moments estimation A form of method of moments estimation in which moments are weighted by the density and/or cumulative distribution functions, useful when moments themselves do not exist. (Hosking, Wallis and Wood, 1985).

probable error An older measure of sampling variability now almost superseded in statistics by the **standard error**. The probable error is 0.6745 times the standard error, the reason for the choice of the numerical coefficient being

321

that the quartiles of a normal distribution with variance σ^2 are distant $0.6745\,\sigma$ from the mean, so that one-half of the distribution lies within the range: mean $\pm 0.6745\,\sigma$.

probit The **normal equivalent deviate** increased by five in order to make negative values very rare. The word was suggested by Bliss (1934) as a contraction of 'probability unit'.

probit analysis The analysis of quantal response data using the **probit** transformation.

probit regression line In the analysis of quantal response data the percentages or proportions of the subjects reacting to the doses of stimulus can be converted into probits and plotted as ordinates against the logarithms of the doses. A line through this scatter of points, fitted by freehand methods or by an arithmetical process, is the probit regression line. The usual arithmetic procedure for obtaining it is an iterative method of successive approximation by means of a weighted linear regression of working probits on the logarithms of the doses.

process average fraction defective The average of the proportion of defective items in samples from a manufacturing process; the probability that an item from a process which is statistically under control is defective.

process with independent increments See **additive (random walk) process**.

processing error A type of error which can occur in the processing of statistical data. In survey data, for example, processing errors may include errors of transcription, errors of coding, errors of data entry and errors of arithmetic in tabulation.

procrustes methods Any modelling procedures in which data are forced to fit a preassigned pattern. Procrustes transformation refers to the rotation of a dispersion matrix to conform to a given factor model. [See **confirmatory factor analysis**.]

producer's risk In acceptance inspection, the risk which a producer takes that a batch will be rejected by a sampling plan even though it conforms to requirements. It is related to the probability of an error of the first kind in the theory of testing hypotheses in that it corresponds to the probability of rejecting a hypothesis when it is, in fact, true. [See also **consumer's risk**.]

product binomial model Representation used for the systematic study of categorical data in which a particular sampling model is assumed for the data.

product integral A mathematical construct bearing the same relation to products as the standard definition of integrals has to sums. Its role in summarizing the analysis of censored survival data was indicated by Cox (1972).

product-limit estimator Another term used particularly by actuaries for the **Kaplan–Meier estimator**.

product moment If the density function of n random variables $X_1, X_2, \ldots, X_n$ is given by $f(x_1, \ldots, x_n)$ the product moment, joint or multivariate moment of order $r, s, \ldots, u$ is the mean value of $X_1^r X_2^s \ldots X_n^u$:

$$\int \ldots \int_{\mathbb{R}^n} x_1^r x_2^s \ldots x_n^u f(x_1, \ldots, x_n) dx_1 \ldots dx_n.$$

product moment correlation A product moment correlation coefficient is so termed because its numerator is the first **product moment** or covariance of the two variables concerned. It is defined as

$$\rho = \frac{\text{covariance}(X, Y)}{\{\text{var}(X)\,\text{var}(Y)\}^{1/2}}.$$

profile analysis A method of clustering the attitude questions in a survey. In common with factor analysis and linkage analysis, it starts from the correlations between the pairs of attitude questions, arranged in the usual correlation matrix. The profile of an attitude question is its correlation with the other attitude questions, i.e., its row in the correlation matrix.

profile likelihood For a parametric model with both parameters of interest and nuisance parameters, the likelihood function (or more commonly its logarithm) considered as a function of the parameter of interest and calculated maximized over the nuisance parameter separately for each possible value of the parameter of interest. It has some but not all the properties of an ordinary likelihood function and in particular can be used to calculate approximate confidence regions via the chi-squared distribution.

prognostic factor An explanatory variable effective in predicting an outcome variable, especially in a medical context with survival or death, or success or failure of treatment as the outcome.

progressively censored sampling In life and dosage response studies it is frequently desirable, or practically imposed, that some of the surviving sample units are withdrawn at an initial stage of censoring, others being withdrawn at later stages. This practice facilitates the economic use of test facilities and does provide some data on the longer life spans. This form has also been referred to as 'hypercensored samples' and 'multiple censored samples'.

projection This term is used in several connected senses. (i) In relation to a time series it means a future value calculated according to predetermined changes in the assumptions of the environment. (ii) More recently it has been used in probability theory to denote the conditional expectation of a variable. Since a regression equation gives the expectation of the dependent variable conditional upon values of the predicated ('independent') variables and such equations are used for forecasting or prediction, the usages are connected. (*iii*) In a geometrical approach to the theory of the linear (and other) models, projection has its usual geometrical meaning. [See also **posterior probability**, **regression**.]

projection pursuit An exploratory technique for seeking structures, like clusters or unexpected shapes or separations, of a high p-dimensional data set by searching through projections of the data to lower k-dimensional space.

propensity interpretation of probability An approach to probability theory developed by Popper (1959), principally to remove a subjective element in the role of 'the observer' in quantum theory.

proper distribution See **improper distribution**.

proportional frequency In relation to a frequency distribution, the proportional frequency in any class is the frequency of the class divided by the total frequency of the distribution. The term sometimes occurs in a different sense in relation to bivariate or multivariate frequency arrays. For instance, if, in a table of p rows and q columns, the q frequencies in each row are proportional to the q row totals and, similarly, therefore, for the columns the case is said to be one of proportional frequencies. The term is sometimes used similarly in connection with proportional subclass numbers in analysis of variance.

proportional hazards model A model assuming that factors affecting survival have an additive effect on the log hazard function. The most familiar application is **Cox's regression model**.

proportional odds model A model in which explanatory variables act by scaling odds ratios.

proportional sampling A method of selecting sample numbers from different strata so that the numbers chosen from the strata are proportional to the population numbers in those strata. [See also **uniform sampling fraction**.]

proportional subclass numbers See **proportional frequency**.

prospective study A **cohort study** based on retrospective records is sometimes referred to as a historical prospective study.

protocol The aims and design of **clinical trials** are usually set out beforehand in a document known as the protocol. 'Protocol departures' refer to instances in which the protocol is not followed, as, for example, changes of treatment necessitated by side-effects. The exclusion of such cases in analysis can lead to serious bias.

proximity analysis A method used in numerical taxonomy for arranging items in a line or a plane, or in a space of higher dimensions, so that like is adjacent to like and far from unlike. The method is based upon rank ordering of interpoint distances and inverse ordering of similarities. [See **cluster analysis**.]

pseudo-factor An artificial or dummy factor used in the design of factorial experiments, generally to render the number of factors a convenient one for the application of some specified balanced design.

pseudo-inverse An alternative name for **generalized inverse**.

pseudo-likelihood A function calculated for a relatively complex model, e.g., with complex dependencies among errors, by ignoring certain features, for instance by assuming the dependency to have simple Markovian form. The resulting maximum likelihood estimates usually have reasonable properties but their precision cannot be found directly from the observed information matrix. Besag (1977) introduced the idea in the context of a spatial model.

pseudo-random numbers Numbers generated by a deterministic process (in a computer or calculator) that have many of the properties of random numbers. They are widely used, particularly for **simulation models**, but are sometimes distrusted because of the possibility of unsuspected periodicities.

pseudo-spectrum A somewhat misleading term denoting the mathematical expectation of the estimates of **spectral density function** obtained without regard to the stationarity of the process.

pseudo-values of the jackknife See **jackknife**.

psychological probability See **probability**.

psychometrics Statistical methods of particular relevance to psychology especially to educational and similar testing procedures.

publication bias The systematic error arising in combining evidence from different studies (**overviews** or meta-analysis) if some studies showing no significant effect are not available.

pure birth process See **birth process**.

pure random process The simplest example of a stationary process where, in discrete time, all the random variables are mutually independent. In continuous time the process is sometimes referred to as **white noise**, relating to the energy characteristics of certain physical phenomena.

pure strategy See **strategy**.

purposive sample A sample in which the individual units are selected by some purposive method. It is therefore subject to biases of personal selection and for this reason is now rarely advocated in its crude form. [See **quota sample, balanced sample**.]

Q

Q–Q plot A quantile–quantile graph in which the theoretical quantiles for a distribution (e.g., the standardized normal) are plotted against the empirical quantiles from the data. A visual judgement of linearity is used to decide if the data appear consistent with the theoretical distribution (Wilk and Gnanadesikan, 1968).

Q-technique A method of analysis of similarities or relationships in which, given a matrix of n observations on m individuals, n rows and m columns, the correlations or other statistical measure are sought between the m columns down the n rows, i.e., between individuals, as distinct from the R-technique which looks for relations between variables, namely between rows along the columns.

quad A square-shaped **basic cell**; also, the area of such a cell.

quadrant dependence If the probability of any quadrant $X \leqslant x, Y \leqslant y$ under the distribution $F(x, y)$ is compared with its independence probability, then the pair (x, y) or its distribution as denoted by Lehmann (1966) is positively quadrant dependent if

$$\Pr(X \leqslant x, Y \leqslant y) \geqslant \Pr(X \leqslant x) \Pr(Y \leqslant y).$$

The negative dependence holds with the inequality reversed.

quadrat A sampling device in the form of a square lattice. It may be a framework which can be placed on the ground, e.g., for dividing a plot into subplots, or a square grid of some transparent material for superposition on a map. More loosely the term is sometimes used to denote (i) a mesh which is rectangular, not necessarily square, and (ii) one unit of the lattice.

quadratic estimator An estimator which is based upon some quadratic function of sample values. For example, the standard deviation may be estimated from the square root of the variance, a quadratic estimator, or from the mean deviation or the range, which are linear estimators.

quadratic exponential model A representation of the joint distribution of a set of binary variables in exponential family form with only linear and quadratic terms (Cox and Wermuth, 1992).

quadratic form A homogeneous quadratic function of the form

$$Q = \sum_{i=1}^{n} \sum_{j=1}^{n} a_{ij} x_i x_j = \boldsymbol{x}' \mathbf{A} \boldsymbol{x}$$

where $\mathbf{A}$ is the matrix of the quadratic form and generally taken to be symmetric. The quadratic form is important in multivariate analysis and, in particular, **analysis of variance**. [See also **Cochran's theorem**.]

quadratic mean See **mean square error**.

quadratic programming A major development of **linear programming** in which some or all of the constraints and the objective function are quadratic in the variables.

quadrature spectrum The form of spectrum which measures the covariance between the sine and cosine components, or out of phase components, of a sample time series. If the quadrature spectrum is denoted by $Q_{12}(f)$ then

$$Q_{12}(f) = -A_{12}(f)\sin F_{12}(f)$$

where $A_{12}(f)$ and $F_{12}(f)$ are the **phase spectrum** and **cross amplitude spectrum** of the two series.

quadri-normal distribution See **modified normal distributions**.

qualitative data See **quantitative data, attribute**.

qualitative interaction An interaction between two treatment variables or a treatment variable and an explanatory variable in which a reversal of direction of effect occurs.

quality adjusted life year Health status scored between 0 and 1 (ideal health) and then summed for an individual over subsequent life span (possibly with future values discounted). Used by health economists in an attempt to suggest rational allocation of resources. Closely related to **disability adjusted life years** used by demographers and others to compare societies and assess time trends.

quality control The statistical analysis of process inspection data for the purpose of controlling the quality of a manufactured product which is produced in large numbers. It aims at tracing and eliminating systematic variations in quality, or reducing them to an acceptable level, leaving the remaining variation to chance. The process is then said to be statistically under control.

quality control chart See **control chart**.

quality of life Term used in both sociology and in medical studies for attempts to measure via suitable questionnaires individuals' attitude to their life. In the medical case, health status is a more neutral term.

quality of official statistics A number of features that reflect user needs. The features vary between organizations but Eurostat's ($2000a$, $2000b$) quality vector has the following components: (1) Relevance of statistical concept. A statistical product is relevant if it meets users' needs. (2) Accuracy of estimates, which is the difference between the estimate and the true parameter value. (3) Timeliness and punctuality in disseminating results. (4) Accessibility and clarity of information. (5) Comparability, which is a prerequisite for harmonized statistics. (6) Coherence. Statistics are coherent if elementary concepts can be combined in more complex ways or, if they come from different sources, they are based on common definitions, classifications and methodological standards. (7) Completeness. Domains for which statistics are available should reflect the needs and priorities expressed by users as a collective.

quantal response The response of a subject to a stimulus is said to be quantal when the only observable, or the only recorded, consequence of applying the stimulus is the presence or absence of a certain reaction, e.g., death. A quantal response may be expressed as a two-valued variable taking values 0 and 1. [See also **binary data**.]

quantile A class of $(n - 1)$ partition values of a variable which divide the total frequency of a population or a sample into a given number n of equal proportions. For example, if $n = 4$ then the $n - 1$ values are the **quartiles** although the central variable value is generally termed the **median**. [See also **deciles, quintiles**.]

quantitative data Strictly, this term, in contrast to qualitative data, should relate to data in the form of numerical quantities such as measurements or counts. It is sometimes, less exactly, used to describe material in which the variables concerned are quantities, e.g., height, weight or price, as distinct from data deriving from qualitative attributes, e.g., sex, nationality or commodity.

quantitative response A reaction, by an experimental unit to a given stimulus, which may be measured on a variable scale. For example, the response may be measured in terms of weight, size or reaction time: in particular the survival time.

quantity relative The ratio of the quantity of a commodity in the given period to the corresponding quantity in the base period; such ratios enter into quantity index numbers of the Laspeyres or Paasche form. [See **price relative**.]

quantum hypothesis A hypothesis in which the possible values of a parameter are discrete and hence increase by quantum jumps.

quantum index An index number which purports to show the changes in quantity, usually of goods or services produced, purchased or sold, independently of changes in prices or money values. One such index is of the Laspeyres type obtained by weighting quantities in the given and base period by prices in the base period. Quantum index numbers do not necessarily measure changes in volume or weight. [See **Laspeyres' index**.]

quartile There are three values which separate the total frequency of a distribution into four equal parts. The central value is called the **median** and the other two the lower (first) and upper (third) quartiles respectively. They are a particular set of **quantiles**.

quartile deviation A measure of dispersion based upon the distance between certain representative values of the variable. In this case the representative values are the upper and lower quartiles and the quartile deviation is defined by

$$\text{QD} = \frac{1}{2}(Q_3 - Q_1).$$

An alternative name for the quartile deviation is the semi-interquartile range.

quartile direction See **circular quartile deviation**.

quartile measure of skewness If the lower quartile is Q_1, the upper quartile is Q_3 and the median is M, the quartile measure of skewness of a frequency distribution is

$$\frac{(Q_3 - M) - (M - Q_1)}{(Q_3 - Q_1)}.$$

quartile variation An alternative to the **standard deviation** as a measure of variation. If the lower quartile is Q_1, and the upper quartile is Q_3, the coefficient of quartile variation, denoted V_Q, is given by

$$V_Q = \frac{100(Q_3 - Q_1)}{Q_3 + Q_1}.$$

quartimax See **factor rotation**.

quartimin See **factor rotation**.

quasi-compact cluster See **cluster sampling**.

quasi-experiments Term used for observational studies in which, while the allocation of treatments is not under the investigator's control, efforts are made to come as close as possible to the conditions of an experiment.

quasi-factorial design An experimental design for which a formal correspondence may be set up between the treatments and their combinations and the combinations of the treatments of a factorial set. Thus, for example, four treatments can be put in correspondence with the four combinations of two factors, each at two levels, and the design for the four treatments derived from one or more of those appropriate to the 2^2 **factorial experiment**. This class of design is useful for treatments which do not have a factorial structure and, especially with the recovery of **information** between blocks, is in general more efficient than designs using **randomized blocks** or 'control plot' techniques. Although terminology is still somewhat fluid 'quasi-factorial' is usually synonymous with 'lattice' in relation to a design. The so-called lattice designs owe their name to the fact that the treatments are allocated in some systematic way according to a pattern which can be represented diagrammatically on a lattice.

quasi-independence A term introduced by Goodman (1968). Suppose that p_{ij} is the proportion of individuals in a population that fall in the ith row and the

jth column of an $R \times C$ contingency table, and that S is a given subset of the cells (i, j) of the table. For the subset S, the row and column classifications of the table are defined as quasi-independent if the proportions p_{ij} can be written as

$$p_{ij} = a_i b_j \qquad \text{for all cells } (i, j) \text{ in } S$$

for a set of positive constants a_i and b_j.

quasi-Latin square A term proposed by Yates (1937b) for certain kinds of factorial designs in the form of Latin squares. In this experimental design certain of the interactions are confounded with the rows and columns of the squares. These designs eliminate the variations due to differences between the rows and the columns from the experimental error of those effects which are not confounded but, unlike the Latin square itself, each treatment does not appear once to every row and column.

quasi-likelihood If a response vector Y has mean vector μ and matrix **variance function** $V(\mu)$, then the log quasi-likelihood Q (Wedderburn, 1974) is determined by the equation

$$\partial Q / \partial \mu = V^{-1}(\mu)(Y - \mu).$$

If Q exists, it has many of the properties of a log-likelihood function, and provides a form of semi-parametric inference that extends the domain of **generalized linear models**.

quasi-Markov chain A stochastic process over $N + 1$ states identified by the integers 0 to N, for which the Markovian property holds between any two states i and j for which $1 \leqslant i, j \leqslant N$. Such a quasi-Markov chain is said to be of order N, and arises in the theory of regenerative phenomena.

quasi-maximum likelihood estimator A set of least squares estimators for multiple equation models where the asymptotic distributional properties of the estimators are almost independent of the form of the distributions of errors in the equation, so that least squares solutions are virtually equivalent to maximum likelihood solutions. [See **quasi-likelihood**.]

quasi-median A generalization of the concept of a sample median to the arithmetic mean of any pair of symmetrically ranked order statistics; thus for a sample of size n they are defined by

$$\bar{\theta}_i = \begin{cases} \frac{1}{2} \left(X_{(k+1-i)} + X_{(k+1+i)} \right) & \text{if } n = 2k + 1 \\[2ex] \frac{1}{2} \left(X_{(k-i)} - X_{(k+1-i)} \right) & \text{if } n = 2k \end{cases}$$

where $X_{(j)}$ is the ith order statistic. The usual definition of a sample **median** is obtained by setting i to zero.

quasi-Newton methods Methods for solving equations, or finding maxima and minima, using the **Newton–Raphson method**, but without supplying the functional form of the derivatives. These are instead estimated from differences.

quasi-normal equations If the normal equations of least squares estimation have the parameter estimates based upon **instrumental variables** rather than the ordinary model variables, then the equations are said to be quasi-normal; 'normal' in this sense has nothing to do with the normal (Gaussian) distribution.

quasi-random sampling Under certain conditions, largely governed by the method of compiling the **sampling frame** or list, a systematic sample of every nth entry from a list will be equivalent for most practical purposes to a random sample. This method of sampling is sometimes referred to as quasi-random sampling.

quasi-range A term proposed by Mosteller (1946) for the difference $x_s - x_r$ where $1 \leqslant r \leqslant s \leqslant n$, in the ascending order statistics of a sample of n observations. Common practice is to take the range of $n - 2i$ observations, omitting the i largest and i smallest.

Quenouille's test A test proposed by Quenouille (1947) for the goodness of fit of an autoregressive model to a time series. The test was extended by Wold (1949) to the case of a moving averages model; and by Walker (1950) for an autoregressive model with error terms comprising moving averages of independent variables. Quenouille (1958) extended it further to a pair of time series or two lengths of one series.

questionnaire A group or sequence of questions designed to elicit information upon a subject, or sequence of subjects, from an informant. [See also **schedule**.]

Quetelet, Adolphe (1796–1874) Active in many areas of science, notably mathematics, astronomy and meteorology, Adolphe Quetelet owes his celebrity to the international blossoming under his impetus of the study of populations emanating from sophisticated statistics organized systematically and treated probabilistically. He wrote popular works on science, like *Instructions populaires sur le calcul des probabilités* published in 1828, and was the tutor of the Princes Ernest and Albert of Saxe-Coburg and Gotha, in 1836. The lessons which Quetelet gave were published in 1846 under the title *Lettres à S.A.R. le Duc régnant de Saxe-Cobourg et Gotha, sur la théorie des probabilités appliquées aux sciences morales et politiques* (Quetelet, 1832).

queueing problem The problem of queues, or congestion, arises in a variety of fields where there is a service to be offered and accepted rather than a product to be made. In general the problem is concerned with the state of a system, e.g., the length of the queue or queues at a given time, the average waiting time, queue discipline and the mechanism for offering and taking the particular service. The analysis of queueing problems makes extensive use of the theory of stochastic processes.

quintiles The set of four values which divide the total frequency into five equal parts. [See **quantile**.]

quota sample A sample, usually of human beings, in which each investigator is instructed to collect information from an assigned number of individuals (the quota) but the individuals are left to the investigator's personal choice. In practice this choice is severely limited by 'controls', e.g., the investigator is instructed to secure certain numbers in assigned age groups, equal numbers of the two sexes, certain numbers in particular social classes, and so forth. Subject to these controls, which are designed to make the sample as representative as possible, the investigator is not restricted to the contacting of assigned individuals as in most forms of probability sampling.

quotient regression A regression of the form

$$Y_t = \frac{\alpha_0 + \alpha_1 X_{1t} + \ldots + \alpha_n X_{nt}}{\beta_0 + \beta_1 Z_{1t} + \ldots + \beta_m Z_{mt}} + \varepsilon_t$$

(Wold, 1966), where X and Z are regressor variables and ε is a random residual. The coefficients α and β are usually estimated by least squares in an iterative procedure.

R

R A flexible and powerful software environment for statistical data analysis and graphical display (Ihaka and Gentleman, 1996). It is a freely available clone of **S-PLUS**, and has replaced the latter as the primary statistical programming tool at many research and academic institutions around the world. R can be downloaded from the Comprehensive R Archive Network (CRAN) at http://cran.r-project.org or mirror sites and is available for most computing systems. R has an extensive collection of tools for statistical modelling and analysis and a great strength is that new functions can be built and implemented relatively easily. For example, the Bioconductor Project which provides bioinformatics software is built primarily on R (see http://bioconductor.org).

***R*-estimator** A **robust estimator** of a type derived from a test based on ranks.

***R*-process** A procedure introduced by Birkes, Dodge and Seely (1976) applied to the incidence matrix N of a two-way classification which determines what cell expectations are estimable.

***R*-technique** See ***Q*-technique**.

racial likeness, coefficient of A coefficient proposed by K. Pearson (1921). It was designed for the testing of homogeneity of two multivariate distributions but was extended to the measurement of distance between them (Fisher, 1936a). For the latter purpose it has certain disadvantages and has been replaced by the D^2 **statistic** of Mahalanobis.

radix A **life table** may show the numbers, surviving at different ages, of an initial number of individuals, e.g., 10,000, which is the radix of that particular tabulation.

Raikov's theorem This theorem, first stated by Raikov (1938), shows that if X_1 and X_2 are independent and $X = X_1 + X_2$ has a Poisson distribution, then each of the random variables X_1 and X_2 has a Poisson distribution. This result can be generalized to any finite number of independent random variables.

raising factor Apart from its ordinary significance this term is used in the following special sense. The coefficients of a linear function of the values of the sample units used to estimate population, stratum, or higher-stage unit totals are called raising, multiplying, weighting or inflation factors of the corresponding sample units. If the raising factors of all the sample units are equal, the common raising factor is called the raising factor of the sample, and the sample itself is called self-weighting. It should be noted that the raising factors depend not only on the sampling plan but also on the method of estimation.

random This word is used in senses ranging from 'non-deterministic' (as in **random process**) to 'purely by chance, independently of other events'.

random allocation designs An alternative name for **random balance design** which emphasizes a principle of construction involving random sampling from the array of treatment combinations in a full k-factorial design.

random balance design A factorial design intended to deal with the case where the number of factors exceeds the number of observations. It is constructed by selecting treatment combinations at random from the possible set, subject to constraints to ensure balance. The method was proposed by Satterthwaite (1959) and developed and criticized by other writers (Dempster, 1960).

random bifurcation The manner of growth of a tree or rooted graph when the free end that bifurcates on each occasion is a random choice from those available. Among many applications, the distribution of the different possible shapes of tree is of interest in population genetics, polymer science and in the study of tributary systems of rivers.

random coefficient model A regression model in which some or all of the regression coefficients are random variables, each with a specified distribution.

random component If a magnitude consists of a number of parts compounded in some way, e.g., by addition or multiplication, any such part as is a **variable** is a random component of the magnitude.

random distribution This expression is sometimes used for a probability distribution. It is also sometimes employed to denote a distribution of probability which is uniform in the range concerned, i.e., a rectangular distribution. It seems better to avoid the term altogether, or, in such expressions as 'points randomly distributed over an area', to specify clearly the law of distribution involved.

random effects model An alternative name for **model II** analysis of variance. [See also **variance components**.]

random error An error, that is to say a deviation of an observed from a true value, which behaves like a variable in the sense that any particular value occurs as though chosen at random from a probability distribution of such errors.

random event An event with a probability of occurrence determined by some probability distribution. The term is used somewhat loosely to denote either an event which may or may not happen at a given trial, such as the throwing of a 6 with an ordinary die, or an event which may or may not happen at any given moment of time such as an industrial accident to an individual.

random impulse process A stochastic process describing the linear motion of a particle subject to a series of small impulses which are random.

random linear graph The formation of lines joining pairs of points independently and randomly selected from a group of, say, N points.

random normal scores test See **normal scores tests**.

random number generator An ideal device to generate sequences of random numbers. In practice, any generator of random numbers on computers uses a deterministic procedure, so it generates **pseudo-random numbers**. A natural random number generator, as proposed by Dodge (1996), is the billions of decimals of π.

random order An order of a set of objects when the ordering process is carried out in such a way that every possible order is equally probable. Tests of random order are freely used to test the hypothesis that there are systematic elements present which would prevent the observed order from being random.

random orthogonal transformations A device used in multivariate analysis (Wijsman, 1957) whereby transformations are performed with orthogonal matrices the elements of which depend on a random vector.

random process In a general sense this term is synonymous with the more usual and preferable **stochastic process**. It is sometimes employed to denote a process in which the movement from one state to the next is determined by a variable which is independent of the initial and final state. It is better to denote such a process as a **pure random process**.

random sample A sample which has been selected by a method of **random selection** as contrasted with one chosen by some method of purposive selection.

random sampling error A **sampling error** in cases where the sample has been selected by a random method. It is common practice to refer to random sampling error simply as 'sampling error' where the random nature of the selective process is understood or assumed.

random sampling numbers Sets of numbers used for the drawing of random samples. They are usually compiled by a process involving a chance element and in their simplest form consist of a series of digits 0 to 9 occurring (so far as can be ascertained) at random with equal probability. [See also **pseudo-random numbers**.]

random selection A sample selected from a finite population is said to be random if every possible sample has equal probability of selection. This applies to sampling without replacement; a random sample with replacement is such that each item is independently selected with equal probability (and so each possible ordered sample is chosen with equal probability).

random series A series the numbers of which may be regarded as drawn at random from a fixed distribution. [See **irregular kollectiv**.]

random start In selecting a **systematic sample** at intervals of n from an ordered population, it is sometimes desirable to select the first sample unit by a random drawing from the first n units of that population. The sample is then said to have a random start.

random tessellation The random subdivision of the plane into non-overlapping convex polygons. [See **Dirichlet tessellation.**]

random variable A real-valued function defined on a **sample space**. [See **discrete random variable, continuous random variable.**]

random walk The path traversed by a particle which moves in steps, each step being determined by chance either in regard to direction or in regard to magnitude or both. Cases most frequently considered are those in which the particle moves on a lattice of points in one or more dimensions, and at each step is equally likely to move to any of the nearest neighbouring points. The theory of random walks has many applications, e.g., to the migration of insects, sequential sampling and, in the limit, to diffusion processes.

randomization A set of objects is said to be randomized when arranged in a random order by some objective method; and, by a slight extension, a set of treatments applied to a set of units is said to be randomized when the treatment applied to any given unit is chosen at random from those available and not already allocated.

randomization tests Any test characterized by the test statistic being referred, under the null hypothesis, to its distribution on permuting the observations in every equally likely way that could have occurred under that hypothesis, especially where explicit randomization is involved in the design. In certain symmetrical situations where each observation could equally well have been positive or negative under the null hypothesis, all possible allocations of plus or minus signs to the absolute values of the observations are also used in producing a similar type of distribution.

randomized blocks An experimental design in which each **block** contains a complete replication of the treatments, which are allocated to the various units within the blocks in a random manner and hence allow unbiased estimates of error to be constructed.

randomized decision function A decision function which is selected from a set of possible decision functions with the help of a chance mechanism, telling the experimenter, before the outcome of an experiment is known, exactly what action to take as a result of the experiment.

randomized fractional factorial designs A class of designs proposed by Ehrenfeld and Zacks (1961) which yield unbiased estimators, valid tests and confidence intervals for the parameters of interest without the usual assumptions concerning the confounding of higher-order interaction. Two methods of randomization are used: (*i*) based upon blocks of treatment combinations, a type of **cluster sampling**; (*ii*) random selection of treatments from every block, a form of stratified sampling. [See **stratified sample.**]

randomized model A statistical model, generally an experimental design, where the treatment combinations are assigned to experimental units by some random arrangement.

randomized response A technique suggested by Warner (1965) for eliminating evasive answer bias in sample surveys of human populations. The respondent selects a question on a probability basis from two or more questions presented to him or her without revealing to the interviewer which has been chosen, and records the answer. A batch of such replies provides useful information for estimating the proportion of the population that has the 'sensitive characteristic'.

randomized significance tests It is generally impossible to define a **critical region** to give an exact **type I error** for discrete variables. The problem may be overcome by attaching fixed probabilities of lying in the critical region to certain values of the test statistic, and declaring such results significant or non-significant according to the value of an unrelated random number. This device is of no practical value, but may be used to compare the power of two tests with exactly the same type I error.

range The largest minus the smallest of a set of variable values. The range is of itself an elementary measure of dispersion but, in terms of the **mean range** in repeated sampling, it may afford a reasonable estimate of the population standard deviation.

range chart A chart used in statistical quality control on which the recorded quality criterion is the range of samples. This **control chart** is used to maintain a check upon the variability of the quality of the particular product or processes. The range is a less sensitive criterion for changes in variability than is the sample variance but is much more easily calculated.

rank This term occurs in statistical work in at least three contexts. (*i*) In the theory of order relations, the rank of a single observation among a set is its ordinal number when the set is ordered according to some criterion such as values of a variable. (*ii*) In matrix theory the term occurs in its usual mathematical sense, being the greatest number r of linearly independent rows or columns which can be found in it. (*iii*) Derived from the previous usage, the rank of a multivariate distribution is the rank of its dispersion matrix, and is thus the number of variables which are independent in the sense of not being connected by linear equations. [See also **singular distribution**.]

rank correlation Rank correlation measures the intensity of correlation between two sets of rankings or the degree of correspondence between them. There are two principal coefficients of rank correlation: **Kendall's** τ (1938) and **Spearman's** ρ (1904).

rank order statistics Statistics based only on the rank order of the sample observations, e.g., the rank correlation coefficients. 'Rank order statistics' are distinguishable from 'order statistics', e.g., the median, the range which make use of the metric values of the observations.

rank-randomization tests Randomization tests with the observations replaced by their size ranks before computation of the test statistic. The sample space and null distribution of the test statistic are thereby standardized, so the null distribution may profitably be tabulated. Rank-randomization tests are usually of the same high order of efficiency as their ordinary randomization counterparts, but are much easier to apply.

rank scores A set of scores attached to ranks, also known as rank transformation, to improve the efficiency of tests based upon them. The original scores given by Fisher and Yates (1938) assumed an underlying normal distribution. Other transformations appropriate for survival data or for situations when only a proportion of items respond are asymmetric, with large differences between the high scores, based for example on an underlying exponential distribution.

rank transformation See **rank scores**.

rank-weighted mean The **L-estimator** (linear combination of order statistics) of the form (Sen, 1964)

$$T_{n,k} = \binom{n}{2k+1}^{-1} \sum_{i=k+1}^{n-k} \binom{i-1}{k} \binom{n-i-1}{k} X_{n:i},$$

where $0 \leqslant k \leqslant (n-1)/2$. Note that $T_{n,0} = \bar{X}$, while $T_{n,k}$ is the median for $k = [(n-1)/2]$.

rankit A transform of quantal response data based on ranks. [See **probit**.]

Rao–Blackwell theorem A theorem concerning minimum variance estimation stated by Rao (1945) and Blackwell (1947). If a **minimum variance** estimator exists it is always a function, e.g., the conditional expected value, of the sufficient estimator. In particular a minimum variance unbiased estimator can be found as the conditioned expectation of any unbiased estimator given the value of a complete sufficient statistic, a process known as Rao–Blackwellization. [See **sufficiency**.]

Rao–Kupper model A generalization of the **Bradley–Terry model** of paired comparisons to allow for tied observations, proposed by Rao and Kupper (1967).

Rao's scoring test A large-sample test of a hypothesis proposed by Rao (1948, 1965) using a scoring system based on likelihood derivatives.

Rasch model A model for a two-way layout of binary response variables in which the logit of the success probability is the sum of row and column parameters.

ratio estimator An estimator which involves the ratio of two variables, i.e., a ratio whose numerator and denominator are both subject to sampling errors. The term occurs particularly in sample survey theory. If the members of a population each bear the values of two characteristics, x and y, and the total of x, say X, is known for the population, the corresponding total of y, say Y, can be estimated by multiplying X by a sample ratio consisting of the sample total of y divided by the sample total of x.

ratio scale A graphical scale in which equal absolute variations correspond to equal proportional variations in the data. The most common form of chart employing the ratio is the logarithmic or **semi-logarithmic chart**.

raw moment A moment of a frequency distribution calculated about some origin other than the arithmetic mean. The usage is not universal and some authors use this term to denote moments either about the mean or not before corrections for grouping are applied. [See **Sheppard's corrections**.]

raw score The score as originally obtained in some psychological, educational or other test. [See also **T-score, z-score**.]

Rayleigh distribution A χ^2 distribution with two degrees of freedom, so called because it was considered by Rayleigh in some physical situations.

Rayleigh tests Tests of uniformity on the circle and on the sphere, due to Rayleigh (1919). Both tests have critical regions of the form

$$\bar{R} > C$$

where $\bar{R}$ is the *mean resultant*. The circular test is the **uniformly most powerful test** invariant under rotations for alternatives in the family of **von Mises distributions** (Ajne, 1968), whilst the spherical test is the UMP test invariant under rotations for alternatives in the family of **Fisher's distributions** (Watson, 1967; Beran, 1968).

realization A realization of a stochastic process $\{X_t\}$ is a particular series of values $(\ldots x_{-2}, x_{-1}, x_0, x_1, \ldots)$, possibly observed from the process. The realization may be regarded as a 'member' of the process in the same way that an individual observation is regarded as the member of a population. In general the realization is of infinite extent, but a finite observed section of it is also sometimes referred to as a realization.

reciprocal scaling Another name for **correspondence analysis**, used particularly in ecological statistics.

record linkage The problem of matching pairs of records held in different files, either to obtain more complete records about specific individuals or to compare records on two sets of characteristics from individuals matched on the basis of a third set.

records tests Distribution-free tests for trend in time series based on the breaking of record values. An observation is called an upper/lower record if it is greater/smaller than all previous observations in the series. Foster and Stuart (1954) have described two such tests.

recovery of information The standard analysis of experiments designed in the form of **incomplete blocks** fails to use information about treatment effects which can be obtained from comparisons among block totals. More refined methods of analysis to 'recover' this information were proposed by Yates (1940).

rectangular association scheme If N is the **design matrix** of a partially balanced incomplete block design with three associate classes, the two modes of classification (of treatments in N) for the relation of first and second association can be superimposed, in the form of a rectangular array (Vartak, 1959).

rectangular distribution Strictly a continuous distribution of type

$$dF(x) = dx/k, \qquad \alpha \leqslant x \leqslant \alpha + k$$

where k is a constant. The expression is also sometimes used to denote a discontinuous distribution for which all variable values have the same probability.

rectangular lattice An experimental design introduced by Harshbarger (1949) as an extension of the **square lattice**. It is appropriate to $k(k+1)$ treatments, which are regarded as corresponding to the points of a $k \times (k+1)$ lattice, the blocks consisting of k units.

rectified index number An index number formula which is obtained by taking the geometric mean of two other index numbers of opposite bias. The two index numbers are sometimes said to be geometrically 'crossed', e.g., the **'ideal' index number**. The object of rectification is usually to make the resulting index satisfy either, or both of, the **factor reversal test** or the **time reversal test**.

rectifying inspection Inspection of a product which aims at removing any defective units found and replacing them by effective units. In this way the quality of a batch of product may be considerably improved and, in any case, a batch is never rejected. This type of inspection is not applicable when the inspection test is destructive. [See also **average outgoing quality limit**.]

rectilinear trend An alternative name for a **linear trend**.

recurrence game A sequence of trials of an event which are conducted as a game, in that a 'reward' is received, or a 'fee' incurred, under certain recurrences.

recurrence time This concept occurs in the analysis of stochastic processes in two ways. In connection with a renewal process it is the time, or number of steps, between two similar (recurrent) states. With reference to a **point process** and in particular a stationary point process where the time origin is arbitrary, the forward recurrence is the interval from the arbitrary time origin to the next point event. The backward recurrence time is formed by direct analogy.

recurrent Markov chain A periodic **Markov chain**.

recurrent state A state k in a Markov chain is said to be recurrent if, with probability one, the Markov chain will eventually return to k, having started at k. A state k is said to be transient or non-recurrent if this probability is less than one.

recursive algorithm A rule defining a function in terms of the function itself. Thus $n! = n \times (n - 1)!$ is a recursive algorithm for calculating an integral factorial function; it must be terminated by defining $0! = 1$.

recursive residual Residuals used in sequential setup; the kth recursive residual is usually the difference of the kth observation and its prediction for this valued based on the observations obtained before the kth observation; more generally, the kth recursive residual is a functional of the kth observation and its prediction.

recursive system The word 'recursive' has been used by some writers to denote relations which are 'recurrent', presumably under the impression that the former is the correct adjectival form derived from the verb 'to recur'. A purist will avoid the word in this sense. Wold (1953) has proposed to describe as 'recursive' systems of equations in econometrics with three properties: (i) they are recurrent in the sense that if the values of the variables are known up to time $t-1$ the equations give the values for time t; (ii) the values of the variables at time t are obtainable one by one in some order or other; (iii) each equation of the system expresses a unilateral causal dependence.

reduced design A device in experiment design for restricting the large size of block required by a full design. For example, in cyclic rotation experiments Patterson (1964) proposed dividing the crop sequences into groups according to comparability.

reduced equations A method of estimation in econometrics whereby the original equations are modified so that each endogenous variable is expressed as a function of the set of exogenous variables and, possibly, the errors.

reduced form method In econometrics, a method of estimating the parameters in a stochastic system which relies on the expression of the **endogenous variables** individually in terms of **predetermined variables**. [See also **limited information methods**.]

reduced inspection See **normal inspection**.

reduced sample A term sometimes applied to the involuntary censoring of a sample when some items are no longer observable (Kaplan and Meier, 1958).

reduced variable A **standardized variable** using location and scale statistics other than the mean and the standard deviation.

reduction of data The process whereby a large number of observations are brought within manageable compass for convenience of handling and interpretation. Particularly in the physical sciences it is an older term for statistical methods.

Reed–Frost model A chain binomial model. If, at any stage, there are r infectives, the probability of infection is taken as $1 - (1 - p)^r$ (instead of p, as in Greenwood's model).

Reed–Münch method A method proposed by Reed and Münch (1938) for the rapid assessment of the equivalent doses of standard and test preparations which would produce a 50 per cent quantal response. The method is strictly applicable only to **tolerance distributions** which are symmetrical.

reference period In one sense this is synonymous with base period. It may also refer to the length of time, e.g., week or year, for which data are collected.

reference prior distribution Forms of prior distribution intended for use in standard Bayesian inference procedures. A modern, more specific use is for the prior chosen to maximize the expected value of perfect information, defined as the maximum of the limiting utility gain when the sample size tends to infinity, compared with having no data (Bernardo and Smith, 1994).

reference set An alternative term for **fundamental probability set**.

reflecting barrier Certain additive or **random walk** processes represent the motion of a particle in one or more dimensions and, in certain cases, limitations may be imposed on the motion in the form of barriers (constraints) which, once reached, reflect the particle and continue the motion, as distinct from absorbing it.

refusal rate In the sampling of human populations, the proportion of individuals who, though successfully contacted, refuse to give the information sought. The proportion is usually and preferably calculated by dividing the number of refusals by the total number of the sample which it was originally desired to achieve. Where, however, there are other causes of non-achievement, e.g., persons have died or left the area, the refusal rate is sometimes calculated as the number of refusals divided by the number of persons contacted, i.e., by the number of refusals plus the number of successful or partially successful contacts.

regenerative process A class of stochastic process attributed to Palm (1943) characterized by possessing regeneration points. These points are those epochs where the occurrence of the state S is sufficient to 'regenerate' the process in the sense that probabilities are no longer dependent on past history. The Markov process is a special case which requires every time to be a regeneration point.

regressand A synonym for dependent or response variable in a regression relation.

regression This term was originally used by Galton to indicate certain relationships in the study of heredity but it has come to mean the statistical method developed to investigate those relationships. If a variable Y consists of two components, the first one depending on a variable X and the latter being a random error, i.e., if $Y = f(X) + \varepsilon$, then the regression of Y on X is the equation $Y = f(X)$ where it is supposed that ε has zero expectation. The definition remains valid if X, instead of being a single variable, refers to a set of variables $X_1, X_2, \ldots$. The X's are called 'explanatory', 'independent', 'predicated' variables, 'predictors' or 'regressors'. Y is called the 'dependent variable', 'predictand' or 'regressand' (Birkes and Dodge, 1993).

regression coefficient The coefficient of a variable in a regression equation.

regression curve A diagrammatic exposition of a regression equation. For two variables this can be shown on a plane with the 'independent' variable X as abscissa and Y as ordinate; and for three variables it is possible to construct solid models or reduce the representation to a plane surface by use of the **isometric chart** and the **stereogram**. The term is sometimes interpreted to mean a regression equation of a degree higher than the first, the emphasis then lying on the word 'curve' as opposed to a straight line.

regression dependence If, for two stochastically dependent variables X, Y, $\Pr(Y \leqslant y | X = x)$ is non-increasing in x, y is said to be positively regression dependent (Lehmann, 1966). Likewise the dependence is negative regression if the probability statement is non-decreasing in x.

regression diagnostics Techniques, graphical and otherwise, for examining fitted multiple regression equations to investigate the effects of particular data points or small groups of data points.

regression equivariant estimator An estimator $\mathbf{T}_n(Y_1, \ldots, Y_n)$ of β in the linear regression model $Y_i = \boldsymbol{x}_i'\boldsymbol{\beta} + e_i$, $i = 1, \ldots, n$, $\boldsymbol{\beta} \in \mathbb{R}^p$, is regression equivariant or affine equivariant if

$$\mathbf{T}_n(Y_1 + \boldsymbol{x}_1'\boldsymbol{b}, \ldots, Y_n + \boldsymbol{x}_n'\boldsymbol{b}) = \mathbf{T}_n(Y_1, \ldots, Y_n) + \boldsymbol{b}$$

for every $\boldsymbol{b} \in \mathbb{R}^p$.

regression estimate In general, an estimate of the value of a dependent variable Y obtained from substituting the known values of the explanatory variables X in a regression equation connecting Y and X. The term has a particular application in sample surveys. If the regression of A on B may be estimated from a sample and the total of B is known for the population, the total of A may be estimated from the regression equation. It is then called a regression estimate.

regression line A straight line representing the dependence on X of the conditional expectation of Y given X.

regression quantile An extension of the empirical (sample) α-quantile ($0 < \alpha < 1$) to the linear regression model

$$Y_i = \boldsymbol{x}'_i\boldsymbol{\beta} + e_i, \qquad x_{i1} = 1, \qquad i = 1, \ldots, n, \qquad \boldsymbol{\beta} \in \mathbb{R}^p,$$

defined as a solution of the minimization

$$\sum_{i=1}^{n} \left\{ \alpha(Y_i - \boldsymbol{x}'_i\boldsymbol{b})^+ (1 - \alpha)(Y_i - \boldsymbol{x}'_i\boldsymbol{b})^- \right\} := \min, \qquad \boldsymbol{b} \in \mathbb{R}^p.$$

Here $a^{\pm}$ denotes the positive or negative part of a, respectively (Koenker and Bassett, 1978).

regression rank scores Introduced by Gutenbrunner and Jurečková (1992) these represent a straightforward extension of the ranks of observations to the linear regression model

$$Y_i = \boldsymbol{x}'_i\boldsymbol{\beta} + e_i, \qquad x_{i1} = 1, \qquad i = 1, \ldots, n, \qquad \boldsymbol{\beta} \in \mathbb{R}^p.$$

They are defined as vector $\hat{a}_n(\alpha) = (\hat{a}_{n1}(\alpha), \ldots, \hat{a}_{nn}(\alpha))$, $0 \leqslant \alpha \leqslant 1$, which is a solution of the parametric linear programming problem

$$\sum_{i=1} Y_i \hat{a}_{ni}(\alpha) : \max$$

$$\sum_{i=1}^{n} \hat{a}_{ni} = n(1 - \alpha)$$

$$\sum_{i=1}^{n} x_{ij}\hat{a}_{ni} = (1 - \alpha) \sum_{i=1}^{n} x_{ij}, \qquad j = 2, \ldots, p$$

$$0 \leqslant \hat{a}_{ni}(\alpha) \leqslant 1, \qquad i = 1, \ldots, n.$$

The regression rank scores test of Wilcoxon type for the hypothesis $\boldsymbol{\beta}^* = \boldsymbol{0}$ in the linear regression model

$$Y_i = \boldsymbol{x}'_i\boldsymbol{\beta} + \boldsymbol{x}^{*\prime}_i\boldsymbol{\beta}^* + e_i, \qquad i = 1, \ldots, n, \qquad \boldsymbol{\beta}^* \in \mathbb{R}^q,$$

is based on the quadratic form of the vector of statistics $\sum_{i=1}^{n} x^{*}_{ij}\hat{b}_{ni}$, $j = 1, \ldots, q$, where $\hat{b}_{ni} = \int_0^1 \hat{a}_{ni}(\alpha)d\alpha$ (Gutenbrunner *et al.*, 1993).

regression surface See **regression curve**.

regressor A synonym for explanatory variable in a regression relation. [See **regression**.]

regret The loss function which arises in decision theory generally involves two terms the second of which represents the difference between total loss and unavoidable loss. It is this excess loss, the 'regret', which has to be minimized (van der Waerden, 1960).

regular best asymptotically normal estimator A class of estimator initially proposed by Neyman (1949a) and for which the limiting distribution is normal with the correct asymptotic mean and minimum asymptotic variance (van der Vaart, 1998).

regular estimator An estimator for which there hold certain regularity conditions, principally concerning the differentiability of the estimator with respect to the variable values on which it depends and of the frequency distribution with respect to its parameters. [See **Cramér–Rao inequality**.]

regular group divisible incomplete block design A group divisible **incomplete block** design is regular (Bose and Connor, 1952) if $r > \lambda_1$ and $rk > \lambda_2 v$ where r is number of replicates, v the product of m groups each of n treatments; λ_1 number of blocks that treatments in the same group occur and λ_2 the blocks in which treatments in different groups occur.

regular Markov renewal process See **regular state**.

regular state An initial state of a Markov renewal process from which there is probability zero of the process undergoing an infinite number of transitions in a finite time. The process is said to be regular if all the states are regular.

regular stationary point process A point process $\{N(t), t \geqslant 0\}$ for which

$$\lim_{t \to 0} \frac{\Pr\left(N(t) \geqslant 2\right)}{t} = 0.$$

Such a point process is sometimes alternatively described as orderly.

reification In multivariate analysis, the substantive interpretation of factors, principal components, etc., defined by the mathematical procedures used.

rejectable quality level The level of quality as determined by percentage defective, for example, for which a buyer would wish to have only a low probability of accepting. This probability is the **consumer's risk**.

rejection error See **α-error, error of the first kind**.

rejection line See **acceptance line**.

rejection number See **acceptance number**.

rejection region In the theory of testing hypotheses, a region of the **sample space** such that if a sample point falls within it the hypothesis under test is rejected. [See **critical region**.]

rejection sampling Monte Carlo simulation in which variables generated from a proposal or envelope distribution are rejected with probabilities that ensure that those accepted have the target distribution. The Metropolis–Hastings algorithm is an important example. Also called acceptance–rejection sampling or the envelope method.

rejective sampling Sampling with unequal probabilities with replacement in which the whole sample is rejected as soon as any individual is selected a second time (Hajek, 1964).

relative efficiency (of an estimator) A measure of comparative efficiency of two estimators of the same parameter. If estimator t_1 has, for sample size n_1, the same precision, in the sense of same sampling variance, as estimator t_2 for sample size n_2, the relative efficiency of t_1, with respect to t_2, is n_2/n_1.

relative efficiency (of a sample design) In the design of experiments this is equivalent to **efficiency factor**. In sample survey work the usage takes into account the cost of the survey and the relative efficiency is the ratio of the 'cost per unit of information', where **information** is used in the sense of Fisher. A third usage concerns sampling plans and is based upon the ratio of the cost of the optimum plan to the plan in operation.

relative efficiency (of a test) The ratio of sample sizes concerned with two tests of statistical hypothesis necessary to yield the same **power** against the same alternative hypothesis. The concept is due to Cochran (1937) and Pitman (1948). An alternative formulation due to Blomqvist (1950) specifies equal slopes for the power curves at the parameter point rather than equal power.

relative frequency The frequency in an individual group of a frequency distribution expressed as a fraction of the total frequency.

relative information A term introduced by Yates (1939) in connection with the **partial confounding** of experimental effects in factorial experimental designs. Where an effect is partially confounded the relative amount of information is the ratio of the amount actually available to what would be available if there were no confounding.

relative potency The relationship of two estimated stimuli, one of which acts as a standard, which produce the same response. In biological assay the relative potency of a test preparation compared with a standard preparation is generally obtained from the inverse ratio of doses which result in identical responses, i.e., equally effective doses.

relative precision A term which is frequently used to denote the ratio of the error variances of two sample designs which are different but which are based upon the same sampling unit and the same size of sample. The usage is not universal, however, since some writers use the term **relative efficiency (of a test)** for this concept. The relative efficiency and the relative precision are equal in the case of simple random sampling for the mean of a large population, but not necessarily otherwise.

relative risk A term used mainly in medical statistics for the ratio of the proportions suffering from a disease in certain classes (e.g., smokers and non-smokers).

relative variance A term sometimes used to denote the square of the coefficient of variation. [See **variation, coefficient of**.]

reliability This term is used in three different contexts. (*i*) In connection with biological assay, Finney (1947) has defined reliability of an assay as the reciprocal of a function of the confidence interval of the estimate of potency of the stimulus. (*ii*) The term is also used in factor analysis, especially in connection with the statistical analysis of psychological and educational tests. The 'reliability' of a result is conceived of as that part which is due to permanent systematic effects, and therefore persists from sample to sample, as distinct from error effects which vary from one sample to another. The term has not spread to other sciences. In a slightly more specialized sense the noun 'reliability' sometimes means a **reliability coefficient**. (*iii*) The term is now also used in the context of the life of industrial components and equipment as the probability of survival, after time t_n, that is to say $1 - F(t)$ where $F(t)$ is the distribution function of the lifetimes. [See also **factor analysis**.]

reliability coefficient A coefficient introduced by Spearman (1910) into psychology. Its object is to assess the systematic component of a variable (test) as distinct from error components. In psychology it is usually measured by the correlation between the results of two administrations of the same test. The 'reliability' as a quantity is the complement of the error variance of the test but this usage requires care in view of the widespread use of the term in connection with industrial equipment in the form $1 - F(t)$ where $F(t)$ is the distribution function of the lifetimes. [See also **factor analysis**.]

REML Reduced maximum likelihood. A maximum likelihood estimate calculated not directly from the full observations but after transformation to remove the influence of some irrelevant aspects. The essential idea is due to Bartlett (1937) and the development into an important method for estimating components of variance from unbalanced data is due to Patterson and Thompson (1974).

remote sensing Data collection by artificial satellite. Such data present special statistical problems associated particularly with atmospheric interference and imprecise location on the surface.

renewal density The derivative of the **renewal function**.

renewal distribution See **general renewal process**.

renewal function For a **renewal process** in continuous time, the expected number of renewals in the time interval $(0, t]$.

renewal process A class of stochastic point process in which times between events are independently and identically distributed. [See **general renewal process**.]

renewal theorem An important theorem in renewal theory (Smith, 1958), from which a number of other results can be deduced. It states that if $m(t)$ is the expected number of renewals for a general renewal process in the interval $[0, t)$, and $Q(t)$ is any non-increasing non-negative function on the interval $[0, \infty)$ satisfying $\int_0^\infty Q(t)dt < \infty$, then provided that the **renewal distribution** is not a **lattice distribution** it follows that

$$\lim_{t \to 0} \int_0^t Q(t-s)dm(s) = \frac{1}{\mu} \int_0^\infty Q(s)ds$$

where μ is the mean renewal time.

renewal theory Originally this term meant the analysis of recurrent events to problems concerning the duration of life in aggregates of physical equipment. Such aggregates are sometimes referred to as self-renewing when the failure of any unit results in its replacement. Currently it is the theoretical study of the **renewal process**.

repeated measures design A design in which each individual gives responses to treatment on different occasions. Analysis as a split-plot experiment is appropriate only if the residuals on different occasions are equally correlated. If the correlation structure is more complex, the appropriate analysis is either a multivariate analysis of variance, or one that assumes a defined time-series model.

repeated survey A sample survey which is performed more than once with essentially the same **questionnaire** or **schedule** but not necessarily with the same sample units. [See also **sampling on successive occasions**.]

repetition A term denoting the execution of a statistical enquiry at different points in space or time, usually as part of a coordinated programme, as distinct from **replication**.

repetitive group sampling plan A set of sampling schemes introduced by Sherman (1965) for acceptance inspection based upon the proportion of defective units. According to a fixed criterion, a sample proportion indicates accept, reject or disregard and repeat the process until a decision to accept or reject is achieved: no account is taken of the number of intermediate samples or their proportion defective.

replacement See **sampling with replacement**.

replacement process A sequential control process for which, at various points, the system can be returned to some initial state.

replication The execution of a measurement or other operation or survey more than once so as to increase precision and to obtain a closer estimation of sampling error. Replication should be distinguished from repetition by the fact that replication of an experiment denotes repetition carried out at one place and, as far as possible, one period of time. Current usage on this point is often rather loose. [See **duplicate sample**.]

representative sample In the widest sense, a sample which is representative of a population. Some confusion arises according to whether 'representative' is regarded as meaning 'selected by some process which gives all samples an equal chance of appearing to represent the population'; or, alternatively, whether it means 'typical in respect of certain characteristics, however chosen'. On the whole, it seems best to confine the word 'representative' to samples which turn out to be so, however chosen, rather than apply it to those chosen with the object of being representative.

reproducibility An experiment or survey is said to be reproducible if, on **repetition** or **replication** under similar conditions, it gives the same results; that is to say, if the variation between experiments is small and negligible. For a similar idea in psychological tests see **reliability**.

resampling procedure An inference procedure involving reuse of observed data, often sampling from them, examples being the bootstrap and jackknife, cross-validation, balanced repeated sampling, permutation testing, etc.

residual A general term denoting a quantity remaining after some other quantity has been subtracted. It occurs in a variety of particular contexts. For example, if the true value of a variable is subtracted from an observed value then the difference may be called a residual; it is also frequently called an error. Similarly, if a mathematical model is fitted to data, the values by which the observations differ from the model values are called residuals. In a slightly wider and less satisfactory sense the word is used to denote a stochastic element which is associated with the predicated or **independent variables** in a regression, e.g., in the linear regression

$$y = \beta X + \varepsilon$$

the variable ε is sometimes called a residual error term, and if the value of β is estimated from the data as, say, b, the difference between an observed value of y and the so-called 'predicted' value bX is also called a residual.

residual sum of squares See **error sum of squares**.

residual treatment effect In experiments which are continued over several consecutive periods of time on the same individual it is important to consider whether the effect of the experimental treatments administered during one period is carried over into the next and subsequent periods. Any such 'carried-over' effects are known as residual treatment effects and, if they are likely to be present, appropriate precautions have to be taken in the design of the experiment and the analysis of the results.

residual variance That part of the variance of a set of data which remains after the effect of certain systematic elements such as treatments is removed. It measures the variability due to unexplained causes or experimental error.

residual waiting time See **waiting time**.

resistant techniques Any statistical procedure that is insensitive to changes in a small part of the data. Such techniques typically use **robust estimators** and are concerned with minimizing the effect of **outliers**.

resolvable balanced incomplete block design An incomplete block design with parameters v (treatments), b (blocks) of size $k < v$ and r (number of blocks in which every treatment occurs) is said to be α-resolvable if the blocks can be divided into t sets each of β blocks so that in each set every treatment is replicated α times. We then have the relationships

$$v\alpha = k\beta; \qquad r = \alpha t; \qquad b = \beta t.$$

An α-resolvable incomplete block design is called affine α-resolvable if any pair of blocks of the same set intersects in q_1 treatments and any pair of blocks from different sets intersects in q_2 treatments.

resolvable designs An incomplete block design is said to be resolvable (Bose, 1942) if the blocks can be grouped in a way such that each group contains every treatment once, so as to form a complete replicate. This property has considerable practical advantages in allocating experimental effort or avoiding premature termination of an experiment.

response The reaction of an individual unit to some form of stimulus. It may be reaction to a drug, as in bioassay, or the reaction to a request for information, as in sample surveys of human beings. Also used more broadly for the dependent variable in a regression analysis. [See **non-response.**]

response error See **non-sampling error.**

response metameter The transformed measurement of the response to a given stimulus. The transformation is made, for example, in biological assay, in order to facilitate computations and diagrammatic representation. [See also **metameter.**]

response surface If a response η depends upon an unknown function ϕ of k quantitative factors $\xi_1, \ldots, \xi_k$, the values of η for varying ξ's may be viewed as a surface in $k + 1$ dimensions. One object of experimentation is to approximate to this surface in some domain of interest, especially where η is a maximum.

response surface designs Experimental designs of factorial type intended to give approximately uniformly efficient estimates of the response surface over the range of the factors. Such a design is usually called **rotatable design.**

response time distribution Where, in biological assay, the response to the stimulus is measured in terms of the time that elapses before a given reaction appears, the different reaction times for different individuals may be put into the form of a distribution of response time.

restricted chi-squared test A modification of the chi-squared test proposed by Neyman (1949a) where the restriction is on the alternative hypotheses as derived from an explicit model. The use of this restriction enables one to obtain higher power in some directions than in others.

restricted maximum likelihood See **REML.**

restricted randomization Special schemes of randomization in which some special configurations are not allowed but the theoretical properties of the randomization scheme are retained (Grundy and Healy, 1950).

restricted sequential procedure A class of closed sequential procedure proposed by Armitage (1957) incorporating the principle of truncation in order to reduce the variability of sample number.

retrospective change point problem Change point problem when all data are available at the beginning of statistical analysis.

retrospective study See **case–control study**.

return level An alternative name for the quantile of a distribution, used in extreme-value statistics. The m-year return level in an annual series is the $1 - 1/m$ quantile, for which the return period is m years.

return period An alternative name for **return time**.

return states In the theory of Markov chains, a state which **communicates** with itself is called a return state. A state which communicates with no state, not even itself, is called a non-return state.

return time In time series, the interval of time taken by the series to return to some assigned value, usually an extreme value, as, for example, 'the return period of flooding' in a river.

reversal design Alternative name for **switchback design**.

reversal test This term occurs in two quite different connections: (i) in certain tests of consistency for index numbers: **factor reversal test** and **time reversal test**; (ii) in the analysis of time series, where one of the tests for random order is based upon 'reversals' in the series. A reversal occurs if, in the first differences of the series, a positive sign follows a negative sign or vice versa. A so-called 'reversal' test for randomness may be formed by considering the proportion of reversals in a given series.

reversible jump Markov chain Monte Carlo A modification of Markov chain Monte Carlo methods for use in statistical inference when parameter spaces of different dimensions are contemplated (Green, 1995).

Rhodes' distribution A unimodal bivariate distribution due to Rhodes (1923). It takes the form

$$h(x,y) = h_0 e^{-lx - my} \left(1 - \frac{x}{a} + \frac{y}{b}\right)^p \left(1 + \frac{x'}{a} - \frac{y'}{b}\right)^{p'}$$

where the region of nonzero density is defined by

$$S = \left\{(x,y) : 1 - \frac{x}{a} + \frac{y}{b} > 0, \ 1 + \frac{x'}{a} - \frac{y'}{b} > 0\right\}$$

and $a, b, a', b', p, p', l, m$ and the resulting h_0 are all positive constants, with $ab' > a'b$ to ensure that the two lines forming the boundaries of the distribution intersect in the third quadrant, and that the region consists of that part of the plane enclosed by the lines that also contains the origin.

Rice, Stuart Arthur (1889–1969) American sociologist and statistician who set up and headed the Office of Statistical Standards in the US Bureau of the Budget in the 1930s. He made major efforts to improve the statistical services of countries around the world, particularly Japan and Korea. He went on to found Stuart A. Rice Associates (later Surveys & Research Corporation), a statistical consulting firm (Rice, 1931).

ridge regression A method proposed by Hoerl and Kennard (1970) to overcome shortcomings of standard least squares methods of estimating regression coefficients arising from the more or less close association between variables. While the least squares estimators are minimum variance unbiased, they are not in general minimum mean square error. The ridge estimators are biased but with smaller mean square error; the approach is essentially one of choosing a constant to achieve a satisfactory balance between bias and variance. [See **penalized least squares**.]

ridit analysis A method proposed by Bross (1958) for analysing subjectively categorized or poorly recorded measurement data. It consists of allocating scores relative to the identified distribution of the data based upon a transformation to the uniform distribution rather than the normal distribution.

Riemann distribution Zipf (1932) suggested the Riemann distribution

$$\phi(n) = \frac{1}{\zeta(s)n^s}$$

where $s > 1$, $n = 1, 2, 3, \ldots$, as a suitable mathematical representation of word frequency distributions.

right angular design An association scheme for **partially balanced incomplete block designs**, proposed by Tharthare (1963), defined four-associate class designs. Designs with this type of association scheme are termed right angular designs; there are $2Sl$ treatments arranged in l right angles of equal arms of length S.

right-censoring This occurs if it is possible that instead of observing the random variable X we observe only that $X > C$. This situation is most often encountered in survival (or failure time) analysis. The assumption of **independent censoring** is often made for a simple treatment.

right-truncation If we observe T only if $T < B$, there is right-truncation. This is different from right-censoring in the sense that if $T \geqslant B$ the individual does not enter into the sample; this is a selection of the sample and typically occurs in epidemiology when treating data coming from registers. As for **censoring** assumptions have to be made about how the truncation mechanism was generated.

Ripley's statistics Two sets of statistics for studying the second-order properties of spatial point processes. For a point process of intensity λ, $\lambda K(t)$ is the expected number of further points within t of an arbitrary point of the process, and $p(t)$ is the probability that there is no point of the process within i of a randomly chosen point. Ripley (1976) introduced these functions, and suggested unbiased estimators for them.

risk factor A characteristic that is of value in predicting risk. A characteristic or activity known to contribute to a specific pathogenesis. For example, high blood pressure and smoking tobacco are both known risk factors for cardiovascular disease. It is to be distinguished from a cause, an imposed change in which, other things being equal, is likely to induce a change in outcome.

risk function The word 'risk' occurs in statistics in its ordinary sense and, apart from occurrence in actuarial statistics, has one special use in the theory of **decision functions**. Where a number of possible decisions have a loss function attached, the risk function $R(\theta, d)$ is the expected cost of the experimentation plus the expected value of the loss function for the state of nature θ and the decision function d.

risk ratio The risk ratio is the factor by which 'risks' for two comparison groups differ, where risk is measured as a proportion or probability (density). It is the ratio of two probabilities (or probability densities), which typically describe the rate of an event of interest in each group. The ratio is a value within a range of zero to infinity with one being neutral. For example, suppose that the proportion of persons suffering from complications after traditional surgery is 0.10, while the proportion suffering from complications after alternative surgery is 0.125. Then the risk ratio is $0.10/0.125 = 0.8$, indicating that 20 per cent fewer patients treated by the traditional method suffer from complications.

Robbins–Munro process A **stochastic approximation procedure** for finding the parameter of a regression equation, or a quantile where assumptions are not made on the distribution, proposed by Robbins and Munro (1951). The concept has been developed by other writers.

robust estimators Estimators, usually of location and scale parameters, robust against the presence of **outliers**. Various types have been defined; among the most important are (i) L-estimators, which are linear functions of order statistics; (ii) M-estimators, which maximize some function of the data points and the parameter; (iii) R-estimators, which are derived from tests based on ranks (Staudte and Sheather, 1990). [See **Huber estimator**.]

robustness Many test procedures involving probability levels depend for their exactitude on assumptions concerning the generating mechanism, e.g., that the parent variation is normal (Gaussian). If the inferences are little affected by departure from those assumptions, e.g., if the significance points of a test vary little if the population departs quite substantially from the normality, the test on the inferences is said to be robust. In a rather more general sense, a statistical procedure is described as robust if it is not very sensitive to specified departures from the assumptions on which it depends.

Room's squares A type of experiment design in square form proposed by Room (1955). It consists of a square of $2n - 1$ rows and columns so that in each row and column there are n symbols ($n - 1$ blanks) which contain all $2n$ digits; ($n[2n - 1]$) symbols.

root estimator (of a mean) When sampling from a highly skewed population, the root estimator, $(1 - c)\bar{x} + c\bar{u}^2$ where $u_i = \sqrt{x_i}$, can provide substantial gain in efficiency over $\bar{x}$ as an estimator of the population mean (Jenkins, Ringer and Hartley, 1973).

root mean square deviation The square root of the second moment of a set of observations taken about some arbitrary origin, that is to say the square root of the **mean square deviation** or mean square error. The minimum value of the root mean square deviation occurs when the origin coincides with the arithmetic mean – it is then called the **standard deviation**. [See also **variance**.]

root mean square error An alternative name for **root mean square deviation**.

rose diagram A form of diagrammatic representation for grouped angular data. Corresponding to each angular interval a sector is constructed with apex at the origin, radius proportional to the class frequency and arc subtending the interval. The disadvantage of this form of diagram is that the areas of sectors vary as the square of the frequency. To overcome this, the sectors may be constructed with radii proportional to the square roots of the class frequencies, to produce what has been called an equi-areal rose diagram. [See **polar-wedge diagram**.]

Rosenbaum's test A non-parametric test for the equality of the scale parameters of two populations known to have equal medians, proposed by Rosenbaum (1953, 1954). The test statistic is the total number of values in the sample from the first population that are either smaller than the smallest or larger than the largest values in the sample from the second population.

rotatable design A k-factor experiment is defined by Box and Hunter (1957) to have a dth-order rotatable design if the expected yield is a polynomial of degree d in the k factor levels and the variance of the estimated response is a function of the distance from the centre of the design. The concept of rotatable designs was generalized by Herzberg (1966, 1967) to **cylindrically rotatable designs** that are rotatable for all factors except one; and by Das and Dey (1967) to **group-divisible rotatable designs** which are those for which the factors can be split into two groups in such a way that the design is rotatable for each group of factors when the levels of the factors in the other group are held constant.

rotation sampling A term suggested by Wilks and developed by Eckler (1955) for the situation of sampling on successive occasions with some (optimum) proportion of units common to successive pairs of samples.

roughness penalty Non-parametric estimates of density or of regression curves are usually required to be smooth; maximum likelihood estimation typically leads to discrete distributions or step functions. Smoothing may be achieved by reducing the likelihood by a function that takes high values for irregular curves and small values for smooth curves. Such a function is known as a roughness penalty. [See **maximum likelihood method**.]

Round Robin design In the case where the same n objects have to be used repeatedly throughout an experiment we have a special form of **resolvable design** which may or may not be a **cyclic design**. A tournament of n (even) players who meet in r rounds of $n/2$ matches is an example and known generally as the Round Robin design.

rounding The process of approximating to a number by omitting certain of the end digits, replacing by zeros if necessary, and adjusting the last digit retained so that the resulting approximation is as near as possible to the original number. If the digit is increased by unity the number is said to be rounded up; if left unchanged it is rounded down. When both are under consideration the process is said to be one of rounding off.

route sampling A procedure similar to **line sampling** and used in surveys of crop acreage in districts which are well provided with roads. A route which adequately covers the area is chosen and the roadside lengths of the different crops recorded. Since the location of roads is unlikely to be random, estimates of acreage so obtained are likely to be biased but changes in acreages may be estimated by using the same route for a number of years. The method of route sampling as a form of systematic sampling can also be applied to crop estimation.

Roy's maximum root criteria The maximum characteristic root or eigen-value statistic for testing the following: (i) equality of k p-variate normal distributions with the same but unknown covariance matrix; (ii) independence between two sets of variables jointly distributed as a normal distribution with unknown mean vector; (iii) equality of covariance matrices of two p-variate normal distributions with unknown mean vectors; (iv) whether the covariance matrix of a p-variate normal distribution with unknown mean vector equals a specified matrix.

ruin problems See **gambler's ruin, absorbing barrier**.

runs In a series of observations of attributes the occurrence of an uninterrupted series of the same attribute is called a run. In particular, a single isolated occurrence may be regarded as a run of one. In a series of variable values, a consecutive set which are monotonically increasing or decreasing are said to provide runs 'up' or 'down' respectively. The theory of runs has been developed in connection with a number of distribution-free tests.

Rutherford's contagious distribution A discrete probability distribution proposed by Rutherford (1954) where the probability of success at any trial depends linearly upon the number of previous successes.

S

S_B, S_U **distributions** Two bounded systems of frequency distributions proposed by Johnson (1949) based on a variable transformation of type

$$z = \gamma + \delta f \left(\frac{x - \xi}{\lambda} \right),$$

where z is a unit normal variable, γ, δ, ξ and λ are constants and f is some convenient function. The S_B distributions use the function

$$\log \left(\frac{x - \xi}{\xi + \lambda - x} \right).$$

The S_U family uses the function

$$\sinh^{-1} \left(\frac{x - \xi}{\lambda} \right).$$

[See also **Johnson's system**.]

Sacks' theorem A theorem due to Sacks (1958) which states under very general conditions that when the **Robbins–Munro process** is used with constants $a_n = c/n$ (c a constant), the distribution of $(X_n - \theta)$ is asymptotically normal with zero mean. The importance of this theorem is that, in conjunction with **Dvoretzky's stochastic approximation theorem**, it gives a theoretical basis for employing the Robbins–Munro process under a very wide range of conditions.

saddle point A stationary point on a surface which is a maximum along some paths and a minimum along others, as at the top of a mountain pass.

saddle point approximation A method of approximating a complex integral by expanding about a saddle point in the complex plane. In statistics, it is mainly used to approximate distribution functions when the moments are known, using the inversion formula for the **characteristic function**.

saddle point expansion A mathematical technique for producing asymptotic expansions via the theory of functions of a complex variable distorting a path of integration to pass through a saddle point and down the direction of steepest descent. The technique was introduced into statistical problems by Daniels (1954*b*). The resulting expansions which often have unexpectedly high accuracy can be derived by other probabilistic methods and so in a statistical context the term is applied more widely than in other fields.

sample A part of a population, or a subset from a set of units, which is provided by some process or other, usually by deliberate selection with the object of investigating the properties of the parent population or set.

sample census If 'census' is taken to mean the examination of each member of a population this term is self-contradictory. If, however, census refers to the kind of material collected then it is possible to use a sample.

sample design The usage is not uniform as regards the precise meaning of this and similar terms like 'sample plan', 'survey design', 'sampling plan' or 'sampling design'. These cover one or more parts constituting the entire planning of a sample survey inclusive of processing etc. The term 'sampling plan' may be restricted to mean all steps taken in selecting the sample; the term 'sample design' may cover in addition the method of estimation; and 'survey design' may cover also other aspects of the survey, e.g., choice and training of interviewers, tabulation plans, etc. 'Sample design' is sometimes used in a clearly defined sense, with reference to a given frame, as the set of rules or specifications for the drawing of a sample in an unequivocal manner.

sample moment See **sampling moment**.

sample plan See **sample design**.

sample point A sample of n variable values $x_1, x_2, \ldots, x_n$ can be represented as a point or vector in an n-dimensional space, usually Euclidean, in which the values of the x's are taken as coordinates. A 'point' in this space corresponding to an observed set of sample values is the sample point. The idea generalizes in a straightforward manner to p-way multivariate variation, the sample then being regarded as defining a point in p_n dimensions or p vectors in n dimensions or n vectors in p dimensions.

sample reuse A general term for methods in which estimates are adjusted or assessed by further examination of the sample values, without distributional assumptions. [See **bootstrap methods, cross-validation, jackknife**.]

sample size The number of sampling units which are to be included in the sample. In the case of a **multi-stage sampling** this number refers to the number of units at the final stage in the sampling. Sometimes used more broadly to denote the number of (independent) observations available.

sample space The set of **sample points** corresponding to all possible samples. The permissible domain of variation of a sample point. Sometimes referred to as sample description space or event space.

sample survey A **survey** which is carried out using a sampling method, i.e., in which a portion only, and not the whole population, is surveyed.

sample unit This term is often synonymous with **sampling unit** but would be better confined to the denotation of any one of the units constituting a specified sample.

sample variance The variance estimated from a sample in the simplest case. For a sample of observations $x_1, \ldots, x_n$ the sampling variance is defined by

$$s^2 = \frac{1}{n-1} \sum_{i=1}^{n} (x_i - \overline{x})^2,$$

where $\overline{x}$ is the sampling mean $(x_1 + \ldots + x_n)/n$.

sampling distribution The distribution of a **statistic** or set of statistics in all possible samples which can be chosen according to a specified sampling scheme. The expression almost always relates to a sampling scheme involving random selection, and most usually concerns the distribution of a function of a fixed number n of independent variables. In a frequency-based theory of statistical inference, the sampling distribution of a statistic is used as a basis for assessing uncertainty and comparing alternative procedures.

sampling error That part of the difference between a population value and an estimate thereof, derived from a random sample, which is due to the fact that only a sample of values is observed, as distinct from errors due to imperfect selection, bias in response or estimation, errors of observation and recording, etc. The totality of sampling errors in all possible samples of the same size generates the sampling distribution of the statistic which is being used to estimate the parent value.

sampling fraction The proportion of the total number of sampling units in the population, stratum, or higher-stage unit within which simple random sampling, with multiple counting of sample units when sampled with replacement, is made. There are thus sampling fractions corresponding to different strata and different stages of sampling. Exactly the same definition is sometimes loosely applied to other sampling schemes, e.g., in sampling with variable probability, or multi-stage sampling (ratio of total number of ultimate units included in the sample to total units in the population). However, for general application it appears desirable to define it as the reciprocal of the **raising factor** of the sample when it exists, i.e., when the sample is a **self-weighting sample**. The term sampling ratio or rate is also used. [See also **overall sampling fraction**, **variable sampling fraction**.]

sampling frame A list of all members of a population used as a basis for sampling. Without such a frame, or its equivalent, methods of sampling with assured properties such as unbiasedness are not available. The frame in effect defines the study population (Thompson, 1992).

sampling inspection The evaluation of the quality of material or units of a product by the inspection of a part, rather than the whole; in contradistinction to total inspection or **screening inspection**.

sampling interval See **systematic sample**.

sampling moment A moment of a sampling distribution, as distinct from a moment of a set of sample observations.

sampling on successive occasions The carrying out of a sampling process at successive intervals of time. Various methods of doing so are employed in sample surveys, e.g., by selection of a new sample on each occasion, by the partial replacement of the sample and by subsampling the initial sample.

sampling ratio See **sampling fraction**.

sampling structure A specification which defines a class of completely specified sample or survey designs. In problems of optimum design the optimization is restricted to a given class of designs, and not to all conceivable possibilities.

sampling unit One of the units into which an aggregate is divided or regarded as divided for the purposes of sampling, each unit being regarded as individual and indivisible when the selection is made. The definition of unit may be made on some natural basis, e.g., households, persons, units of product, tickets, etc., or upon some arbitrary basis, e.g., areas defined by grid coordinates on a map. In the case of **multi-stage sampling** the units are different at different stages of sampling, being 'large' at the first stage and growing progressively smaller with each stage in the process of selection. The term **sample unit** is sometimes used in a synonymous sense; but refer to that term for a different meaning.

sampling with replacement When a sampling unit is drawn from a finite population and is returned to that population, after its characteristic(s) have been recorded, before the next unit is drawn, the sampling is said to be 'with replacement'. In the contrary case the sampling is 'without replacement'. A different usage occurs in sample surveys when samples are taken on successive occasions. If the same members are used for successive samples there is said to be no replacement; but if some members are retained and others are replaced by new individuals there is 'partial replacement'.

sampling zeros See **structural zeros**.

sandwich estimate An estimate of the precision of a formal maximum likelihood estimate obtained not directly from the information matrix but via the variability of individual values of the score function thus allowing for some forms of departure from the original model.

SAS Acronym for Statistical Analysis System. This is an advanced statistical package for data management, graphics and statistical analysis. SAS's data handling and management capabilities especially for large and complex data sets are a great strength, and SAS is widely used in the pharmaceutical and finance industries, and in government agencies. In addition to a broad range of statistical procedures, e.g., for time series, survival and mixed models, SAS offers programming capabilities and a graphical user interface. SAS is generally considered to have more advanced statistical capabilities than most other mainstream packages. SAS Enterprise Miner provides data mining tools and is only one of many specialist products; see http://www.sas.com for a complete listing.

Satterthwaite's test An approximation to the variance ratio test when the denominator is a combination of different estimates of variance (Satterthwaite, 1946). [See **Behrens–Fisher problem, Welch's test.**]

saturated model A model in which the number of independent observations is equal to the number of adjustable parameters. A special case is one full replicate of a factorial experiment with all contrasts to be estimated (Federer, Hedayat and Ratkoe, 1981).

saturation In the factor analysis of multivariate material the correlation between a common factor and a variable is called the saturation of that particular variable. It measures the extent to which the factor 'appears' in the variable or the extent to which the variable is 'saturated' with the factor. [See **factor loading.**]

scale equivariant estimator An estimator $T_n(Y_1, \ldots, Y_n)$ of β in the linear regression model $Y_i = x_i'\beta + e_i$, $i = 1, \ldots, n$, $\beta \in \mathbb{R}^p$, is scale equivariant, if it satisfies $T_n(cY_1, \ldots, cY_n) = cT(Y_1, \ldots, Y_n)$ for every $c > 0$ and $(Y_1, \ldots, Y_n)$.

scale parameter A parameter of a frequency distribution which is functionally related to the scale of the variable, e.g., the standard deviation in a normal distribution.

scan statistic The maximum number of events in a window (usually of time) of predetermined width. Its most common application is testing for clustering conditional on the total number of events observed.

scatter coefficient A term proposed by Frisch (1929) to indicate a property of a multivariate distribution. It is the square root of the determinant whose elements are the intercorrelations r_{ik} between the pairs of variables; that is to say, it is the square root of the correlation determinant. For the case of two variables the scatter coefficient is the same as the coefficient of alienation. [See **alienation, coefficient of.**]

scatterplot A graph that shows the relationship between two variables by plotting all pairwise values in two dimensions. The 'shape' is used to describe the relationship between the two variables.

scatterplot matrix An array of scatterplots relating every pair of variables.

scatterplot smoother A smoothing procedure in which the conditional mean, median, etc., of the abscissa variable of a scatterplot given a value of the ordinate variable is estimated, typically by locally weighted averaging.

scedastic curve See **regression.**

scedasticity A little-used word denoting dispersion, especially as measured by variance. In a bivariate distribution, the graph of the variance of arrays of one variable against the corresponding values of the other variable is called a scedastic curve (see also **kurtosis**). If the variance of one variable is the same for all fixed values of the other, the distribution is said to be homoscedastic in the first variable; in the contrary case it is heteroscedastic.

Schach's two-sample tests A class of two-sample non-parametric tests for the equality of circular populations, proposed by Schach (1969). The test statistics are rotationally invariant functions of the ranks of the members of one of the samples within the combined sample, and they provide a means of deriving a corresponding two-sample test of equality from any given single-sample test of uniformity.

schedule Apart from its customary connotation of 'list', this word occurs in the theory of sample surveys in the specialized sense of a group, or sequence, of questions designed to elicit information upon a subject. It is then synonymous with 'questionnaire'. Usually it is completed by an investigator on the basis of information supplied by the particular member of the population chosen for inclusion in the sample; but sometimes it is completed by that member him- or herself, as in postal enquiries.

Scheffé, Henry (1907–1977) An American mathematical statistician. Scheffé obtained his PhD in mathematics in 1935 from the University of Wisconsin. He made contributions in **analysis of variance**. One of his most important papers appeared in 1953 on the S-method of simultaneous confidence intervals for estimable functions in the subspace of the parameter space (Scheffé, 1959).

Scheffé test A **multiple comparisons** test (Scheffé, 1959) for means in an **analysis of variance**. It is based on the multivariate confidence region for the means, and is appropriate for testing any contrast among them. When used for comparisons in pairs, it is extremely conservative.

Schuster periodogram An alternative name for the unqualified term '**periodogram**' so called because it was introduced by Schuster (1898). [See also **periodogram**.]

Schwarz's inequality An inequality which in its probabilistic form states that for two arbitrary random variables X and Y

$$\{E(XY)\}^2 \leqslant E(X^2)E(Y^2)$$

whenever the mathematical expectations exist. A generalization of this result for non-negative random variables is **Hölder's inequality**.

score Generally, a numerical value assigned to an observation (usually where no measurable variable exists or none can conveniently be measured) as a substitute for a variable considered as varying over a scale. The word is also used to describe the first derivative of the natural logarithm of the likelihood function.

score equation If U is the **score statistic**, the score equation is $U(\theta) = 0$. The solution of this equation is, in well-behaved situations, the maximum likelihood estimator.

score statistic The score or score statistic, often denoted U, is defined as the derivative of the log-likelihood relatively to the parameters:

$$U(\theta) = \frac{\partial L}{\partial \theta}.$$

Under the measure defined at θ it has zero expectation and its variance is the **Fisher information matrix**. Asymptotically it has a normal distribution.

score test Test based on the score statistic for testing a particular value of a parameter $\theta = \theta_0$. Under the null hypothesis its expectation is zero, its variance can be obtained from the Fisher **information matrix** and one most often uses the asymptotic normality of the distribution. Many simple tests, such as the χ^2 test or the **log-rank test** are score tests. The advantage of this type of test is that the statistic often takes a simple form and its distribution is often relatively simple since it is computed entirely under the null hypothesis. In situations where the alternative hypotheses are more complex than the null hypothesis (such as in **homogeneity** testing) this is a great advantage.

score vector The vector of first derivatives of the log-likelihood function with respect to a vector parameter. In many cases the maximum likelihood estimator is obtained as the solution of the score equation, which equates the score vector to zero.

scoring, method of An iterative process useful for solving nonlinear maximum likelihood equations (Rao, 1973).

scree test A criterion for the number of factors to be fitted in factor analysis. It is based on the ordered eigenvalues of the correlation matrix; if these show a sharp drop, the number of factors is taken as the number of eigenvalues before the drop (Cattell, 1978).

screening design The statistical design for a programme of experiments which has the object of selecting a promising subset of treatments for further development. The selection process can be optimum according to a number of criteria but is likely to include balancing the errors of the first and second kind. The approach has been used, for example, in the fields of plant breeding, drug screening and educational selection.

screening inspection The complete inspection of a block of material or units of a product, and the rejection of all items or portions found defective. It is also known as 'total inspection' or '100 per cent inspection'. [See also **sampling inspection**.]

seasonal variation In time series, that part of the movement which is assigned to the effect of the seasons on the year, e.g., seasonal variation in rainfall.

Sometimes the term is used in a wider sense relating to oscillations generated by periodic external influences, e.g., daily variations in temperature might be described as 'seasonal'.

second limit theorem A theorem which, broadly speaking, states that if the moments of a sequence of distribution functions F_n tend to the moments of a distribution function F, moments of all orders existing, then F_n tends to F, provided that the latter is uniquely determined by its moments. [See also **first limit theorem**.]

second-order analysis Any method of analysis dependent on moments of the first and second order only. The term is used in particular in connection with the analysis of spatial stochastic processes.

second-order stationary See **covariance stationary process**.

secondary process In many types of stochastic process it is possible that the events or occurrences constituting the process may also be characterized by one or more additional random variables. Each of these may be regarded as a secondary process.

secondary unit In sampling, a synonym of second-stage unit. [See **multi-stage sampling**.]

secretary problem A classical problem of sequential **decision theory**. One of n items is to be chosen; they are examined in a random sequence and each is given a score or rank. At each step, an irrevocable decision to accept or reject is made, and the problem is to maximize the chance of choosing the best, or to maximize the expected score.

secular trend An alternative name for **trend** in time series which is sometimes reserved for a trend extending over a long period of years, say, centuries, as against 'trends' extending over decades.

selection bias The error introduced when a sample is not representative of the population about which an investigator wishes to make inferences.

selected points, method of A method of fitting a curve to a large number of points whereby a small number of points is selected as representative, more or less subjectively, and a curve fitted to them. The number of points chosen depends on the type of curve which it is intended to fit; for the fitting of a polynomial of degree n at least $(n + 1)$ points are necessary.

selection with arbitrary (variable) probability A procedure for selecting a sample in which the probabilities of selection for the sampling units in the population are allocated in advance in a purposive but arbitrary manner. When the sample units are selected one by one, as is usually done, a different set of probabilities may be associated with each drawing. [See also **selection with equal probability**.]

selection with equal probability Fundamentally, selection of a single element from a set of such elements in such a way that selection probabilities of all elements are equal. There is, however, no uniform usage in respect of selection of a sample of more than one element; it has then reference to (*i*) the actual operation of selection, of any one of them individually and/or collectively when two or more operations are involved; (*ii*) or the final product: the entire sample, obtained by all such operations with or without particular reference to the component sample units. Thus, for example, in stratified simple random sampling, with different sampling fractions in different strata, the entire sample is sometimes referred to as being selected with unequal probability even though the actual operation of selection within a stratum is basically with equal probability.

selection with probability proportional to size A sampling procedure under which the probability of a unit being selected is proportional to the size of the unit. Generally this probability has reference to each drawing separately when sample units are selected one by one. Thus the procedure known commonly as sampling with probability proportional to size, with replacement, ensures such a probability of any particular drawing, but considered in its entirety, the series of drawings does not make the probability of inclusion in the sample of any specified unit proportional to its size unless the units are all of the same size.

self-avoiding random walks A random walk in which the units have physical size and no two units may occupy the same region of space; or, on a lattice, a walk which never visits a site for a second time.

self-conjugate Latin square A Latin square which remains the same if its rows and columns are interchanged, i.e., it is symmetrical about its main diagonal.

self-correlation coefficient An alternative term for a **reliability coefficient**, but one which is better avoided.

self-renewing aggregate See **renewal theory**.

self-similar process A stochastic process whose joint distributions are invariant to rescaling of the time variable, except for change of scale. The simplest examples are the **Poisson process** and **Brownian motion process**.

self-similarity, coefficient of See **fractional Brownian motion**.

self-weighting sample If the **raising factors** of the sample units are all equal the sample is self-weighting, of course with respect to the particular linear estimator under consideration; but it may not be a self-weighting sample for another estimator. A self-weighting sample, usually in respect of the total of the entire population, is generally incorporated in a sample design to simplify tabulation work, because the population total is easily estimated from the sample total. In two-stage (multi-stage) sampling the number or proportion of second-stage sample units is sometimes fixed in such a manner that the sample becomes self-weighting.

semi-averages, method of A particular case of the method of selected points in which the data are divided into two equal groups and a straight line drawn through the means of the groups or two other representative points, one in each group. This method is used to provide a rapid estimate of a linear regression line. [See **selected points, method of.**]

semi-interquartile range An alternative name for **quartile deviation**; that is to say, one-half of the distance between the two quartiles of a sample or a distribution.

semi-invariant In older usage this term, introduced by Thiele (1889), related to what is now called **cumulant**. The words 'semi-invariant' or 'seminvariant' are now better confined to statistics which are independent of the origin and are multiplied by a scale factor under transformations of scale. The moments about the sample mean and the cumulants are both seminvariant in this sense and other symmetric functions of the observations exist with seminvariant properties. The term is not, but could be, used to describe statistics such as the range, which are not symmetric functions of the observations.

semi-Latin square An experimental design for $2k$ treatments arranged in the form of a rectangle with k rows of $2k$, each row being an arrangement of the $2k$ treatments, each pair of columns 1, 2, 3, 4, etc., being also an arrangement of the $2k$ treatments; no row or column therefore contains the same treatment more than once. It may also be regarded as a $k \times k$ Latin square with each plot split. The design has been criticized on the grounds that it leads to biased estimates of error.

semi-logarithmic chart A form of graphic presentation in which one axis only is scaled in terms of logarithms. The logarithms may be based upon any suitable number although in the case of specially printed chart papers they are usually to base 10, i.e., are common logarithms.

semi-Markov process A stochastic process first defined by Lévy (1954), and Smith (1955b) as a regenerative stochastic process, and applied by Takacs (1954) to counter problems. This type of process is essentially a Markov chain with randomly distributed lengths of time in any one stage having arbitrary distribution.

semi-martingale See **martingale**.

semi-normal distribution An alternative but little-used name for a **half-normal distribution**.

semi-range A statistic equal to one-half of the range. [See also **mid-range**.]

semi-stable law See **stable law**.

semi-stationary process A somewhat illogical but convenient word to describe a process which is largely stationary in the sense that its non-stationary characteristics are 'slowly' time-dependent (Priestley, 1965).

semivariogram In **geostatistics**, suppose Z is a random variable defined at points $x, x + h$ separated by a vector h, and

$$\gamma = \frac{1}{2}E\left\{Z\left(x\right) - Z\left(x + h\right)\right\}^{2}.$$

In general, γ is a function of x and h. In a stationary process it is a function of h only, and if the process is also isotropic it is independent of the direction of h. A plot of γ against h is a semivariogram; for stationary systems it is essentially equivalent to a correlogram, since γ is then proportional to $(1 - r_h^2)$.

sensitivity analysis Reassessment of the fitted model to detect whether changing any of the assumptions leads to different interpretations.

sensitivity and specificity If a test, particularly a diagnostic test for a disease, gives a proportion α of false positives and β of false negatives, $1 - \beta$ and $1 - \alpha$ are called respectively the sensitivity and specificity of the test.

sensitivity data A term which is sometimes used as an alternative to **quantal response** data to describe data consisting of measured reactions at various levels of a stimulus. This particular term has been largely used in connection with tests of explosives.

sensitivity function A function proposed by Nyquist (1992) for assessing the sensitivity of the outputs to small changes in the inputs of a given statistical process. Let M_ϕ be a parametric model indexed by a parameter ϕ, $T(M_\phi)$ be a parametric function of interest (e.g., an estimate of the model based on the data), and $P_T(M_\phi)$ be a specified property of T under the model M_ϕ (e.g., an observed value of T or the sampling distribution of T). The sensitivity of a specific property of T with respect to changes in ϕ can be measured by the sensitivity function

$$SF(M, T, P) = \lim_{\phi \to 0} d(P_T(M_\phi), P_T(M_0))/\phi,$$

where $d(\cdot, \cdot)$ is some suitable metric. Several statistical applications of the sensitivity function are given in Hadi and Nyquist (2002).

separation theorem Any result in the theory of graphical models that expresses a condition for the independence of two sets of variables A, B given a third set C via the notion that the set C in some sense separates the sets A and B.

sequential analysis The analysis of material derived by a sequential method of sampling; that is to say, it is the data, not the analysis, which are sequential.

sequential change point problem Change point problem when data arrive sequentially and after each new observation one has to decide whether the observations obtained so far indicate a change in the model or not; observations continue only in the latter case.

sequential estimation   Estimation from data obtained by a sequential sampling process.

sequential probability ratio test   A sequential test for a hypothesis H_0 against an alternative hypothesis H_1, due to Wald (1947). At the end of each stage in the sampling the probability ratio p_1/p_0 is computed where the suffixes 0 and 1 refer to the null and alternative hypotheses respectively and p is the known probability function of all sample members so far drawn. Then if $B < p_1/p_0 < A$ the sampling is continued another stage. But if $p_1/p_0 \leqslant B$ the null hypothesis (H_0) is accepted, and if $p_1/p_0 \geqslant A$ the null hypothesis is rejected and the alternative hypothesis (H_1) accepted. The two constants A and B are determined by reference to prescribed requirements concerning the two types of errors to be made in testing hypotheses, the rejection of H_0 when it is true and the acceptance of H_1 when it is false.

sequential sampling   A sampling in which the members are drawn one by one or in groups in order, and the results of the drawing at any stage decide whether sampling is to continue. The sample size is thus not fixed in advance but depends on the actual results and varies from one sample to another. The sampling terminates according to predetermined rules which are decided by the degree of precision required.

sequential test   A test of significance for a statistical hypothesis which is carried out by using the methods of sequential analysis. An example is the **sequential probability ratio test**. These tests are particularly important (*i*) in industrial statistics, where a small average sample number is desirable, and (*ii*) in medical statistics, where the procedure ensures that no patient receives a treatment that has been shown to be inferior, so limiting ethical problems.

sequential tolerance region   A **statistical tolerance region** where the sample for the purpose of determining the boundaries proceeds sequentially (Saunders, 1960) and is terminated when those boundaries are unchanged after a predetermined small number of observations.

serial cluster   A type of cluster in which the actual demarcation of a cluster or listing of units constituting a cluster is avoided by means of a rule which makes use of the serial numbers already assigned to the units in the frame. [See **entry plot**.]

serial correlation   The correlation between members of a time series (or space series) and those members lagging behind or leading by a fixed distance in time (or space). Thus, if the series is $u_1, u_2, \ldots$ the serial correlation of order k is the correlation between the pairs (u_1, u_{1+k}), (u_2, u_{2+k}), etc. An analogous definition may be framed for continuous time. In this sense the serial correlation is the sample value of the parent autocorrelation. Some writers, however, use 'autocorrelation' to denote the correlation of members of a series with themselves whether of sample or parent, and 'serial correlation' to denote the correlations of members of two different series. [See **lag covariance**.]

serial design An experimental design (Thompson and Seal, 1964) which uses overlapping effects in time in order (i) to secure estimates of quantities for which an evaluation would otherwise be impossible and (ii) to secure a basis for continuing the experiment for a long period. [See also **evolutionary operation**.]

serial dilution assay A method for estimating the density of organisms distributed at random in a medium. It uses only binary (presence, absence) data. The medium is tested perhaps in its original form and also diluted by factors $d, d^2, \ldots$, where a common value of d is 2. At each dilution presence or absence of organisms in a small sample is recorded.

serial linear rank statistics Denote by $R_{n;t}$ the rank of $X_{n;t}$ in the n-tuple $(X_{n;1}, \ldots, X_{n;n})$. A serial linear rank statistics is a statistic of the form

$$S_n := \sum_{t=p+1}^{n} a_n(R_{n;t}, R_{n;t-1}, \ldots, R_{n;t-p}),$$

where $a_n : \{1, \ldots, n\}^{p+1} \to \mathbb{R}$ is a collection of **scores**.

serial sampling inspection schemes These schemes assume that batches of items produced sequentially in industrial situations will be positively correlated. Schemes can be constructed by several methods which include those derived from Bayes' theorem applied after setting up a stochastic process representing the system and thereby obtaining a sentencing rule for particular batches.

serial variation A statistic for studying short-term variations in time series proposed by Jowett (1952). It is the mean semi-square difference

$$d_{(x)s} = \frac{1}{2(n-s)} \sum_{t=1}^{n-s} (x_t - x_{t+s})^2$$

and is a simple function of the **serial correlation** coefficient. [See also **variogram**.]

serially balanced sequence An experimental design proposed by Finney and Outhwaite (1956) for the situation where a single experimental subject has to receive a number of treatments during a period of time. The treatments are arranged in a series of complete blocks so that residual effects of any treatment occur the same number of times in association with the direct effect. A somewhat similar procedure for the Latin square design was suggested by Bradley (1958).

series queues A queueing system wherein each arrival unit is served in turn by facilities $1, 2, \ldots, k$ which are parts of one system. A somewhat different concept is expressed by the term 'tandem queues'.

Shannon–Wiener index An index of **diversity** based on information theory. If a large population is divided into s groups with proportions p_i, the index is

$$- \sum_{i=1}^{s} p_i \log p_i.$$

shape Aspect of a set, e.g., a two- or three-dimensional region invariant under changes of scale and location. See Dryden and Mardia (1998).

shape parameter A term used for a parameter of a frequency distribution which is associated with **skewness** or **kurtosis**, or more broadly for any measure describing the broad form of a distribution.

Shapiro–Wilk test A type of analysis of variance test of normality for a complete sample where the test statistic is the ratio of the square of a linear combination of the sample order statistics to the usual estimate of variance (Shapiro and Wilk, 1965).

Sheppard's corrections The calculation of moments from a grouped frequency distribution introduces certain errors as a result of assuming that frequencies are concentrated at the central values of the class intervals. Sheppard (1898) proposed a set of corrected moments $(\bar{\mu})$ which, for moments about the mean, are given by

$$\bar{\mu}_2 = \mu_2 - \frac{1}{12}h^2,$$
$$\bar{\mu}_4 = \mu_4 - \frac{1}{2}\mu_2 h^2 + \frac{7}{240}h^4,$$

and so on, where h is the grouping interval. Similar corrections have been given by various authors to cover factorial moments, the multivariate case, discontinuous variation and cumulants. These corrections are appropriate only for bell-shaped distributions, with high-order contact with the axis at the tails. [See also **correction for grouping**.]

Sherman's test statistic A test proposed by Sherman (1950) on the null hypothesis that data from an observed series of events arise from a **Poisson process** against the alternative hypothesis that the data arise from a **renewal process**. [See also **Moran's test statistic**.]

Shewhart control chart See **control chart**.

shock and error model A system of equations which contains both stochastic elements associated with specific variables (errors in variables) and elements associated with specific equations in the system, i.e., shocks (errors in equations). [See also **error in equations, errors-in-variables model**.]

shock model In econometric analysis, a system of equations which contains random disturbances, as opposed to one in which the variables are subject to **errors of observation**. [See also **error in equations**.]

Short distribution The convolution of a Neyman **type A distribution** with parameters λ, θ, and a **Poisson distribution** with parameter ϕ.

short-term fluctuation A fluctuation in a time series which has a short duration; a continuing set of such fluctuations. [See **trend**.]

shortest confidence intervals See **most selective confidence intervals**.

shorth estimator An estimator of location that is defined as the arithmetic mean of the 'shortest half' of a sample, i.e.,

$$\frac{1}{[n/2]+1} \sum_{i=l}^{l+[n/2]} x_{(i)}$$

where l is chosen to minimize $x_{(l+[n/2])} - x_{(i)}$. It is a measure that is relatively insensitive to **outlier** observations (Andrews *et al.*, 1972).

shot noise See **noise**.

shrinkage estimator When several parameters can be thought of as issuing from a population, their estimation is often improved by combining natural estimators of the individual parameters, e.g., **maximum likelihood estimators**, with an overall estimator of the population mean. The combined estimator is often a weighted linear combination of the natural and overall estimators, with weights depending on their relative precision, and lies between them; one can view the natural estimator as shrunk towards the overall one. Shrinkage arises in empirical Bayes and hierarchical Bayes models, and in many **smoothing** contexts (Carlin and Louis, 2000).

Siegel–Tukey test A distribution-free test of whether or not two populations known to be identical in every respect except scale are also identical with regard to scale parameter (Siegel and Tukey, 1960*a*; Siegel and Tukey, 1960*b*). It is a modification of the **Wilcoxon rank sum test**, the test statistic of both tests having the same null distribution. The earlier Freund–Ansari test (1957) is slightly different although more logically appealing in the manner of assigning ranks to observations, but suffers the minor disadvantage that because the null distribution of its test statistic is different, special tables are required to use the test. Similar remarks apply to the David–Barton test (Barton, David and Mallows, 1958; Barton and David, 1958).

sieve estimator An estimator whose complexity increases with sample size, though typically more slowly.

sigmoid curve A curve lying between two horizontal asymptotes representing a function which increases monotonically and has a point of inflexion somewhere near a point half-way between them; hence a curve somewhat resembling a letter S. In statistical work this type of curve is met in connection with, among others, the distribution function of unimodal distributions, **growth curves**, such as the logistic curve, and a particular **dose–response relationship** in biological assay.

sign test A test of significance depending on the signs of certain quantities and not on their magnitude; for example, one possible test for trend in a time series is based on the ratio of positive to negative signs of the first differences.

signed likelihood ratio statistic The square root of the likelihood ratio statistic for a null hypothesis about a one-dimensional parameter with a sign attached depending on the sign of the difference between the maximum likelihood estimate and the null value. It allows the determination of one-sided tests and confidence limits by contrast with the essentially two-sided character of the likelihood ratio statistic itself (Barndorff-Nielsen and Cox, 1994).

signed rank test See **Wilcoxon signed rank test**.

significance An effect is said to be significant if the value of the statistic used to test it lies outside acceptable limits, so that there is strong evidence against the hypothesis that the effect is not present. A test of significance is one which, by use of a test statistic, purports to provide a test of the hypothesis that the effect is absent. By extension the critical values of the statistics are themselves called significant. [See **level of significance**.]

significance level See **level of significance**.

similar action The name given to the action of mixtures of stimuli, e.g., the toxic effect of a mixture of poisons, when the stimuli are statistically independent and additive. The effect of a mixture is then predictable from the relative proportions of the constituents and the known response of each.

similar regions In the theory of testing hypotheses a region in the sample space is said to be similar to another if a correspondence can be set up between them such that the probability of a sample point falling in a part of one is proportional to the probability that a sample point falls into the corresponding part of the other. For example, the distribution of n independent normal variables with zero mean and unit variance is spherically symmetric in the sample space. It is possible to set up a correspondence between the whole space and the surface of a hypersphere of unit radius centred at the origin. The probability that a sample point falls in any cone with vertex at the origin is proportional to the probability that a point on the sphere falls in the area which that cone cuts off on it; the surface of the sphere is thus similar to the sample space.

similarity index If two individuals each bear the values of $p(0,1)$ variables, and in m cases they both exhibit, or do not exhibit the same variable value, the ratio m/p is a similarity index. The complementary quantity $1 - (m/p)$ is a dissimilarity index, and is used particularly in **cluster analysis** as a distance function.

simple hypothesis A statistical hypothesis which completely specifies the distribution function of the variables concerned in the hypothesis. [See also **composite hypothesis**.]

simple lattice design See **square lattice**.

simple random sampling Sampling in which every member of the population has an equal chance of being chosen and successive drawings are independent as, for example, in sampling with replacement.

simple sample A random sample in which the probabilities of selection of members are all equal and are constant throughout the drawing.

simple structure In factor analysis, the concept, suggested by Thurstone (1931), that each factor should affect only a small number of the observed variables. Many numerical rotation procedures, such as **varimax**, are based on the idea of simple structure. It conflicts with Spearman's postulate of a single general factor. [See **size and shape**.]

simplex algorithm A general technique for maximizing or minimizing a function of p variables. The function is evaluated at the vertices of a simplex, and the point with the 'worst' value is reflected in the centroid of the remainder. If this gives an improvement, the procedure is repeated with the new simplex until no further improvement is possible. The simplex is contracted, and the procedure repeated until effective convergence to a point.

simplex centroid design A design of an experiment with mixtures proposed by Scheffé (1963) wherein if there are m components and all mixtures are of equal proportions, the design involves observations on all subsets of mixtures using from 1 to m components.

simplex designs A simplex is the n-dimensional analogue of the triangle in two, or the tetrahedron in three, dimensions. In experimental design the simplex usually consists of the $n+1$ vertices which are thought of as equidistant, like an equilateral triangle, surrounding the domain of interest; the object is to reach an optimal point by some such method as steepest ascents. If the design is applied to investigating the composition of multi-component systems, the sum of proportions of components being unity, it is sometimes described as a simple lattice. The superposition of two or more simplexes to provide certain kinds of rotatable designs is known as a simplex sum design (Box and Behnken, 1960).

simplex method An **algorithm** to solve **linear programming** problems due to Dantzig (1949). The linear constraints, in general, define a feasible region within which the optimum solution must lie. This is a simplex, and the method, in brief, seeks a point on the simplex and finds its way to the optimum by traversing a path along the edges of the simplex.

simplex models Models for the analysis of relationships among variables that can be arranged according to a logical ordering that may be known beforehand or has to be inferred from the data (Jöreskog, 1970).

Simpson's paradox See **Yule–Simpson paradox**.

simulation model A model of a dynamic or other system which is too complicated for explicit analytical solution, but whose behaviour can be simulated in a variety of conditions by starting from suitable numerical initial circumstances.

simultaneous confidence intervals The setting of confidence intervals for several parameters which are simultaneously under estimate. The problem of locating the parameters separately in confidence intervals when they are not independent is possibly unsolvable without some further assumption, but they may jointly be located in a confidence region. [See **joint prediction interval**.] Similar conditions apply to simultaneous tolerance intervals and simultaneous discrimination intervals.

simultaneous equations model A model representing a stochastic situation in which the relations between the variables are expressed by a set of simultaneous equations containing them. This interdependent system was first proposed by Haavelmo (1943).

simultaneous estimation The estimation of two or more parameters from the same data.

simultaneous tolerance interval See **simultaneous confidence intervals**.

simultaneous variance ratio test A test of equality of variances of $(k + 1)$, $k \geqslant 2$, univariate normal populations, proposed by Gnanadesikan (1959), for the case where one variance is chosen as the standard and the other k variances are compared with it. The alternative hypothesis in this case is that not all the other k variances are equal to the standard; this distinguishes the test from the **maximum F-ratio** test.

single-blind A **clinical trial** in which the patient is kept unaware of which treatment he or she is receiving is called single-blind. [See **double-blind**.]

single factor theory A representation of multivariate data, introduced into factor analysis by Spearman (1904), in which there is only one single common factor. There is some ambiguity of terminology since the 'Single (Common) Factor Method' is equivalent to the two-factor method of Spearman which uses one common factor and one specific or unique factor for each test. [See also **general factor**, **hierarchy**.]

single-linkage clustering A method of cluster analysis in which the distance between two clusters is defined as the least distance between a pair of items in the respective clusters.

single sampling A type of sampling inspection where the decision to accept or reject the hypothesis that the material concerned accords with some specification is taken after the inspection of a single sample. [See also **double sampling**.]

single sampling plan In quality control, a procedure which provides for the drawing of one sample, on the basis of which a lot is accepted or rejected, as distinct from a double sampling plan, which may allow for indecision on the first sample to be resolved by a further sample.

single tail test An alternative term for a **one-sided test**.

singly linked block design A class of incomplete block design proposed by Youden (1951a) in which every pair of blocks has one treatment in common. For example, for 10 treatments in five blocks of four plots each design is as follows:

$$
\begin{array}{cccc}
1 & 1\,2 & 3\,4 \\
2 & 5\,5 & 6\,7 \\
3 & 6\,8 & 8\,9 \\
4 & 7\,9 & 10\,10
\end{array}
$$

It is a particular case of the **triangular design**.

singular distribution A multivariate normal distribution is singular if the rank of the correlation matrix or, equivalently, the dispersion matrix is less than the number p of variables. It is then possible to express the frequency in terms of fewer than p variables, linearly related to the original set. More generally a continuous distribution is singular if a set of measure zero has probability one.

singular value decomposition Any real $m \times n$ matrix $\mathbf{A}$ with $m > n$ may be expressed in the form $\mathbf{A} = \mathbf{USV}'$, where $\mathbf{U}$ and $\mathbf{V}'$ are orthogonal matrices of eigenvectors and $\mathbf{S}$ is an $n \times n$ diagonal matrix containing the non-negative square roots of the eigenvalues of $\mathbf{A}'\mathbf{A}$, the singular values of $\mathbf{A}$. The singular value decomposition is the basis of **correspondence analysis**.

singular weighing design A weighing design where the matrix $\mathbf{S} = \mathbf{X}'\mathbf{X}$ is singular; $\mathbf{X} = \{x_{ij}\}$ is the design matrix (Raghavarao, 1964).

sinusoidal limit theorem A theorem stated by Slutsky (1937) to the effect that if a random series x is subject to n iterated summations of pairs of items, followed by the calculation of the mth differences, and if the ratio m/n is kept constant, any arbitrary section of the resulting series will tend (with probability 1 as $n \to \infty$) to a sine curve of period

$$
T = \frac{2\pi}{\arccos r}, \quad \text{where} \quad r = \frac{1 - m/n}{1 + m/n}.
$$

The result has subsequently been generalized.

size and shape When a number of variables are all positively correlated, the first principal component is a weighted sum of the variables, and is said to represent 'size'. Other components are orthogonal to it, and are contrasts between groups of variables. They correspond to 'shape'. The distinction is important in **allometry**, and extends by analogy to variables of other types, as in factor analysis.

size of a region or test In the theory of testing statistical hypotheses, the size of a **critical region** is a measure of probability and is the same as probability of an **α-error** or **error of the first kind**. For composite hypotheses it has sometimes been used to denote the limits of the α-error where no **similar regions** exist.

skew distribution A distribution which is not symmetrical; a distribution for which a measure of **skewness** has some value other than zero.

skew-normal distribution A skew distribution obtained by conditioning on positivity of one component of a bivariate normal variable.

skew regression An obsolete term for **curvilinear regression**.

skewness Asymmetry, in relation to a frequency distribution or a measure of that asymmetry. If a unimodal distribution has a longer tail extending towards lower values of the variable it is said to have negative skewness; in the contrary case, positive skewness.

skip free process A term proposed by Keilson (1962) for a class of random walk where in passing from x_1 to $x_2 > x_1$ all intervening states must be encountered at least once. As stated the process would be positive skip free; the opposite mode, negative skip free, requires the passage from x_2 to $x_1 < x_2$.

Skitovich–Darmois theorem This states (Darmois, 1953; Skitovich, 1954), that if $a_i, b_i, i = 1, \ldots, n$, are constant coefficients and $X_1, \ldots, X_n$ are independent, but not necessarily identically distributed, scalar random variables, then if

$$\sum_{i=1}^{n} a_i X_i \qquad \text{and} \qquad \sum_{i=1}^{n} b_i X_i$$

are independent, it follows that the random variables for which $a_j b_j \neq 0$ are all normal.

slash distribution A heavy-tailed distribution used in robustness studies and obtained from division of a normal by an independent uniform random variable.

slippage test A significance test of k samples in which the hypothesis is one of homogeneity in the means, as against the alternative that one member or set of members has 'slipped' away from the others. For example, where the samples are observations on an industrial process at successive time points and it is suspected that the magnitude of the variable is systematically moving away from the intended value.

slope ratio assay A general class of biological assay where the dose–response lines for the standard test stimuli are not in the form of two parallel regression lines but of two lines with different slopes intersecting the ordinate corresponding to zero dose of the stimuli. The relative potency of these stimuli is obtained by taking the ratio of the estimated regression coefficients, i.e., the ratio of the two slopes, hence the same 'slope ratio' assay. The slope ratio assay generally employs an odd number of points, and is called a $(2k + 1)$-point design. This compares with the $2k$-point design of the parallel line assay, although this general class of design can be adapted for the slope ratio assay by omitting the test at the common zero dose.

slowly varying function A function L such that $\lim_{t \to \infty} L(tx)/L(t) = 1$ for all $x > 0$.

Slutsky process A synonym for the **moving average** process.

Slutsky–Yule effect An effect in the averaging of random series studied independently by Slutsky and Yule. If a moving average is applied to such a series the averaged series contains undulations of an apparently systematic kind. Further averaging enhances the effect and under certain types of repeated moving average the resulting series approaches a sine wave. [See **sinusoidal limit theorem**.]

Slutsky's lemma Results giving convergence properties of rational functions of random variables. If all the variables converge in probability to constants, then a rational function of them converges in probability to the function of the constants, if finite. This is true more generally for continuous functions, implying that convergence in probability is invariant under continuous functional transforms. If some variables converge jointly in distribution and the remainder in probability to constants, then convergence to the function of constants and the limiting variable holds (van der Vaart, 1998).

small numbers, law of A term suggested by von Bortkiewicz (1898) to describe the behaviour of rare events obeying a **Poisson distribution**. The term is not antithetical to the **law of large numbers**, and in fact is itself related to the behaviour of large numbers in which only small proportions are events of the kind under study.

Smirnov tests See **Kolmogorov–Smirnov test**, **Cramér–von Mises test**.

Smirnov–Birnbaum–Tingey distribution The probability distribution $P_n(\varepsilon)$ that if an empirical distribution function $F_n(x)$ is constructed from a random sample of size n from a continuous probability density with distribution function $F(x)$, then $F(x)$ will be nowhere greater than $F_n(x) + \varepsilon$. The earliest attempts at deriving the distribution were by Smirnov (1939); the explicit form

$$P_n(\varepsilon) = 1 - \varepsilon \sum_{j=0}^{[n(1-\varepsilon)]} \binom{n}{j} \left(1 - \varepsilon - \frac{j}{n}\right)^{n-j} \left(\varepsilon + \frac{j}{n}\right)^{j-1},$$

where $0 < \varepsilon \leqslant 1$, and $[x]$ means the integral part of x, is due to Birnbaum and Tingey (1951).

smooth regression analysis A concept proposed by Watson (1964) extending the basic principles of smoothing frequency or probability density functions to the bivariate case of regression analysis. [See **kernel smoothing in regression.**]

smooth test A test of goodness of fit between data and hypothesis in which the alternative hypotheses are regarded as moving away from the null hypothesis 'smoothly' in the parameters, i.e., with high continuity and differentiability in them (Neyman, 1937*b*). It has the important property of taking into account the nature and the order of the signs of deviations between observation and expectation as well as the size of these deviations.

smoother Any procedure whereby data are smoothed to obtain a local estimate. See **smoothing.**

smoothing The process of removing or reducing fluctuations in data in order to produce a result that is visually smooth. Typically this is performed by some form of local averaging, e.g., by a **nearest-neighbour method**, a **kernel estimator**, a **spline estimate**, or by **local likelihood estimation**. An older term for smoothing is graduation (Simonoff, 1996; Wand and Jones, 1995). [See also **error reducing power.**]

smoothing power A term used in connection with the smoothing of a time series. The 'smoothness' of a series may be tested by examining the order as well as the size of differences between successive observations. There are a number of bases upon which a smoothing index can be constructed, the conventional one involving the use of differences of the third order. [See also **error reducing power.**]

snap reading method A simple method of time sampling proposed by Tippett (1935) for determining the proportions of time a system spends in various states. The term derives from 'snap (shot = time) reading'.

Snedecor, George Waddel (1881–1974) George Snedecor is widely known for his text *Statistical methods* (Snedecor, 1937). Snedecor is best remembered as a pioneer in making statistical tools accessible to experimenters in agriculture, biology, and others areas of application. He was the founder of the Iowa State College Statistical Laboratory in 1933, the first of its kind in the United States.

Snedecor's *F*-distribution See *F*-distribution.

snowball sampling A form of sampling proposed by Goodman (1961) whereby a random sample of n individuals is drawn from a finite population. Each individual is asked to name k further individuals, and so on for s stages. The initial sample solution and its size govern the kind of analysis and inferences that can be drawn from the data.

sojourn time The length of uninterrupted time that a stochastic system spends in a given state.

space–time clustering In investigating whether there is an infectious element in the aetiology of a disease from its incidence in space and time, the spatial distribution is likely to be extremely irregular, but an excess of cases close together in space that occurred at about the same time may be evidence of infection. The problem was first studied by Bartlett and Knox (1964), and more refined procedures were introduced by Mantel (1967) and Pike and Smith (1968). [See **Knox's test**]

spatial statistics Statistical methods applied to data spatially referenced, or tied to a specific region (Cressie, 1993).

spatial systematic sample A systematic sample taken in two dimensions; also termed 'plane sampling' by Quenouille (1949*b*).

Spearman, Charles Edward (1863–1945) An English psychologist. His initial choice of career was the army. In 1897 he went to Leipzig to study psychology. After jobs in various German universities he returned to London in 1907 as professor of psychology at University College, retiring in 1931. His first paper on correlation, introducing what is now called **Spearman's** ρ, appeared in 1904. In the same year a second paper laid the foundations for **factor analysis**. Spearman was elected to the National Academy of Sciences in 1943.

Spearman–Brown formula A formula for the estimation of the **reliability coefficient** of a psychological or educational test which is n times as long as a basic test for which the reliabilities are known. If r_1 is the reliability of a test of unit length, the reliability of a test of length n (not necessarily integral) is

$$\frac{nr_1}{1 + (n-1)r_1}.$$

[See also **split half method**.]

Spearman estimator The estimator, non-parametric, of the mean effective dose relevant to the **Spearman–Kärber method** of bioassay or other stimulus experiment. If the doses (k) are equally spaced the mean is $x_k + \frac{1}{2}d - d\sum p_i$ where d is the common difference between doses and p_i the individual dose/stimuli responses.

Spearman–Kärber method A method for estimating **equivalent doses** of stimuli which generate **quantal responses**. In general, this method estimates the average logarithmic tolerance, i.e., the mean effective dose, rather than the **median effective dose** and requires an unlimited range of doses for its successful application.

Spearman two-factor theorem An alternative term for the basic result underlying the two-factor theory, due largely to the work of Spearman. [See also **single factor theory**.]

Spearman's footrule A coefficient of rank correlation proposed by Spearman (1906) and defined through the scaled sum of the absolute values of the differences between the ranks ascribed when the same individuals are given two different rankings.

Spearman's ρ A coefficient of rank correlation proposed by Spearman (1906) and defined through the scaled sum of the squared values of the differences between the ranks ascribed when the same individuals are given two different rankings.

species abundance A sample of items falling into different classes may be presented as a table showing the number of classes represented by r individuals. The species abundance problem is the problem of estimating the total number of classes in the population, or equivalently the number of classes unrepresented in the sample. There are applications in ecology, linguistics and numismatics.

species of Latin square In the enumeration of Latin squares by combinatorial methods certain distinguishable types appear from which other squares may be obtained by permutation of the letters or of rows or of columns and also by interchange of the three categories. These types have been called species.

specific factor See **common factor**.

specific rate A rate which is based upon some homogeneous subgroups of a population instead of the whole population. For example, death rates may be specific for age, i.e., may be calculated separately for a number of age groups of the population.

specification bias A term suggested by Anderson and Hurwicz (1949) for the bias which arises from incorrect specification of the model under analysis, e.g., by the use of an **errors-in-variables model** instead of one with errors in equations.

specificity In multivariate, and particularly in factor, analysis, the specificity of a variable is the proportion of its total variance attributable to a specific factor. [See **common factor**, **factor analysis**, **sensitivity and specificity**.]

spectral average An average over a **spectral density function** $f(\omega)$ of the form

$$J(A) = \int A(\omega)f(\omega)d\omega$$

where $A(\omega)$ is a suitably chosen function. It occurs particularly in the smoothing of the spectral density function and $A(\omega)$ is sometimes known as the **spectral window**. Special forms of such a window are known by the names of Daniell

(1946), Bartlett (1963), Blackman and Tukey (1959) and Parzen (1969). [See also **spectral weight function**.]

spectral decomposition A square matrix **A** may be expressed in the form $\mathbf{SDS}^{-1}$, where **D** is a diagonal matrix of eigenvalues and **S** is a matrix of right-hand eigenvectors. This form is important, in particular, in the theory of **Markov chains**.

spectral density function The derivative of the **spectral function**.

spectral distribution function See **spectral function**.

spectral function A necessary and sufficient condition for $\rho(\tau)$, $\tau = 0, 1, 2, \ldots$, to be an autocorrelation function of a discrete time stationary stochastic process, is that it is expressible in the form

$$\rho(\tau) = \frac{1}{\pi} \int_0^\pi \cos(\tau\omega) dF(\omega),$$

where $F(\omega)$ is a non-decreasing function called the spectral distribution function with $F(0) = 0, F(\pi) = \pi$. For a continuous time process the corresponding condition is that

$$\rho(\tau) = \int_0^\infty \cos(\tau\omega) dF(\omega),$$

with $F(0) = 0$, $F(\infty) = 1$. The function $F(\omega)$ is variously called the spectral distribution function, integrated spectral function, power spectrum, or integrated power spectrum. Similarly $dF(\omega)/d\omega$ is called the spectral density. Both spectral function and spectral density can be defined, without invoking the concept of autocorrelation, in terms of the intensities given by harmonic analysis, i.e., by a decomposition of the process rather than via the autocorrelation.

spectral weight function A weight function used in the estimation of the spectral density; individual specifications have been proposed, for example, by Daniell (1946), Bartlett (1948), Blackman and Tukey (1959), Parzen (1961) and Bartlett (1963). The function is sometimes known as a 'spectral window'.

spectral window See **spectral weight function**.

spectrum A term which is applied by physical analogy (i) to the graphical representation of the spectral function; (ii) to the graphical representation of the spectral density; (iii) to the spectral function itself; and (iv) to the spectral density function itself. Usage varies, but it would seem convenient to speak of 'spectral function' and 'spectral density function' for the mathematical functions; of the 'integrated spectrum' as the graph of the spectral function against frequency as abscissa; and of the 'spectrum' as the graph of the spectral density against frequency as abscissa. As for ordinary frequency functions, the spectral density may not exist, the spectral values condensing at certain points to provide a discrete spectrum. The concept of spectrum for a single series can be extended to the relations between a pair of series.

Spencer formula A moving-average-type graduation formula by Spencer (1904) in two forms, one using 15 terms and the other 21 terms, both reproducing cubic trends. The 15-point formula has weights $[4]^2$ $[5]$ $[-3, 3, 4, 3, -3]/320$ and the 21-point formula the expression $[5]^2$ $[7]$ $[-1, 0, 1, 2, 1, 0, -1]/350$, where $[r]$ means a simple moving average of extent r.

spent waiting time See **waiting time.**

spherical distribution A probability density on a sphere, representing the relative frequencies of directions in three dimensions.

spherical mean direction The direction determined by taking the arithmetic mean of the direction cosines of a sample of directions in three dimensions, or equivalently of points on the sphere. For a population of such directions, the arithmetic mean is replaced by expectation.

spherical normal distribution A trivariate normal distribution with zero correlation between pairs of variables, zero means and unit variance, i.e., standardized. [See **Fisher's (spherical normal) distribution.**]

spherical standard error A measure of the inaccuracy of the estimated mean direction of a unimodal spherical distribution.

spherical variance A measure of the variability of a sample or distribution of directions in three dimensions, relative to a specified direction.

spherical variance function See **variance function.**

spherically symmetric distributions See **elliptically symmetric distributions.**

sphericity, tests for In multivariate analysis, tests of the hypothesis that the variables are uncorrelated and have equal variance.

sphering Linear transformation of multivariate data to new variables with equal variances and zero correlation.

Spitzer's identity An identity (Spitzer, 1956a) connecting a sequence of distributions from random walks and sums of variables. Those from random walk concern the maximum distance achieved by a particle during n steps.

splicing In an index number it may become necessary at certain times to make provision for the appearance of new items or the disappearance of items previously in use, e.g., in price index numbers, when commodities go off the market. The method of effecting the change is known as splicing. For example, if the index at period k based on period 0 is I_{0k} and a change then occurs in the content of the index, and if a new index for period l on period k as base is I'_{kl}, then one form of spliced index relating period k to period 0 is $I_{0k} \times I'_{kl}$ divided by 100 if necessary. [See also **chain index.**]

spline estimate Spline curves consist of polynomial segments smoothly joined at points known as knots. Often the polynomials are cubic, and are constrained to have continuous first and second derivatives at the knots. They are commonly used in **smoothing** estimators of regression curves, and in **generalized additive models** (Green and Silverman, 1994).

split half method A method used, mainly in psychology, to estimate the reliability of a test. Two scores are obtained from the same test, either from alternate items, the so-called odd–even technique, or from parallel sections of items. The correlation between the halves is usually raised to the reliability expected for the test as a whole by the **Spearman–Brown formula**. An analogous use of this term occurs in connection with the design of sample surveys. If there is a question which permits of two formulations, the sample can be split into two halves, and one version given to each half. In this way it is possible to determine the appropriate wording of the question or the more general interpretation of the replies.

split plot confounding Confounding in a design embodying split plots. There are two different kinds of possible confounding: (*i*) the effects of whole plots may be confounded just as they would be if no splitting were present; (*ii*) interactions between the factors represented in the split plots and certain differences between whole plots may be confounded. The object is to ensure that a limited number of important comparisons can be made within plots while less important comparisons are made between plots.

split plot design An experimental method in which units are divided into subunits that receive different treatments or combinations of treatments. Such designs originated in agricultural experimentation, where the units were plots of land, but are now widespread in many applications, including **longitudinal survey**. Subunits may themselves be further split. Analysis must take into account variation at all levels.

split test method An alternative term for the **split half method**.

S-PLUS An advanced environment for statistical graphics and data analysis (http://www.insightful.com/products/product.asp?PID=3). It is an enhanced version of S, which is itself a powerful and flexible programming language. S-PLUS provides an extensive suite of tools for statistical modelling, analysis and graphical display. It is more flexible than standard statistical packages in that functions implementing new statistical methods can be easily built in S-PLUS. S-PLUS for Windows is a popular version. For many details and illustrative examples, see Venables and Ripley (2002). Associated software is available from http://www.stats.ox.ac.uk/pub/MASS4/Software.html. See also **R**, a freeware version of S-PLUS.

spread See **concordant sample**.

spreadsheet In computer technology, a two-way table, with entries which may be numbers or text. Facilities include operations on rows or columns. Entries may also give references to other entries, making possible more complex operations.

SPSS Acronym for Statistical Package for the Social Sciences. This offers a wide range of statistical data analysis and graphical procedures including generalized linear models and time series. SPSS also supports a programming syntax and a polished graphical user interface, although it does not quite match the powerful general data handling capabilities of **SAS**. SPSS is popular among science and social science researchers. SPSS Clementine provides data mining tools. Further details are available from http://www.SPSS.com/.

spurious correlation A term proposed by K. Pearson (1897) for the case where correlation is found to be present between ratios or indices in spite of the original values being random observations on uncorrelated variables. More generally, correlation may be described as spurious if it is induced by the method of handling the data and is not present in the original material. It is to be distinguished from **illusory association or correlation.**

square contingency See **contingency.**

square lattice An experimental design for testing treatments which are a perfect square, say k^2 in number. If the treatments in arbitrary order are denoted by the number pairs (i,j), $i,j = 1, \ldots, k$, they may be arrayed in a square:

$$
\begin{array}{cccc}
(1,1) & (1,2) & \ldots & (1,k) \\
(2,1) & (2,2) & \ldots & (2,k) \\
\vdots & \vdots & \vdots & \vdots \\
(k,1) & (k,2) & \ldots & (k,k)
\end{array}
$$

Various designs can be constructed from this array; for example (trivially), k blocks of k by selecting rows; a set of $2k$ blocks by selecting rows and columns (a simple lattice design); a set of $3k$ blocks by taking in addition the members corresponding to identical letters in a $k \times k$ Latin square superposed on this square (a triple lattice design), and so on. There are in general $(k+1)k$ blocks of k providing orthogonal comparisons. From the point of view of factorial experiments the k^2 treatments are regarded as the combination of two factors each at k levels (see **quasi-factorial design**). If k is prime it is possible to find $k+1$ replicates of the square such that the $k^2 - 1$ degrees of freedom assignable to treatment effects are divided into $k+1$ orthogonal sets of $k-1$ degrees of freedom. If each of these is confounded with the rows of one replicate and the columns of another, the $k+1$ replicates are called a completely balanced (or balanced) lattice square; every treatment then occurs with every other in one row and in one column. For non-prime k such a design may not exist but certain designs possessing a kind of balance may sometimes be found. If fewer than the $k+1$ replicates are employed the design is said to be partially balanced.

square root transformation A variable transformation which is used to 'stabilize the variance' of sample data drawn from Poisson populations; that is to say, to give variables whose variance will be nearly independent of their means. The square root transformation bears the same relation to the Poisson distribution as the **arc sine transformation** to the binomial distribution. It avoids the difficulty of testing homogeneity in variance analysis where variances in different classes differ; but in cases other than homogeneity tests it distorts the model. [See **stabilization of variance**.]

squariance A term proposed by Pitman (1938) in place of the phrase 'sum of squares about the mean' for the sake of brevity. [See also **deviance**.]

St Petersburg paradox A gambler offers to toss a coin, and pay 2^k dollars if the first head appears on the kth toss. The 'fair price' to play the game is infinite. The 'paradox', first discussed in the eighteenth century, is an early example of a distribution with infinite mean.

stabilization of variance The process of transforming a variable whose distribution is dependent on a parameter, in order to make the variance of the transformed variable as insensitive as possible to the values of the parameter. The transformation is used mainly in order to provide tests of significance, the same test being then approximately valid over a fairly wide range of parameter values; or to ensure, in tests of homogeneity of means in variance analysis, that the variances of the variables are approximately equal, as is required by standard tests.

stable law A distribution function F is called stable if, to any positive a_1 and a_2 and any real b_1 and b_2, there correspond constants $a > 0$ and b such that the relation

$$F(a_1 x + b) * F(a_2 x + b_2) = F(ax + b)$$

is satisfied, where $*$ denotes convolution. A symmetric stable law is a stable law whose characteristic function $\phi(t)$ is of the form

$$\log \phi(t) = -c|t|^\lambda, \qquad c \leqslant 0, \quad 0 < \lambda \leqslant 2.$$

A semi-stable law is a distribution for which

$$\phi(t) = \phi(\beta t)^\gamma, \qquad \gamma > 0, \qquad 0 < |\beta| < 1,$$

for all real t. A generalized stable law is a distribution for which $\phi(t)$ is non-vanishing and satisfies for all real t an equation of the form

$$\phi(\beta_1 t)^{\alpha_1} \dots \phi(\beta_k t)^{\alpha_k} = \phi(\beta_{k+1})^{\alpha_{k+1}} \dots \phi(\beta_n t)^{\alpha_n},$$

where $a_i > 0$ for $i = 1, \dots, n$. The concepts are useful in the characterization of some probability distributions.

stable Paretian distributions A class of distributions discussed by Lévy (1925) and Mandelbrot (1960) which follow the 'weak' or asymptotic form of the Pareto law with distribution function of the form

$$1 - F(x) \sim (x/k)^{-a} \qquad \text{as} \quad x \to \infty,$$

where $0 < a < 2$. The distributions are also known as stable non-Gaussian distributions.

stable process (distribution) An alternative term for **stationary** or **stochastic process**. The term is to be avoided because of confusion with a 'loi stable' in the sense used by French authors. [See **stable law**.]

stable state A state of a **Markov chain** in continuous time for which the expected **sojourn time** is greater than zero.

Stacy's distribution A form of **generalized gamma distribution**.

staircase design An experiment design proposed by Graybill and Pruitt (1958) extending the basic **randomized block** to the case where all blocks do not contain the same number of experimental units.

staircase method An alternative name for the **up and down method** especially when used in connection with fatigue tests.

standard deviation The most widely used measure of dispersion of a frequency distribution introduced by K. Pearson (1894b). It is equal to the positive square root of the **variance**.

standard error The positive square root of the variance of the sampling distribution of a statistic. It includes the precision with which the statistic estimates the relevant parameter as contrasted with the standard deviation that describes the variability of primary observations.

standard Latin square A Latin square in which the first row and column are in the natural order of letters, A, B, C, ..., or numbers, $1, 2, 3, \ldots$. All Latin squares of a given order can be obtained by permutations of the rows and columns of the standard Latin squares of that order. For example, the 576 squares of order 4 are obtained in this way from four standard squares.

standard measure If X is a variable with mean μ and standard deviation σ the transformed variable $Y = (X - \mu)/\sigma$ is said to be in standard measure. It has zero mean and unit standard deviation.

standard normal variable Normal random variable with mean zero and unit variance.

standard population The population in a given period or area which can be used as a basis for comparison with that at another period or in another area, e.g., in constructing standardized rates of birth or death.

standard score An alternative name for **z-score**. [See also **T-score**.]

standardized deviate The value of a deviate reduced to standardized form (zero mean, unit variance) by subtracting the parent mean and then dividing by the parent standard deviation. Standardization of the sample values is often carried out by a similar process with the sample mean and standard deviation, where the parent values are unknown.

standardized mortality ratio An index number in the Paasche form used in the analysis of vital statistics. The ratios of the age-specific death rates for the given year to similar rates for the base or standard year are weighted by the 'expected' deaths in the given year. [See also **Paasche's index, comparative mortality figure**.]

standardized regression coefficients The parameters of a linear model in which the regressor variables have been standardized to zero mean and unit variance; they correspond to correlations if the response variable has also been so standardized. The term is sometimes also used when the regressor variables have unit variance but nonzero mean, in which case the intercept absorbs the difference.

standardized variable A variable in **standard measure**.

Stata A powerful statistical package which is especially popular amongst epidemiologists and medical statisticians. It offers functions and graphics for a wide range of statistical procedures and has some programming capabilities. Stata's documentation is excellent, as are the on-line updating and help facilities. *The Stata Journal* is a useful contribution to the statistical software literature. Further information can be obtained from http://www.stata.com/.

stationary process A stochastic process is said to be strictly stationary if the joint distribution of any finite subset of its constituent random variables is invariant to changes in the location of the subset. It is called stationary in the wide sense or second-order stationary if the mean and covariances of the subset exist and are invariant to such changes in location.

stationary stochastic process See **stationary process**.

statistic A summary value calculated from a sample of observations, usually but not necessarily as an estimator of some population parameter; a function of sample values and hence observable.

statistical decision function See **decision function**.

statistical graphics The field of using graphs to analyse and/or present data. Common forms of statistical graphs include **histograms, pie charts** and **scatterplots**.

statistical literacy The ability to understand and critically evaluate statistical results coupled with the ability to appreciate the contributions that statistical thinking can make in public and private, professional and personal decisions (Wallman, 1993).

statistical numeracy Statistical numeracy requires a feel for numbers, an appreciation of levels of accuracy, the making of sensible estimates, a common-sense approach to data in supporting an argument, the awareness of the variety of interpretation of figures and a judicious understanding of widely used concepts such as mean and percentages (Cockcroft, 1982).

statistical population See **population**.

statistical tolerance interval The interval between an upper and a lower **statistical tolerance limit**.

statistical tolerance limit A quantity, computed on the basis of data, above or below which the value of a future random variable will lie with a specified probability; it is also sometimes known as a prediction limit. It should be contrasted with a **confidence limit**, which has the same property but for an unknown parameter (Wilks, 1942).

statistical tolerance region A generalization of the concept of a **statistical tolerance interval** between an upper and lower **statistical tolerance limit** to the case of two or more dimensions.

statistically equivalent block A term proposed by Tukey (1947) for the multivariate analogue of the **statistical tolerance interval** based upon order statistics.

statistics Numerical data relating to an aggregate of individuals; the science of collecting, analysing and interpreting such data.

statistics anxiety The feeling of confusion and indecision that some students experience when faced with more than one possible path to solving a statistics or probability problem, or the anxiety that some people feel at the very mention of the word 'statistics' or 'probability', at the sight of a table of data, or at the thought of having to deal with uncertainty.

steepest ascent, method of A method introduced into experimentation by Box and Wilson (1951) to find the maximum value of a **response surface**, i.e., to select from experimental data the treatment or set of treatments which are, in some sense, best. The method uses an initial two-level factorial experiment to determine the direction near the starting point in which the response surface rises most steeply. It then proceeds in a standard fashion until a three-level experiment is used to investigate the shape of the response surface near the optimum. An analogous technique is also widely used in numerical analysis.

Stein–Chen method A Poisson approximation using systems of estimating functions to establish bounds on differences among expectations (Barbour, Holst and Janson, 1992).

Stein's paradox The sample mean as an estimate of a p-dimensional mean is admissible under squared error loss if $p \leqslant 2$, but not if $p > 2$.

Stein's two-sample procedure A method of sampling proposed by Stein (1945) to overcome a difficulty in sampling from a normal population that, for fixed sample size n, no similar test of the mean exists which is independent of the variance. The method chooses a sample of fixed size n_0 and then proceeds to a further sample of $n - n_0$, n being determined by the observations of the first sample.

Steiner's triple systems Steiner proposed the problem of arranging w different objects in triplets such that every pair occurs in one and only one triplet (Steiner, 1853). It is a special case of a balanced **incomplete block** design. In generalization, a balanced incomplete block design arranging v elements in m blocks of size k so that each set of β elements occurs exactly once is known as a Steiner system.

stem and leaf display A method of displaying data due to Tukey (1977). The 'stem' corresponds to a range of values, the 'leaf' to the next digit of the data. The display has the shape of a **histogram**, but contains information on the distribution of values within each cell.

step-down procedure A procedure (Ray, 1958) whereby differences are examined in one observed variable first and then in another observed variable after eliminating by regression the effect of the first, and so on. The variables are, in some sense, arranged in a descending order of importance.

Stephan's iterative process An approach to fitting a **log-linear model** to a **contingency table**, closely related to **iterative proportional fitting** (Stephan, 1942).

stepwise regression A method of selecting the 'best' set of regressor variables for a regression equation. It proceeds by introducing the variables one at a time (forward selection) or by beginning with the whole set and rejecting them one at a time (backwards elimination). The criterion for accepting or rejecting a variable can depend on the extent to which it affects the multiple correlation coefficient, the residual variance, or some measure of overall fit such as **Akaike's information criterion** (AIC), **Bayes information criterion** (BIC), etc. Some programs consider more than one variable at a time, or reconsider some which have already been accepted or rejected.

STER distribution A distribution derived from a non-negative integer-valued random variable Y having a probability function $P_Y(y)$. The derived probability function is defined by Sums which are successively Truncated from the Expectation of the Reciprocal of the (zero-truncated) random variable Y. [See also **Bissinger distributions**.]

stereogram A general class of diagram which purports to show a three-dimensional figure on a plane surface. In particular, the name is given to the three-dimensional form of the **histogram**, namely the diagram showing the frequencies of a bivariate distribution.

stereology Inference about three-dimensional structures based on two- or one-dimensional samples, sections or probes.

Stevens–Craig distribution The number of colours represented in a random sample of size n drawn with replacement from an urn containing ka balls, a being of each of k different colours, has a Stevens–Craig distribution with parameters k and n and probability mass function

$$\frac{1}{k^n} \binom{k}{x} \Delta^x 0^n, \qquad x = 1, \dots, k.$$

Stirling distribution A Stirling distribution of the first type is an m-fold convolution of the **logarithmic series distribution**. The name appears to be derived from the use of Stirling numbers of the first kind in the numerator of the density function.

stochastic The adjective 'stochastic' implies the presence of a random variable, e.g., stochastic variation is variation in which at least one of the elements is a variable and a stochastic process is one wherein the system incorporates an element of randomness, usually evolving in time, as opposed to a deterministic system. The word derives from Greek στόχος, a target, and στοχαστιχης was a person who forecast a future event in the sense of aiming at the truth. In this sense it occurs in sixteenth-century writings. Bernoulli in the *Ars conjectandi* (1713) refers to the '*ars conjectandi sive stochastice*'. The word passed out of usage until revived in the twentieth century.

stochastic abundance models Models for the 'unobserved species problem'. If a sample of individuals includes types (species, or words used by an author) the aim is inference about the number of types in the population (genus, or total vocabulary).

stochastic approximation procedure A non-parametric method of iterative estimation for a functional or regression relationship which incorporates random elements (Derman, 1956). It may be contrasted with the Newton–Raphson method of approximation and compared with the **up and down method**. An extension to take account of time changes was made by Dupač (1965). [See also **Kiefer–Wolfowitz process**, **Robbins–Munro process**.]

stochastic comparison of tests If two or more allowable tests of significance of a single hypothesis are made on the one set of data the critical levels obtained may be regarded as random variables. In this sense Bahadur (1960b) referred to the stochastic comparison of tests: when applied to a comparison of two tests the concept was first explicitly suggested by Anderson and Goodman (1957).

stochastic convergence Different modes of convergence of a sequence of random variables $\{X_n\}$ to a limiting variable X can be defined. The most important are: (i) convergence in probability, weak convergence, i.e.,

$$\lim_{n \to \infty} \Pr(|X_n - X| > \varepsilon) = 0,$$

for every $\varepsilon > 0$; (ii) almost sure, or strong, convergence, i.e.,

$$\lim_{n \to \infty} \{\Pr(|X_n - X| > \varepsilon \text{ for any } m \geqslant n)\} = 0;$$

(iii) convergence in mean square, i.e.,

$$\lim_{n \to \infty} E\left\{(X_n - X)^2\right\} = 0;$$

and (iv) convergence in distribution, or in law, i.e.,

$$\lim_{n \to \infty} \Pr(X_n \leqslant x) = \Pr(X \leqslant x)$$

at every x for which the probability distribution function of X is continuous. Either (ii) or (iii) implies (i) which in turn implies (iv), but no other implications hold in general.

stochastic differentiability See **stochastic process**.

stochastic disturbance A disturbance which possesses a probability distribution. [See also **shock model**, **shock and error model**.]

stochastic integrability See **stochastic process**.

stochastic kernel A sophisticated way of referring to a probability distribution dependent on a parameter. A stochastic kernel K is a function of two variables, a point and a set, such that $K(x\Gamma)$ is (i) for a fixed x, a probability distribution in Γ, and (ii) for any interval F, a particular closed set of continuous functions in x. If the probability distribution in (i) is a **defective probability distribution** the kernel is referred to as a substochastic kernel.

stochastic matrix A matrix of transition probabilities in stochastic processes, the matrix being called stochastic if the sum of the entries in each row is equal to one. If, in addition, the sum of the entries in each column is equal to one the matrix is said to be doubly stochastic.

stochastic model A **model** which incorporates some stochastic elements.

stochastic process A family of random variables $\{X_t\}$ where t assumes values in a certain set, in applications typically a subset of the integers, the real line, the plane, and so on (Grimmett and Stirzaker, 2001).

stochastic programming A form of **linear programming** where the parameters of the objective function, derived from the constraints, are not fixed but comprise sample items from a given distribution of values.

stochastic transitivity The condition that, given two probabilities Π_{ij} and Π_{ik} where Π_{ij} is the probability that a binary random variable X_{ij} has unit value, $\Pi_{ij} \geqslant \frac{1}{2}$ and $\Pi_{jk} \geqslant \frac{1}{2}$ imply $\Pi_{ik} \geqslant \frac{1}{2}$. If the final term is replaced by $\max(\Pi_{ij}, \Pi_{jk})$ we have a more stringent condition known as strong stochastic transitivity.

stochastic variable An alternative name for a **variable** or random variable.

stochastic volatility The property that the **volatility**, i.e., variability, of a time series itself varies unpredictably in time. It is the focus of much interest in financial contexts (Shephard, 1996). [See **autoregressive conditional heteroscedasticity (ARCH) model**.]

stochastically larger or smaller A variable with distribution function F is said to be stochastically larger than a variable with distribution function G if $F(x) \leqslant G(x)$ for all real x.

stopping rule A procedure in sequential sampling for dividing a sample space into two regions: one in which further observations are taken, and the other in which sampling is terminated.

strata chart A chart upon which two or more time series are plotted with the vertical scales arranged so that the curves do not cross. The bands, or strata, between successive curves may be distinctively coloured or hatched. This kind of chart is valuable in connection with the presentation of time-series data in which a total can be broken down into its constituent parts.

strategy In the theory of games, a schedule giving the possible courses of action open to an individual according to the state of the game and possibly to previous action by the opponents. If the strategy lays down a single course of action for each situation it is said to be pure. If there are choices which are determined by a chance mechanism the strategy is mixed. In the theory of sampling an estimation procedure is sometimes referred to as a strategy (Pathak, 1967). The term is also used to describe general considerations about how statistical methods are to be applied.

stratification The division of a population into parts, known as strata, especially for the purpose of drawing a sample, an assigned proportion of the sample then being selected from each stratum. The process of stratification may be undertaken on a geographical basis, e.g., by dividing up the sampled area into sub-areas on a map; or by reference to some other quality of the population, e.g., by dividing the persons in a town into two strata according to sex or into three strata according to whether they belong to upper, middle or lower income groups. The term stratum is sometimes used to denote any division of the population for which a separate estimate is desired, i.e., in the sense of a **domain of study**. It is also used sometimes to denote any division of the population for which neither separate estimates nor actual separate sample selection is made.

stratification after selection It sometimes happens that the proportional numbers lying in certain strata are known but that it is impossible to identify in advance the stratum to which a chosen member belongs. The sample selection then has to be made without reference to the strata, e.g., by simple random sampling. The resulting sample may, however, be stratified after selection and treated as an ordinary stratified sample. The term post-stratification is also used. The procedure is almost as efficient as sampling with a **uniform sampling fraction**.

stratified sample A sample selected from a population which has been stratified, part of the sample coming from each stratum.

stratum See **stratification**.

strength of a test See **power**.

stress In **multi-dimensional scaling**, the stress is a statistic indexing the distortion of the data caused by representation in fewer dimensions. Various definitions have been used. The problem then becomes one of minimizing the stress among possible representations in a space of given dimension.

strictly dominated estimator In decision theory an estimator (**decision function**) d_1 is said to be strictly dominated by another estimator (decision function) d_2 if the risk function of d_2 never exceeds that of d_i and assigns smaller risk values to d_2 at some points. Any estimator strictly dominated by another estimator is said to be inadmissible and every estimator which is not dominated by any other estimator is called admissible.

strictly stationary process See **stationary process**.

strong completeness The property, for a family of probability measures, that any **unbiased estimator** of zero for a subfamily of the probability measures is identically zero **almost everywhere** with respect to the full family provided the measures omitted form a set having measure zero.

strong convergence See **stochastic convergence**.

strong law of large numbers Let $\{X_i\}$, $i = 1, 2, \ldots$, be a sequence of variables with expectations μ_i. In its classical form the strong law of large numbers gives conditions under which

$$\Pr\left\{ \lim_{n\to\infty} \frac{1}{n} \sum_{i=1}^{n} (X_i - \mu_i) = 0 \right\} = 1,$$

i.e., convergence happens almost surely. The weak law gives conditions under which

$$\Pr\left\{ \left| \frac{1}{n} \sum_{i=1}^{n} (X_i - \mu_i) \right| > \varepsilon \right\} \longrightarrow 0$$

for any given $\varepsilon > 0$. Modern versions of both laws are concerned with conditions under which these statements hold for more general normalizing constants. For independent and identically distributed random variables, the mean $\overline{X}_n$ converges almost surely to μ if and only if $E(X_i)$ exists and equals μ.

strongly consistent estimator An estimator which converges strongly to its limit in the probabilistic sense. [See also **stochastic convergence**.]

structural changes A class of statistical procedures developed particularly for detection and identification of change(s) in econometric models.

structural equation An expression which is usually employed to denote a mathematical relation among variables entering into the specification of a model and hence expressing its structure. In statistics and economics the variables are frequently stochastic in character, and the term has been used to differentiate between such relations and those of a functional kind in the ordinary mathematical sense.

structural equation models General statistical models for multi-equation systems where observed variables can be continuous, ordered, dichotomous or censored. The models can be viewed as an interrelated system of regression equations where some of the variables have multiple measures, and where measurement error is taken into account when estimating relationships. They can be expressed as factor analysis models in which some factor loadings are restricted to fixed constants, like zero, and factors may affect each other, directly and indirectly.

structural parameters The parameters appearing in structural equations. [See **partially consistent observations**.]

structural time-series models These time-series models have components such as trends, seasonal variation and cycles that have a direct interpretation to describe the main features of the system.

structural zeros Cells with zero frequency in cross-classifications of categorical variables that occur because it is impossible to observe a given combination of categories, i.e., occur with probability one. Sampling zeros are those that can occur with probability less than one.

structure The structure of a model is the pattern of relationship between its constituent variables as distinct from their values or coefficients associated with them. In factor analysis the structure expresses the pattern of relationship between the variables and the underlying common factors. In the special case where each variable does not depend on all the factors common to the system the structure is called simple. An equation appearing in the explicit formulation of a model is called a **structural equation**; the estimation of any parameters appearing in it is called (not very happily) structural estimation; a coefficient in such an equation is called a structural coefficient. Strictly speaking, perhaps, the adjective 'structure' should be applied only to those variables which appear in the system more than once and hence knit the structure together; but this requirement is not always observed. [See **factor pattern**, **partially consistent observations**.]

Student See **Gosset, William Sealy (1876–1937)**.

studentization The process of removing complications due to the existence of an unknown parent scale parameter by constructing a statistic whose sampling distribution is independent of it, especially by dividing a statistic which is of a certain degree in the observations by another statistic of the same degree. The expression is derived from the *nom de plume* of W.S. Gosset, who first introduced the process in 1907 by discussing the distribution of the mean divided by the sample standard deviation. [See also **Student's hypothesis**.]

studentized confidence interval Bootstrap confidence interval based on an approximate pivot obtained by studentizing an estimator.

studentized M-estimator The M-estimator of β in the linear regression model $Y_i = x_i'\beta + e_i$, $i = 1, \ldots, n$, $\beta \in \mathbb{R}^p$, defined as a solution of the minimization

$$\sum_{i=1}^{n} \rho(Y_i - x_i'\beta) := \min,$$

with an appropriate function ρ, is generally not scale equivariant. To make it scale equivariant, we redefine the estimator by means of the minimization

$$\sum_{i=1}^{n} \rho\left(\frac{Y_i - x_i'\beta}{S_n}\right) := \min,$$

where $S_n = S_n(Y_1, \ldots, Y_n)$ is a scale statistic, regression invariant and scale equivariant in the sense that $S_n(cY_1 + x_1'b, \ldots, cY_n + x_n'b) = cS_n(Y_1, \ldots, Y_n)$ for every $c > 0$ and $b \in \mathbb{R}^p$. This procedure is called **studentization**. [See **M-estimate, regression equivariant estimator**.]

studentized maximum absolute deviate Given a sample of independent values $x_1, \ldots, x_k$ this quantity is defined as

$$d = \max_{i=1,\ldots,k} \frac{|x_i - \bar{x}_i|}{s}$$

where s is the sample standard deviation.

studentized range The range of a sample of n observations divided by the sample standard deviation, usually based on $n - 1$ degrees of freedom.

Student's distribution See **t-distribution**.

Student's hypothesis A **composite hypothesis** asserting that the mean of a sample drawn from a normal population has a certain value or lies in a certain range. For this hypothesis the **t-test** based on Student's distribution has certain optimal properties.

Sturges' rule An empirical rule for assessing the desirable number of frequency groups into which the distribution of observed data should be classified (Sturges, 1926). If N is the number of items and k the number of groups, then

$$k = 1 + \log_2 N.$$

For example, a distribution of 100 items should have not less than eight frequency groups according to this particular rule.

subexponential distribution A property of a distribution function proposed by Tukey (1958*b*) and defined as follows: the distribution is subexponential to the right if

$$\frac{F(z+h) - F(h)}{1 - F(h)} = 1 - \frac{1 - F(z+h)}{1 - F(h)}$$

is monotonically decreasing for fixed $z > 0$ as h increases. The property of being subexponential to the left, or in both directions, is defined analogously.

subgroup confounded In certain kinds of experimental design the comparisons made among the observations may be regarded as a group in the mathematical sense. When certain sets of high-order interactions are **confounding** they form a subgroup, also in the mathematical sense. The design is then said to have a subgroup of comparisons confounded.

subjective Bayesian inference Bayesian inference in which the prior distributions are defined subjectively, typically by selecting from a family of prior distributions one that the investigator considers best to represent his or her prior beliefs.

subjective probability This expression can be used in at least three senses: (*i*) to describe the intensity of belief in a proposition held by an individual (not necessarily quantifiable); (*ii*) to denote a theory of (quantifiable) probability founded on such intensity of belief with the help of axioms; (*iii*) to describe any theory which is not objective in the sense of being based either on generally accepted axioms or on observable frequencies.

sub-martingale See **martingale**.

subnormal dispersion A term proposed by Lexis (1877) to denote the case where the **Lexis ratio** is less than unity. Lexis referred to data giving rise to such a situation as 'constrained'. [See also **Poisson variation, Lexis variation.**]

sub-Poisson distribution A form of the **hyper-Poisson distribution**.

subsample A sample of a sample. It may be selected by the same method as was used in selecting the original sample, but need not be so. [See also **multi-phase sampling.**]

subsampling The selection of elements or sub-elements of an existing sample, generally at random. More specialized meanings include, in multi-stage sampling, the selection of second-stage units from any selected first-stage unit. In **multi-phase sampling**, the part of the first-phase sample taken up for the second-phase enquiry is called a subsample of the first-phase sample, and the process of selection at the second phase is called subsampling.

substitution In sampling enquiries it is sometimes difficult to make contact with, or obtain information from, a particular member of the sample. In such cases it is sometimes the practice to substitute a more conveniently examined member of the population in order to maintain the size of sample. Any such substitution should, however, be carried out upon a strictly controlled plan in order to avoid bias.

successive difference statistic A group of statistics related to observations which have an order in space or, more usually, in time. They depend on the difference between successive members of the series, or the differences of those differences. [See **variable difference method**.]

sufficiency A property of an estimator defined by Fisher (1922b). A statistic T is said to be sufficient for a parameter θ if the distribution of a sample $X_1, \ldots, X_n$ given $T = t$ does not depend on θ. The distribution of T then contains all the information in the sample relevant to the estimation of θ and knowledge of T and its sampling distribution is 'sufficient' to give that information, so long as the model is used as the basis of the analysis. Generally a set of estimators or statistics $T_1, \ldots, T_k$ is 'jointly sufficient' for parameters $\theta_1, \ldots, \theta_l$ if the distribution of sample values given $T_1, \ldots, T_k$ does not depend on these θ's. If $k > l$ the set is exhaustive and if $k = l$ the set is minimal sufficient (Fisher, 1956). [See also **minimal sufficient statistic**.]

Sukhatme d-statistic A little-used expression for the quantity d occurring in the **Behrens–Fisher test**.

Sukhatme's test A non-parametric test for the equality of the variances of two populations, proposed by Sukhatme (1957). The test suffers from the disadvantage that it can only be applied when the medians of both populations are known. It is asymptotically equivalent to the **Freund–Ansari test**. [See also **Siegel–Tukey test**.]

superefficiency In most contexts of parametric inference the **maximum likelihood estimator** has asymptotically the smallest variance among estimators of the parameter, but in certain cases, particularly **non-regular** ones, estimators exist which have smaller variance than the maximum likelihood estimator. Typically this occurs only when the parameter belongs to a certain set of values of measure zero, this being known as the set of superefficiency.

superfluous variable In regression analysis, an independent variable which does not add anything to the goodness of fit of a regression line to data. In **bunch map analysis** a variable is deemed to be superfluous if its inclusion in the analysis does not make the 'bunch' tighter.

super-martingale See **martingale**.

supernormal dispersion See **Lexis variation**.

super-Poisson distribution A form of the hyper-Poisson distribution.

superpopulation models An approach to the theory of survey sampling in which the population of values under study is treated as a random sample from a (usually hypothetical) superpopulation, rather than a set of fixed values on which a sampling distribution is induced by randomization.

superposed process A stochastic point process consisting of the superimposition of p individual processes (Cox and Isham, 1980).

superposed variation Variation which is additive to the variation under discussion but is not part of the generative scheme, e.g., errors of observation, as contrasted with variation, like the error terms in an autoregressive equation where the occurrence of any particular value is followed by its incorporation into the motion of the system.

supersaturated design A design is said to be saturated when the number of observations equals the number of parameters of interest, and supersaturated when it is less. The term is used in particular in connection with **random balance designs**, factorial designs (Satterthwaite, 1959) with more factors than observations.

supplementary information In sample survey design, information about the sampling units which is supplementary to the characteristics under investigation in the survey. Such information may be used for stratification, for the determination of the probabilities of selection, or in estimation. When used in estimation it provides the basis for estimators based on ratios or regression. For example, in a survey of business firms, supplementary information on, say, gross turnover provided by a previous census may be used either in the sample design or to improve the efficiency of sample estimates.

supplemented balance A feature of experiment design, originally proposed by Hoblyn, Pearce and Freeman (1954) and developed by Pearce (1960), whereby one particular treatment (control) has additional replications.

support (*i*) A specialized use of the word adopted by Jeffreys (1936) and again by Edwards (1972) to mean the natural logarithm of the **likelihood ratio**. The term support function is defined as the natural logarithm of the **likelihood** function itself. Also used by some authors for the likelihood ratio, or the **Bayes' factor**. (*ii*) If $f(x)$ is a probability density, the set of values of x for which $f(x) > 0$ is called the support of $f(x)$.

surprise index A device proposed by Weaver (1948) to provide a basis for the 'surprise' experienced when an event governed by a set of probabilities is observed; see also Good (1956).

survey An examination of an aggregate of units, usually human beings or economic or social institutions. Strictly speaking, perhaps, 'survey' should relate to the whole population under consideration and to material collected in considerable detail. However, it is often used to denote a **sample survey**, i.e., an examination of a sample made in order to draw conclusions about the whole.

survey design See **sample design**.

survey methodology Survey methodology concerns methodology for selecting a sample of population objects, measuring the sample units and then estimating population parameters based on the measurements. A survey can be defined in the following way (Dalenius, 1985): (1) A survey concerns a set of objects comprising a population. (2) The population under study has one or more measurable properties. (3) One wants to describe the population by one or more parameters defined in terms of these properties. This requires observing (a sample of) the population. (4) In order to get observational access to the population a frame is needed, i.e., an operational representation such as a list of the population objects or a map of a geographical area. (5) A sample of objects is selected from the frame in accordance with a sampling design which specifies a probability mechanism and a sample size. (6) Observations are made on the sample in accordance with a measurement process, i.e., a measurement method and a prescription as to its use. (7) Based on the measurements an estimation process is applied to compute estimates of the parameters when making inference from the sample to the population.

survey weight A value associated with a survey response that indicates the number of persons (or elements) in the population that the surveyed individual represents. A basic survey weight is the reciprocal of the individual's probability of selection. Further adjustments to the basic weights are often made to compensate for non-response, to calibrate weight totals to externally known totals, etc.

survival analysis The analysis of data relating to failure times, or times to the occurrence of a specified event. The aim is inference about the distribution of survival times, usually from censored data and its dependence on explanatory variables.

survivor function The survivor function corresponding to a distribution function $F(x)$ is $1 - F(x)$. For a positive random variable X this gives the probability of the event $X > x$, and is key in lifetime data analysis, reliability and renewal theory.

swindle In computer simulation, any device that reduces effort or improves precision, as compared with naïve techniques. An example is the use of **antithetic variables**.

switchback design An experiment design which relates to a three-period sequence for two treatments rather than a two-, or multiple of two-, period sequence, i.e., ABA, rather than AB. Denoting periods as suffixes, the use of $A_1 B_2 A_3$ and $B_1 A_2 B_3$, for example, would deal with the case where trend effects contribute to experimental error independently of treatment effects.

symmetric circular distribution See **circular distribution**.

symmetric sampling A symmetric sampling procedure is one where all elements in the population are treated on the same footing during the drawing procedure. For example, sampling with probability proportional to size is not symmetric.

symmetric stable law See **stable law**.

symmetrical distribution A frequency distribution for which the variable values equidistant from a central value are equally frequent. In a symmetrical distribution all odd-order moments about the mean and all odd-order cumulants, where they exist, are zero.

symmetrical factorial design A factorial design is said to be symmetrical if, in the experiment to which it relates, the number of levels of each factor is the same. In the contrary case it is said to be asymmetrical.

symmetrical test See **two-tailed test**.

symmetrical unequal block arrangement A device proposed by Kishen (1941) to overcome the difficulty in incomplete block design if naturally defined blocks comprise different numbers of plots. The symmetric unequal block arrangements preserve the property of complete balance.

sympathy effect A term used in connection with the sampling of human populations to describe the situation where a member of the sample responds to a question from the investigator in the way which is believed to please the investigator, rather than giving an accurate reply. The respondent is often not conscious of deliberate falsehood, which usually adds to the difficulty of countering the bias to which the effect gives rise.

synergy If an effect associated with several exposures exceeds what would have been expected based on their separate effects, the phenomenon is described as synergy.

synthesis of variance The estimation of a variance in a new random system derived by first estimating components of variance and then combining them appropriately. For example, the variance of a sample mean may be estimated for a sampling scheme different from that used for the data under analysis (Cox and Solomon, 2002).

systematic This word is frequently used in statistics in contrast to 'random' or 'stochastic'. Thus, a variable y consisting of a constant m plus a variable x with zero mean is sometimes said to have a systematic component m and a stochastic component x, although it might equally well be regarded as a stochastic component y with mean m. More typically, however, the systematic part would be specified by a linear or nonlinear relation expressing dependence on explanatory variables. Similarly, an error variable is said to be a systematic error if it has a nonzero mean; and a sampling process is 'systematic' if it is not random. The usage is convenient but occasionally gives rise to difficulty. Many processes embody both 'systematic' and 'stochastic' elements and should not properly be described by either adjective alone; for example, the so-called systematic sampling of a list may begin from a randomly chosen point, and a random sample may be chosen from systematically determined strata. The basic difficulty is that even a random event may be the outcome of a systematic procedure and it is not to be resolved by substituting some other word for 'systematic'.

systematic design An experimental design laid out without any randomization. The term is difficult to define exactly because in one sense every design is systematic; it usually refers to a situation where experimental observations are taken at regular intervals in time or space. [See **systematic**.]

systematic error An error which leads to biased inferences, e.g., due to uncorrected non-random sampling of a population.

systematic sample A sample which is obtained by some systematic method, as opposed to random choice; for example, sampling from a list by taking individuals at equally spaced intervals, called the sampling intervals, or sampling from an area by determining a pattern of points on a map.

systematic square Early attempts at creating experimental designs for the elimination of variability in two directions orthogonal to each other did not generally use the principle of randomization. Thus the allocation of treatment to the rows and columns proceeded in some 'systematic' way. For example, a square design could be laid out as follows with systematic arrangement in the NW–SE diagonals:

$$A \ B \ C \ D$$
$$D \ A \ B \ C$$
$$C \ D \ A \ B$$
$$B \ C \ D \ A$$

[See also **Knut Vik square**.]

systematic statistic A term proposed by Mosteller (1946) for a statistic consisting of a linear combination of order statistics.

systematic variation A term used in two slightly different senses: (i) to denote variation which is deterministic, as opposed to stochastic, and which can therefore be represented by a deterministic mathematical expression; and (ii) to denote variation in observations resulting from experimental or other situations as a result of factors which are not under statistical control. [See also **assignable variation**.]

T

t-distribution This distribution, originally due to Student (Gosset, 1908b), is usually written in the form, as modified by Fisher (1925b)

$$f(t) = \frac{\Gamma\left\{\frac{1}{2}(\nu+1)\right\}}{\Gamma\left(\frac{1}{2}\nu\right)\sqrt{\nu\pi}} \left(1 + \frac{t^2}{\nu}\right)^{-(\nu+1)/2}, \qquad -\infty < t < \infty, \qquad \nu = 1, 2, \ldots,$$

where ν is called the number of **degrees of freedom**. The distribution is, among other things, that of the ratio of a sample mean, measured from the parent mean, to its estimated standard error, in samples from a normal population. It is thus independent of the parent scale parameter and can be used to set confidence intervals to the mean independently of the parent variance. [See also **studentization**.]

T-distribution An alternative name for **Hotelling's T^2** distribution.

t-ratio distribution The distribution of the ratio of two random variables which follow the bivariate t-distribution, developed by Press (1969). The distribution arises in **Bayesian inference** in econometrics. It can be shown that the distribution has no convergent moments.

T-score A value of a random variable obtained by a method of rescaling marks, or scores, in a test, proposed by McCall (1922). The method is essentially one of transforming the scores into deviates of a normal distribution which has a mean of 50 and standard deviation of 10 units. Hence the range of the T-score of 0 to 100 is equivalent to one of five times the standard deviation on each side of the mean in a normal distribution. [See z-**score**.]

T-square test A test for randomness of spatial point processes, introduced by Besag and Gleaves (1973). The test compares the distance of the nearest event of the process from a randomly chosen point with the nearest-neighbour distance between randomly chosen events. The latter is estimated from the distance of the nearest neighbour to the first event outside the T-square defined by the normal to the vector from the original random point.

t-test A test based on the **t-distribution**.

T-test There are three tests of significance which may be encountered under this name. One is the test using **Hotelling's T^2** distribution. Another is a rank order test of trend in a time series introduced by Mann (1945) and the third is a non-parametric test for comparing two variances proposed by Sukhatme (1957). In this last case the statistic is

$$T = \frac{1}{mn} \sum_{i=1}^{m} \sum_{j=1}^{n} \psi(x_i, y_j)$$

which is a modified form of the **Wilcoxon–Mann–Whitney** test; $\psi(x_i, y_j) = 1$ if $y_j < x_i < 0$ or $0 < x_i < y_j$ and zero otherwise.

taboo state See **taboo probability**.

taboo probability For a **Markov chain** in discrete time, the probability that the chain starts from one state and ends in another after a given number of steps, without passing through a specified third state. The notion is important in the study of queueing systems.

Taguchi methods In quality control, Taguchi is mainly known for statistical techniques aimed at optimizing industrial design. The phrase is used rather loosely for studies concentrating on the development, production and inspection of a product rather than on the last stage only. It is also used to refer to designs and methods of analysis proposed by Taguchi (1987). Occasionally it is used as a synonym for techniques of experimental design in an industrial context.

tail area (of a distribution) The portion of the area under a **frequency curve** which lies between the start of the distribution and some point lying between the start and the mode; or symmetrically, between some point lying between the mode and the end of the range of the random variable and the end of the distribution.

Tajima's D statistic A standardized version of $\pi - \hat{\theta}$, where π is the nucleotide diversity and $\hat{\theta}$ is Watterson's estimate of mutation rate. D measures deviations from the neutral **infinitely-many-sites model**, where its expectation is zero.

Takacs process A **Markov process** in continuous time obtained by Takacs (1955) by considering a single server queueing system with Poisson input and a general service-time distribution. If $X(t)$ is defined to be the time which a customer who arrived at time t would have to wait until the commencement of service, $X(t)$ is called the Takacs or 'virtual waiting time' process.

tandem queues A situation where the output from the service stage of one queueing system is the direct input to a service stage of a second system (Reich, 1957). This is not the same concept as **series queues**.

tandem tests A term proposed by Abramson (1966) for a sequence of two Wald-type **sequential probability ratio tests** the first carried out on one of two variables and the second, which may depend upon the outcome of the first, performed on the other variable.

Tchebychev (or Chebyshev), Pafnutii Lvovich (1821–1894) Tchebychev is regarded as the founder of the St Petersburg School of Mathematics, which encompassed pathbreaking work in probability theory. The **Tchebychev**

inequality carries his name; he initiated rigorous work on a general version of the **central limit theorem**.

Tchebychev–Hermite polynomials Polynomials based upon derivatives of the **normal distribution**

$$\phi(x) = \frac{1}{\sqrt{2\pi}} e^{-x^2/2}.$$

The polynomial of order r, $H_r(x)$, is defined by

$$\left(\frac{-d}{dx}\right)^r \phi(x) = H_r(x)\phi(x).$$

These polynomials have important orthogonal properties. It appears that they were originally derived by Laplace but are known in statistical work as Hermite or Tchebychev–Hermite polynomials. The first four are

$$H_1 = x, \qquad H_2 = x^2 - 1, \qquad H_3 = x^3 - 3x, \qquad H_4 = x^4 - 6x^2 + 3.$$

Tchebychev inequality If $g(x)$ is a non-negative function of a variable x, Tchebychev's (1874) inequality states that for every $k > 0$

$$\Pr\{g(X) > k\} \leqslant \frac{E\{g(X)\}}{k}.$$

If $g(x) = (x - m)^2$, m being the mean of x and $k = t^2\sigma^2$, σ^2 being the variance of x, this reduces to the **Bienaymé–Tchebychev inequality**. More general inequalities of a similar kind involving moments higher than the second are sometimes known as inequalities 'of the Tchebychev type'. A multivariate inequality was given by Olkin and Pratt (1958).

temporally continuous process An expression sometimes used to denote a stochastic process which is dependent upon a continuous time parameter. The terminology is not recommended.

temporally homogeneous process A stochastic process for which the **transition probabilities** are the same for any time interval of given length t.

terminal decision In sampling schemes of a sequential type, a decision which involves terminating the sampling process. For example, under a single sampling scheme for acceptance inspection there are two possible decisions which are both terminal decisions: to accept or reject the lot under inspection. If the scheme made provision for a third type of decision, to continue sampling, this third type would not be terminal.

Terry–Hoeffding test An expected **normal scores test** of the identity of two populations sensitive to unequal locations (Hoeffding, 1951).

Terry's test A rank order test most powerful against a location-shift alternative in underlying normal populations. The transformation (Terry, 1952) is based on the methods of Hoeffding (1951), and is closely related to the rank transformation of Fisher and Yates.

test coefficient In factor analysis, a synonym for **factor loading**.

test of normality A test of a set of observations to see whether they could have arisen by random sampling from a normal population. Such tests may be carried out by a comparison of the sample distribution function with a normal distribution function. Certain other tests are said to be tests of normality when, in fact, they are only tests of agreement of certain sample statistics with the values of the corresponding population parameters; for example, a test of the sample moment ratio $b_1 = m_3/m_2^{3/2}$, where m_r is the rth sample moment, against the normal value of zero, or $b_2 = m_4/m_2^2$ against the normal value 3 are spoken of as tests of normality.

test statistic A function of a sample of observations which provides a basis for testing a statistical hypothesis.

testing the constancy of regression relationships over time Tests for detection and identification of change(s) in regression models where the design matrix can be random fixed or mixed. The considered changes concern either regression parameters and/or variance.

tetrachoric correlation An estimate of the parameter ρ, equivalent to the product moment correlation between two normally distributed random variables, obtained from the information contained in a 2×2 table or double dichotomy of their bivariate distribution. The term is almost entirely confined to a particular estimator of rather complicated form developed by Pearson (1913).

tetrachoric function A function which is related to the **Tchebychev–Hermite polynomials** and which is used in the calculation of the **tetrachoric correlation** coefficient. The function of order r may be defined as

$$\tau_r = \frac{(-1)^{r-1} D^{r-1}\phi(x)}{(r!)^{1/2}} = \frac{H_{r-1}(x)\phi(x)}{(r!)^{1/2}}$$

where $D^{r-1}\phi(x)$ is the $(r-1)$th derivative of $\phi(x) = e^{-x^2/2}/\sqrt{2\pi}$, the standard normal distribution, and $H_{r-1}(x)$ is the Tchebychev–Hermite polynomial of order $r-1$.

tetrad difference See **hierarchy**.

Theil's inequality coefficient See **inequality coefficient**.

Theil's mixed regression estimator Theil (1963) obtained an estimator for a regression coefficient by combining sample information with stochastic a priori information. For the linear regression model $y = x\beta + u$ where β is a vector of regression coefficients and u is an error vector with $E(u) = 0, E(uu') = \sigma^2 I$, prior information about β in the form $r = \mathbf{R}\beta + v$ is known, where $\mathbf{R}$ is a matrix of known constants, r is a prior estimate of $\mathbf{R}\beta$, and v is an error vector with $E(v) = \mathbf{0}$, $E(vv') = \psi$ and $E(uv) = 0$. The estimator of β is then given by

$$\hat{\beta} = \left(\frac{x'x}{s^2} + \mathbf{R}'\psi^{-1}\mathbf{R}\right)^{-1}\left(\frac{\hat{x}y}{s^2} + \mathbf{R}\psi^{-1}r\right)$$

where s^2 is an unbiased estimator of σ^2.

theoretical frequencies The frequencies which would fall into assigned ranges of the random variable if some theoretical distribution law were exactly followed, as distinct from the actual frequencies which may be observed in a sample.

Thiessen polygon See **Dirichlet tessellation**.

Thomas distribution A distribution equivalent to the **double Poisson distribution**. The number of parents in an area follows a Poisson distribution with mean m_1, and the number of offspring per parent follows a Poisson distribution with mean m_2. Thus the number of offspring per area follows a **type A distribution**, and the total of parents and offspring per area follows a Thomas distribution.

Thompson's rule A studentized rejection rule for outlying observations proposed by Thompson (1935). The criterion, determined from F- or t-tables, provides the basis for rejecting all observations whose studentized residual is larger than the critical value. This rule is not particularly suitable for a single outlier, or two outliers which lie at opposite extremes of the sample observations.

three-dimensional lattice A general class of lattice design of which the **cubic lattice** is a particular case. For example, 120 treatments could be laid out as a $4 \times 5 \times 6$ three-dimensional lattice. This would be regarded as equivalent to a factorial design with three factors at 4, 5 and 6 levels.

three-series theorem A theorem due to Kolmogorov (1928) concerning the sums of mutually independent variables. Let $\{X_k\}$ be a sequence of independent random variables and let $\{a_k\}$ be a bounded sequence of positive numbers. Define

$$Y_k = \begin{cases} X_k & \text{if } |X_k| \leqslant a_k \\ 0 & \text{if } |X_k| > a_k. \end{cases}$$

Then the series $\sum X_k$ converges with probability one if, and only if, all the following three series converge:

$$(i) \quad \sum \Pr(X_k \neq Y_k),$$

$$(ii) \quad \sum E(Y_k),$$

$$(iii) \quad \sum E\{Y_k - E(Y_k)\}^2.$$

three-stage least squares A method for estimating the coefficients in a **simultaneous equations model** as a whole rather than as a set of single equations in isolation. [See also **two-stage least squares**.]

threshold model Any model involving a change at some threshold value of a regressor variable (see **two-phase regression**), particularly models for the effects of drugs that imply zero effect below a critical level.

threshold theorems In epidemic theory, results showing the impossibility or extreme improbability of an epidemic unless the density of susceptibles reaches a certain level. The first (deterministic) result of this sort was given by Kermack and McKendrick (1927).

ticket sampling A method of selecting a random sample in which the characteristics of each member of the population are noted on a separate card and the required number of cards are drawn from the resulting pack. Also known as 'lottery sampling'.

tied double changeover design An experimental design proposed by Federer and Ferris (1956) for situations where treatments (t) are applied in sequence and the effects persist into the succeeding period. Both direct and residual effects may be estimated and the design makes use of $(t-1)$ orthogonal Latin squares in its construction.

tied ranks When a set of objects have to be ranked it may happen that certain of them are indistinguishable as regards their order and are therefore placed together in a group. To complete the ranking equal rank numbers are allotted to each member of the group, which are then said to be 'tied' and to exhibit 'tied ranks'. The most common method is to allocate to each member the mean of the ranks which the tied members would have if they were ordered. This is called the mid-rank method.

tight A random variable X is tight if for any ε there is an M such that $\Pr(|X| > M) < \varepsilon$. A sequence of random variables is uniformly tight if it is bounded in probability.

time antithesis An index number formula derived from another formula by interchanging the subscripts denoting the base period and the given period and then taking the reciprocal.

time comparability factor In the analysis of vital statistics it sometimes happens that the standard population used to construct index numbers of mortality becomes out of date. In order to make comparisons between periods during which this has occurred an adjusting factor known as the time comparability factor is used. A common form for this factor is derived as the average of the age-specific death rates for the new time period weighted by the mean populations at the specific ages in the base period, divided by a similar average of rates weighted by the mean populations at the specific ages in the new time period. [See also **area comparability factor**.]

time-dependent covariates Particularly in **Cox's regression model**, covariates that change during the survival time of the subjects.

time-dependent Poisson process See **inhomogeneous Poisson process**.

time domain The analysis of time series in terms of autocorrelation and related properties. [See **autocorrelation, frequency domain**.]

time lag The difference in time by which one observation lags behind or is later than another. [See also **lag, lag covariance**.]

time reversal test One of the criteria proposed by Irving Fisher (1922a) for a 'good' index number. The time reversal test is satisfied when an index number satisfies the following relationship:

$$I_{0n} I_{n0} = 1$$

where the base and given periods are designated by '0' and 'n' respectively. The advantage of index numbers obeying this test is that the comparison of two periods is symmetric and consistent results are obtained whichever is regarded as the base.

time series A set of ordered observations on a quantitative characteristic of an individual or collective phenomenon taken at different points of time. Although it is not essential, it is common for these points to be equidistant in time. The essential quality of the series is the order of the observations according to the time variable, as distinct from those which are not ordered at all, e.g., in a random sample chosen simultaneously, or are ordered according to their internal properties, e.g., a set arranged in order of magnitude.

Tobit model A regression model proposed by Tobin (1958) for a dependent variable that is continuous apart from an atom at zero. Applications are mainly in economic statistics.

tolerance In computing stepwise regression, the tolerance of a regressor variable is the proportion of its sum of squares about the mean not accounted for by other variables already included in the regression equation.

tolerance distribution The distribution among a number of individuals of the critical level of intensity at which a stimulus will just produce a reaction in each individual.

tolerance factor In quality control, the difference between the upper and lower tolerance limits divided by some measure of the variability of the product, usually the standard deviation. Sometimes one-half of this quantity is taken as the tolerance factor, especially where the distribution of the variable under measurement is symmetrical.

tolerance limits In quality control, the limiting values between which measurements must lie if an article is to be acceptable, as distinct from **confidence limits**. [See also **statistical tolerance limit**.]

tolerance number of defects An expression which might be better expressed as 'tolerable number of defects'. It is obtained by multiplying the **lot tolerance per cent** or fraction defective by the size of the **lot** or batch submitted for inspection.

Tong's inequality Tong (1970) showed that, if a random variable X is non-negative with probability one, then

$$E(X^k) \geqslant \left(E(X^{k/s})\right)^s \geqslant (E(X))^k + \left[E(X^{k/s}) - (E(X))^{k/s}\right]^s.$$

total determination, coefficient of In regression analysis, the square of the coefficients of multiple correlation: R^2. It represents the proportion of the total variance of the dependent variable which is accounted for by the variation of the independent variables in the multiple correlation. In this respect it is a generalization of the **coefficient of determination** and is sometimes called the coefficient of multiple determination. Similarly, a coefficient of total non-determination, or multiple non-determination, can be written $K = 1 - R^2$.

total time on test transformation The transformation

$$\int_0^{F^{-1}(t)} [1 - F(u)]du, \qquad 0 \leqslant t \leqslant 1$$

of the distribution function F of component lifetimes. The transformation arises in the theory of life testing.

trace correlation See **vector alienation coefficient**.

Tracy–Widom distribution The distribution of the limiting law of the largest eigenvalue of an $n \times n$ Gaussian symmetric matrix (Tracy and Widom, 1994).

traffic intensity A critical ratio of importance in the analysis of congestion problems; it is generally given as

$$\rho = \frac{\text{mean rate of arrival (at a queue)}}{\text{mean rate of service}}$$

but may also be stated as the mean service time divided by the mean inter-arrival time.

transfer function Alternatively called the **frequency response function**.

transformation set of Latin squares If the rows, columns and letters of a Latin square are permuted, the resulting set of Latin squares is known as the transformation set. In the case of squares of certain sizes, e.g., 6×6 squares, not all the squares of a transformation set will be different.

transient state Alternative name for **non-recurrent state**.

transition intensity The probability density corresponding to a change in the state of a **stochastic process** to a specified new state at a given time, conditional on its history thus far.

transition matrix A matrix of **transition probabilities** p_{ij}, representing either the probability of transition from state i to j, or the probability of state i conditional on j at the previous instant. According to the definition, the rows or the columns sum to unity.

transition probability In the theory of stochastic processes, the **conditional probability** that a system in a given state will be in another given state at some specified later time.

translation equivariant estimator An estimator $T_n = T_n(X_1, \ldots, X_n)$ of the shift parameter θ in the model $X_i = \theta + e_i$, $i = 1, \ldots, n$, is translation equivariant if it satisfies $T_n(X_1 + c, \ldots, X_n + c) = T_n(X_1, \ldots, X_n) + c$ for every $c \in \mathbb{R}$ and for every $(X_1, \ldots, X_n)$.

translation parameter A parameter of location, i.e., the parameter θ, in a density of the form $g(y - \theta)$.

travelling salesman problem A classical problem in **operational research**, that of identifying the shortest path passing through a set of points.

treatment In experimentation, a stimulus which is applied in order to observe the effect on the experimental situation, or to compare its effect with those of other treatments. In practice, 'treatment' may refer to a physical substance, a procedure or anything which is capable of controlled application according to the requirements of the experiment.

treatment mean square A mean square in an **analysis of variance** assignable to differences among the effects of the experimental treatments.

tree-pruning The reduction of the complexity of a classification or regression tree.

tree regression A class of non-parametric regression procedures in which the covariate space is recursively split into regions, within each of which the fitted response is typically constant. These are generally first overfitted by taking too many regions, which are subsequently combined until some measure of prediction accuracy has been optimized; this is known as pruning the tree. The method applies to continuous and to categorical responses, the latter case being known as classification trees (Breiman *et al.*, 1984).

tree structured statistical methods Such methods include classification and regression trees, survival trees, and trees for correlated data such as longitudinal data and multiple discrete responses. All are based on the recursive partitioning idea as described in **tree regression** (Zhang and Singer, 1999).

trend A long-term movement in an ordered series, say a time series, which may be regarded, together with the oscillation and random component, as generating the observed values. An essential feature of the concept of trend is that it is smooth over periods that are long in relation to the unit of time for which the series is recorded. 'Long' for this purpose is somewhat arbitrarily defined so that a movement which is a trend for one purpose may not be so for another, e.g., a systematic movement in climatic conditions over a century would be regarded as a trend for most purposes, but might be part of an oscillatory movement taking place over geological periods of time. In practice trend is usually represented by some smooth mathematical function (analytic trend) such as polynomials in the time variable or logistic form; but graduation procedures by **moving averages** or other smoothing procedures are also common.

trend fitting The general process of representing the trend component of a time series. A trend may be represented by a particular curve form, e.g., the logistic, or by a particular form of the general class of polynomial in time, or by a **moving average**. [See also **variable difference method**.]

trial In probability theory a 'trial' is a realization of a random variable or of an event under a probabilistic scheme, e.g., the tossing of a coin is a 'trial', the outcome being one of two possible events, a head or a tail. More generally, a 'trial' is any controlled study with an outcome of an uncertain kind.

triangle test A test in which three objects, two of which are alike, are presented to a judge who attempts to select the dissimilar object. [See also **duo–trio test**.]

triangular association scheme This type of association scheme for **partially balanced incomplete block designs** states that, for the $n(n-1)/2$ treatments, there exist sets (S_j) such that (i) each S_j consists of $n-1$ treatments, (ii) any treatment is in precisely two sets, and (iii) any two sets S_i, S_j have exactly one common treatment.

triangular design A class of experimental design in which $\frac{1}{2}n(n-1)$ treatments are arranged in n incomplete blocks of $n-1$ according to a pattern which may be illustrated as follows for the case $n = 4$.

$$
\begin{array}{cccc}
X & 1 & 2 & 3 \\
1 & X & 4 & 5 \\
2 & 4 & X & 6 \\
3 & 5 & 6 & X
\end{array}
$$

The diagonals of a 4×4 table are eliminated and the six treatments filled in as shown. Each row then constitutes an incomplete block. The phrase 'triangular design', due to Bose (1939), is also applied to more general incomplete block designs based on the above scheme. [See also **linked blocks**.]

triangular distribution A probability distribution a graph of whose density is triangular in shape, e.g., the density $f(x) = \max(0, 1 - |x|)$.

triangular multiply linked block design For a **triangular design** to be multiply linked, an extension of singly and doubly linked, either $r = 2\lambda_1 - \theta_2$ with $b = n$ or $r = (n-3)\lambda_2 - (n-4)\lambda_1$ and $b = \frac{1}{2}(n-1)(n-2)$ with the letters denoting replicates (r), blocks (b) and number of treatments (n) in a group.

411

triangular (singly or doubly) linked blocks A subtype of **triangular designs**.

trimmed least squares estimator A straightforward extension of the trimmed mean to the linear regression model reducing the influence of outliers in the errors e_i of measurement. The α-trimmed least squares estimator $(0 < \alpha < \frac{1}{2})$ trims off the observations Y_i such that either $Y_i \leqslant x_i'\widehat{\beta}(\alpha)$ or $Y_i \geqslant x_i'\widehat{\beta}(1-\alpha)$, $i = 1, \ldots, n$, where $\widehat{\beta}(\alpha)$ and $\widehat{\beta}(1-\alpha)$ are the respective α- and $(1-\alpha)$-regression quantiles, and then calculates the ordinary least squares estimator from the remaining observations (Koenker and Bassett, 1978). [See also **least squares estimator, regression quantile, trimmed mean**.]

trimmed mean A method of reducing the sensitivity of the **arithmetic mean** to extreme observations. It involves discarding a proportion of the outer observations on either side and averaging the remainder. If $0 \leqslant \alpha \leqslant 0.5$ then the trimmed mean $\bar{x}_\alpha$ is the same as the arithmetic mean when $\alpha = 0$ and converges to the median when α converges to 0.5.

trinomial distribution The **multinomial distribution** for three exhaustive and mutually exclusive classes.

triple comparisons The statistical model for paired comparisons can be extended to the $\binom{t}{3}$ triple comparisons that may be formed from t treatments.

triple lattice See **square lattice**. Generally for any lattice design, if three replications are selected from those possible under the design the resultant is said to be triple.

trough An observation in a discontinuous time series which is lower than each of the two neighbouring observations; or, in the continuous case, a point where the series has a minimum.

truncation A truncated distribution is one formed from another distribution by cutting off and ignoring the part lying to the right or left of a fixed variable value. A truncated sample is likewise obtained by ignoring all values greater than or less than a fixed value. In this sense truncation is to be distinguished from **censoring**. The word also occurs in a different sense to denote the cessation of a sampling process. For example, in **sequential analysis** the successive drawing of members of a sample may have to be stopped before a decision has been reached under the terms of the sequential scheme. This cutting off with respect to time may be called truncation but is different from cutting off with respect to a variable value. [See also **cut-off**.]

Tukey's gap test A test for comparing individual mean values in analysis of variance suggested by Tukey (1949). It is based upon excessive gaps occurring between individual or groups of mean values.

Tukey's pocket test An alternative name for **Tukey's quick test**.

Tukey's quick test A simple and compact **slippage test** for two populations, proposed by Tukey (1959). It is based upon the overlap of the sample values, i.e., upon the number of observations in one sample which are greater than all those in the other sample plus the number of observations in the other sample which are less than all those in the one; the test can only be used if both these numbers are greater than zero. The test is quick to execute and the critical values for the common percentage points are easy to memorize. It is sometimes called **Tukey's pocket test**.

Tukey's test A multiple comparison test of mean values (Tukey, 1953) arising in analysis of variance based upon the **studentized range**. It is a step-by-step procedure but when all the sample sizes are equal can be modified to a simultaneous procedure. [See **Gabriel's test**.]

turning point In an ordered series, an observation which is a **peak** or a **trough**. When several contiguous values are equal, and greater than or less than the neighbouring values, a convention is required to determine which is regarded as the turning point, e.g., the middle one may be chosen.

TWIST Acronym for time without symptoms and toxicity.

two-armed bandit A device delivering rewards at random at different rates according to which of two levers is pressed. The optimal strategy for obtaining rewards is studied in **bandit problems**.

two-factor theory See **single-factor theory**.

two-phase regression A class of statistical procedures for detection and identification of change(s) in regression models. If there are additional restrictions on possible change one speaks about restrictive two-phase regression.

two-phase sampling See **multi-phase sampling**.

two-sided test An alternative name for **two-tailed test**.

two-stage least squares A **limited-information method** which can be used to estimate the coefficients of any identifiable equation (see **identifiability**) in isolation from the other equations in a **simultaneous equations model**. [See also **three-stage least squares**.]

two-stage sample A simple case of **multi-stage sampling**. In this case the population to be sampled is first classified into primary units, each of which consists of a collection of the basic sampling unit, the secondary unit. A sample of these primary units is taken, constituting the first stage, and these are then subsampled with respect to their secondary units: this constitutes the second stage.

two-tailed test A test for which the **rejection region** comprises areas at both extremes of the sampling distribution of the test function. It is usual but not essential to allot one-half of the probability of rejection to each extreme, giving a symmetrical test.

two-way classification The classification of a set of observations according to two criteria of classification as, for example, in a double dichotomy or a **correlation table**.

type A distribution A form of compound Poisson distribution proposed by Neyman (1939) for use as a **contagious distribution**. It is to be distinguished from the Gram–Charlier type A expansion of a frequency function. The frequency of the discontinuous random variable X (which takes values $0, 1, 2, \ldots$) at x is the coefficient of t^x in

$$\exp\left\{-m_1(1 - e^{m_2(t-1)})\right\}.$$

It is thus the distribution of the number in the second generation, when the parents follow a Poisson distribution with mean m_1, and the number of offspring per parent follows a Poisson distribution with mean m_2.

type B distribution A distribution formed by the limit of the sum of a large number of independent Neyman **type A distributions** wherein the second parameter has a uniform distribution.

type C distribution A distribution formed in a manner similar to the Neyman type B with a beta distribution for the second parameter.

type I distribution One of the three main types of the Pearson system of frequency distributions. In general it is a unimodal distribution with limited range. With the origin at the mode, it was written by Pearson (1894a, 1895) in the form

$$f(x) \propto \left(1 + \frac{x}{a_1}\right)^{m_1} \left(1 - \frac{x}{a_2}\right)^{m_2}, \qquad -a_1 \leqslant x \leqslant a_2, \quad m_1, m_2 > -1,$$

where $m_1/a_1 = m_2/a_2$. With a suitable choice of origin and scale, it is equivalent to the **beta distribution**.

type II distribution A particular case of the **type I distribution**. Its density function may be written

$$f(x) \propto \left(1 - \frac{x^2}{a^2}\right)^m, \qquad -a \leqslant x \leqslant a, \quad m \geqslant -1,$$

and is symmetrical, usually platykurtic in shape and with limited range. An alternative way of expressing the density is

$$\frac{1}{B(p, p)} x^{p-1}(1 - x)^{p-1}, \qquad 0 \leqslant x \leqslant 1, \quad p > 0.$$

In the special case when $m = 0$ or $p = 1$ it becomes the **rectangular distribution**.

type III distribution This distribution, in the Pearson system of frequency distributions, has unlimited range in one direction and it is generally unimodal. Its density may be written

$$f(x) \propto \left(1 + \frac{x}{a}\right)^{\gamma a} e^{-\gamma x}, \qquad -a \leqslant x < \infty, \quad \gamma > 0, \quad a > 0,$$

or, more simply, by choice of a suitable origin and scale:

$$f(x) = \frac{1}{\Gamma(p)} x^{p-1} e^{-x}, \qquad 0 \leqslant x < \infty, \quad p > 0.$$

The **chi-squared distribution** is of this form, as is the **gamma distribution**.

type IV distribution One of the types of the Pearson system of frequency distributions. Its general shape is that of a unimodal skew distribution with unlimited range in both directions. Its density has the form

$$f(x) \propto \left(1 + \frac{x^2}{a^2}\right)^{-m} e^{-\mu \arctan(x/a)}, \qquad -\infty < x < \infty, \quad a > 0, \quad \mu > 0.$$

type V distribution A unimodal distribution of special type in the Pearson system with origin at the start of the distribution. Its density is usually written in the form

$$f(x) \propto x^{-p} e^{-\gamma/x}, \qquad 0 \leqslant x < \infty, \quad \gamma > 0, \quad p > 1.$$

A transformation of type $y = \gamma/x$ turns it into a **type III distribution**.

type VI distribution The third of the three main types in the Pearson system of frequency curves. Its density is generally unimodal and skew with unlimited range in one direction and may be written as

$$f(x) \propto x^{-q_1} (x - a)^{q_2}, \qquad a \leqslant x < \infty, \quad q_1 > q_2 + 1.$$

The substitution $y = a/x$ reduces this distribution to the **type I** form.

type VII distribution A unimodal symmetrical distribution of special kind in the Pearson system. It has unlimited range in both directions and its density may be written

$$f(x) \propto \left(1 + \frac{x^2}{a^2}\right)^{-m}, \qquad -\infty < x < \infty, \quad m > \frac{1}{2}.$$

The **t-distribution** is a special case of this type.

type VIII distribution A member of the less important group of Pearson curves. Its density may be written

$$f(x) \propto \left(1 + \frac{x}{a}\right)^{-m}, \qquad -a \leqslant x \leqslant 0, \quad 0 \leqslant m \leqslant 1.$$

type IX distribution A member of the less important group of Pearson curves. Its density may be written

$$f(x) \propto \left(1 + \frac{x}{a}\right)^m, \qquad -a \leqslant x \leqslant 0, \quad m > -1.$$

type X distribution A distribution of the Pearsonian system which is the same as the **exponential distribution**.

type XI distribution A J-shaped distribution in the Pearson system whose density may be written

$$f(x) \propto x^{-m}, \qquad b \leqslant x < \infty, \quad m > 0.$$

The start of the distribution is at a suitable $x = b$. [See **Pareto curve**.]

type XII distribution A special distribution in the Pearson system which has a twisted J-shape and constitutes a particular case of the **type I distribution**. The form of the density is

$$f(x) \propto \left(\frac{1 + (x/a_1)}{1 - (x/a_2)}\right)^m, \qquad -a_1 \leqslant x \leqslant a_2, \quad |m| > 1.$$

type I error An alternative term for **error of the first kind**, i.e., rejection of a hypothesis when it is true, or α-**error**. [See also **producer's risk**.]

type II error An alternative term for **error of the second kind** or β-**error**. [See also **consumer's risk**.]

type III error See **error of the third kind**.

type I and II probabilities See **type I sampling**.

type A region In the theory of testing statistical hypotheses, a locally unbiased critical region for testing a simple hypothesis specifying one parameter. Regions of this kind are obtained by maximizing the curvature of the power curve at $\theta = \theta_0$ subject to conditions of local unbiasedness and control of errors of the first kind. If a type A region does not suffer from the restriction of being merely locally unbiased, but is, in effect, unbiased everywhere, then it is known as a type A_1 region.

type B region An extension of the concept of the **type A region** to the case of a composite hypothesis.

type C region An extension of the concept of the unbiased critical region of type A proposed by Neyman and Pearson (1933) to cover a simple hypothesis specifying two parameters. A critical region of this class must be of a given size, unbiased and of best local power in the neighbourhood of the null values of the parameters, say θ_1^0, θ_2^0. The exact determination of type C regions rests upon knowledge of the errors of the second kind and in order to overcome the general absence of this information Isaacson (1951) proposed the region of type D. [See **Neyman–Pearson lemma**, **Neyman–Pearson theory**.]

type D region An unbiased critical region proposed by Isaacson (1951) for testing simple hypotheses specifying the values of several parameters. The type D region, which is a generalization of the type A region, is one which maximizes the curvature of the power surface subject to conditions of size and unbiasedness.

type E region A development of the **type D region** but stated to be of doubtful existence for a wide class of multivariate tests (Giri and Kiefer, 1964).

type I sampling A term sometimes used in Bayesian analysis to denote ordinary sampling from a given population. If this population is itself the result of sampling from a superpopulation the term type II sampling is applied to the selection of the population. The probabilities that govern the two types of sampling are called type I and type II probabilities respectively. The hierarchy can be extended to types III, IV, etc.

type II sampling See **type I sampling**.

type A series A term introduced by Charlier to denote the expansion of a continuous density function as a series of derivatives of the normal density function. It is more usually known as the **Gram–Charlier series**. [See also **Edgeworth expansion**.]

type B series A term introduced by Charlier to denote the expansion of a frequency function in terms of derivatives of a Poisson distribution. [See **Gram–Charlier series, type B**.]

type C series An expansion of a frequency function proposed by Charlier as an alternative to his type A. The latter can give rise to negative densities in the tails of the distribution and type C purports to remove this anomaly. It has not come into general use. [See **Gram–Charlier series, type C**.]

U

U-shaped distribution A frequency distribution shaped more or less like a letter U, though not necessarily symmetrical, i.e., with the maximum frequencies at the two extremes of the range of the random variable.

U-statistic A class of statistic first occurring in the construction of rank order tests for the two-sample problem proposed by Wilcoxon (1945) and Mann and Whitney (1947). U-statistics are of the form

$$U = \sum_{i=1}^{n_1} \sum_{j=1}^{n_2} Z_{ij}$$

where

$$Z_{ij} = \begin{cases} 1 & \text{if } X_i < Y_j \\ 0 & \text{if } X_i \geqslant Y_j \end{cases}$$

and $(X_1, \ldots, X_{n_1}), (Y_1, \ldots, Y_{n_2})$ are ordered observations from the two samples. A much more general notion due to Hoeffding (1948a) defines for independent and identically distributed random variables $Y_1, \ldots, Y_n$ a statistic of the form $\sum Y_{i_1} \ldots Y_{i_r}$, where the sum is over all subsets $i_1, \ldots, i_r$ of $1, \ldots, n$.

U_N^2 test A goodness-of-fit test, analogous to the **Cramér–von Mises test** based on the W_N^2 statistic, introduced by Watson (1961, 1962) and developed by Pearson and Stephens (1962). The test may be extended to the case where the two samples are not necessarily the same size. [See **Watson's U_N^2 test**.]

ultimate cluster The aggregate of ultimate or final stage units included in a primary unit.

ultrametric inequality An inequality implying that all triangles are isosceles with the equal sides longer than the other. It applies to the distances at which items combine in **hierarchical cluster analysis**, and to the geometry of the **dendrogram**.

unadjusted moment A moment of a frequency distribution before any adjustment is made for the effect of grouping the observations, e.g., before the application of **Sheppard's corrections**. [See also **raw moment**.]

unbiased confidence intervals A system of confidence intervals I which covers the true value θ of the parameter with the assigned probability α and the other values θ' as little as possible is said to be unbiased if

$$\Pr_\theta(\theta \in I) = \alpha \geqslant \Pr_{\theta'}(\theta \in I),$$

where the subscript denotes the value of θ under which the probabilities are calculated. This use of 'unbiased' is analogous to that for tests of significance.

unbiased critical region See **critical region**.

unbiased error An error which may be regarded as drawn at random from an error population with zero mean. Thus in the long run positive and negative errors tend to cancel out.

unbiased estimating equation An equation for the estimation of a parameter in which the terms are unbiased estimators of the corresponding parent values. It does not follow that the estimator of the parameter is then unbiased itself. For example, if the estimator t of a parameter θ is given by $A - Bt = 0$, where A and B are random variables, A and B may be unbiased and hence the equation is unbiased, but the ratio A/B may still give a biased estimator of θ (Godambe, 1976).

unbiased estimator An estimator whose expectation equals the estimand; it is sometimes called absolutely unbiased if this is true for all values of the estimand.

unbiased minimum variance estimator See **minimum variance linear unbiased estimator**.

unbiased sample A sample drawn and recorded by a method which is free from bias. This implies not only freedom from bias in the method of selection, e.g., random sampling, but freedom from any bias of procedure, e.g., wrong definition, non-response, design of questions, interviewer bias, etc. An unbiased sample in these respects should be distinguished from unbiased estimating processes which may be employed upon the data.

unbiased test A test is unbiased if the probability of rejecting the null hypothesis is a minimum when the null hypothesis is true.

underdispersion See **overdispersion**.

unequal subclasses See **disproportionate subclass numbers**.

uniform distribution An alternative term for the **rectangular distribution**.

uniform sampling fraction If a sample is selected from a population which has been grouped into strata, in such a way that the number of units selected from each stratum is proportional to the total number of units in that stratum, the sample is said to have been selected with a uniform sampling fraction.

uniform scores test A non-parametric test of the equality of two circular populations, proposed by Wheeler and Watson (1964). For independent random samples of size n_1 and n_2 from the respective populations the test statistic, which is invariant under rotations, may be written in the form

$$R_1^2 = n_1 + \sum_{i=1}^{n_1} \sum_{j=1}^{n_2} \cos\left\{2\pi(r_i - r_j)/n\right\}$$

where r_i is the rank order of the ith order statistic of the first sample within the combined sample, and $n = n_1 + n_2$.

uniform spectrum A spectrum in which all ordinates have the same value. It may arise from **white noise**, a term used for a completely random process.

uniformity trial An experiment, or set of trials, in which each experimental unit receives exactly the same treatment. The object, *inter alia*, may be to estimate some standard characteristic of that treatment or to investigate some aspects of the experimental technique, e.g., plot size or layout in agricultural trials.

uniformly best constant risk estimator Of the class of constant risk estimators, i.e., those for which the **risk function** is constant, the estimator that minimizes the expected risk with reference to an a priori distribution is termed a uniformly best constant risk estimator. It is usual to obtain this restricted class of estimators indirectly through the theorem that any constant risk estimator which is also a minimax estimator is also a uniformly best constant risk estimator. [See also **minimax principle**.]

uniformly best distance power test If the hypotheses alternative to a specified null hypothesis can be specified by parameters with continuous variation, then it is possible to determine a region which yields the same **power** for a given subset of the alternative hypotheses, which are said to be equidistant from the null hypothesis. A test of the null hypothesis taking this region as the acceptance region is a uniformly best distance power test if the region has the desired **size** and if for any specified alternative hypothesis the power is no smaller than the power for any other region of the same kind.

uniformly better decision function The merit of a decision function may be judged by reference to its **risk function**. A decision function δ_1 is said to be a uniformly better decision function than δ_2 if the risk function for δ_1 is never greater than the corresponding function for δ_2 and is smaller for some values.

uniformly minimum risk A characteristic of some multiple decision procedures, where the risk, in relation to a loss function, is minimized over a range of hypotheses.

uniformly most accurate A family of confidence intervals for a parameter is called uniformly most accurate if the probability that it contains any value of the parameter is no larger than the corresponding probability for any other family of confidence intervals, uniformly for all values of the true parameter.

uniformly most powerful (UMP) test A test of a hypothesis against a family of alternative hypotheses which is most powerful for each of the alternative hypotheses. In most cases a uniformly most powerful test only exists when the alternative hypotheses are restricted in some fashion; for instance, if the hypothesis is that some parameter $\theta = 0$, the alternatives might be $\theta > 0$ or

$\theta < 0$ but not both. If the test is uniformly most powerful for either of these alternative sets it is said to be the uniformly most powerful one-sided test.

unimodal An adjective describing a frequency distribution which has a single **mode**.

union–intersection tests A class of significance tests introduced by Roy (1957). A composite null hypothesis can be regarded as the intersection of an infinite set of simple hypotheses; the corresponding rejection region is the union of their separate rejection regions. The tests often coincide with likelihood ratio tests, but not in all cases. In canonical analysis, for example, the likelihood ratio test is based on a determinant, and the union–intersection test on an extreme eigenvalue.

unique factor In factor analysis this term sometimes occurs in the sense of **specific factor** and should be avoided in that sense. More usually, in psychology, it refers to a clearly identifiable trait which forms a factor common to the tests under discussion.

uniqueness See **factor analysis**. The uniqueness of a random variable (test) is the complement of the communality.

unit normal random variable A random variable which is normally distributed with zero mean and unit standard deviation. The corresponding distribution is often written $N(0, 1)$. [See **normal distribution**.]

unit stage sampling See **multi-stage sampling**.

unitary sampling Sampling in which the ultimate units are directly chosen, as contrasted with a multi-stage sampling where primary groups are first chosen.

univariate distribution A distribution of one random variable only as contrasted with bivariate, trivariate or multivariate distributions.

universe An alternative term for **population** derived from the 'universe of discourse' of classical logic.

unobserved confounder A variable, especially in an observational study, distorting an effect of interest. The nature of the variable may be unknown to the investigator or may be known but in fact unmeasured. See **confounding**.

unreduced designs A particular kind of balanced **incomplete block** design derived by taking all possible sets of k-treatments from the full set t, each set of k forming a block. The design is completed by permuting the blocks and arranging the treatments randomly.

unreliability See **reliability**.

unrestricted random sample A sample which is drawn from a population by a random method without any restriction; that is to say, all possible samples have the same chance of being selected.

unweighted mean A mean of a set of observations in which no weights are attached to them, except in the trivial sense that each has weight unity.

unweighted means method (in analysis of variance) In analysis of variance, a simple method for the analysis of a set of results for which the subgroup frequencies are unequal. It involves taking the mean values for each subclass and carrying out an ordinary variance analysis on those means.

up and down method A method of estimating the 50 per cent response point of quantal response data. It is essentially a unit sequential process of testing. If the first object to be tested reacts to a given stimulus the next is subjected to a decreased stimulus. If this reacts then the level is again reduced but if it fails to react the object is retested at the previous high level. This progressive testing at levels of the stimulus which are put 'up and down' according to each result accounts for the name of the method (Anderson, McCarthy and Tukey, 1946). The method has been developed for responses less than 50 per cent by calculating the proportion of zeros rather than positive responses, and for responses greater than 50 per cent in Wetherill, Chen and Vasudeva (1966).

up cross See **down cross**.

upper control limit See **control chart**.

upper quartile See **quartile**.

urn model A family of distributions on the non-negative integers motivated by sampling an urn containing balls of different colours. A ball is drawn at random and depending on its colour new balls of specified colours are added to the urn. The process is repeated a given number of times and the numbers of sampled balls of various colours recorded. The idea has a long history and there are many special variants (Johnson and Kotz, 1977).

Uspensky's inequality An inequality developed from the **Bienaymé–Tchebychev inequality** for use in cases where large positive deviations from the mean are important. It may be written

$$\Pr(X - \mu \leqslant t) = \begin{cases} \leqslant \sigma^2/(\sigma^2 + t^2), & \text{if } t < 0, \\ \geqslant 1 - \{\sigma^2/(\sigma^2 + t^2)\}, & \text{if } t \geqslant 0. \end{cases}$$

This can also be applied to the sum of random variables.

utility theory A formulation of **decision theory**, in which emphasis is placed on the numerical values attached to the 'utilities' of different outcomes.

V

validation A procedure which provides, by reference to independent sources, evidence that an enquiry is free from bias or otherwise conforms to its declared purpose. It may be applied to a sample investigation with the object of showing that the sample is reasonably representative of the population and that the information collected is accurate. Validity is to be contrasted with consistency, which is concerned with the internal agreement of data or procedures among themselves. Validation of a measuring instrument involves establishing near-equivalence to a gold-standard method, should one exist. Otherwise one relies on face-validity and construct validity. For an instrument formed from an item-based questionnaire, face-validity checks that the items cover the aspects required on subject-matter grounds, whereas construct validity requires that differences between distinct groups of individuals are in the expected direction.

value at risk (VaR) The quantile of a profit and loss distribution of a financial portfolio over a fixed period, used to set levels of risk for banks etc.

value index An index number formed from the ratio of aggregate values in the given period to the aggregate values in the base period. Strictly speaking this is not an index number as ordinarily understood but a value relative. [See also **price relative**.]

van der Waerden's test A distribution-free test proposed by van der Waerden (1952, 1953), which is sensitive to differences in location for two samples which are from otherwise identical populations. The test statistic is

$$V = \sum_{i=1}^{N} \Phi^{-1}\left(\frac{R_i}{N+1}\right)$$

where Φ^{-1} are inverse normal scores, N is the combined sample size and R_i the ranks of the ith element of one sample in the combined sample.

variable Generally, any quantity which varies. More precisely, a variable in the mathematical sense, i.e., a quantity which may take any one of a specified set of values. It is convenient to apply the same word to denote non-measurable characteristics, e.g., 'sex' is a variable in this sense since any human individual may take one of two 'values', male or female. It is useful, but far from being the general practice, to distinguish between a variable as so defined and a **random variable**.

variable difference method A method of analysis of time series which consist of a systematic and a random component. It is based essentially on the consideration that if the systematic part of a series can be represented by a polynomial, then successive differencing will eliminate this element and hence allow for the isolation of the random element, or at least the estimation of its variance.

variable lot size plan A sampling plan for acceptance inspection which combines the useful features of a lot-by-lot plan with the flexibility of continuous plans. It is intended to deal with continuous production; an essential feature is that screening of rejected 'lots' does not have to be done immediately.

variable sampling fraction If from a stratified population a simple random sample is selected from each stratum in such a way that the proportion of units sampled in each stratum varies from stratum to stratum, the sample is said to be selected with variable sampling fraction. Applicability of the term to other sampling schemes rests upon the general definition of **sampling fraction**.

variable selection The problem of selecting subsets of variables, in regression or multivariate statistics, that contain most of the relevant information in the full data set. [See **stepwise regression, all-possible-subsets regression**.]

variables inspection An **acceptance inspection** where the criteria for classifying or judging a sample submitted for inspection are quantitative rather than qualitative. In this sense 'variable' relates to a measurable quantity, as distinct from an attribute, and is not used in the broader sense noted in the definition of variable.

variance The second moment of a frequency distribution taken about the arithmetic mean as the origin, namely

$$\int_{-\infty}^{\infty} (x - \mu_1')^2 f(x)dx$$

where μ_1' is the mean and f the density function (Fisher, 1918). It is a quadratic mean in the sense that it is the mean of the squares of variations from the arithmetic mean. It may also be regarded as one-half of the mean square of differences of all possible pairs of variable values. [See **sample variance**.]

variance analysis See **analysis of variance**.

variance component One of the objects of the **analysis of variance** is to split up the sum of squares of observations about their mean into portions which can be assigned to variation between the classes or subclasses according to which the data are classified. If the variables defining the classes are 'fixed', that is to say if all the classes under consideration actually appear, these constituent parts of the sums of squares indicate through mean squares the magnitude of class differences, and the extent to which they differ from the residual mean square affords a test of the hypothesis that such differences are governing the situation. A second generating model often considered in variance analysis regards the classificatory variables observed as themselves variables, i.e., as samples chosen from a wider classification. The expected values of the mean squares derived from the variance analysis can then be used to obtain estimates of the variances of the classifying variables. For example, in a two-way classification with r rows and c columns and k members in each cell, one possible model expressing additive row and column effects is that the observations x are given by

$$x_{ij} = a_i + b_i + (ab)_{ij} + e_{ij}, \qquad i = 1, 2, \ldots, r, \qquad j = 1, 2, \ldots, c.$$

The expected mean squares in an analysis of variance are

Rows: $\sigma_e^2 + k\sigma_{ab}^2 + kc\sigma_a^2$
Columns: $\sigma_e^2 + k\sigma_{ab}^2 + kr\sigma_b^2$
Interaction: $\sigma_e^2 + k\sigma_{ab}^2$
Residual: σ_e^2

where σ_a^2, for example, is the variance of a_i. The various σ^2 are called the variance components of x and can be estimated by equating the estimated and theoretical mean squares.

variance–covariance matrix See **covariance matrix**.

variance function A function expressing how the variance of a random variable depends upon its mean. The Poisson distribution, for example, has equal mean and variance, and so its variance function is the identity. The concept is important in **generalized linear models**, in **quasi-likelihood** theory, and in a different sense in **response surface** designs where the dependence is on explanatory variables (Box and Hunter, 1957; McCullagh and Nelder, 1989).

variance ratio distribution See F-**distribution**.

variance ratio test See F-**test**.

variance stabilizing transformation A transformation to give approximate independence between mean and variance as a preliminary to analysis of variance. For distributions with a known relationship $\nu = f(\mu)$ between mean and variance, transformations $y = \int^y f^{-1/2}(x)dx$, or modifications of this form, are often used. Sometimes the relationship must be inferred from the data; see **Box–Cox transformations**.

variate An alternative name for **variable**.

variation, coefficient of The standard deviation of a random variable divided by the mean. Its dimensionless form makes it convenient for summarization.

variation flow analysis A technique of evaluating the transfer of variations in stock, where the product from several machines at one processing stage is fed randomly to several machines of the succeeding stage.

varimax See **factor rotation**.

variogram See **semivariogram** (the variogram plots 2γ instead of γ).

425

vector alienation coefficient In canonical analysis, suppose two sets of variables have dispersion matrices $\mathbf{V}_{11}(p \times p)$ and $\mathbf{V}_{22}(q \times q)$, and $\mathbf{V}_{12}$ is the matrix of covariances between them, with $p \leqslant q$. Then if λ_i are the roots of $\lambda\mathbf{V}_{11} - \mathbf{V}_{12}\mathbf{V}_{22}^{-1}\mathbf{V}_{21} = 0$ (the squared canonical correlations), Hotelling (1936) defined the vector alienation coefficient as $\prod(1 - \lambda_i)$. With the same notation, two correlation coefficients may be defined. The trace correlation is given by $r_T^2 = \sum \lambda_i/p$, and the determinant correlation by $r_D^2 = \prod \lambda_i$.

Venn diagram A graphical method of representing operations on sets, of great use in illustrating problems in probability (Venn, 1891).

virtual waiting time process An alternative name, derived from the particular variable considered, for the **Takacs process**.

vital statistics Data or statistics that describe basic life-cycle events, such as birth, death, stillbirth, marriage and divorce microdata, counts and rates.

volatility The variability of a time series of financial returns.

von Mises, Richard (1883–1953) Austrian mathematician and aerodynamicist who made valuable contributions to statistics and the theory of probability, in which he emphasized the idea of random distribution. Von Mises was born in Lvov, Ukraine, and educated at Vienna. He was professor at the University of Strasbourg, 1909–1918. In 1920 he was appointed director of the Institute for Applied Mathematics at Berlin. His main contributions on the theory of probability are based on the idea that a probability cannot be simply the limiting value of a relative frequency, and that any event should be irregularly or randomly distributed in the series of occasions in which its probability is measured. Von Mises's ideas were contained in two papers which he published in 1919. Little noticed at the time, they have come to influence all modern statisticians.

von Mises distribution The circular normal distribution as derived by von Mises (1918).

von Mises expansion Functional expansion of statistical parameter using distribution functions, analogous to Taylor series expansion. The first derivative is related to the **influence curve** (van der Vaart, 1998).

von Neumann's ratio The ratio of the **mean square successive difference** to the variance of a series was proposed by von Neumann (1941) as a statistic for testing the independence of successive observations in an ordered series for which the underlying distribution is normal. In large samples from a random series the distribution of the ratio tends to be normal with mean 2 and variance $4(n - 2)/(n^2 - 1)$ where n is the number of observations. The use of the ratio to test independence in a series of observations is equivalent to the use of the older **Abbe criterion**.

Voronoi polygon See **Dirichlet tessellation**.

W

W_N^2 **test** See **Cramér–von Mises test**.

W **test for normality** See **Shapiro–Wilk test**.

waiting time In the theory of queues this term is self-explanatory, but in **renewal theory** residual waiting time is the period from a given point in time t to the next renewal point. Spent waiting time is the time elapsed since the last renewal. The sum of the two waiting times is the length of recurrence interval. In the theory of **random walks** residual waiting time is called 'point of first entry' or 'hitting point' for the interval (t, ∞).

Wald statistic An alternative name for the standardized **maximum likelihood estimator**, particularly when used as a basis for tests.

Wald test Test based on the asymptotic distribution of the maximum likelihood estimator: in the regular case the distribution is normal with expectation θ and variance given by the inverse of the Fisher **information matrix**. Under the null hypothesis $\theta = \theta_0$ the expectation is θ_0 and it is easy to form a statistic which has a standard normal distribution. These tests are easy to compute once the likelihood has been maximized.

Wald–Wolfowitz runs test A non-parametric test of the identity of the distribution functions of two continuous populations against general alternative hypotheses, proposed by Wald and Wolfowitz (1940). A sample of independent values from the first population is combined with a sample of independent values from the second population, the combined sample is ranked, and the test statistic is computed as the number of runs, where a run is defined to be a succession of adjacently ranked values from the same population which are followed and preceded in rank by values from the other population or no values at all.

Wald–Wolfowitz test A large-sample distribution-free test of randomness based upon serial covariance proposed by Wald and Wolfowitz (1943). For a series of observations $X_1, \ldots, X_n$ measured about their mean the test statistic is $\sum_{t=1}^{n} X_t X_{t+k}$, with $X_{n+j} = X_j$, i.e., the formula is **circular**.

Wald's classification statistic A statistic suggested by Wald (1945) which is effectively the same as Fisher's discriminant function (Fisher, 1936c).

Wald's fundamental identity A result due to Wald (1944) which plays an important role in the theory of the **sequential probability ratio test**. It is also important in the theory of **first passage times**.

Ward's method A hierarchical method of **cluster analysis**. Clusters are joined to give the smallest possible increase in the residual sum of squares, defined as the sum of squares of the distances of each point from its cluster mean.

Waring's distribution See **factorial distribution**.

Watson–Williams test In its single-sample form, a test of the hypothesis that the **mean direction** of a **Fisher distribution** is equal to a specified direction against the alternative that it is not. In its two-sample form a test of the hypothesis of equality of two mean directions against the alternative of unequal mean directions. The tests were proposed by Watson and Williams (1956).

Watson's distribution See **Dimroth–Watson's distribution**.

Watson's U_N^2 test I Either of two non-parametric tests proposed by Watson (1961, 1962). For a single random sample of n observations, with order statistics $X_{(1)}, X_{(2)}, \ldots, X_{(n)}$, the test of the hypothesis that a population has a specific distribution function $F_0(x)$ has a test statistic that may be written in the form

$$U^2 = \sum_{i=1}^{n} \left(U_i - \bar{U} - \frac{2i-1}{2n} + \frac{1}{2} \right)^2 + \frac{1}{12n}$$

where the U_i are the uniform order statistics defined by

$$U_i = F_0(X_{(i)}), \qquad i = 1, \ldots, n, \quad \text{and} \quad \bar{U} = \frac{1}{n} \sum_{i=1}^{n} U_i.$$

Watson's U_N^2 test II For the test of the hypothesis that two independent random samples, of sizes m and n, come from the same unknown population, the test statistic may be written in the form

$$U_{m,n} = \frac{mn}{(m+n)^2} \sum_{i=1}^{m+n} (D_i - \bar{D})^2$$

where D_i is the difference between the two sample distribution functions at the ith order statistic of the combined samples (Watson, 1961; Watson, 1962). Because for angular data these statistics are invariant under rotations, they are particularly suited to testing hypotheses concerning **circular distributions**.

Watterson's mutation estimate An estimate of the scaled mutation rate under the **infinitely-many-sites model** of mutation. If S_n is the number of segregating sites in a sample of DNA sequences then the estimate is $\hat{\theta} = S_n / \sum_{j=1}^{n-1} j^{-1}$.

wavelets A system of orthogonal functions each with finite support. They can thus be used to decompose a time series or image into components, the coefficient of each depending only on a finite (often quite small) range. The wavelet coefficients can then be used, for example, to study relatively transient effects. There is an extensive literature and applications in statistics and numerical analysis (Percival and Walden, 2000).

weak convergence See **stochastic convergence**.

Weibull distribution A distribution whose density function has the form

$$f(x) = \alpha\lambda x^{\alpha-1}\exp(-\lambda x^{\alpha}), \quad x > 0, \quad \alpha, \lambda > 0,$$

proposed by Weibull (1939) to describe data arising from life and fatigue tests. It arises as a limiting distribution for minima as a transformation of the **generalized extreme-value distribution**.

weighing design An experimental design, due to Hotelling (1944), for the efficient weighing of N objects using a two-pan balance. Various measures of efficiency have been proposed each being optimum for a set of conditions on the experiment.

weight The importance of an object in relation to a set of objects to which it belongs; a numerical coefficient attached to an observation, frequently by multiplication, in order that it shall assume a desired degree of importance in a function of all the observations of the set.

weight bias Bias, usually in an index number, due to the use of incorrect or undesirable weights. Since the true value of the complete quantity which an index purports to measure is not in general capable of direct measurement, bias in this sense is to some extent an arbitrary quantity.

weight function A non-negative function used for weighting purposes, especially in the theory of decision functions, where the word is often used synonymously with **loss function**.

weighted average An average of quantities to each of which has been attached a **weight** intended to account for its relative importance or precision. The weighted arithmetic mean of $x_1, \ldots, x_n$ with weights $w_1, \ldots, w_n$ equals $\sum w_j x_j / \sum w_j$, for example.

weighted battery A group of educational or psychological tests wherein the relative importance of each test is determined by attaching a weight to the score obtained in that test.

weighted index number An index number in which the component items are weighted according to some system of weights reflecting their relative importance. In one sense nearly all index numbers are weighted by implication; for example, an index number of prices amalgamates prices per unit of quantity and the size of these units may vary from one commodity to another in such a way as to constitute weighting. It is, however, usual to describe an index as 'weighted' only when weighting coefficients enter explicitly into its definition and calculation.

weighted least squares A modification of ordinary least squares based formally on a representation of each response observation as the sum of a systematic part depending on unknown parameters and an error term uncorrelated as between different individuals and having for the jth individual variance $w_j^{-1}\sigma^2$, where the w_j are regarded as known constants. Efficient estimation then proceeds by minimizing a sum of squares of deviations in which the contribution of the jth individual receives weight w_j. [See also **generalized least squares estimator**.]

weighted regression Any form of regression analysis in which values for different individuals are given different weights (or measures of importance), for example because they are based on different numbers of underlying data values or because the intrinsic variances are different. [See also **weighted least squares**.]

weighting coefficient The coefficient attached to an observation as its **weight** in a procedure involving weighting. [See also **raising factor**.]

Welch's test An approximate solution, based on a Taylor expansion, of the **Behrens–Fisher problem** in the framework of the frequency theory of probability. It uses the same test statistic as the **Behrens–Fisher test**, but gives narrower confidence intervals. The intervals may, in fact, be narrower than those obtained by the t-test when variances are assumed equal, and this has led to some criticism (Welch, 1936).

white noise By analogy with the continuous energy distribution in white light from an incandescent body, a covariance stationary stochastic process which has equal power in all frequency intervals over a wide frequency range is called white noise. It is essentially a completely random sequence. The sequence $\{X(t), t \geqslant 0\}$ is said to be a white noise process if it possesses a constant **spectral density function**.

Whittaker periodogram See **periodogram**.

Whittle distribution The distribution of the number of transitions from one state to another among consecutive transitions in a **Markov chain** (Whittle, 1955).

Whittle estimation Estimation of time-series parameters using the approximate exponential distribution of periodogram ordinates.

Wicksell's equations A set of equations (Wicksell, 1925) relating the moments of the diameters of a population of spheres to the moments of the diameters of random sections of the spheres. The results were later extended to moments relating to ellipsoids and their elliptical sections. They are important in **stereology**.

wide sense stationary See **covariance stationary process**.

Wiener–Khintchine theorem The theorem stating that the **covariance function** of a stationary stochastic process is positive definite and thus a Fourier transform of the power spectrum.

Wiener process A diffusion process, or Brownian motion process, characterized by independent increments satisfying

$$E(dX) = \mu dt, \qquad E(dX^2) = \sigma^2 dt.$$

The process is Gaussian; the increment in time t is normally distributed with mean μt and variance $\sigma^2 t$.

Wilcoxon, Frank (1892–1965) A notable American chemist who had a significant role in developing modern insecticides. He and Jack Youden were two of a three-person study group meeting to study R.A. Fisher's newly published book, *Design of experiments*. Both Wilcoxon and Youden became statisticians and made important contributions. Wilcoxon was interested in ranking methods, tabling, individual comparisons by ranking methods, nomograms, and method of dose and time response (Wilcoxon, 1945).

Wilcoxon–Mann–Whitney test A non-parametric test for comparing two samples using only ranked values.

Wilcoxon rank sum test A distribution-free test of the equality of the location parameters of two otherwise identical populations proposed by Wilcoxon (1945). Given a random sample from both populations, the test statistic is formed by combining the samples, ranking the observations in the combined sample, and summing the ranks of the observations belonging to one of the samples.

Wilcoxon signed rank test A distribution-free test of the difference between two treatments using matched samples. If the differences $|x_i - y_i|$ of n pairs of observations are ranked according to size, and each rank given the sign of the original difference, the sum of positive ranks is the test statistic proposed by Wilcoxon (1945) and developed by other writers.

Wilks–Rosenbaum tests Distribution-free tests of location and dispersion proposed by Rosenbaum (1953, 1954) based upon the concept of **statistical tolerance limits** due to Wilks (1942).

Wilks's criterion A criterion of general use in multivariate analysis for testing hypotheses concerning multivariate normal populations, especially hypotheses of homogeneity in means or dispersions. The criterion, which is essentially the ratio of maximized log likelihoods, depends on the ratio of the determinants of two matrices of sums of squares and products; the numerator corresponding to a sum-within-classes and the denominator to a total sum. It occurs in various forms. The Λ criterion was derived by Wilks (1932b) and has subsequently been extended by him and other authors. He also derived a test of significance of the criterion, sometimes known as Wilks's test.

Wilks's empty cell test A distribution-free test (Wilks, 1961) of the null hypothesis of identity of two continuously distributed populations. A simple random sample of n values is drawn from one population, dividing the real line into $n + 1$ intervals or 'cells'; each member of a simple random sample from the second population can then be considered to have fallen into a particular cell. There is clearly a tendency for the test statistic, the number of empty cells, to be smallest when the null hypothesis is true.

Wilks's internal scatter A term introduced by Wilks (1932a) for the generalized sample variance, i.e., the determinant of the sample variance–covariance matrix.

Willcox, Walter Francis (1861–1964) An American demographer and statistician who founded the statistical research office in the US Census Office, and was a major innovator in demographic analysis and apportionment theory and methodology in the United States (Willcox, 1897).

Wilson–Hilferty transformation A transformation of χ^2 proposed by Wilson and Hilferty (1931) for the purpose of ascertaining its distribution function approximately from the normal distribution. If the number of degrees of freedom is ν, the transformed quantity $(\chi^2/\nu)^{1/3}$ is distributed approximately normally with mean $1 - 2/9\nu$ and variance $2/9\nu$.

window See **spectral weight function**.

Winsorized estimation A resistant method of estimating the mean of a relatively small sample of observations by using **linear systematic statistics** and replacing extreme observations by those next in magnitude. This process is associated with the name of Winsor who proposed this approach and it can apply to symmetric or non-symmetric censoring and adjustment (Hastings *et al.*, 1947).

Wishart distribution The distribution of Wishart (1928) is a p-dimensional generalization of the χ^2 distribution (the distribution of $\sigma^2 \chi^2$). Let $\boldsymbol{X}_1, \ldots, \boldsymbol{X}_n$ be independent, $\boldsymbol{X}_i \sim \boldsymbol{N}_p(\boldsymbol{\mu}_i, \boldsymbol{\Sigma})$, $\boldsymbol{\Sigma} > 0$, i.e., $\boldsymbol{X}_i$ has a p-dimensional multivariate normal distribution with mean vector $\boldsymbol{\mu}_i$ and covariance matrix $\boldsymbol{\Sigma}$, and let $\boldsymbol{W} = \sum \boldsymbol{X}_i \boldsymbol{X}_i'$, $\boldsymbol{\delta} = \sum \boldsymbol{\mu}_i \boldsymbol{\mu}_i'$. If $\boldsymbol{\delta} = 0$, $\boldsymbol{W}$ has a p-dimensional Wishart distribution with n **degrees of freedom** on the **covariance matrix** $\boldsymbol{\Sigma}$ given by

$$K|\boldsymbol{\Sigma}|^{-n/2}|\boldsymbol{W}|^{(n-p-1)/2}\exp\{-\tfrac{1}{2}\operatorname{tr}(\boldsymbol{\Sigma}^{-1}\boldsymbol{W})\}, \qquad \boldsymbol{W} > 0,$$

where

$$K^{-1} = 2^{-np/2}\pi^{p(p-1)/4}\prod_{i=1}^{p}\Gamma((n+1-i)/2).$$

[See also Mardia *et al.* (1979).]

Wishart's modal analysis A technique of cluster analysis based on a direct search for modes, introduced by Wishart (1969), and incorporated into the program CLUSTAN.

within-group variance See **intraclass variance**.

Wold's decomposition theorem A theorem which asserts that any univariate non-deterministic stochastic process can be decomposed into a one-sided moving average process and a deterministic process.

Wold's Markov process of intervals A generalization of the **renewal process** due to Wold (1948) which assumes that the sequence of intervals between events in the process $X_1, X_2, \ldots$ forms a time homogeneous Markov sequence.

Wolfowitz minimum distance method A method of estimation developed by Wolfowitz (1957). In many stochastic structures where the distribution function depends continuously upon the parameters and distribution functions of the random variables in the structure, those parameters and distribution functions which are uniquely determined by the distribution function of the structure can be strongly consistently estimated.

Woodbury distribution A form of the **Pólya–Eggenberger distribution** derived by Woodbury (1949) which differs from that distribution in that the probability of success depends only on the number of previous successes and not on the number of previous failures, and so not on the total number of previous trials.

working mean An alternative term for **arbitrary origin**.

working probit The iterative calculations for the maximum likelihood estimation of a **probit regression line** are usually performed by finding the weighted linear regression of the working probit on the dose metameter. The working probit is a quantity compounded of the **empirical probit** and the **expected probit**, with which it coincides if the empirical value lies exactly on the provisional line.

wrapped Cauchy distribution The probability distribution (Lévy, 1939) resulting from wrapping a **Cauchy distribution** on the real line around the unit circle. If the Cauchy density is

$$f(x; a) = \frac{1}{\pi} \frac{a}{a^2 + x^2}, \qquad -\infty < x < \infty, \qquad a > 0,$$

then the corresponding wrapped Cauchy distribution function is

$$p_\rho(\theta) = \frac{1}{2\pi} \frac{1 - \rho^2}{1 + \rho^2 - 2\rho \cos\theta}, \qquad 0 \leqslant \theta < 2\pi, \qquad 0 \leqslant \rho \leqslant 1,$$

where $\rho = e^{-a}$. [See also **wrapped distribution**.]

wrapped distribution The name for a distribution on the real line that has been wrapped around the circle of unit radius. If X is a random variable on the line with distribution function $F(x)$, the random variable X_w of the corresponding wrapped distribution is given by

$$X_w = X \ (\mathrm{mod} \ 2\pi)$$

and the distribution function of X_w is given by

$$F_w(\theta) = \sum_{k=-\infty}^{\infty} \{F(\theta + 2\pi k) - F(2\pi k)\}, \qquad 0 \leqslant \theta < 2\pi.$$

If X has a discrete distribution concentrated on the points $X = 2\pi k/m$, for $k = \ldots, -1, 0, 1, \ldots$, and m is an integer, then

$$p_w\left(\theta = \frac{2\pi r}{m}\right) = \sum_{k=-\infty}^{\infty} p\left(\frac{2\pi r}{m} + 2\pi k\right), \qquad r = 0, 1, \ldots, m-1,$$

where p and p_w are the probability mass functions of X and X_w respectively. If X has a probability distribution function $f(x)$, the corresponding probability distribution function of X_w is

$$f_w(\theta) = \sum_{k=-\infty}^{\infty} f(\theta + 2\pi k).$$

wrapped normal distribution The probability distribution resulting from wrapping a **normal distribution** on the real line around the unit circle. If the normal distribution has mean zero and variance σ^2, the corresponding wrapped normal distribution has probability density function

$$f_w(\theta) = \frac{1}{\sigma\sqrt{2\pi}} \sum_{k=-\infty}^{\infty} \exp\left\{-\frac{1}{2} \frac{(\theta + 2\pi k)^2}{\sigma^2}\right\}, \qquad 0 < \theta \leqslant 2\pi.$$

The functional form of $f_w(\theta)$ with $\sigma^2 = ct$ was first discovered by Hass-Lorentz (1913) as the probability density function of the position on a circle at time t, after undergoing Brownian motion, of a particle that starts at $\theta = 0$ at time zero. [See also **wrapped distribution**.]

wrapped Poisson distribution The distribution defined by

$$\Pr(\theta = 2\pi r/m) = \sum_{k=0}^{\infty} p(r + km; \lambda), \qquad r = 0, 1, \ldots, m-1,$$

where m is an integer and p is the Poisson probability mass function

$$p(x; \lambda) = e^{-\lambda}\lambda^x/x!, \qquad x = 0, 1, \ldots, \qquad \lambda > 0.$$

[See also **wrapped distribution**.]

Y

Yates, Frank (1902–1994) A British statistician. After reading mathematics at Cambridge, he worked for some years in Africa on surveying. He was then appointed to Rothamsted Experimental Station, initially to work with R.A. Fisher, and he remained there until his retirement. He made major contributions to the design of experiments, to sampling methods and to the systematic use of computers in statistics.

Yates–Grundy–Sen estimator An unbiased estimator of the variance of an estimated mean or total based on a sampling design for which the inclusion probability of any two members of the population appearing together in the sample is positive and known (Yates and Grundy, 1953; Sen, 1953; Thompson, 1997).

Yates's algorithm A simple procedure for estimating main effects and interactions in a two-level factorial experiment by repeated addition and subtraction of pairs of values. It is related to the fast Fourier transform and mathematically connected with the inversion of Kronecker products of matrices.

Yates's correction An adjustment proposed by Yates (1934) in the χ^2 calculation for a 2×2 table and also known as χ^2-correction. It consists of subtracting $\frac{1}{2}$ from one cell in the table and adjusting the other cells so that row and column totals remain constant and working on the value of χ^2 computed from the resulting table. The general effect is to bring the distribution based on discontinuous frequencies nearer to the continuous χ^2 distribution from which the published tables for testing χ^2 are derived.

Youden, William John (1900–1971) Beginning with the balanced incomplete block designs known as **Youden squares**, W.J. Youden created many designs tailored to the needs of the physical and engineering sciences (Youden, 1951*b*). His contributions in scientific research were so impressive that the statistical societies gave his name to three series of awards and prizes.

Youden square An experimental design proposed by Youden (1937). It is not a square, and would be better known as a 'Youden design'. For example, a design for seven treatments could be laid out as follows:

$$A \ B \ C \ D \ E \ F \ G$$
$$B \ C \ D \ E \ F \ G \ A$$
$$D \ E \ F \ G \ A \ B \ C$$

which may be regarded as three rows of a **Latin square**, hence the alternative name of **incomplete Latin square**.

Yule, George Udny (1871–1951) A Scottish statistician known, among other things, for the **coefficient of association**, Yule's paradox (later called Simpson's paradox or the **Yule–Simpson paradox**) and the **Yule process** as well as for contributions to Mendelian theory and time-series analysis (Yule, 1911).

Yule distribution A distribution proposed by Yule (1925*a*) for biological species investigations of the form $f(i) = AB(i, \rho + 1)$ where A and ρ are constants and $B(i, \rho + 1)$ is a beta function. Its use was extended by Simon (1955). An entirely different form of distribution is also associated with Yule (1925*b*), for which see **factorial distribution**.

Yule process A stochastic birth process used by Yule (1925*b*). It is essentially equivalent to the **Furry process**. The name is sometimes applied to an **autoregressive process** of the second order.

Yule–Simpson paradox The effect whereby a contingency table may show a positive dependence between two classification factors at each level of a third factor, but a negative dependence between them when merged over the levels of the third factor.

Yule–Walker equations A set of p linear relationships between the parameters $\alpha_1, \ldots, \alpha_p$ of the pth-order autoregressive process

$$u_t = \alpha_1 u_{t-1} + \alpha_2 u_{t-2} + \ldots + \alpha_p u_{t-p} + \varepsilon_t$$

in terms of the autocorrelations $\rho_1, \ldots, \rho_p$ of the process. The equations were introduced by Yule (1927) and Walker (1931).

Yule's equation A name sometimes given to an autoregressive equation of the second order, e.g.,

$$u_t + a u_{t-1} + \beta u_{t-2} = \varepsilon_t.$$

Yule's hyperbolic distribution A modified form of the **Yule distribution** (1925*a*) proposed by Powell (1955) in connection with the generation time of bacteria.

Yule's notation A notation for variables, regression coefficients, correlation coefficients, variances, etc., in the general linear least squares regression model, introduced by Yule (1912) and now standard. Subscripts following the dot refer to variables in the model for which adjustment has been made. For example, $b_{ij.kl}$ is the regression coefficient of x_j in the regression equation for x_i, when x_k and x_l also appear in the equation.

436

Z

Z-chart A form of graphic presentation of a time series consisting of three lines which usually take the shape of a letter 'Z'. The lower line is a plot of original data in the form of a time series; the centre line is a cumulative total; the upper line consists of a moving total of the original data.

z-distribution The distribution of a logarithmic transformation of a variance ratio, defined as $Z = \frac{1}{2}\log(S_1/S_2)$, where S_1 and S_2 are two independent estimates of a variance (Fisher, 1924). This **variance stabilizing transformation** was formerly used to avoid interpolation in tables of significance points. [See also **beta distribution, F-distribution**.]

z-score A term used by some writers in connection with educational and psychological testing as an alternative to standardized scores. A z-score for an observation is the score expressed as a deviation from the sample mean value in units of the sample standard deviation.

z-test A significance test based upon the **z-distribution** (Lin and Mudholkar, 1980). In most cases it is tantamount to an **F-test**, but also is used as an approximation to tests with more complicated distributions, in which case variance ratios may not be involved. In elementary textbooks, it is sometimes used for a test based on the normal distribution, with z representing a standard normal deviate.

z-transformation See **Fisher's transformation**.

Zelen's inequality A one-sided inequality of the Tchebychev type stated explicitly by Zelen (1954),

$$\Pr\left[\xi - E(\xi) \geqslant t\sigma\right] \leqslant \left[1 + t^2 + \frac{(t^2 - t\alpha_3 - 1)^2}{\alpha_4 - \alpha_3^2 - 1}\right]^{-1},$$

for $t \geqslant \frac{1}{2}[\alpha_3 + (\alpha_3^2 + 4)^{1/2}]$, where α_3 and α_4 are standardized third and fourth moments. When t is at its lower limit the inequality reduces to the one-sided version of the **Bienaymé–Tchebychev inequality**. At its upper limit, it reduces to **Cantelli's inequality**.

zero-sum game A game played by a number of persons in which the winner takes all the stakes provided by the losers so that the algebraic sum of gains at any stage is zero. It has been argued that many decision problems may be viewed as zero-sum games between two persons.

Zipf's law A general law proposed by Zipf (1949) to approximate to the distribution of individuals bearing variable values which are either integral or grouped in intervals of equal width. The simplest form is $f(x) = k/x^p$, $x = 1, \ldots, n-1$, where p is some constant. A special case occurs when $p = 1$ and x takes the range $1, 2, 3, \ldots$, and hence can be regarded as a rank; in this case the product of the frequency and its rank is a constant. There are various later generalizations of this simple form. See, for instance, **factorial distribution**. The term 'harmonic distribution' has been used for Zipf's distribution: this should be avoided as inaccurate and likely to lead to confusion with the distribution of harmonics in spectral analysis. [See also **Riemann distribution**.]

zonal polynomial Certain homogeneous symmetric polynomials in the characteristic roots of a symmetric matrix, of use in the study of multivariate distributions (James, 1960).

zonal sampling A term used sometimes to indicate sampling by zones, zone in this context denoting a stratum determined on a geographical basis.

zone of indifference See **zone of preference**.

zone of preference In connection with the test of a hypothesis, the zone of indifference is defined as the region in the sample space, if any, which is left after the removal of a region of acceptance and a region of rejection. These latter two together are sometimes called a zone of preference. It is more customary, and seems better practice, to use the word 'region' instead of 'zone'.

REFERENCES

Abbe, E. (1906). *Gesammelte Abhandlungen*, Volume 2. Fisher, Jena.

Abelson, R.P. and Tukey, J.W. (1963). Efficient utilization of non-numerical information in quantitative analysis: general theory and the case of simple order. *Annals of Mathematical Statistics*, **34**, 1347–1369.

Abramson, L.R. (1966). Asymptotic sequential design of experiments with two random variables. *Journal of the Royal Statistical Society. Series B*, **28**, 73–87.

Aitchinson, J. (1955). On the distribution of a positive random variable having a discrete probability mass at the origin. *Journal of the American Statistical Association*, **50**, 901–908.

Aitchinson, J. (1986). *The statistical analysis of compositional data*. Chapman & Hall, London.

Aitken, A.C. (1934). The normal form of compound and induced matrices. *Proceedings of the London Mathematical Society, II*, **38**, 354–376.

Aitken, A.C. (1935). On least squares and linear combination of observations. *Proceedings of the Royal Society of Edinburgh. Series A*, **55**, 42–48.

Aitken, A.C. (1948). On the estimation of many statistical parameters. *Proceedings of the Royal Society of Edinburgh. Series A*, **62**, 369–377.

Aitken, A.C. and Gonin, H.T. (1935). On fourfold sampling with and without replacement. *Proceedings of the Royal Society of Edinburgh. Series A*, **55**, 114–125.

Aitkin, M.A. (1974). Simultaneous inference and the choice of variable subsets in multiple regression. *Technometrics*, **16**, 221–227.

Ajne, B. (1968). A simple test for uniformity of a circular distribution. *Biometrika*, **55**, 343–354.

Akaike, H. (1969). Fitting autoregressive models for prediction. *Annals of the Institute of Statistical Mathematics, Tokyo*, **21**, 243–247.

Allan, F.E. and Wishart, J. (1930). A method of estimating the yield of a missing plot in field experimental work. *Journal of Agricultural Science*, **20**, 399–406.

Andersen, P.K., Borgan, O., Gill, R.D., and Keiding, N. (1993). *Statistical methods based on counting processes*. Springer-Verlag, New York.

Andersen, P.K., Hansen, L.S., and Keiding, N. (1991). Non- and semi-parametric estimation of transition probabilities from censored observation of a non-homogeneous Markov process. *Scandinavian Journal of Statistics*, **18**, 153–167.

Anderson, E. (1960). A semigraphical method for the analysis of complex problems. *Technometrics*, **2**, 387–391.

Anderson, O. (1935). *Einführung in die mathematische Statistik*. Julius Springer,

Wien.

Anderson, T.W. (1951). Estimating linear restrictions on regression coefficients for multivariate normal distributions. *Annals of Mathematical Statistics*, **22**, 327–351.

Anderson, T.W. and Darling, D.A. (1952). Asymptotic theory of certain 'goodness of fit' criteria based on stochastic processes. *Annals of Mathematical Statistics*, **23**, 193–212.

Anderson, T.W. and Darling, D.A. (1954). A test of goodness of fit. *Journal of the American Statistical Association*, **49**, 765–769.

Anderson, T.W. and Goodman, L.A. (1957). Statistical inference about Markov chains. *Annals of Mathematical Statistics*, **28**, 89–110.

Anderson, T.W. and Hurwicz, L. (1949). Errors and shocks in economic relationships. *Econometrica, Chicago*, **17 Suppl.**, 23–24.

Anderson, T.W., McCarthy, P.J., and Tukey, J.W. (1946). *'Staircase' method of sensitivity testing*, Volume 65-46. Naval Ordnance Report, Princeton, NJ.

Andrews, D.F. (1972). Plots of high-dimensional data. *Biometrics*, **28**, 125–136.

Andrews, D.F., Bickel, P.J., Hampel, F.R., Huber, P.J., Rogers, W.H., and Tukey, J.W. (1972). *Robust estimation of location: survey and advances*. Princeton University Press, Princeton, NJ.

Andrews, D.F., Gnanadesikan, R., and Warner, J.L. (1973). *Methods for assessing multivariate normality in multivariate analysis III*. Academic Press, New York.

Andrews, F.C. and Chernoff, H. (1955). A large-sample bioassay design with random doses and uncertain concentration. *Biometrika*, **42**, 307–315.

Ansari, A.R. and Bradley, R.A. (1960). Rank-sum tests for dispersions. *Annals of Mathematical Statistics*, **31**, 1174–1189.

Anscombe, F.J. and Tukey, J.W. (1963). Analysis of residuals. *Technometrics*, **5**, 141–160.

Aranda-Ordaz, F.J. (1983). An extension of the proportional-hazards model for grouped data. *Biometrics*, **39**, 109–117.

Arfwedson, G. (1951). A probability distribution connected with Stirling's second class numbers. *Skandinavien Aktuarie Tidskrift*, **34**, 121–132.

Armitage, P. (1957). Restricted sequential procedures. *Biometrika*, **44**, 9–26.

Armitage, P. (1959). The comparison of survival curves. *Journal of the Royal Statistical Society. Series A*, **122**, 279–300.

Armitage, P. and Colton, T. (1998). *Encyclopedia of biostatistics*, Volume 1. Wiley, New York.

Arnold, K.J. (1941). *Spherical probability distribution*. Ph.D. thesis, Massachusetts Institute of Technology.

Arthanari, T.S. and Dodge, Y. (1981). *Mathematical programming in statistics*.

Wiley, New York.

Aspin, A.A. and Welch, B.L. (1949). Tables for use in comparisons whose accuracy involves two variances, separately estimated. *Biometrika*, **36**, 290–296.

Athreya, K.B. and Ney, P.E. (1972). *Branching processes*. Springer-Verlag, New York, Berlin.

Bagai, O.P. (1962). Statistics proposed for various tests of hypotheses and their distributions in particular cases. *Sankhyā*, **24**, 409–418.

Bahadur, R.R. (1960a). On the asymptotic efficiency of tests and estimates. *Sankhyā*, **22**, 229–252.

Bahadur, R.R. (1960b). Stochastic comparison of tests. *Annals of Mathematical Statistics*, **31**, 276–295.

Bailey, N.T.J. (1953). The use of chain-binomials with a variable chance of infection for the analysis of intra-household epidemics. *Biometrika*, **40**, 279–286.

Bailey, N.T.J. (1956). Significance tests for a variable chance of infection in chain-binomial theory. *Biometrika*, **43**, 332–336.

Ballot, B. (1847). *Les changements périodiques de température*. Utrecht.

Barbour, A.D., Holst, L., and Janson, S. (1992). *Poisson approximation*. Oxford University Press, Oxford.

Bardwell, G.E. and Crow, E.L. (1964). A two-parameter family of hyper-Poisson distributions. *Journal of the American Statistical Association*, **59**, 133–141.

Barlow, R.E. (1968). Some recent developments in reliability theory. *European Meeting 1968, Selected Statistics Papers*, **2**, 49–66.

Barlow, R.E., Marshall, A.W., and Proschan, F. (1963). Properties of probability distributions with monotone hazard rate. *Annals of Mathematical Statistics*, **34**, 375–389.

Barlow, R.E. and Proschan, F. (1965). *Mathematical theory of reliability*. Wiley, New York.

Barnard, G.A. (1947). Significance tests for 2×2 tables. *Biometrika*, **34**, 123–138.

Barnard, G.A. (1949). Statistical inference. *Journal of the Royal Statistical Society. Series B*, **11**, 115–139.

Barnard, G.A. (1959). Control charts and stochastic processes. *Journal of the Royal Statistical Society. Series B*, **21**, 239–257.

Barnard, G.A. (1963a). Discussion of paper by M.S. Bartlett. *Journal of the Royal Statistical Society. Series B*, **25**, 294.

Barnard, G.A. (1963b). The logic of least squares. *Journal of the Royal Statistical Society. Series B*, **25**, 124–127.

Barndorff-Nielsen, O.E. (1983). On a formula for the distribution of the maximum likelihood estimator. *Biometrika*, **70**, 343–365.

Barndorff-Nielsen, O.E. and Cox, D.R. (1989). *Asymptotic techniques for use in statistics.* Chapman & Hall, London.

Barndorff-Nielsen, O.E. and Cox, D.R. (1994). *Inference and asymptotics, monographs on statistics and applied probability,* Volume 52. Chapman & Hall, London.

Bartels, J. (1935a). Random fluctuations, persistence, and quasi-persistence in geophysical and cosmical periodicities. *Terrestrial Magnetism and Atmospheric Electricity,* **40**, 1–60.

Bartels, J. (1935b). Zur Morphologie geophysikalischer Zeitfunktionen. *Sitzungsberichte der Preussischen Akademie Wissenschaften, Physikalische-Mathematische Klasse,* **30**, 504–522.

Bartlett, M.S. (1933). On the theory of statistical regression. *Proceedings of the Royal Society of Edinburgh. Series A,* **53**, 260–283.

Bartlett, M.S. (1935a). Contingency table interactions. *Supplement to the Journal of the Royal Statistical Society,* **2**, 248–252.

Bartlett, M.S. (1935b). The statistical estimation of g. *British Journal of Psychology,* **26**, 199–206.

Bartlett, M.S. (1937). Properties of sufficiency and statistical tests. *Proceedings of the Royal Society of London. Series A,* **160**, 268–282.

Bartlett, M.S. (1946). *Stochastic process.* Note of a course, University of North Carolina.

Bartlett, M.S. (1948). A note on the statistical estimation of supply and demand relations from time series. *Econometrica,* **16**, 323–329.

Bartlett, M.S. (1949). Some evolutionary stochastic processes. *Journal of the Royal Statistical Society. Series B,* **11**, 211–229.

Bartlett, M.S. (1951). The goodness of fit of a single hypothetical discriminant function in the case of several groups. *Annals of Eugenics,* **16**, 199–214.

Bartlett, M.S. (1955). *An introduction to stochastic processes.* Cambridge University Press, Cambridge.

Bartlett, M.S. (1963). The spectral analysis of point processes (with discussion). *Journal of the Royal Statistical Society. Series B,* **25**, 264–296.

Bartlett, M.S. and Diananda, P.H. (1950). Extension of Quenouille's test for autoregressive schemes. *Journal of the Royal Statistical Society. Series B,* **12**, 108–115.

Bartlett, M.S. and Kendall, D.G. (1946). The statistical analysis of variance-heterogeneity and the logarithmic transformation. *Supplement to the Journal of the Royal Statistical Society,* **8**, 128–138.

Bartlett, M.S. and Knox, E.G. (1964). The detection of space-time interactions. *Applied Statistics,* **13**, 25–30.

Barton, D.E. and David, F.N. (1958). Non-randomness in a sequence of two alternatives. I: Run test. *Biometrika,* **45**, 253–256.

Barton, D.E., David, F.N., and Mallows, C.L. (1958). Non-randomness in a sequence of two alternatives. I: Wilcoxon's and allied test statistics. *Biometrika*, **45**, 166–180.

Basu, A.P. (1967). On two *k*-sample rank tests for censored data. *Annals of Mathematical Statistics*, **38**, 1520–1535.

Basu, D. (1955). On statistics independent of a complete sufficient statistic. *Sankhyā*, **15**, 377–380.

Bates, G.E. and Neyman, J. (1952). Contributions to the theory of accident proneness. *University of California Publications in Statistics*, **1**, 215–276.

Bayes, T. (1763). An essay towards solving a problem in the doctrine of chances. *Philosophical Transactions*, **53**, 370–418.

Beall, B. and Rescia, G. (1953). A generalization of Neyman's contagious distributions. *Biometrics*, **5**, 207–212.

Bechhofer, R.E. (1954). A single-sample multiple decision procedure for ranking means of normal populations with known variances. *Annals of Mathematical Statistics*, **25**, 16–39.

Becker, A. and Cleveland, W.S. (1987). Brushing scatterplots. *Technometrics*, **29**, 127–142.

Behrens, B.V. (1929). Ein Beitrag zur Fehlerberechnung bei weinige Beobachtungen. *Landwirtschaft Jahrbuch*, **68**, 807–837.

Bellman, R. (1957). *Dynamic programming*. Princeton University Press, Princeton, NJ.

Bellman, R. and Harris, T. (1948). On the theory of age-dependent stochastic branching processes. *Proceedings of the National Academy of Sciences of the USA*, **34**, 601–604.

Bellman, R. and Harris, T. (1952). On age-dependent binary branching processes. *Annals of Mathematics, II Series*, **55**, 280–295.

Belson, W. (1959). Matching and prediction on the principle of biological classification. *Applied Statistics*, **8**, 65–75.

Bennett, B.M. (1965). On multivariate signed rank tests. *Annals of the Institute of Statistical Mathematics, Tokyo*, **17**, 55–61.

Benoît, Commandant (1924). Note sur une méthode de résolution des équations normales provenant de l'application de la méthode des moindres cárres à un système d'équations linéares en nombre inférieur à celui des inconnues. — Application de la méthode à la résolution d'un système déini d'équations linéares (Procédé du Commandant Cholesky). *Bullétin Géodésique (Toulouse)*, **2**, 5–77.

Beran, R.J. (1968). Testing for uniformity on a compact homogeneous space. *Journal of Applied Probability*, **5**, 177–195.

Beran, R.J. (1969). Asymptotic theory of a class of tests for uniformity of a circular distribution. *Annals of Mathematical Statistics*, **40**, 1196–1206.

Berge, P.O. (1932). Über das Theorem von Tchebycheff und andere Grenzen einer Wahrscheinlichkeitsfunktion. *Skandinavien Aktuarie Tidskrift*, **15**, 65–77.

Berge, P.O. (1938). A note on a form of Tchebycheff's theorem for two variables. *Biometrika*, **29**, 405–406.

Berkson, J. (1944). Application of the logistic function to bio-assay. *Journal of the American Statistical Association*, **39**, 357–365.

Berkson, J. (1950). Are there two regressions? *Journal of the American Statistical Association*, **45**, 164–180.

Berkson, J. (1955a). Estimate of the integrated normal curve by minimum normit chi-square with particular reference to bio-assay. *Journal of the American Statistical Association*, **50**, 529–549.

Berkson, J. (1955b). Maximum likelihood and minimum χ^2 estimates of the logistic function. *Journal of the American Statistical Association*, **50**, 130–162.

Bernardo, J.M. and Smith, A.F.M. (1994). *Bayesian theory*. Wiley, New York.

Bernoulli, J. (1713). *Ars conjectandi*. Basel.

Bernstein, S. (1927). Sur l'extension du théorème du calcul des probabilités aux sommes de quantités dépendantes. *Mathematisches Annalen*, **97**, 1–59.

Berry, A.C. (1941). The accuracy of the Gaussian approximation to the sum of independent variates. *Transactions of the American Mathematical Society*, **49**, 122–136.

Besag, J. (1974). Spatial interaction and statistical analysis of lattice systems (with discussion). *Journal of the Royal Statistical Society. Series B*, **36**, 192–236.

Besag, J. (1977). Efficiency of pseudolikelihood estimation for simple Gaussian fields. *Biometrika*, **64**, 616–618.

Besag, J. and Gleaves, J.T. (1973). On the detection of spatial pattern in plant communities. *Bulletin of the International Statistical Institute*, **45**, 153–158.

Beverton, R.J.H. and Holt, S.J. (1957). *On the dynamics of exploited fish populations*. Chapman & Hall, London.

Bhat, B.R. and Nagnur, B.N. (1965). Locally asymptotically most stringent tests and Lagrangian multiplier tests of linear hypotheses. *Biometrika*, **52**, 459–468.

Bhattacharya, K.N. (1943). A note on two-fold triple systems. *Sankhyā*, **6**, 313–314.

Bhattacharyya, A. (1946). On some analogues of the amount of information and their use in statistical estimation. I. *Sankhyā*, **8**, 1–14.

Bhattacharyya, B.C. (1942). The use of McKay's Bessel function curves for graduating frequency distributions. *Sankhyā*, **6**, 175–182.

Biemer, P. and Lyberg, L. (2003). *Introduction to survey quality*. Wiley, New York.

Bienaymé, I. (1853). Considérations à l'appui de la découverte de Laplace sur la loi des probabilités dans la méthode des moindres carrés. *Comptes-Rendus de l'Académie des Sciences de Paris*, **37**, 309–324.

Billor, N., Hadi, A.S., and Velleman, P.F. (2000). BACON: Blocked Adaptive Computationally-Efficient Outlier Nominators. *Computational Statistics and Data Analysis*, **34**, 279–298.

Bingham, C. (1974). An antipodally symmetric distribution on the sphere. *Annals of Statistics*, **2**, 1201–1225.

Birkes, D. and Dodge, Y. (1986). The number of minimally connected block designs. *Computational Statistics and Data Analysis*, **4**, 269–275.

Birkes, D. and Dodge, Y. (1993). *Alternative methods of regression*. Wiley, New York.

Birkes, D., Dodge, Y., and Seely, J. (1976). Spanning sets for estimable contrasts in classification models. *Annals of Statistics*, **4**, 86–107.

Birnbaum, A. (1961*a*). Confidence curves: An omnibus technique for estimation and testing statistical hypotheses. *Journal of the American Statistical Association*, **56**, 246–259.

Birnbaum, A. (1961*b*). On the foundations of statistical inference. Binary experiments. *Annals of Mathematical Statistics*, **32**, 414–435.

Birnbaum, A. (1969). Concepts of statistical evidence. In *Philosophy science and method* (ed. S. Morgenbesser, P. Suppes and M. White), pp. 112–143. St Martin's Press, New York.

Birnbaum, Z.W., Esary, J.D., and Saunders, S.C. (1961). Multi-component systems and structures and their reliability. *Technometrics*, **3**, 55–77.

Birnbaum, Z.W., Raymond, J., and Zuckerman, H.S. (1947). A generalization of Tchebyshev's inequality to two dimensions. *Annals of Mathematical Statistics*, **18**, 70–79.

Birnbaum, Z.W. and Saunders, S.C. (1969). A new family of life distributions. *Journal of Applied Probability*, **6**, 319–327.

Birnbaum, Z.W. and Tang, V.K.T. (1964). Two simple distribution-free tests of goodness of fit. *Reviews of the International Statistical Institute*, **32**, 2–13.

Birnbaum, Z.W. and Tingey, F.H. (1951). One-sided confidence contours for probability distribution functions. *Annals of Mathematical Statistics*, **22**, 592–596.

Bissinger, B.H. (1963). A type-resisting distribution generated from considerations of an inventory decision model. In *Proceedings of the International Symposium on Discrete Distributions*, pp. 14–17. Montreal.

Black, F. and Scholes, M. (1973). The pricing of options and corporate liabilities. *Journal of Political Economy*, **81**, 637–659.

Blackman, R.B. and Tukey, J.W. (1959). *The measurement of power spectra. From the point of view of communications engineering. Unabridged and corrected republ.* Dover Publications, New York.

Blackwell, D. (1947). Conditional expectation and unbiased sequential estimation. *Annals of Mathematical Statistics*, **18**, 105–110.

Blackwell, D. (1948). A renewal theorem. *Duke Mathematical Journal*, **15**, 145–150.

Bliss, C.L. (1934). The method of probits. *Science*, **79**, 38–39; 409–410.

Blom, G. (1958). *Statistical estimates and transformed beta variables*. Wiley, New York.

Blom, G. (1960). Hierarchical birth and death processes. I: Theory. II: Applications. *Biometrika*, **47**, 235–251.

Blomqvist, N. (1950). On a measure of dependence between two random variables. *Annals of Mathematical Statistics*, **21**, 593–600.

Bloomfield, P. and Steiger, W.L. (1983). *Least absolute deviations. Theory, applications, and algorithms*. Birkhauser, Boston.

Blum, J.R. (1954). Multidimensional stochastic approximation methods. *Annals of Mathematical Statistics*, **25**, 737–744.

Blum, J.R., Kiefer, J., and Rosenblatt, M. (1961). Distribution free test of independence based on the sample distribution function. *Annals of Mathematical Statistics*, **32**, 485–498.

Blum, J.R. and Rosenblatt, J. (1963). On multistage estimation. *Annals of Mathematical Statistics*, **34**, 1452–1458.

Bock, R.D. (1958). Remarks on the test of significance for the method of paired comparisons. *Psychometrika*, **23**, 323–334.

Bogert, B.P. and Healy, M.J.R. (1963). Fortran subroutines for time series analysis. *Communications of the ACM*, **6**, 32–34.

Bonferroni, C.E. (1927). *Elementi di statistica generale*. Università Bocconi, Milano.

Boole, G. (1854). *An investigation of the laws of thought on which are founded the mathematical theories of logic and probabilities*. Macmillan, London.

Borel, E. (1909). Les probabilités dénombrables et leurs applications arithmétiques. *Rendus Circulaire de Mathématique Palermo*, **27**, 247–271.

Borel, E. (1942). Sur l'emploi du théorème de Bernoulli pour faciliter le calcul des coefficients. Applications au problème de l'attente aux guichets. *Comptes-Rendus de l'Académie des Sciences de Paris*, **214**, 452–456.

Borges, R.Z. (1970). Eine Approximation der Binomialverteilung durch die Normalverteilung der Ordnung $1/n$. *Zeitschrift für Wahrscheinlichkeitstheorisch Verwaltung Gebiet*, **14**, 189–199.

Boscovich, R.J. (1757). De literaria expeditione per pontificiam ditionem et synopsis amplioris operis... *Bononiensi Scientiarum et Artum Instituto atque Academia Commentarii, Reprint with a Croatian translation by N. Cubranić, Institute of Higher Geodesy, Zagreb*, **4** (1961), 353–396.

Bose, R.C. (1934). On the application of the hyperspace geometry to the theory of multiple correlation. *Sankhyā*, **1**, 338–344.

Bose, R.C. (1939). On the construction of balanced incomplete block designs. *Annals of Eugenics*, **9**, 353–399.

Bose, R.C. (1942). A note on the resolvability of balanced incomplete designs. *Sankhyā*, **6**, 105–110.

Bose, R.C. (1947). Mathematical theory of the symmetric factorial designs. *Sankhyā*, **8**, 107–166.

Bose, R.C. (1949). *Least square aspects of analysis of variance*. Institute of Statistics Mimeo Series 9, Chapel Hill, NC.

Bose, R.C. (1956). Paired comparison designs for testing concordance between judges. *Biometrika*, **43**, 113–121.

Bose, R.C. and Connor, W.S. (1952). Combinatorial properties of group divisible incomplete block designs. *Annals of Mathematical Statistics*, **23**, 367–383.

Bose, R.C. and Nair, K.R. (1939). Partially balanced block designs. *Sankhyā*, **4**, 337–372.

Bose, R.C., Shrikhande, S.S., and Parker, E.T. (1960). Further results on the construction of mutually orthogonal Latin squares and the falsity of Euler's conjecture. *Canadian Journal of Mathematics*, **12**, 203.

Bowley, A.R. (1928). Bilateral monopoly. *Economic Journal*, **38**, 216–237.

Box, G.E.P. (1953). Non-normality and tests on variance. *Biometrika*, **40**, 318–335.

Box, G.E.P. and Behnken, D.W. (1960). Some new three-level designs for the study of quantitative variables. *Technometrics*, **2**, 455–475.

Box, G.E.P. and Cox, D.R. (1964). An analysis of transformations. *Journal of the Royal Statistical Society. Series B*, **26**, 211–243.

Box, G. and Draper, N.R. (1998). *Evolutionary operation. A statistical method for process improvement*. Wiley, New York.

Box, G.E.P. and Hunter, J.S. (1957). Multi-factor experimental designs for exploring response surfaces. *Annals of Mathematical Statistics*, **28**, 195–241.

Box, G.E.P. and Jenkins, G.M. (1962). Some statistical aspects of adaptive optimization and control. *Journal of the Royal Statistical Society. Series B*, **24**, 297–343.

Box, G.E.P. and Jenkins, G.M. (1970). *Time series analysis: Forecasting and control*. Holden-Day Series in Time Series Analysis. Holden-Day, San Francisco.

Box, G.E.P. and Jenkins, G.M. (1976). *Time series analysis: Forecasting and control* (Rev. edn). Holden-Day Series in Time Series Analysis. Holden-Day, San Francisco.

Box, G.E.P. and Muller, M.E. (1958). A note on the generation of random normal deviates. *Annals of Mathematical Statistics*, **29**, 610–611.

Box, G.E.P. and Tiao, G.C. (1975). Intervention analysis with applications to economic and environmental problems (in theory and methods). *Journal of the American Statistical Association*, **70**, 70–79.

Box, G.E.P. and Wilson, K.B. (1951). On the experimental attainment of optimum conditions. *Journal of the Royal Statistical Society. Series B*, **13**, 1–45.

Bradford, S.C. (1948). *Documentation.* Crosby Lockwood, London.

Bradford Hill, A. (1965). The environment and disease: Association or causation? *Proceedings of the Royal Society of Medicine*, **58**, 295–300.

Bradley, J.V. (1958). Complete counterbalancing of immediate sequential effects in a Latin square design. *Journal of the American Statistical Association*, **53**, 525–528 and 1030–1031.

Bradley, J.V. (1968). *Distribution-free statistical tests.* Prentice Hall, Englewood Cliffs, NJ.

Bradley, R.A. (1953). Some statistical methods in taste testing and quality evaluation. *Biometrics*, **9**, 22–38.

Bradley, R.A. and Terry, M.E. (1952). Rank analysis of incomplete block designs, I: The method of paired comparisons. *Biometrika*, **39**, 324–345.

Breiman, L. (1994). *Bagging predictors.* Technical Report, University of California, Berkeley.

Breiman, L., Friedman, J., Olshen, R., and Stone, C. (1984). *Classification and regression trees.* Chapman & Hall, New York.

Breslow, N. and Day, N. (1980). *The analysis of case-control studies.* IARC Scientific Publications, Lyons.

Broadbent, S.R. (1955). Quantum hypotheses. *Biometrika*, **42**, 45–57.

Brookmeyer, R. and Gail, M.H. (1988). A method for obtaining short-term projections and lower bounds on the size of the aids epidemic. *Journal of the American Statistical Association*, **83**, 301–308.

Bross, I.D.J. (1958). How to use ridit analysis. *Biometrics*, **14**, 18–38.

Brown, G.W. (1947). On small-sample estimation. *Annals of Mathematical Statistics*, **18**, 582–585.

Brown, G.W. and Mood, A.M. (1951). On median tests for linear hypotheses. In *Proceedings of the Berkeley Symposium on Mathematical Statistics and Probability, California, July 31 – August 12, 1950*, pp. 159–166.

Brown, M. (1959). The structure of stochastic difference equation models. *Econometrica*, **27**, 116–120.

Brown, R.G. (1963). *Smoothing, forecasting and prediction of discrete time series.* Prentice Hall, Englewood Cliffs, NJ.

Brown, R.G. and Meyer, R.F. (1961). The fundamental theorem of exponential smoothing. *Operations Research*, **9**, 673–685.

Brownlee, K.A. (1965). *Statistical theory and methodology in science and engineering* (2nd edn). Wiley, New York.

Brunk, H.D. (1962). On the range of the difference between hypothetical distribution function and Pyke's modified empirical distribution function. *Annals of Mathematical Statistics*, **33**, 525–532.

Buffon, L. (1777). *Essai d'arithmétique morale. Supplément à l'histoire naturelle*, Volume 4.

Burg, J.P. (1975). *Maximum entropy spectrum analysis*. Ph.D. thesis, Stanford University.

Burke, P.J. (1956). The output of a queueing system. *Operations Research*, **4**, 699–704.

Burkholder, D.L. (1956). On a class of stochastic approximation processes. *Annals of Mathematical Statistics*, **27**, 1044–1059.

Burns, A.F. and Mitchell, W.C. (1946). *Measuring business cycles*. National Bureau of Economic Research, New York.

Burr, I.W. (1942). Cumulative frequency functions. *Annals of Mathematical Statistics*, **13**, 215–232.

Burt, C. (1940). *The factors of the mind*. University of London Press, London.

Bury, A.K. (1975). *Statistical methods in applied science*. Wiley, New York.

Butler, C.C. (1969). A test for symmetry using the sample distribution function. *Annals of Mathematical Statistics*, **40**, 2209–2210.

Camp, B.H. (1951). Approximation to the point binomial. *Annals of Mathematical Statistics*, **22**, 130–131.

Cantelli, F.P. (1910). Intorno ad un teorema fondamentale della teoria del rischio. *Bollettino Associazione Attuari Italia*, 1–23.

Cantelli, F.P. (1917). Sulla probabilità come limite della frequenza. *Rendiconti della R. Accademia dei Lincei. Classe di scienze fisiche matematiche e naturali*, Ser. 5a , **26**, 39–45.

Capon, J. (1961). Asymptotic efficiency of certain locally most powerful rank tests. *Annals of Mathematical Statistics*, **32**, 88–100.

Carleman, T. (1926). *Les fonctions quasi-analytiques*. Gauthier-Villars, Paris.

Carlin, B.P. and Louis, T.A. (2000). *Bayes and empirical Bayes methods for data analysis* (2nd edn). Chapman & Hall, London.

Carlstein, E., Muller, H.G., and Siegmund, D. (1994). Change-point problems. In *IMS lecture notes* (ed. E. Carlstein *et al.*), Volume 23, pp. 130–144. Institute of Mathematical Statistics, Hayward, CA.

Carothers, A.D. (1971). An examination and extension of Leslie's test of equal catchability. *Biometrics*, **27**, 615–630.

Carroll, J.D. and Arabie, P. (1996). Multidimensional scaling. In *Handbook of perception and cognition. Measurement, judgement and decision*, Volume 3, pp.

179–250. Academic Press, New York.

Carver, H.C. (1919). On the graduation of frequency distribution. *Proceedings of Casualty, Actuarial Society of America*, **6**, 52–72.

Castillo, E., Gutiérrez, J.M., and Hadi, A.S. (1997). *Expert systems and probabilistic network models*. Springer-Verlag, New York.

Cattell, R.B. (1978). *The scientific use of factor analysis*. Plenum Press, New York.

Cauchy, M.A.L. (1821). *Cours d'analyse de l'Ecole Royale Polytechnique*. L'Imprimerie Royale, Paris.

Chakraborti, S. and van der Laan, P. (1996). Precedence tests and confidence bounds for complete data: An overview and some results. *The Statistician*, **45**, 351–369.

Chakravarti, I.M. (1956). Fractional replication in asymmetrical factorial designs and partially balanced arrays. *Sankhyā*, **17**, 143–164.

Champernowne, D.G. (1952). The graduation of income distribution. *Econometrica*, **20**, 591–615.

Charlier, C.V.L. (1905a). Über das Fehlergesetz. *Arkiv für Matematik Astronomie und Physik*, **2(8)**, 1–9.

Charlier, C.V.L. (1905b). Über die Darstellung willkuerlicher Funktionen. *Arkiv für Matematik Astronomie und Physik*, **2(20)**, 1–35.

Charlier, C.V.L. (1928). *A new form of the frequency function*. Gleerup, Lund.

Chatterji, S. and Hadi, A. (1988). *Sensitivity analysis in linear regression*. Clarendon Press, Oxford.

Chauvenet, W. (1864). *A manual of spherical and practical astronomy, 2*. Lippincott, Philadelphia.

Cherian, K.C. (1941). A bivariate correlated gamma-type distribution function. *Journal of the Indian Mathematical Society*, **5**, 133–144.

Chernoff, H. (1973). The use of faces to represent points in k-dimensional space graphically. *Journal of the American Statistical Association*, **68**, 361–368.

Chernoff, H. and Savage, I.R. (1958). Asymptotic normality and efficiency of certain nonparametric test statistics. *Annals of Mathematical Statistics*, **29**, 972–994.

Choi, E. and Hall, P. (1999). Data sharpening as a prelude to density estimation. *Biometrika*, **87**, 941–947.

Christaller, W. (1933). *Die zentralen Orte in Süddeutschland*. Gustav Fischer, Jena.

Chung, K.L. and Feller, W. (1949). On fluctuations in coin tossing. *Proceedings of the National Academy of Sciences of the USA*, **35**, 605–608.

Chung, K.L. and Fuchs, W.H.J. (1951). On the distribution of values of sums of random variables. *Memoirs of the American Mathematical Society*, **6**, 1–12.

Clark, W. (1922). *Gantt chart*. Ronald Press, New York.

Clatworthy, W.H. (1967). Some new families of partially balanced designs of the Latin square type and related designs. *Technometrics*, **9**, 229–244.

Clayton, D. and Schifflers, E. (1987). Models for temporal variation in cancer rates II. Age-period-cohort models. *Statistics in Medicine*, **6**, 469–481.

Cliff, A.D. and Ord, J.K. (1981). *Spatial processes, models, inferences and applications*. Pion, London.

Cochran, W.G. (1934). The distribution of quadratic forms in a normal system, with applications to the analysis of covariance. *Proceedings of the Cambridge Philosophical Society*, **30**, 178–191.

Cochran, W.G. (1937). Problems arising in the analysis of a series of similar experiments. *Supplement to the Journal of the Royal Statistical Society*, **4**, 102–118.

Cochran, W.G. (1941). The distribution of the largest of a set of estimated variances as a fraction of their total. *Annals of Eugenics*, **11**, 47–52.

Cochran, W.G. (1950). The comparison of percentages in matched samples. *Biometrika*, **37**, 256–266.

Cochran, W.G. (1965). The planning of observational studies of human populations. *Journal of the Royal Statistical Society. Series A*, **128**, 234–266.

Cochran, W.G. and Cox, G.M. (1957). *Experimental designs*. Wiley, New York.

Cockcroft, W.H. (1982). *Mathematics counts: Report of the committee of enquiry into the teaching of mathematics*. HMSO, London.

Coles, S.G. (2001). *An introduction to the statistical modeling of extreme values*. Springer-Verlag, New York.

Connor, W.S. and Youden, W.J. (1954). New experimental designs for paired observations. *Journal of Research of the National Bureau of Standards*, **53**, 191–196.

Cook, R.D. (1979). Influential observations in linear regression. *Journal of the American Statistical Association*, **74**, 169–174.

Cornfield, J. (1962). Joint dependence of risk of coronary heart disease on serum cholesterol and systolic blood pressure: a discriminant function analysis. *Federation Proceedings*, **21**, 58–61.

Cornish, E.A. and Fisher, R.A. (1937). Moments and cumulants in the specification of distribution. *Reviews of the International Statistical Institute*, **5**, 307–320.

Cowden, D.J. (1952). The multiple-partial correlation coefficient. *Journal of the American Statistical Association*, **47**, 442–456.

Cox, D.R. (1952*a*). A note on the sequential estimation of means. *Proceedings of the Cambridge Philosophical Society*, **48**, 447–450.

Cox, D.R. (1952*b*). Sequential tests for composite hypotheses. *Proceedings of the*

Cambridge Philosophical Society, **48**, 290–299.

Cox, D.R. (1955). Some statistical methods connected with series of events. *Journal of the Royal Statistical Society. Series B*, **17**, 129–164.

Cox, D.R. (1958*a*). The regression analysis of binary sequences (with discussion). *Journal of the Royal Statistical Society. Series B*, **20**, 215–242.

Cox, D.R. (1958*b*). Some problems connected with statistical inference. *Annals of Mathematical Statistics*, **29**, 357–372.

Cox, D.R. (1958*c*). Two further applications of a model for binary regression (in Miscellanea). *Biometrika*, **45**, 562–565.

Cox, D.R. (1962). *Renewal theory*. Methuen, London.

Cox, D.R. (1970). *The analysis of binary data*. Methuen, London.

Cox, D.R. (1972). Regression models and life-tables. *Journal of the Royal Statistical Society. Series B*, **34**, 187–220.

Cox, D.R. (1984). Interaction. *International Statistical Review*, **52**, 1–25.

Cox, D.R. and Hinkley, D.V. (1974). *Theoretical statistics*. Wiley, New York.

Cox, D.R. and Isham, V. (1980). *Point processes*. Chapman & Hall, London.

Cox, D.R. and Lewis, P.A.W. (1966). *The statistical analysis of series of events*. Wiley, New York.

Cox, D.R. and Miller, H.D. (1965). *The theory of stochastic processes*. Chapman & Hall, London.

Cox, D.R. and Reid, N. (1989). On the stability of maximum-likelihood estimators of orthogonal parameters. *Canadian Journal of Statistics*, **17**, 229–233.

Cox, D.R. and Snell, E.J. (1989). *Analysis of binary data* (2nd edn). Chapman & Hall, London.

Cox, D.R. and Solomon, P.J. (2002). *Components of variance*. Chapman & Hall, London.

Cox, D.R. and Stuart, A. (1955). Some quick sign tests for trend in location and dispersion. *Biometrika*, **42**, 80–95.

Cox, D.R. and Wermuth, N. (1992). Response models for mixed binary and quantitative variables. *Biometrika*, **79**, 441–461.

Cox, D.R. and Wermuth, N. (1996). *Multivariate dependencies: Models, analysis and interpretation*, Volume Monographs on Statistics and Applied Probability 67. Chapman & Hall, London.

Craig, A.T. (1933). Variables correlated in sequence. *Bulletin of the American Mathematical Society*, **39**, 129–136.

Cramér, H. (1925). On some classes of series used in mathematical statistics. *Transactions of the 6th Congress of Scandinavian Mathematics*, 399–425.

Cramér, H. (1928). On the composition of elementary errors. *Skandinavien Aktuarie Tidskrift*, **11**, 13–74; 141–180.

Cramér, H. (1936). Sur une propriété de la loi de Gauss. *Comptes-Rendus de l'Académie des Sciences de Paris*, **202**, 615–616.

Cramér, H. (1946). *Mathematical methods of statistics.* Princeton University Press, Princeton, NJ.

Cramér, H. and Wold, H. (1936). Some theorems on distribution functions. *Journal of the London Mathematical Society*, **11**, 290–294.

Cressie, N. (1993). *Statistics for spatial data* (Rev. edn). Wiley, New York.

Crofton, M.V. (1869). On the theory of local probability, applied to straight lines drawn at random in a plane; the methods used being also extended to the proof of certain new theorems in the integral calculus. *Philosophical Transactions of the Royal Society of London. Series A*, **158**, 181–199.

Cronbach, L. (1951). Coefficient alpha and the internal structure of tests. *Psychometrika*, **16**, 297–334.

D'Agostino, R.B. (1971). An omnibus test of normality for moderate and large size samples. *Biometrika*, **58**, 341–348.

Dalenius, T. (1950). The problem of optimum stratification. *Skandinavien Aktuarie Tidskrift*, **33**, 203–213.

Dalenius, T. (1957). Possibilities and limits of sampling in regional inquires. *Bulletin of the Institute of International Statistics*, **35**, 337–355.

Dalenius, T. (1985). *Elements of survey sampling.* Swedish Agency for Research Cooperation with Developing Countries.

Daniel, C. (1959). Use of half normal plots in interpreting factorial two-level experiments. *Technometrics*, **1**, 311–341.

Daniell, P.J. (1946). Discussion on symposium on autocorrelation in time series. *Supplement to the Journal of the Royal Statistical Society*, **8**, 86–88.

Daniels, H.E. (1944). The relation between measures of correlation in the universe of sample permutations. *Biometrika*, **33**, 129–135.

Daniels, H.E. (1950). Rank correlation and population models. *Journal of the Royal Statistical Society. Series B*, **12**, 171–181.

Daniels, H.E. (1954a). A distribution-free test for regression parameters. *Annals of Mathematical Statistics*, **25**, 499–513.

Daniels, H.E. (1954b). Saddlepoint approximations in statistics. *Annals of Mathematical Statistics*, **25**, 631–650.

Dantzig, G.B. (1949). Programming of interdependent activities: II. Mathematical model. *Econometrica, Chicago*, **17**, 200–211.

Dantzig, G.B. (1957). *Les fonctions génératrices liées à quelques tests non-paramétriques.* Mimeographed course, Mathematisch Centrum, Amsterdam.

Darmois, G. (1951). Sur une propriété caractéristique de la loi de probabilité de Laplace. *Comptes-Rendus de l'Académie des Sciences de Paris*, **232**, 1999–2000.

Darmois, G. (1953). Analyse générale des liaisons stochastiques. Etude particulière de l'analyse factorielle linéaire. *Revue Institut International de Statistique*, **21**, 2–8.

Darwin, C. (1876). *The effects of cross and self fertilization in the vegetable kingdom.* John Murray, London.

Das, M.N. and Dey, A. (1967). Group-divisible rotatable designs. *Annals of the Institute of Statistical Mathematics, Tokyo*, **19**, 331–347.

David, F.N. (1947). A power function for tests of randomness in a sequence of alternatives. *Biometrika*, **34**, 335–339.

David, F.N. (1950). Two combinatorial tests of whether a sample has come from a given population. *Biometrika*, **37**, 97–110.

David, F.N. and Neyman, J. (1938). Extension of the Markoff theorem on least squares. *Statistical Research Memoirs*, **2**, 105–116.

David, H.A. and Wolock, F.W. (1965). Cyclic designs. *Annals of Mathematical Statistics*, **36**, 1526–1534.

Davison, A.C. and Hinkley, D.V. (1997). *Bootstrap methods and their application.* Cambridge University Press, Cambridge.

Davison, A.C. and Smith, R.L. (1990). Models for exceedances over high thresholds (with discussion). *Journal of the Royal Statistical Society. Series B*, **52**, 393–442.

Dayhoff, E. (1964). On the equivalence of polykays of the second degree and σ's. *Annals of Mathematical Statistics*, **35**, 1663–1672.

de Finetti, B. (1937). Problemi di optimum. *Giornale d'Istituto di Italiano Attuari*, **8**, 48–67.

de Finetti, B. (1974). *Theory of probability. A critical introductory treatment*, Volume 1. Wiley Series in Probability and Mathematical Statistics. Wiley, New York.

de Moivre, A. (1718). *Doctrine of chance.* Millar, London.

de Moivre, A. (1733). *Approximatio ad summam terminorum binomii* $(a + b)^n$, *in seriem expansi.* Supplement to Miscellanea Analytica, London.

DeGroot, M.H. (1970). *Optimal statistical decision.* McGraw-Hill, New York.

DeGroot, M.H. (1975). *Probability and statistics.* Addison-Wesley, Reading, MA.

DeGroot, M.H. (1987). *Bayesian analysis and uncertainty in economic theory.* Rowman and Littlefield, Lanham, MD.

Delaunay, B. (1963). Theorie der regulaere Dirichletschen Zerlegungen des n-dimensionalen euklidischen Raumes. *Schrift. Int. Math. Deutsch Akad. Wiss. Berlin*, **13**, 27–31.

Deming, W.E. (1950). *Some theory of sampling.* Wiley, New York.

Deming, W.E. and Stephan, F.F. (1940). On a least squares adjustment of a sampled frequency table when the expected marginal totals are known. *Annals of Mathematical Statistics*, **11**, 427–444.

Dempster, A.P. (1960). Random allocation designs I: In general classes of estimation methods. *Annals of Mathematical Statistics*, **31**, 885–905.

Dempster, A.P., Laird, N.M., and Rubin, D.B. (1977). Maximum likelihood from incomplete data via the EM algorithm (with Discussion). *Journal of the Royal Statistical Society. Series B*, **39**, 1–38.

Derman, C. (1956). An application of Chung's lemma to the Kiefer–Wolfowitz stochastic approximation procedure. *Annals of Mathematical Statistics*, **27**, 532–536.

Derman, C., Johns, M.V., and Lieberman, G.J. (1959). Continuous sampling procedures without control. *Annals of Mathematical Statistics*, **30**, 1175–1191.

D'Esopo, D.A. (1961). A note on forecasting by the exponential smoothing operator. *Operations Research*, **9**, 686–687.

Diewert, W.E. (1987). 'Index number'. In *The new Palgrave: A dictionary of economics* (ed. J. Eatwell, M. Milgate and P. Newmann), Volume 1, pp. 767–780. Macmillan, London.

Diggle, P., Liang, K., and Zeger, S. (1994). *Analysis of longitudinal data*. Clarendon Press, Oxford.

Dimroth, E. (1963). Fortschritte der Gefuegestatistik. *Neues Jahrbuch der Mineralogie, Monatsheft, 1963*, 186–192.

Divisia, F. (1925). L'indice monétaire et la théorie de la monnaie. *Revue d'Economie Politique*, **39**, 842–861, 980–1008, 1121–1151. Also published in 1926 by the Société Anonyme du Recueil Sirey, Paris.

Dixon, W.J. (1950). Analysis of extreme values. *Annals of Mathematical Statistics*, **21**, 488–506.

Dixon, W.J. (1951). Ratios involving extreme values. *Annals of Mathematical Statistics*, **22**, 68–78.

Dodge, H.F. (1943). A sampling inspection plan for continuous production. *Annals of Mathematical Statistics*, **14**, 264–279.

Dodge, H.F. (1970). Note on the evolution of acceptance sampling plans. IV. *Journal of Quality Technology*, **2**, 1–8.

Dodge, Y. (1985). *Analysis of experiments with missing data*. Wiley, New York.

Dodge, Y. (1996). A natural random number generator. *International Statistical Review*, **64**, 329–344.

Dodge, Y. and Jurečkovà, J. (2000). *Adaptive regression*. Springer-Verlag, New York.

Dodge, Y. and Majumdar, D. (1979). An algorithm for finding least square generalized inverses for classification models with arbitary patterns. *Journal of Statistical Computation and Simulation*, **9**, 1–17.

Dodge, Y. and Rousson, V. (1999). Multivariate L_1-mean. *Metrika*, **49**, 127–134.

Dodge, Y. and Rousson, V. (2001). On asymmetric properties of the correlation coefficient in the regression setting. *Annals of Statistics*, **55**, 51–54.

Dodge, Y. and Thomas, D. (1980). On the performance of the normal theory and nonparametric multiple comparison procedures. Part 1 and 2. *Sankhyā*, **42**, 1–17.

Donner, A. and Klar, N. (eds) (2000). Cluster randomization trials. *Statistical Methods in Medical Research*, **9**, 79–179.

Donoho, D.L. and Huber, P.J. (1983). The notion of breakdown point. In *Festschrift for Erich L. Lehmann* (ed. P.J. Bickel, K.A. Doksum and J.L. Hodges), pp. 157–184. Wadsworth, Belmont, CA.

Donsker, M.D. (1951). *An invariance principle for certain probability limit theorems*, Volume 6. Memoirs of the American Mathematical Society.

Doolittle, M.H. (1878). Method employed in the solution of normal equations and the adjustment of a triangulation. *US Coast and Geodesic Survey Report*, pp. 115–120.

Downton, F. (1966). Linear estimates with polynomial coefficients. *Biometrika*, **53**, 129–141.

Dowson, D.C. and Landau, B.V. (1982). The Fréchet distance between multivariate normal distributions. *Journal of Multivariate Analysis*, **12**, 450–455.

Dryden, I.L. and Mardia, K.V. (1991). General shape distributions in a plane. *Advances in Applied Probability*, **23**, 259–276.

Dryden, I.L. and Mardia, K.V. (1998). *Statistical shape analysis*. Wiley, New York.

Duncan, A.J. (1962). Bulk sampling: Problems and lines of attack. *Technometrics*, **4**, 319–344.

Duncan, A.J. (1986). *Quality control and industrial statistics*. Irwin-Dorsey, Homewood, IL.

Duncan, D.B. (1952). On the properties of the multiple comparison test. *Virginia Journal of Science*, **3**, 50–67.

Duncan, D.B. (1955). Multiple range and multiple *F*-tests. *Biometrics*, **11**, 1–42.

Dunn, O.J. and Clark, V. (1971). Comparison of tests of the equality of dependent correlation coefficients. *Journal of the American Statistical Association*, **66**, 904–908.

Dupač, V. (1965). A dynamic stochastic approximation method. *Annals of Mathematical Statistics*, **36**, 1695–1702.

Durbin, J. (1961). Some methods of constructing exact tests. *Biometrika*, **48**, 41–55.

Durbin, J. (1967). Design of multi-stage surveys for the estimation of sampling errors. *Applied Statistics*, **16**, 152–164.

Durbin, J. and Watson, G.S. (1950). Testing for serial correlation in least squares regression. I. *Biometrika*, **37**, 409–428.

Dvoretzky, A. (1956). On stochastic approximation. *Proceedings of the 3rd Berkeley Symposium on Mathematical Statistics and Probability*, **1**, 39–55.

Dwass, M. (1956). The large-sample power of rank order tests in the two-sample problem. *Annals of Mathematical Statistics*, **27**, 352–374.

Dwass, M. (1960). Some k-sample rank-order tests. In *Contributions to probability and statistics, Essays in honor of H. Hotelling* (ed. I. Olkin), pp. 198–202. Stanford University Press, Stanford, CA.

Dwass, M. (1964). Extremal processes. *Annals of Mathematical Statistics*, **35**, 1718–1725.

Eckler, A.R. (1955). Rotation sampling. *Annals of Mathematical Statistics*, **26**, 664–685.

Edgeworth, F.Y. (1885). *Methods of statistics*, Volume Jubilee. Royal Statistical Society, London.

Edgeworth, F.Y. (1898). Miscellaneous applications of the calculus of probabilities. *Journal of the Royal Statistical Society*, **61**, 119–131.

Edgeworth, F.Y. (1905). The law of error. *Cambridge Philosophical Transactions*, **20**, 36–66, 113–141.

Edgeworth, F.Y. (1925). *Papers relating to political economy*, 3 Volumes. Macmillan, London.

Edwards, A.W.F. (1972). *Likelihood. An account of the statistical concept of likelihood and its application to scientific inference.* Cambridge University Press, Cambridge.

Efron, B. (1979). Bootstrap methods: Another look at the jackknife. *Annals of Statistics*, **7**, 1–26.

Ehrenfeld, S. and Zacks, S. (1961). Randomization and factorial experiments. *Annals of Mathematical Statistics*, **32**, 270–297.

Ehrenfest, P. and Ehrenfest, T. (1907). Über zwei bekannte Einwände gegen das Boltzmannsche H-Theorem. *Physikalische Zeitschrift*, **8**, 311–314.

Eisenhart, C. (1947). The assumptions underlying the analysis of variance. *Biometrics*, **3**, 1–21.

Elfving, G. (1947). The asymptotical distribution of range in samples from a normal population. *Biometrika*, **34**, 111–119.

Engen, S. (1974). On species frequency models. *Biometrika*, **61**, 263–270.

Erlang, A.K. (1917). Solution of some problems in the theory of probabilities of significance in automatic telephone exchanges. *Post Office Electrical Engineer's Journal*, **10**, 189–197.

Esscher, F. (1924). On a method of determining correlation from the ranks of the variates. *Skandinavien Aktuarie Tidskrift*, **7**, 201–219.

Esseen, C.G. (1945). Fourier analysis of distribution functions. A mathematical study of the Laplace–Gaussian law. *Acta Mathematica*, **77**, 1–125.

Euler, L. (1782). Recherches sur une nouvelle espèce de carrés magiques. *Verhandlungen der Zeeuw. Gen. Weten. Vlissengen*, **9**, 85–239.

Eurostat (2000*a*). *Definition of quality in statistics*. Eurostat Working Group on Assessment of Quality in Statistics, Luxembourg, 4–5 April 2000.

Eurostat (2000*b*). *Standard quality report*. Eurostat Working Group on Assessment of Quality in Statistics, Luxembourg, 4–5 April 2000.

Fan, J. and Gijbels, I. (1996). *Local polynomial modelling and its applications*. Chapman & Hall, London.

Federer, W.T. and Ferris, G.E. (1956). *Least squares estimates of effects and sum of squares for a tied double change-over design*. Mimeo. No. BU-73-M in Biometrics Unit Series, Cornell University, Ithaca, NY.

Federer, W.T., Hedayat, A.S., and Ratkoe, B.L. (1981). *Factorial designs*. Wiley, New York.

Fellegi, I.P. (1963). Sampling with varying probabilities without replacement: Rotating and non-rotating samples. *Journal of the American Statistical Association*, **58**, 183–201.

Feller, W. (1968). *An introduction to probability theory and its applications*, Volume I. Wiley, New York.

Feller, W. (1990). *An introduction to probability theory and its applications*, Volume II. Wiley, New York.

Ferreri, C. (1964). Una nuova funzione di frequenza per l'analisi delle variabili statistiche semplici. *Statistica*, **24**, 223–251.

Fieller, E.C. (1940). The biological standardization of insulin. *Supplement to the Journal of the Royal Statistical Society*, **7**, 1–64.

Fieller, E.C., Hartley, H.O., and Pearson, E.S. (1957). Tests for rank correlation coefficients. I. *Biometrika*, **44**, 470–481.

Finney, D.J. (1941). On the distribution of a variate whose logarithm is normally distributed. *Supplement to the Journal of the Royal Statistical Society*, **7**, 155–161.

Finney, D.J. (1945). The fractional replication of factorial arrangements. *Annals of Eugenics*, **12**, 291–301.

Finney, D.J. (1947). The estimation from individual records of the relationship between dose and quantal response. *Biometrika*, **34**, 320–334.

Finney, D.J. (1949). The choice of a response metameter in bio-assay. *Biometrics*, **5**, 261–272.

Finney, D.J. (1978). *Statistical method in biological assay* (3rd edn). Macmillan, New York.

Finney, D.J. and Outhwaite, A.D. (1956). Serially balanced sequences in bioassay. *Proceedings of the Royal Society of London. Series B*, **145**, 493–507.

Fisher, I. (1922a). *The making of index numbers.* Houghton-Mifflin, Boston.

Fisher, I. (1925a). Our unstable dollar and the so-called business cycle. *Journal of the American Statistical Association,* **20**, 179–202.

Fisher, I. (1927). The total value criterion: A new principle in index number construction. *Journal of the American Statistical Association,* **28**, 1–13.

Fisher, N.I. and Switzer, P. (1985). Chi-plots for assessing dependence. *Biometrika,* **72**, 253–265.

Fisher, R.A. (1918). The correlation between relatives on the supposition of Mendelian inheritance. *Transactions of the Royal Society of Edinburgh,* **52**, 399–433.

Fisher, R.A. (1922b). On the mathematical foundation of the theoretical statistics. *Philosophical Transactions of the Royal Society of London. Series A,* **222**, 309–368.

Fisher, R.A. (1924). On a distribution yielding the error functions of several well known statistics. *Proceedings of the International Congress on Mathematics, Toronto,* **2**, 805–813.

Fisher, R.A. (1925b). Applications of 'Student's' distribution. *Metron,* **5**, 90–104.

Fisher, R.A. (1925c). Theory of statistical estimation. *Proceedings of the Cambridge Philosophical Society,* **22**, 700–725.

Fisher, R.A. (1928a). Moments and product moments of sampling distributions. *Proceedings of the London Mathematical Society,* **30**, 199–238.

Fisher, R.A. (1928b). On a property connecting the chi-square measure of discrepancy with the method of maximum likelihood. *Atti del Congresso Internazionale dei Matematici, Bologna,* **6**, 95–100.

Fisher, R.A. (1930). Inverse probability. *Proceedings of the Cambridge Philosophical Society,* **26**, 528–535.

Fisher, R.A. (1935a). *The design of experiments.* Oliver and Boyd, Edinburgh and London.

Fisher, R.A. (1935b). The fiducial argument in statistical inference. *Annals of Eugenics,* **6**, 391–398.

Fisher, R.A. (1936a). The coefficient of racial likeness and the future of craniometry. *Journal of the Royal Anthropological Institute,* **66**, 57–63.

Fisher, R.A. (1936b). Uncertain inference. *Proceedings of the American Academy of Arts and Sciences,* **71**, 245–258.

Fisher, R.A. (1936c). The use of multiple measurements in taxonomic problems. *Annals of Eugenics,* **7**, 179–188.

Fisher, R.A. (1939). The sampling distribution of some statistics obtained from non-linear equations. *Annals of Eugenics,* **9**, 238–249.

Fisher, R.A. (1941). The negative binomial distribution. *Annals of Eugenics,* **11**, 182–187.

Fisher, R.A. (1956). On a test of significance in Pearson's Biometrika tables. *Journal of the Royal Statistical Society. Series B*, **18**, 56–60.

Fisher, R.A. and Yates, F. (1938). *Statistical tables for biological, agricultural and medical research*. Oliver, London and Edinburgh.

Flury, B. (1997). *A first course in multivariate statistics*. Springer-Verlag, New York.

Foster, F.G. (1953). On the stochastic matrices associated with certain queuing processes. *Annals of Mathematical Statistics*, **24**, 355–360.

Foster, F.G. and Stuart, A. (1954). Distribution-free tests in time-series based on the breaking of records. *Journal of the Royal Statistical Society. Series B*, **16**, 1–22.

Fraser, D.A.S. (1957). Most powerful rank-type tests. *Annals of Mathematical Statistics*, **28**, 1040–1043.

Fréchet, M. (1927). On the probability law of maximum values. *Annales de la Société Polonaise Mathématique de Cracow*, **6**, 93–116.

Fréchet, M. (1940). *Les probabilités associées à un système d'événements compatibles et dépendants. 1. Événements en nombre fini fixe*. Hermann, Paris.

Fréchet, M. (1957). Sur la distance de deux lois de probabilité. *Comptes-Rendus de l'Académie des Sciences de Paris*, **244**, 689–692.

Freedman, D. (1971). *Brownian motion and diffusion*. Holden-Day Series in Probability and Statistics. Holden-Day, San Francisco.

Freeman, M.F. and Tukey, J.W. (1950). Transformations related to the angular and the square root. *Annals of Mathematical Statistics*, **21**, 607–611.

Freund, J.E. and Ansari, A.R. (1957). *Two-way rank-sum test for variances*, Volume 34. Virginia Polytechnic Institute, Technical Report, Department of Statistics.

Friedman, M. (1937). The use of ranks to avoid the assumption of normality implicit in the analysis of variance. *Journal of the American Statistical Association*, **32**, 675–701.

Frisch, R. (1929). Correlation and scatter in statistical variables. *Nordic Statistical Journal*, **8**, 36–102.

Frisch, R. (1934). *Statistical confluence analysis by means of complete regression systems*, Volume 192 S. University of Oslo.

Frisch, R. (1937). Note on the phase diagram of two variates. *Econometrica*, **5**, 326–328.

Furry, W.H. (1937). A symmetry theorem in the positron theory. *Physical Review, II Series*, **51**, 125–129.

Gabriel, K.R. (1964). A procedure for testing the homogeneity of all sets of means in analysis of variance. *Biometrics*, **20**, 459–477.

Gabriel, K.R. (1971). The biplot graphic display of matrices with application to principal component analysis. *Biometrika*, **58**, 453–467.

Gabriel, K.R. and Sen, P.K. (1968). Simultaneous test procedures for one-way ANOVA and MANOVA based on rank scores. *Sankhyā*, **30**, 303–312.

Galambos, J. (1975). Methods for proving Bonferroni type inequalities. *Journal of the London Mathematical Society, II Series*, **9**, 561–564.

Gall, M.D., Borg, W.R., and Gall, J.P. (1996). *Educational research: An introduction* (6th edn). Longman, White Plains, NY.

Galton, F. (1907). Grades and deviates (including a table of normal deviates corresponding to each millesimal grade in the length of an array, and a figure). *Biometrika*, **5**, 400–406.

Galton, F. and Watson, H.W. (1874). On the probability of the extinction of families. *Journal of the Anthropological Institute of Great Britain and Ireland*, **4**, 138–144.

Gart, J.J. (1969). An exact test for comparing matched proportions in crossover designs. *Biometrika*, **56**, 75–80.

Garwood, F. (1940). An application of the theory of probability to the operation of vehicular-controlled traffic signals. *Supplement to the Journal of the Royal Statistical Society*, **7**, 65–77.

Gauss, C.F. (1823). Theoria combinationis observationum erroribus minimis obnoxiae. *Werke*, **4**, 1–108.

Gautschi, W. (1957). Some remarks on systematic sampling. *Annals of Mathematical Statistics*, **28**, 385–394.

Geary, R.C. (1935). The ratio of the mean deviation to the standard deviation as a test of normality. *Biometrika*, **27**, 310–332.

Geary, R.C. (1936). Moments of the radio of the mean deviation to the standard deviation for normal samples. *Biometrika*, **28**, 295–305.

Geary, R.C. (1944). Comparison of the concepts of efficiency and closeness for consistent estimates of a parameter. *Biometrika*, **33**, 123–128.

Geary, R.C. (1954). The contiguity ratio and statistical mapping. *Incorporated Statistician*, **5**, 115–145; 129–145.

Geary, R.C. (1966). The average critical value method for adjudging relative efficiency of statistical tests in time series regression analysis. *Biometrika*, **53**, 109–119.

Gehan, E.A. (1965). A generalized two-sample Wilcoxon test for doubly censored data. *Biometrika*, **52**, 650–653.

Geisser, S. (1957). The distribution of the ratios of certain quadratic forms in time series. *Annals of Mathematical Statistics*, **28**, 724–730.

Geisser, S. (1993). *Predictive inference. An introduction.* Chapman & Hall, London.

Ghosh, M.N. (1955). Simultaneous tests of linear hypotheses. *Biometrika*, **42**,

441–449.

Gibbons, J.D. and Pratt, J.W. (1975). P-values: Interpretation and methodology. *The American Statistician*, **29**, 20–25.

Gini, C. (1910). *Indici di concentrazione e di dipendenza*. UTET, Torino.

Gini, C. (1912). Variabilità e mutabilità. *Studi Economico-Giuridici della R. Univ. Cagliari*, **3**, 3–159.

Giri, N. and Kiefer, J. (1964). Local and asymptotic minimax properties of multivariate tests. *Annals of Mathematical Statistics*, **35**, 21–35.

Girshick, M.A. (1939). On the sampling theory of roots of determinantal equations. *Annals of Mathematical Statistics*, **10**, 203–204.

Gittins, J.C. and Jones, D.M. (1974). A dynamic allocation index for the sequential design of experiments. *Progress in Statistics, European Meeting, Budapest 1972, Colloques de Mathématique Société de Janos Bolyai* (ed. K.J. Gani and I. Vincze), **9**, 241–266. North–Holland, Amsterdam.

Glivenko, V. (1933). Sulla determinazione empirica delle leggi di probabilità. *Giornale d'Istituto di Italiano Attuari*, **4**, 92–99.

Gnanadesikan, R. (1959). Equality of more than two variances and of more than two dispersion matrices against certain alternatives. *Annals of Mathematical Statistics*, **30**, 177–184.

Gnedenko, B.V. and Koroljuk, V.S. (1951). Über die maximale Divergenz zweier empirischer Verteilungen. *Doklady Akademii Nauk SSSR*, **80**, 525–528.

Godambe, V.P. (1966). A new approach to sampling from finite populations. I, II. *Journal of the Royal Statistical Society. Series B*, **28**, 310–328.

Godambe, V.P. (1976). Conditional likelihood and unconditional optimum estimating equation. *Biometrika*, **63**, 277–284.

Gompertz, B. (1825). On the nature of the function expressive of the law of human mortality and on the new mode of determining the value of life contingencies. *Philosophical Transactions of the Royal Society of London*, **115**, 513–585.

Good, I.J. (1956). The surprise index for the multivariate normal distribution. *Annals of Mathematical Statistics*, **27**, 1130–1135.

Goodall, D.W. (1966). A new similarity index based on probability. *Biometrics*, **22**, 882–907.

Goodman, L.A. (1961). Statistical methods for the mover-stayer model. *Journal of the American Statistical Association*, **56**, 841–868.

Goodman, L.A. (1968). The analysis of cross-classified data: Independence, quasi-independence, and interactions in contingency tables with or without missing entries. *Journal of the American Statistical Association*, **63**, 1091–1131.

Goodman, L.A. and Kruskal, W.H. (1954). Measures of association for cross-classification. *Journal of the American Statistical Association*, **49**, 732–764.

Goodman, N.R. (1963). Statistical analysis based on a certain multivariate complex Gaussian distribution. An introduction. *Annals of Mathematical Statistics*, **34**, 152–177.

Gosset, S.W. 'Student' (1908a). Probable error of a correlation coefficient. *Biometrika*, **6**, 302–310.

Gosset, S.W. 'Student' (1908b). The probable error of a mean. *Biometrika*, **6**, 1–25.

Gould, S.J. (1966). Allometry and size in ontogeny and phylogeny. *Biological Reviews*, **41**, 587–640.

Gower, J.C. (1971). A general coefficient of similarity and some of its properties. *Biometrics*, **27**, 857–872.

Gram, J.P. (1883). Über die Entwicklung reeler Functionen in Reihen mittelst der Methode der kleinsten Quadrate. *Journal für die Reine und Angewandte Mathematik*, **94**, 41–73.

Granger, C.W.J. and Hatanaka, M. (1964). *Spectral analysis of economic time series*, Volume 1. Princeton Studies in Mathematical Economics. Princeton University Press, Princeton, NJ.

Graybill, F.A. and Pruitt, W.E. (1958). The staircase design: Theory. *Annals of Mathematical Statistics*, **29**, 523–533.

Green, P.J. (1984). Iteratively reweighted least squares for maximum likelihood estimation, and some robust and resistant alternatives. *Journal of the Royal Statistical Society. Series B*, **46**, 149–192.

Green, P.J. (1995). Reversible jump Markov chain Monte Carlo computation and Bayesian model determination. *Biometrika*, **82**, 711–732.

Green, P.J. and Silverman, B.W. (1994). *Nonparametric regression and generalized linear models: A roughness penalty approach*. Chapman & Hall, London.

Green, R.F. (1976). Outlier-prone and outlier-resistant distributions. *Journal of the American Statistical Association*, **71**, 502–505.

Green, W.H. (2000). *Econometric analysis* (4th edn). Prentice Hall, Upper Saddle River, NJ.

Greenwood, M. (1931). On the statistical measure of infectiousness. *Journal of Hygiene, Cambridge*, **31**, 336–351.

Greenwood, M. (1949). The infectiousness of measles. *Biometrika*, **36**, 1–8.

Greiner, R. (1909). Über das Fehlersystem der Kollektivmasslehre. *Zeitschrift für Mathematik und Physik*, **57**, 121, 225, 337.

Grenander, U. (1951). On empirical spectral analysis of stochastic processes. *Arkiv für Matematik*, **1**, 503–531.

Grenander, U. (1956). The theory of mortality measurement. Part II. *Skandinavien Aktuarie Tidskrift*, **39**, 125–153.

Grimmett, G.R. and Stirzaker, D.R. (2001). *Probability and random processes* (3rd edn). Clarendon Press, Oxford.

Groeneboom, P. and Wellner, J.A. (1992). *Information bounds and nonparametric maximum likelihood estimation.* DMV Seminar Band 19, Birkhäuser, Basel.

Grubbs, F.E. (1948). On estimating precision of measuring instruments and product variability. *Journal of the American Statistical Association*, **43**, 243–264.

Grubbs, F.E. (1950). Sample criteria for testing outlying observations. *Annals of Mathematical Statistics*, **21**, 27–58.

Grundy, P.M. and Healy, M.J.R. (1950). Restricted randomization and quasi-Latin squares. *Journal of the Royal Statistical Society. Series B*, **12**, 286–291.

Guldberg, A. (1934). On discontinuous frequency functions of two variables. *Skandinavien Aktuarie Tidskrift*, **17**, 89–117.

Gumbel, E.J. (1935). Les valeurs extrêmes des distributions statistiques. *Annales de l'Institut Henri Poincaré*, **5**, 115–158.

Gumbel, E.J. (1958). *Statistics of extremes.* Columbia University Press, New York.

Gumbel, E.J. (1960). Bivariate exponential distributions. *Journal of the American Statistical Association*, **55**, 698–707.

Gumbel, E.J. and von Schelling, H. (1950). The distribution of the number of exceedances. *Annals of Mathematical Statistics*, **21**, 247–262.

Gupta, S.S. (1956). *On a decision rule for a problem in ranking means*, Volume Mimeograph Series, No. 150. University of North Carolina, Chapel Hill.

Gurland, J. (1958). A generalized class of contagious distributions. *Biometrics*, **14**, 229–249.

Gutenbrunner, C. and Jurečková, J. (1992). Regression rank scores and regression quantiles. *Annals of Statistics*, **20**, 305–330.

Gutenbrunner, C., Jurečková, J., Koenker, R., and Portnoy, S. (1993). Tests of linear hypotheses based on regression rank scores. *Journal of Nonparametric Statistics*, **2**, 307–331.

Guttman, L. (1959). Metricizing rank-ordered or unordered data for a linear factor analysis. *Sankhyā*, **21**, 257–268.

Haavelmo, T. (1943). The statistical implications of a system of simultaneous equations. *Econometrica*, **11**, 1–12.

Hadi, A.S. and Nyquist, H. (2002). Sensitivity analysis in statistics. *Journal of Statistical Studies*, in press.

Haight, F.A. (1965). Counting distributions for renewal processes.

Biometrika, **52**, 395–403.

Hajek, J. (1964). Asymptotic theory of rejective sampling with varying probabilities from a finite population. *Annals of Mathematical Statistics*, **35**, 1491–1523.

Haldane, J.B.S. (1951). A class of efficient estimates of a parameter. *Bulletin of the Institute of International Statistics*, **23**, 231–248.

Hall, P. (1986). On the bootstrap and confidence intervals. *Annals of Statistics*, **14**, 1431–1452.

Hallin, M. (1994). On the Pitman-nonadmissibility of correlogram-based methods. *Journal of Time Series Analysis*, **15**, 607–612.

Hammersley, J.M. and Morton, K.W. (1956). A new Monte Carlo technique: antithetic variates. *Proceedings of the Cambridge Philosophical Society*, **52**, 449–475.

Hamming, R.W. and Tukey, J.W. (1949). Measuring noise color. *Bell Telephone Labs Memorandum*, **MM**, 49–110.

Hampel, F.R. (1971). A general qualitative definition of robustness. *Annals of Mathematical Statistics*, **42**, 1887–1896.

Hanson, D.L. and Koopmans, L.H. (1964). Tolerance limits for the class of distributions with increasing hazard rates. *Annals of Mathematical Statistics*, **35**, 1561–1570.

Harley, B.I. (1957). Relation between the distributions of non-central t and of a transformed correlation coefficient. *Biometrika*, **44**, 219–224.

Harris, T.E. (1952). First passage and recurrence distributions. *Transactions of the American Mathematical Society*, **73**, 471–486.

Harrison, P.J. (1964). Short-term sales forecasting – Developments and methods for seasonal estimation. *Presented in Royal Statistical Society Conference at Cardiff, September 1964*, **14**.

Harshbarger, B. (1949). Triple rectangular lattices. *Biometrics*, **5**, 1–13.

Harter, H.L. (1983). Harter's adaptive robust method. In *Encyclopedia of statistical sciences* (ed. S. Kotz and N.L. Johnson), Volume 3, pp. 576–578. Wiley, New York.

Hartigan, J.A. (1975). *Clustering algorithms*. Wiley, New York.

Hartley, H.O. (1948). The estimation of non-linear parameters by 'internal least squares'. *Biometrika*, **35**, 32–45.

Hartley, H.O. (1950). The maximum F-ratio as a short-cut test for heterogeneity of variance. *Biometrika*, **37**, 308–312.

Hartley, H.O. and Rao, J.N.K. (1962). Sampling with unequal probabilities and without replacement. *Annals of Mathematical Statistics*, **33**, 350–374.

Hass-Lorentz, G.L. de (1913). *Die Brownsche Bewegung und einige verwandte Erscheinungen*. Vieweg, Braunschweig.

Hastie, T.J. and Tibshirani, R.J. (1990). *Generalized additive models*. Chapman

& Hall, London.

Hastings, C., Mosteller, F., Tukey, J.W., and Winsor, C.P. (1947). Low moments for small samples: A comparative study of order statistics. *Annals of Mathematical Statistics*, **18**, 413–426.

Hastings, W.K. (1970). Monte Carlo sampling methods using Markov chains and their applications. *Biometrika*, **57**, 97–109.

Hedayat, A.S., Sloane, N.J.A., and Stufken, J. (1999). *Orthogonal arrays: Theory and applications*. Springer-Verlag, New York.

Heitjan, D.F. and Rubin, D.B. (1991). Ignorability and coarse data. *Annals of Statistics*, **19**, 2244–2253.

Heitler, W. (1937). On the analysis of cosmic rays. *Proceedings of the Royal Society of London. Series A*, **161**, 261–283.

Helmert, F.R. (1876). Die Genauigkeit der Formel von Peters zur Berechnung des wahrscheinlichen Beobachtungsfehlers directer Beobachtungen gleicher Genauigkeit. *Astronomische Nachrichten*, **88**, 113–120.

Herzberg, A.M. (1966). Cylindrically rotatable designs. *Annals of Mathematical Statistics*, **37**, 242–247.

Herzberg, A.M. (1967). A method for the construction of second order rotatable designs in k dimensions. *Annals of Mathematical Statistics*, **38**, 177–180.

Heyde, C.E. and Seneta, E. (1977). *I.J. Bienaymé. Statistical theory anticipated*. Springer-Verlag, New York and Berlin.

Hill, B. (1975). A simple approach to inference about the tail of a distribution. *Annals of Statistics*, **3**, 1163–1174.

Hillier, F.S. (1964). New criteria for selecting continuous sampling plans. *Technometrics*, **6**, 161–178.

Hinkelmann, K. (1964). Extended group divisible partially balanced incomplete block designs. *Annals of Mathematical Statistics*, **35**, 681–695.

Hinkelmann, K. and Kempthorne, O. (1963). Two classes of group divisible partial diallel crosses. *Biometrika*, **50**, 281–291.

Hinkley, D.V. (1970). Inference about the change-point in a sequence of random variables. *Biometrika*, **57**, 1–17.

Hinkley, D.V. (1979). Predictive likelihood. *Annals of Statistics*, **7**, 718–728.

Hirotsu, C. (1986). Cumulative chi-squared statistic as a tool for testing goodness of fit. *Biometrika*, **73**, 165–173.

Hirschfeld, H.O. (1935). A connection between correlation and contingency. *Proceedings of the Cambridge Philosophical Society*, **31**, 520–524.

Hoaglin, D.C. and Welsch, R.E. (1978). The hat matrix in regression and ANOVA. *The American Statistician*, **32**, 17–22.

Hoblyn, T.N., Pearce, S.C., and Freeman, G.H. (1954). Some considerations in the design of successive experiments in fruit plantations. *Biometrics*, **10**,

503–515.

Hodges, J.L. (1955). A bivariate sign test. *Annals of Mathematical Statistics*, **26**, 523–527.

Hodges, J.L. and Lehmann, E.L. (1963). Estimates of location based on rank tests. *Annals of Mathematical Statistics*, **34**, 598–611.

Hoeffding, W. (1948*a*). A class of statistics with asymptotically normal distribution. *Annals of Mathematical Statistics*, **19**, 293–325.

Hoeffding, W. (1948*b*). A non-parametric test of independence. *Annals of Mathematical Statistics*, **19**, 546–557.

Hoeffding, W. (1951). 'Optimum' nonparametric tests. In *Proceedings of the Berkeley Symposium on Mathematical Statistics and Probability, California, July 31 – August 12, 1950*, pp. 83–92.

Hoeffding, W. (1963). Probability inequalities for sums of bounded random variables. *Journal of the American Statistical Association*, **58**, 13–30.

Hoerl, A.E. and Kennard, R.W. (1970). Ridge regression: Biased estimation for nonorthogonal problems. *Technometrics*, **12**, 55–82.

Hogg, R.V. (1967). Some observations on robust estimation. *Journal of the American Statistical Association*, **62**, 1179–1186.

Hogg, R.V. (1975). Estimates of percentile regression lines using salary data. *Journal of the American Statistical Association*, **70**, 56–59.

Holla, M.S. and Bhattacharya, S.K. (1965). On a discrete compound distribution. *Annals of the Institute of Statistical Mathematics, Tokyo*, **17**, 377–384.

Holland, P. (1986). Statistics and causal inference. *Journal of the American Statistical Association*, **81**, 945–970.

Hollander, M. (1970). A distribution-free test for parallelism. *Journal of the American Statistical Association*, **65**, 387–394.

Hollander, M. (1971). A nonparametric test for bivariate symmetry. *Biometrika*, **58**, 203–212.

Hollander, M. and Proschan, F. (1972). Testing whether new is better than used. *Annals of Mathematical Statistics*, **43**, 1136–1146.

Holzinger, K.J. (1937). Note on Professor Kelley's method (in notes). *Journal of the American Statistical Association*, **32**, 360–362.

Holzinger, K.J. (1944). Factoring test scores and implications for the method of averages. *Psychometrika*, **9**, 155–167.

Hooke, R. (1956). Some applications of bipolykays to the estimation of variance components and their moments. *Annals of Mathematical Statistics*, **27**, 80–98.

Horvitz, D.G. and Thompson, D.J. (1952). A generalization of sampling without replacement from a finite universe. *Journal of the American Statistical Association*, **47**, 663–685.

Hosking, J.R.M., Wallis, J.R., and Wood, E.F. (1985). Estimation of the generalized extreme-value distribution by the method of probability-weighted moments. *Technometrics*, **27**, 251–261.

Hosmer, D.W. and Lemeshow, S. (1989). *Applied logistic regression*. Wiley, New York.

Hotelling, H. (1926). Multiple-sheeted spaces and manifolds of states of motion. *Transactions of the American Mathematical Society*, **28**, 479–490.

Hotelling, H. (1931). The generalization of Student's ratio. *Annals of Mathematical Statistics*, **2**, 360–378.

Hotelling, H. (1933). Analysis of a complex of statistical variables into principal components. *Journal of Educational Psychology*, **24**, 417–441 and 498–520.

Hotelling, H. (1936). Relations between two sets of variates. *Biometrika*, **28**, 321–377.

Hotelling, H. (1940). The selection of variates for use in prediction with some comments on the general problem of nuisance parameters. *Annals of Mathematical Statistics*, **11**, 271–283.

Hotelling, H. (1944). Some improvements in weighing and other experimental techniques. *Annals of Mathematical Statistics*, **15**, 297–305.

Hotelling, H. (1951). A generalized t test and measure of multivariate dispersion. In *Proceedings of the Berkeley Symposium on Mathematical Statistics and Probability, California, July 31 – August 12, 1950*, pp. 23–41.

Hsu, P.L. (1939). A new proof of the joint product moment distribution. *Proceedings of the Cambridge Philosophical Society*, **35**, 336–338.

Hsu, P.L. (1941). Analysis of variance from the power function standpoint. *Biometrika*, **32**, 62–69.

Huber, P.J. (1964). Robust estimation of a location parameter. *Annals of Mathematical Statistics*, **35**, 73–101.

Huitema, B.E. (1980). *Analysis of covariance and alternatives*. Wiley, New York.

Huzurbazar, V.S. (1950). Probability distributions and orthogonal parameters. *Proceedings of the Cambridge Philosophical Society*, **46**, 281–284.

Hyrenius, H. (1953). On the use of ranges, cross-ranges and extremes in comparing small samples. *Journal of the American Statistical Association*, **48**, 534–545.

Iglehart, D.L. (1964). Multivariate competition processes. *Annals of Mathematical Statistics*, **35**, 350–361.

Ihaka, R. and Gentleman, R. (1996). R: A language for data analysis and graphics. *Journal of Computational and Graphical Statistics*, **5**, 299–314.

Irwin, J.O. (1935). Tests of significance for difference between percentages based on small numbers. *Metron*, **12**, 83–94.

Irwin, J.O. (1963). The place of mathematics in medical and biological statistics. *Journal of the Royal Statistical Society. Series A*, **126**, 1–45.

Isaacson, S.L. (1951). On the theory of unbiased tests of simple statistical hypotheses specifying the values of two or more parameters. *Annals of Mathematical Statistics*, **22**, 217–234.

Isham, V. (1988). Estimation of the incidence of HIV infection. *Philosophical Transactions of the Royal Society of London. Series B*, **325**, 113–121.

Ising, E. (1925). Beitrag zur Theorie des Ferromagnetismus. *Zeitschrift für Physik*, **31**, 253–258.

Isserlis, L. (1914). On the partial correlation ratio. *Biometrika*, **10**, 391–411.

Ito, K. (1956). Asymptotic formulae for the distribution of Hotelling's generalized T^2. *Annals of Mathematical Statistics*, **27**, 1091–1105.

Ito, K. (1960). Asymptotic formulae for the distribution of Hotelling's generalized t^2 statistic. II. *Annals of Mathematical Statistics*, **31**, 1148–1153.

Jackson, J.R. (1957). Networks of waiting lines. *Operations Research*, **5**, 518–521.

James, A.T. (1960). The distribution of the latent roots of the covariance matrix. *Annals of Mathematical Statistics*, **31**, 151–158.

James, W. and Stein, C. (1961). Estimation with quadratic loss. *Proceedings of the 4th Berkeley Symposium on Mathematical Statistics and Probability*, **1**, 361–380.

Janossy, L., Renyi, A., and Aczel, J. (1950). On composed Poisson distributions. I. *Acta Mathematica Academiae Scientiarum Hungaricae*, **1**, 209–224.

Jeffreys, H. (1936). Further significance tests. *Proceedings of the Cambridge Philosophical Society*, **32**, 416–445.

Jeffreys, H. (1938). The use of minimum χ^2 as an approximation to the method of maximum likelihood. *Proceedings of the Cambridge Philosophical Society*, **34**, 156–157.

Jeffreys, H. (1939). *Theory of probability*. Oxford University Press, Oxford.

Jeffreys, H. (1948). *Theory of probability* (2nd edn). International Series of Monographs on Physics. Oxford University Press, Oxford.

Jenkins, O.C., Ringer, L.J., and Hartley, H.O. (1973). Root estimators. *Journal of the American Statistical Association*, **68**, 414–419.

Jiřina, M. (1952). Sequential estimation of distribution-free tolerance limits. *Czechoslovak Mathematical Journal*, **2**, 221–232.

John, J.A. (1966). Cyclic incomplete block designs. *Journal of the Royal Statistical Society. Series B*, **28**, 345–360.

Johns, M.V. (1957). Non-parametric empirical Bayes procedures. *Annals of Mathematical Statistics*, **28**, 649–669.

Johnson, N.L. (1949). Systems of frequency curves generated by methods of translation. *Biometrika*, **36**, 149–176.

Johnson, N.L. and Kotz, S. (1977). *Urn models and their applications*. Wiley, New York.

Johnson, N.L., Kotz, S., and Balakrishnan, N. (1995). *Continuous univariate distributions* (2nd edn), Volumes 1 and 2. Wiley, New York.

Jonckheere, A.R. (1954). A distribution-free k-sample test against ordered alternatives. *Biometrika*, **41**, 133–145.

Jones, H.E. (1937). The nature of regression functions in the correlation analysis of time series. *Econometrica*, **5**, 305–325.

Jordan, C. (1927). A valoszinuségszamitas alapfogalmai (Les fondements du calcul des probabilités). *Mathematikai és Physikai Lapok*, **34**, 109–136.

Jöreskog, K. (1970). A general method for analysis of covariance structures. *Biometrika*, **57**, 239–251.

Jöreskog, K.G. (1973). A general method for estimating a linear structural equation system. In *Structural equation models in the social sciences* (ed. A.S. Goldberger and O.D. Duncan). Seminar Press, New York.

Jorgensen, B. (1987). Exponential dispersion models. *Journal of the Royal Statistical Society. Series B*, **49**, 127–162.

Jowett, G.H. (1952). A simply constructed adding machine. *Mathematical Gazette*, **36**, 267–269.

Jowett, G.H. (1955). Sampling properties of local statistics in stationary stochastic series. *Biometrika*, **42**, 160–169.

Jowett, G.H. and Wright, W.M. (1959). Jump analysis. *Biometrika*, **46**, 386–399.

Kagan, A.M., Linnik, Y.V., and Rao, C.R. (1965). On a characterization of the normal law based on a property of the sample average. *Sankhyā*, **27**, 405–406.

Kagan, A.M., Linnik, Y.V., and Rao, C.R. (1975). *Characterization problems in mathematical statistics*. Wiley, New York.

Kalbfleisch, J.D. and Prentice, R.L. (2002). *The statistical analysis of failure time data* (2nd edn). Wiley, New York.

Kalman, R.E. (1960). A new approach to linear filtering and prediction problems. *Transactions of the ASME, Series D, Journal of Basic Engineering*, **82**, 35–45.

Kamat, A.R. and Sathe, Y.S. (1962). Asymptotic power of certain test criteria (based on first and second differences) for serial correlation between successive observations. *Annals of Mathematical Statistics*, **33**, 186–200.

Kaplan, E.L. and Meier, P. (1958). Nonparametric estimation from incomplete observations. *Journal of the American Statistical Association*, **53**, 457–481.

Kapteyn, J. (1903). *Skew frequency curves in biology and statistics*. Astronomical Laboratory Noordhoff, Groningen.

Kapteyn, J. and van Uven, M.J. (1916). *Skew frequency curves in biology and statistics*. Hoitsema Brothers, Groningen.

Kärber, J. (1931). Beitrag zur kollektiven Behandlung pharmakologische Reihenversuche. *Archiv für Experimentelle Pathologie und Pharmakologie*, **161**, 480–483.

Katti, S.K. and Rao, A.V. (1970). The log-zero-Poisson distribution. *Biometrics*, **26**, 801–813.

Katz, L. (1950). On the relative efficiencies of ban estimates. *Annals of Mathematical Statistics*, **21**, 398–405.

Keilson, J. (1962). The homogeneous random walk on the half-line and the Hilbert problem. *Bulletin of the Institute of International Statistics*, **39**, 279–291.

Kelley, T.L. (1919). Principles underlying the classification of men. *Journal of Applied Psychology*, **3**, 50–67.

Kelley, T.L. (1925). Measures of correlation determined from groups of varying homogeneity. *Journal of the American Statistical Association*, **20**, 512–521.

Kemp, C.D. and Kemp, A.W. (1965). Some properties of the 'Hermite' distribution. *Biometrika*, **52**, 381–394.

Kendall, D.G. (1948). On the generalized birth-and-death process. *Annals of Mathematical Statistics*, **19**, 1–15.

Kendall, M.G. (1938). A new measure of rank correlation. *Biometrika*, **30**, 81–93.

Kendall, M.G. (1946). *The advanced theory of statistics, II.* Charles Griffin, London.

Kendall, M.G. (1971). The work of Ernst Abbe. *Biometrika*, **58**, 369–373.

Kendall, M.G. and Babington Smith, B. (1940). On the method of paired comparisons. *Biometrika*, **31**, 324–345.

Kendall, M.G. and Dickinson, J. (1990). *Rank correlation methods* (5th edn). Arnold, London.

Kendall, M.G. and Stuart, A. (1958). *The advanced theory of statistics* (4th edn). Charles Griffin, London.

Kermack, W.O. and McKendrick, A.G. (1927). Contributions of the mathematical theory of epidemics. *Proceedings of the Royal Society of London. Series A*, **115**, 700–721.

Kershner, R.P. and Federer, W.T. (1981). Two-treatment crossover designs for estimating a variety of effects. *Journal of the American Statistical Association*, **76**, 612–619.

Kesten, H. (1958). Accelerated stochastic approximation. *Annals of Mathematical Statistics*, **29**, 41–59.

Keuls, M. (1952). The use of the 'studentized range' in connection with an analysis of variance. *Euphytica*, **1**, 112–122.

Khatri, C.G. (1959). On certain properties of power-series distributions. *Biometrika*, **46**, 486–490.

Khintchine, A. (1924). Über einen Satz der Wahrscheinlichkeitsrechnung. *Fundamenta Mathematica*, **6**, 9–20.

Khintchine, A. (1929). Über einen neuen Grenzwertsatz der Wahrscheinlichkeitsrechnung. *Mathematisches Annalen*, **101**, 745–752.

Khintchine, A. (1932). Mathematisches über die Erwartung vor einem oeffentlichen Schalter. *Recueil de la Société Mathématique de Moscou*, **39**, 73–84.

Khintchine, A. (1933). Zum Birkhoff Lösung des Ergodenproblems. *Mathematisches Annalen*, **107**, 485–488.

Khintchine, A. (1934). Korrelationstheorie der stationären stochastischen Prozessen. *Mathematisches Annalen*, **109**, 604–615.

Kiefer, J. and Wolfowitz, J. (1952). Stochastic estimation of the maximum of a regression function. *Annals of Mathematical Statistics*, **23**, 462–466.

Kiefer, J. and Wolfowitz, J. (1959). Optimum designs in regression problems. *Annals of Mathematical Statistics*, **30**, 271–294.

Kiefer, J. and Wolfowitz, J. (1960). The equivalence of two extremum problems. *Canadian Journal of Mathematics*, **12**, 363–366.

Kingman, J.F.C. (1964). The stochastic theory of regenerative events. *Zeitschrift für Wahrscheinlichkeitstheorisch Verwaltung Gebiet*, **2**, 180–224.

Kingman, J.F.C. (1982). The coalescent. *Stochastic Processes and their Applications*, **13**, 235–248.

Kingman, J.F.C. (1993). *Poisson processes*. Clarendon Press, New York.

Kishen, K. (1941). Symmetrical unequal block arrangements. *Sankhyā*, **5**, 329–344.

Kitagawa, T. (1956). Some contributions to the design of sample surveys. *Sankhyā*, **17**, 1–36.

Klotz, J. (1959). Null distribution of the Hodges bivariate sign test. *Annals of Mathematical Statistics*, **30**, 1029–1033.

Klotz, J. (1962). Nonparametric tests for scale. *Annals of Mathematical Statistics*, **33**, 498–512.

Koenker, R. and Bassett, G. (1978). Regression quantiles. *Econometrica*, **46**, 33–50.

Kolmogorov, A.N. (1928). Über die Summen durch den Zufall bestimmter Unabhangiger. *Grossen Mathematisches Annalen*, **99**, 309–319.

Kolmogorov, A.N. (1930). Sur la loi forte des grands nombres. *Comptes-Rendus de l'Académie des Sciences de Paris*, **191**, 910–912.

Kolmogorov, A.N. (1931). Über die analytischen Methoden in der Wahrscheinlichkeitsrechnung. *Mathematisches Annalen*, **104**, 415–458.

Kolmogorov, A.N. (1933). *Grundbegriffe der Wahrscheinlichkeitsrechnung*. Springer-Verlag, Berlin.

Kolmogorov, A.N. (1941). Interpolation and extrapolation of stationary time series. *Bulletin de l'Académie des Sciences de l'URSS, Série Mathématique*, **5**, 3–14.

Konijn, H.S. (1956). On the power of certain tests for independence in bivariate populations. *Annals of Mathematical Statistics*, **27**, 300–323.

Koopman, B.O. (1936). On distributions admitting a sufficient statistic. *Transactions of the American Mathematical Society*, **39**, 399–409.

Koul, H.L. and Saleh, A.K.Md.E. (1995). Autoregression quantiles and related rank-scores processes. *Annals of Statistics*, **23**, 670–689.

Kounias, E.G. (1968). Bounds on the probability of a union, with applications. *Annals of Mathematical Statistics*, **39**, 2154–2158.

Krige, D.G. (1951). *A statistical approach to some mine valuation and allied problems on the Witwatersrand*. Masters Thesis, University of Witwatersrand.

Kruskal, W.H. (1952). A nonparametric test for the several sample problem. *Annals of Mathematical Statistics*, **23**, 525–540.

Kruskal, W.H. and Wallis, W.A. (1952). Use of ranks in one-criterion analysis of variance. *Journal of the American Statistical Association*, **47**, 583–621 and errata, **48**, 907–911.

Krzanowski, W.J. (1975). Discrimination and classification using both binary and continuous variables. *Journal of the American Statistical Association*, **70**, 782–790.

Krzanowski, W.J. (1980). Mixtures of continuous and categorical variables in discriminant analysis. *Biometrics*, **36**, 493–499.

Krzanowski, W.J. (1982). Mixtures of continuous and categorical variables in discriminant analysis: A hypothesis-testing approach. *Biometrics*, **38**, 991–1002.

Kshirsagar, A.M. (1961). The non-central multivariate beta distribution. *Annals of Mathematical Statistics*, **32**, 104–111.

Kshirsagar, A.M. and Gupta, R.P. (1965). The goodness of fit of two (or more) hypothetical principal components. *Annals of the Institute of Statistical Mathematics, Tokyo*, **17**, 347–356.

Kuder, G.F. and Richardson, M.W. (1937). The theory of the estimation of test reliability. *Psychometrika*, **2**, 151–160.

Kullback, S. and Leibler, R.A. (1951). On information and sufficiency. *Annals of Mathematical Statistics*, **22**, 79–86.

Lachenbruch, P.A. (1966). Discriminant analysis when the initial samples are misclassified. *Technometrics*, **8**, 657–662.

Laha, R.G. and Rohatgi, V.K. (1979). *Probability theory*. Wiley, New York.

Lancaster, H.O. (1969). *The chi-squared distribution*. Wiley, New York.

Lance, G.N. and Williams, W.T. (1967). Mixed data classificatory programs. I Agglomerative systems. *Australian Computing Journal*, **1**, 82–85.

Land, A.H. and Doig, A.G. (1960). An automatic method for solving discrete programming problems. *Econometrica*, **28**, 497–520.

Laplace, P.S. (1812). *Théorie analytique des probabilités.* Courcier, Paris.

Larson, R.G. (1977). Construction of Knut Vik designs. *Journal of Statistical Planning and Inference,* **1**, 289–297.

Laspeyres, E. (1871). Die Berechnung einer mittleren Waarenpreissteigerung. *Jahrbücher fur Nationalökonomie und Statistik,* **16**, 296–314.

Lauritzen, S.L. (1974). Sufficiency, prediction and extreme models. *Scandinavian Journal of Statistics, Theory and Applications,* **1**, 128–134.

Lauritzen, S.L. (1996). *Graphical models.* Clarendon Press, Oxford.

Lauritzen, S.L., Thiesson, B., and Spiegelhalter, D.J. (1994). Diagnostic systems created by model selection methods. A case study. In *Selecting models from data: AI and statistics IV* (ed. P. Cheeseman and R.W. Oldford), Volume 89, pp. 143–152. Springer-Verlag, New York.

Lawley, D.N. (1939). A correction to 'A generalization of Fisher's z test'. *Biometrika,* **30**, 467–469.

Lazarsfeld, P.F. (1950). The logical and mathematical foundation of latent structure analysis. *Studies in Social Psychology in World War II, Measurement and Prediction,* **4**, 362–412.

Legendre, A.M. (1805). *Nouvelles méthodes pour la détermination des orbites des comètes.* Courcier, Paris.

Lehmann, E.L. (1951). Consistency and unbiasedness of certain nonparametric tests. *Annals of Mathematical Statistics,* **22**, 165–179.

Lehmann, E.L. (1953). The power of rank tests. *Annals of Mathematical Statistics,* **24**, 23–43.

Lehmann, E.L. (1966). Some concepts of dependence. *Annals of Mathematical Statistics,* **37**, 1137–1153.

Lehmann, E.L. (1986). *Testing statistical hypotheses* (2nd edn). Series in Probability and Mathematical Statistics. Wiley, New York.

Lehmann, E.L. and Scheffé, H. (1950). Completeness, similar regions, and unbiased estimation. *Sankhyā,* **10**, 305–340.

Leipnik, R.B. (1947). Distribution of the serial correlation coefficient in a circularly correlated universe. *Annals of Mathematical Statistics,* **18**, 80–87.

Leslie, P.H. (1945). On the use of matrices in certain population mathematics. *Biometrika,* **33**, 183–212.

Leslie, P.H. (1958). A stochastic model for studying the properties of certain biological systems by numerical methods. *Biometrika,* **45**, 16–31.

Lévy, P. (1925). *Calcul des probabilités.* Gauthier-Villars, Paris.

Lévy, P. (1939). L'addition des variables aléatoires définies sur une circonférence. *Bulletin de la Société Mathématique de France,* **67**, 1–41.

Lévy, P. (1954). Processus semi-markoviens. *Proceedings of the International Congress on Mathematics, Amsterdam,* **3**, 416–426.

Lewis, P.A.W. (1964). A branching Poisson process model for the analysis of computer failure patterns. *Journal of the Royal Statistical Society. Series B*, **26**, 398–456.

Lexis, W. (1875). *Einleitung in die Theorie der Bevölkerungsstatistik.* Karl J. Trübner, Strassburg.

Lexis, W. (1877). *Zur Theorie der Massenerscheinungen in der menschlichen Gesellschaft.* Wagner, Freiburg im Br.

Lexis, W. (1879). Über die Theorie der Stabilität statisticher Reihen. *Jahrbücher für Nationalökonomie und Statistik*, **32**, 60–98.

Liapunov, A.M. (1901). Nouvelle forme du théorème sur la limite des probabilités. *Mémoires de l'Académie Impérial des Sciences de St. Petersburg*, **12**, 1–24.

Lieberman, G.J. (1961). Prediction regions for several predictions from a single regression line. *Technometrics*, **3**, 21–27.

Lieberman, G.J. and Solomon, H. (1955). Multi-level continuous sampling plans. *Annals of Mathematical Statistics*, **26**, 686–704.

Likert, R. (1932). *Some Applications of Behavioural Research.* Paris.

Lilliefors, H.W. (1967). On the Kolmogorov–Smirnov test for normality with mean and variance unknown. *Journal of the American Statistical Association*, **64**, 399–402.

Lin, C.-C. and Mudholkar, G. (1980). A simple test for normality against asymmetric alternatives. *Biometrika*, **67**, 455–461.

Lincoln, F.C. (1930). Calculating waterfowl abundance on the basis of banding returns. *US Department of Agriculture Circular*, **118**, 1–4.

Lindley, D.V. (1957). A statistical paradox. *Biometrika*, **44**, 187–192.

Liu, R. (1990). On a notion of data depth based on random simplices. *Annals of Statistics*, **18**, 405–414.

Lombard, P.B. and Brunk, H.D. (1963). Evaluating the relation of juice composition of mandarin oranges to present unacceptance of a taste panel. *Food Technology*, **17**, 113–115.

Lorenz, M.O. (1905). Methods for measuring concentration of wealth. *Publications of the American Statistical Association*, **9**, 209–219.

Lotka, A.J. (1920). Undamped oscillations derived from the law of mass action. *Journal of the American Chemical Society*, **42**, 1595–1599.

Lowe, J. (1823). *The present state of England in regard to agriculture, trade, and finance* (2nd edn). Longman, Hurst, Rees, Orme and Brown, London.

Lundberg, O. (1940). Ph.D. thesis, University of Stockholm, Uppsala. VII.

Macaulay, F.R. (1931). *The smoothing of time series.* National Bureau of Economic Research, New York.

MacDonald, I. and Zucchini, W. (1997). *Hidden-Markov and other models for discrete-valued time series.* Chapman & Hall, New York.

Madow, W.G. (1945). Note on the distribution of the serial correlation coefficient. *Annals of Mathematical Statistics*, **16**, 308–310.

Mahalanobis, P.C. (1930). On tests and measures of group divergence. Part 1. Theoretical formulae. *Journal of the Asiatic Society of Bengal*, **26**, 541–588.

Mahalanobis, P.C. (1944). On large-scale sample surveys. *Philosophical Transactions of the Royal Society of London. Series B, Biological Sciences*, **231**, 329–451.

Mahalanobis, P.C. (1961). A method of fractile graphical analysis. *Sankhyā*, **23**, 41–64.

Makeham, W.M. (1860). On the law of mortality and the construction of annuity tables. *Journal of the Institute of Actuaries*, **8**, 301–310.

Mandelbrot, B.B. (1960). The Pareto-Lévy law and the distribution of income. *International Economic Review*, **1**, 79–106.

Mandelbrot, B.B. (1975). *Les objets fractals: forme, hasard et dimension.* Flammarion, Paris.

Mann, H.B. (1945). Nonparametric tests against trend. *Econometrica*, **13**, 245–259.

Mann, H.B. and Whitney, D.R. (1947). On a test whether one of two random variables is stochastically larger than the other. *Annals of Mathematical Statistics*, **18**, 50–60.

Mantel, N. (1966). Evaluation of survival data and two new rank order statistics arising in its consideration. *Cancer Chemotherapy Reports*, **50**, 163–170.

Mantel, N. (1967). Ranking procedures for arbitrarily restricted observations. *Biometrics*, **23**, 65–78.

Mardia, K.V. (1962). Multivariate Pareto distributions. *Annals of Mathematical Statistics*, **33**, 1008–1015.

Mardia, K.V. (1972). A multi-sample uniform scores test on a circle and its parametric competitor. *Journal of the Royal Statistical Society. Series B*, **34**, 102–113.

Mardia, K.V., Kent, J.T., and Bibby, J.N. (1979). *Multivariate analysis.* Academic Press, London.

Maritz, J.S. (1966). Smooth empirical Bayes estimation for one-parameter discrete distributions. *Biometrika*, **53**, 417–429.

Markov, A.A. (1913). An example of statistical investigation in the text of 'Eugene Onegin' illustrating coupling of 'tests' in chains. *Proceedings of the Academy of Sciences of St. Petersburg*, **7**, 153–162.

Marshall, A. (1887). Remedies for fluctuations of general prices. *Contemporary Review*, **51**, 355–375.

Marshall, A.W. and Olkin, I. (1967a). A generalized bivariate exponential distribution. *Journal of Applied Probability*, **4**, 291–302.

Marshall, A.W. and Olkin, I. (1967b). A multivariate exponential distribution. *Journal of the American Statistical Association*, **62**, 30–44.

Mather, K. (1949). *Biometrical genetics.* Chapman & Hall, London.

Matheron, G. (1962). *Traité de géostatistique appliquée.* Technip, Paris.

McAlister, D. (1879). The law of the geometric distribution. *Proceedings of the Royal Society of London. Series A*, **52**, 558–577.

McCall, W.A. (1922). *How to measure in education.* Macmillan, New York.

McCullagh, P. (2000). Invariance and factor models (with discussion). *Journal of the Royal Statistical Society. Series A*, **62**, 209–256.

McCullagh, P. and Nelder, J.A. (1989). *Generalized linear models* (2nd edn). Chapman & Hall, London.

McKay, A.T. (1932). A Bessel function distribution. *Biometrika*, **24**, 39–44.

McKay, A.T. (1934). Sampling for batches. *Journal of the Royal Statistical Society. Series B*, **1**, 207–216.

Meijering, J.L. (1953). Interface area, edge length, and number of vertices in crystal aggregates with random nucleation. *Philips Research Reports*, **8**, 270–290.

Merrington, M. and Pearson, E.S. (1958). An approximation to the distribution of non-central *t*. *Biometrika*, **45**, 484–491.

Metropolis, N. and Ulam, S. (1949). The Monte Carlo method. *Journal of the American Statistical Association*, **44**, 335–341.

Mills, J.F. (1926). Table of the ratio: Area to bounding ordinate, for any portion of normal curve. *Biometrics*, **18**, 395–400.

Mitchell, R.B. (1974). *From actuarius to actuary.* Society of Actuaries, Chicago.

Mood, A.M. (1950). *Introduction to the theory of statistics.* McGraw-Hill, New York.

Mood, A.M. (1954). On the asymptotic efficiency of certain nonparametric two-sample tests. *Annals of Mathematical Statistics*, **25**, 514–522.

Moore, G.H. (1950). *Statistical indicators of cyclical revivals and recessions.* National Bureau of Economic Research, New York.

Moran, P.A.P. (1951). Estimation methods for evolutive processes. *Journal of the Royal Statistical Society. Series B*, **13**, 141–146.

Morgan, J.N. and Sonquist, J.A. (1963). Problems in the analysis of survey data, and a proposal. *Journal of the American Statistical Association*, **58**, 415–434.

Morgan, W.A. (1939). A test for the significance of the difference between the two variances in a sample from a normal bivariate population. *Biometrika*, **31**, 13–19.

Morgenstern, D. (1956). Einfache Beispiele zweidimensionaler Verteilungen.

Mitteilungen-Blätter Mathematik und Statistik, **8**, 234–235.

Morse, P.M. (1958). *Queues, inventories and maintenance: The analysis of operational systems with variable demand and supply.* Wiley, New York.

Mortara, G. (1949). *Methods of using census statistics for the calculation of life tables and other demographic measures.* United Nations, New York.

Moses, L.E. (1952). A two-sample test. *Psychometrika*, **17**, 239–247.

Moses, L.E. (1963). Rank tests of dispersion. *Annals of Mathematical Statistics*, **34**, 973–983.

Mosimann, J.E. (1963). On the compound negative multinomial distribution and correlations among inversely sampled pollen counts. *Biometrika*, **50**, 47–54.

Mosteller, F. (1946). On some useful 'inefficient' statistics. *Annals of Mathematical Statistics*, **17**, 377–408.

Mosteller, F. (1948). k-sample slippage test for an extreme population. *Annals of Mathematical Statistics*, **19**, 58–65.

Mosteller, F. and Tukey, J.W. (1949). The uses and usefulness of binomial probability paper. *Journal of the American Statistical Association*, **44**, 174–212.

Mosteller, F. and Tukey, J.W. (1950). Significance levels for a k-sample slippage test. *Annals of Mathematical Statistics*, **21**, 120–123.

Murthy, M.N. (1957). Ordered and unordered estimators in sampling without replacement. *Sankhyā*, **18**, 379–390.

Myers, M.H., Schneiderman, M.A., and Armitage, P. (1966). Boundaries for closed (wedge) sequential t-test plans. *Biometrika*, **53**, 431–437.

Nair, R.C. (1966). On partially linked block designs. *Annals of Mathematical Statistics*, **37**, 1401–1406.

Nelder, J.A. (1966). Inverse polynomials. *Biometrics*, **22**, 128–141.

Nelder, J.A. and Wedderburn, R.W.M. (1972). Generalized linear models. *Journal of the Royal Statistical Society. Series A*, **135**, 370–384.

Nelson, L.S. (1963). Tables for a precedence life test. *Technometrics*, **5**, 491–499.

Newcomb, S. (1859–1861). Notes on the theory of probabilities. *Mathematics Monthly*, **1**: 136–139, 233–235, 331–355; **2**: 134–140, 272–275; **3**: 68, 119–125, 341–349.

Newman, D. (1939). The distribution of range in samples from a normal population, expressed in terms of an independent estimate of standard deviation. *Biometrika*, **31**, 20–30.

Neyman, J. (1934). On the two different aspects of the representative method: The method of stratified sampling and the method of purposive selection. *Journal of the Royal Statistical Society. Series A*, **97**, 558–625.

Neyman, J. (1935). Su un teorema concernente le cosiddette statistiche sufficienti. *Giornale d'Istituto di Italiano Attuari*, **6**, 320–334.

Neyman, J. (1937*a*). Outline of a theory of statistical estimation based on the classical theory of probability. *Philosophical Transactions of the Royal Society of London. Series A*, **236**, 333–380.

Neyman, J. (1937*b*). Smooth test for goodness of fit. *Skandinavien Aktuarie Tidskrift*, **20**, 149–199.

Neyman, J. (1939). On a new class of 'contagious' distributions, applicable in entomology and bacteriology. *Annals of Mathematical Statistics*, **10**, 35–57.

Neyman, J. (1949*a*). Contribution to the theory of the χ^2 test. In *Proceedings of the Berkeley Symposium on Mathematical Statistics and Probability, August 1945 and January 1946*, pp. 239–273.

Neyman, J. (1949*b*). On the problem of estimating the number of school of fish. *University of California Publications in Statistics*, **1**, 21–36.

Neyman, J. (1959). Optimal asymptotic tests of composite statistical hypotheses. In *Probability and statistics, H. Cramér volume* (ed. U. Grenander), pp. 213–234. Wiley, New York.

Neyman, J. and Pearson, E.S. (1928). On the use and interpretation of certain test criteria for purposes of statistical inference. Parts I and II. *Biometrika*, **20A**, 175–240, 263–294.

Neyman, J. and Pearson, E.S. (1933). On the problem of the most efficient tests of statistical hypotheses. *Philosophical Transactions of the Royal Society of London. Series A*, **231**, 289–337.

Neyman, J. and Scott, E.L. (1948). Consistent estimates based on partially consistent observations. *Econometrica, Chicago*, **16**, 1–32.

Neyman, J. and Scott, E.L. (1958). Statistical approach to problems of cosmology. *Journal of the Royal Statistical Society. Series B*, **20**, 1–29.

Neyman, J. and Scott, E.L. (1971). Outlier proneness of phenomena and of related distributions. In *Optimizing methods in statistics* (ed. J.S. Rustagi), pp. 413–430. Academic Press, New York.

Noether, G.E. (1956). Two sequential tests against trend. *Journal of the American Statistical Association*, **51**, 440–450.

Norton, H.W. (1939). The 7×7 squares. *Annals of Eugenics*, **9**, 269–307.

Nyquist, H. (1928). Certain topics in telegraph transmission theory. *AIEE Transactions*, **47**, 617–644.

Nyquist, H. (1992). Sensitivity analysis in empirical studies. *Journal of Official Statistics*, **8**, 167–182.

Ogawa, J. (1963). On the null-distribution of the F-statistic in a randomized balanced incomplete block design under the Neyman model. *Annals of Mathematical Statistics*, **34**, 1558–1568.

Olkin, I. and Pratt, J.W. (1958). A multivariate Tchebycheff inequality. *Annals of Mathematical Statistics*, **29**, 226–234.

Olkin, I. and Tate, R. (1961). Multivariate correlation models with mixed discrete and continuous variables. *Annals of Mathematical Statistics*, **32**, 448–465.

Olmsted, P.S. and Tukey, J.W. (1947). A corner test for association. *Annals of Mathematical Statistics*, **18**, 495–513.

Ord, J.K. (1967). On a system of discrete distributions. *Biometrika*, **54**, 649–656.

Ornstein, L.S. and Uhlenbeck, G.E. (1930). On the theory of Brownian motion. *Physical Review*, **36**, 823–841.

Ott, J. (1999). *Analysis of human genetic linkage*. Johns Hopkins University Press, Baltimore, MD.

Owen, A.B. (1988). Empirical likelihood ratio confidence intervals for a single functional. *Biometrika*, **75**, 237–249.

Paasche, H. (1874). Über die Preisentwicklung der letzten Jahre nach den Hamburger Borsennotirungen. *Jahrbücher fur Nationalökonomie und Statistik*, **23**, 168–178.

Pairman, E. and Pearson, K. (1919). On corrections for the moment-coefficients of limited range frequency distributions when there are finite or infinite ordinates and any slopes at the terminals of the range. *Biometrika*, **12**, 231–258.

Palgrave, I. (1925). *Dictionary of political economy*. Macmillan, London.

Palm, C. (1943). *Intensitätsschwankungen im Fernsprechverkehr*, Volume 44. Ericsson Technics, Stockholm.

Pandy, K.N. (1965). On generalized inflated Poisson distribution. *Journal of Scientific Research of the Banares Hindu University*, **15**, 157–162.

Papadakis, J.S. (1937). Méthode statistique pour des expériences sur champs. *Thessaloniki Plant Breeding Institute Scientific Bulletin*, **23**, 1–30.

Pareto, V. (1897). *Cours d'économie politique*. Rouge, Lausanne.

Parzen, E. (1957). A central limit theorem for multilinear stochastic processes. *Annals of Mathematical Statistics*, **28**, 252–256.

Parzen, E. (1961). Mathematical considerations in the estimation of spectra. *Technometrics*, **3**, 167–190.

Parzen, E. (1967). *Time series analysis papers*. Holden-Day, San Francisco.

Parzen, E. (1969). Multiple time series modelling. In *Multivariate analysis II* (ed. P.R. Krishnaiah), pp. 389–409. Academic Press, New York.

Pathak, P.K. (1967). Asymptotic efficiency of Des Raj's strategy. I. *Sankhyā*, **29**, 283–298.

Patil, V.T. (1957). The consistency and adequacy of the Poisson-Markoff model for density fluctuations. *Biometrika*, **44**, 43–56.

Patnaik, P.B. (1949). The non-central χ^2 and F-distributions and their applications. *Biometrika*, **36**, 202–232.

Patterson, H.D. (1964). Theory of cyclic rotation experiments. *Journal of the Royal Statistical Society. Series B*, **26**, 1–45.

Patterson, H.D. (1969). Queries and notes. *Biometrics*, **25**, 159–164.

Patterson, H.D. and Thompson, R. (1974). Maximum likelihood estimation of components of variance. In *Proceedings of the 8th International Biometric Conference*, pp. 197–207.

Paulson, E. (1942). An approximate normalization of the analysis of variance distribution. *Annals of Mathematical Statistics*, **13**, 233–235.

Pearce, S.C. (1960). A method of studying manner of growth. *Biometrics*, **16**, 1–6.

Pearl, J. (1998). Graphical models of probabilistic and causal reasoning. In *Handbook of defeasible reasoning and uncertainty management systems* (ed. D.M. Gabbay), Volume 1, pp. 367–389. Kluwer Academic Publishers, Dordrecht.

Pearson, E.S. and Stephens, M.A. (1962). The goodness-of-fit tests based on W^2 and U^2. *Biometrika*, **49**, 397–402.

Pearson, K. (1894*a*). Contributions to the mathematical theory of evolution. *Philosophical Transactions of the Royal Society of London. Series A*, **185**, 71–110.

Pearson, K. (1894*b*). On the dissection of asymmetrical frequency curves. *Philosophical Transactions of the Royal Society of London. Series A*, **185**, 719–810.

Pearson, K. (1895). Contributions to the mathematical theory of evolution. II. Skew variation in homogeneous material. *Philosophical Transactions of the Royal Society of London. Series A*, **186**, 343–414.

Pearson, K. (1897). Mathematical contribution to the theory of evolution. On a form of spurious correlation which may arise when indices are used in the measurements of organs. *Proceedings of the Royal Society of London*, **60**, 489–502.

Pearson, K. (1900). On the criterion that a given system of deviations from the probable in the case of a correlated system of variables is such that it can reasonably be supposed to have arisen from random sampling. *Philosophical Magazine, Series 5*, **50**, 157.

Pearson, K. (1905). 'Das Fehlergesetz und seine Verallgemeinerungen durch Fechner und Pearson.' A rejoinder. *Biometrika*, **4**, 169–212.

Pearson, K. (1909). On a new method for determining correlation between a measured character A and a character B of which only the percentage of cases wherein B exceeds (of falls short of) a given intensity is recorded for each grade of A. *Biometrika*, **7**, 96–106.

Pearson, K. (1913). On the probable error of a correlation coefficient as found from a fourfold table. *Biometrika*, **9**, 22.

Pearson, K. (1921). Note on the fundamental problem of practical statistics. *Biometrika*, **13**, 300–301.

Pearson, K. (1923). Notes on skew frequency surfaces. *Biometrika*, **15**, 222–230.

Pearson, K. (1924a). On a certain double hyper-geometrical series and its representation by continuous frequency surfaces. *Biometrika*, **16**, 172–188.

Pearson, K. (1924b). On the moments of the hypergeometrical series. *Biometrika*, **16**, 157–162.

Peek, R.L. (1933). Some new theorems on limits of variation. *Bulletin of the American Mathematical Society*, **39**, 953–959.

Percival, D.B. and Walden, A.T. (2000). *Wavelet methods for time series analysis*. Cambridge University Press, Cambridge.

Perks, W.F. (1932). On some experiments in the graduation of mortality statistics. *Journal of the Institute of Actuaries, London*, **58**, 12–57.

Perry, J.N. and Taylor, L.R. (1985). Adés: New ecological families of species-specific frequency distributions that describe repeated spatial samples with an intrinsic power-law variance-mean property. *Journal of Animal Ecology*, **54**, 931–995.

Peters, C.A.F. (1856). Über die Bestimmung des wahrscheinlichen Fehlers einer Beobachtung aus den Abweichungen der Beobachtungen von ihrem arithmetischen Mittel. *Astronomische Nachrichten*, **44**, 30–31.

Peto, R., Pike, M.C., Armitage, P., Breslow, N.E., Cox, D.R., Howard, S.V., Mantel, N., McPherson, K., Peto, J., and Smith, P.G. (1977). Design and analysis of clinical trials requiring prolonged observation of each patient. II Analysis and examples. *British Journal of Cancer*, **35**, 1–39.

Pike, M.C. and Smith, P.G. (1968). Disease clustering: A generalisation of Knox's approach to the detection of space-time interaction. *Biometrics*, **24**, 541–556.

Pitman, E.J.G. (1936). Sufficient statistics and intrinsic accuracy. *Proceedings of the Cambridge Philosophical Society*, **32**, 567–579.

Pitman, E.J.G. (1937a). The 'closest' estimates of statistical parameters. *Proceedings of the Cambridge Philosophical Society*, **33**, 212–222.

Pitman, E.J.G. (1937b). Significance tests which may be applied to samples from any populations. *Supplement to the Journal of the Royal Statistical Society*, **4**, 119–130.

Pitman, E.J.G. (1938). Significance tests which may be applied to samples from any populations. III. The analysis of variance test. *Biometrika*, **29**, 322–335.

Pitman, E.J.G. (1939a). The estimation of the location and scale parameters of a continuous population of any given form. *Biometrika*, **30**, 391–421.

Pitman, E.J.G. (1939b). Tests of hypotheses concerning location and scale parameters. *Biometrika*, **31**, 200–215.

Pitman, E.J.G. (1948). *Non-parametric statistical inference*. University of North Carolina Institute of Statistics, (Mimeographed Lecture Notes).

Pitman, E.J.G. (1949). *Lecture notes on nonparametric statistical inference*. Columbia University.

Plackett, R.L. (1960). Models in the analysis of variance. *Journal of the Royal Statistical Society. Series B*, **22**, 195–209, Discussion 209–217.

Plackett, R. L. (1965). A class of bivariate distributions. *Journal of the American Statistical Association*, **60**, 516–522.

Poisson, S.D. (1835). *Law of large numbers*. Paris.

Poisson, S.D. (1837). *Recherches sur la probabilité des jugements en matière criminelle et en matière civile, Procédés des règles générales du calcul des probabilités*. Bachelier, Imprimeur-Libraire pour les Mathématiques, Paris.

Politz, A.N. and Simmons, W.R. (1949). An attempt to get 'not at homes' into the sample without callbacks. *Journal of the American Statistical Association*, **44**, 9–31.

Politz, A.N. and Simmons, W.R. (1950). An attempt to get 'not at homes' into the sample without callbacks. *Journal of the American Statistical Association*, **45**, 36–137.

Pollaczek, F. (1930). Über eine Aufgabe der Wahrscheinlichkeitstheorie. I-II. *Mathematische Zeitschrift*, **32**, 64–100.

Pollaczek, F. (1952). Fonctions caractéristiques de certaines répartitions définies au moyen de la notion d'ordre. Application à la théorie des attentes. *Comptes-Rendus de l'Académie des Sciences de Paris*, **234**, 2334–2336.

Pólya, G. (1930). Sur quelques points de la théorie des probabilités. *Annales de l'Institut Henri Poincaré*, **1**, 117–161.

Pólya, G. (1949). Remarks on characteristic functions. In *Proceedings of the Berkeley Symposium on Mathematical Statistics and Probability, August 1945 and January 1946*, pp. 115–123.

Pólya, G. and Eggenberger, F. (1923). Über die Statistik verketteter Vorgänge. *Zeitschrift für Angewandte Mathematische Mechanik*, **3**, 279–289.

Popper, K. (1959). The propensity interpretation of probability. *British Journal of the Philosophy of Science*, **10**, 25–42.

Potthoff, R.F. (1963). Use of the Wilcoxon statistic for a generalized Behrens-Fisher problem. *Annals of Mathematical Statistics*, **34**, 1596–1599.

Powell, E.O. (1955). Some features of the generation times of individual bacteria. *Biometrika*, **42**, 16–44.

Prentice, R.L. (1986). A case-cohort design for epidemiologic cohort studies and disease prevention trials. *Biometrika*, **73**, 1–11.

Press, S.J. (1969). On serial correlation. *Annals of Mathematical Statistics*, **40**, 188–196.

Priestley, M.B. (1962). Basic considerations in the estimation of spectra. *Technometrics*, **4**, 551–571.

Priestley, M.B. (1965). Evolutionary spectra and non-stationary processes. *Journal of the Royal Statistical Society. Series B*, **27**, 204–237.

Pukelsheim, F. (1993). *Optimal design of experiments*. Wiley, New York.

Pyke, R. (1961). Markov renewal processes: Definitions and preliminary properties. *Annals of Mathematical Statistics*, **32**, 1231–1242.

Quenouille, M.H. (1947). A large-sample test for the goodness of fit of autoregressive schemes. *Journal of the Royal Statistical Society. Series A*, **110**, 123–129.

Quenouille, M.H. (1949a). Approximate tests of correlation in time series. *Journal of the Royal Statistical Society. Series B*, **11**, 68–84.

Quenouille, M.H. (1949b). Problems in plane sampling. *Annals of Mathematical Statistics*, **20**, 355–375.

Quenouille, M.H. (1952). *Associated measurements*. Academic Press, New York.

Quenouille, M.H. (1958). The comparison of correlations in time-series. *Journal of the Royal Statistical Society. Series B*, **20**, 158–164.

Quetelet, A. (1832). *Sur la possibilité de mesurer l'influence des causes que modifient les éléments sociaux, Lettre à M. Willermé de l'Institut de France*. Bruxelles.

Rabe-Hesketh, S., Pickles, A., and Skrondal, A. (2001). *GLLAMM manual*. Department of Biostatistics and Computing, Institute of Psychiatry, King's College, University of London, London.

Raghavarao, D. (1964). Singular weighing designs. *Annals of Mathematical Statistics*, **35**, 673–680.

Raikov, D. (1938). On the decomposition of Gauss and Poisson laws. *Bulletin de l'Académie des Sciences de l'URSS, Série Mathématique*, **1**, 91–124.

Raktoe, B.L., Hedayat, A., and Federer, W.T. (1981). *Factorial designs*. Wiley, New York.

Rao, C.R. (1945). Information and the accuracy attainable in the estimation of statistical parameters. *Bulletin of the Calcutta Mathematical Society*, **37**, 81–91.

Rao, C.R. (1946a). Confounded factorial designs in quasi-Latin squares. *Sankhyā*, **7**, 295–304.

Rao, C.R. (1946b). Hypercubes of strength d leading to confounded designs in factorial experiments. *Bulletin of the Calcutta Mathematical Society*, **38**, 67–78.

Rao, C.R. (1948). Large sample tests of statistical hypotheses concerning several parameters with applications to problems of estimation. *Proceedings of the Cambridge Philosophical Society*, **44**, 50–57.

Rao, C.R. (1962). A note on generalized inverse of a matrix with applications to problems in mathematical statistics. *Journal of the Royal Statistical Society. Series B*, **24**, 152–158.

Rao, C.R. (1965). *Linear statistical inference and its applications*. Wiley, New York, London and Sydney.

Rao, C.R. (1971). Estimation of variance and covariance components. MINQUE Theory. *Journal of Multivariate Analysis*, **1**, 257–275.

Rao, C.R. (1973). *Linear statistical inferences and its applications.* Wiley, New York.

Rao, C.R. and Mitra, S.K. (1967). *Generalized inverse of matrices and its applications.* Wiley, New York.

Rao, P.V. and Kupper, L.L. (1967). Ties in paired-comparison experiments: A generalization of the Bradley–Terry model. *Journal of the American Statistical Association*, **62**, 194–204.

Ray, D. (1958). Stable processes with an absorbing barrier. *Transactions of the American Mathematical Society*, **89**, 16–24.

Rayleigh, J.W.S. (1919). On the problem of random vibrations, and of random flights in one, two and three dimensions. *Philosophical Magazine*, **37**, 321–347.

Reed, L.J. and Münch, H.A. (1938). Simple method of determining fifty percent endpoints. *American Journal of Hygiene*, **27**, 494–497.

Reich, E. (1957). Waiting times when queues are in tandem. *Annals of Mathematical Statistics*, **28**, 768–773.

Reuter, G.E.H. (1961). Competition process. *Proceedings of the 4th Berkeley Symposium on Mathematical Statistics and Probability*, **2**, 421–430.

Rhodes, E.C. (1923). On a certain skew correlation surface. *Biometrika*, **14**, 355–377.

Rice, S. (1931). *Methods in social science: A case book.* University of Chicago Press, Chicago.

Ripley, B.D. (1976). The second-order analysis of stationary point processes. *Journal of Applied Probability*, **13**, 255–266.

Ripley, B.D. (1996). *Pattern recognition and neural networks.* Cambridge University Press, Cambridge.

Robbins, H. (1951). Asymptotically subminimax solutions of compound statistical decision problems. In *Proceedings of the Berkeley Symposium on Mathematical Statistics and Probability, California, July 31 – August 12, 1950*, pp. 131–148.

Robbins, H. (1956). An empirical Bayes approach to statistics. *Proceedings of the 3rd Berkeley Symposium on Mathematical Statistics and Probability*, **1**, 157–163.

Robbins, H. (1964). The empirical Bayes approach to statistical decision problems. *Annals of Mathematical Statistics*, **35**, 1–20.

Robbins, H. and Munro, S. (1951). A stochastic approximation method. *Annals of Mathematical Statistics*, **22**, 400–407.

Robins, J.M. (1998). Correction for non-compliance in equivalence trials. *Statistics in Medicine*, **17**, 269–302.

Rojar, B. and White, R.F. (1957). The modified Latin square. *Journal of the Royal Statistical Society. Series B*, **19**, 305–317.

Romanowski, M. (1964). On the normal law of errors. *Bulletin of Geodesy*, **73**, 195–216.

Room, T.G. (1955). A new type of magic square. *Mathematical Gazette*, **39**, 307.

Roos, C.F. (1925). A mathematical theory of competition. *American Journal of Mathematics*, **47**, 163–175.

Rosenbaum, S. (1953). Tables for a nonparametric test of location. *Annals of Mathematical Statistics*, **24**, 663–668.

Rosenbaum, S. (1954). Tables for a nonparametric test. *Annals of Mathematical Statistics*, **25**, 146–150.

Roy, A.D. (1960). A note on prediction from an autoregressive process using pistimetric probability. *Journal of the Royal Statistical Society. Series B*, **22**, 97–103.

Roy, J. (1957). A note on estimation of variant components in multistage sampling with varying probabilities. *Sankhyā*, **17**, 367–372.

Roy, P.M. (1953). A note on the unreduced balanced incomplete block designs. *Sankhyā*, **13**, 11–16.

Roy, S.N. (1939). p-statistics, or some generalizations in analysis of variance appropriate to multivariate problems. *Sankhyā*, **4**, 381–396.

Ruben, H. (1963). The estimation of a fundamental interaction parameter in an emigration-immigration process. *Annals of Mathematical Statistics*, **34**, 238–259.

Ruben, H. (1966). On the simultaneous stabilization of variances and covariances. *Annals of the Institute of Statistical Mathematics, Tokyo*, **18**, 203–210.

Rubin, D.B. (1976). Inference and missing data. *Biometrika*, **63**, 581–592.

Rubin, D.B. (1987). *Multiple imputation for nonresponse in surveys*. Wiley, New York.

Rutherford, R.S.G. (1954). On a contagious distribution. *Annals of Mathematical Statistics*, **25**, 703–713.

Sacks, J. (1958). Asymptotic distribution of stochastic approximation procedures. *Annals of Mathematical Statistics*, **29**, 373–405.

Satterthwaite, F.E. (1941). Synthesis of variance. *Psychometrika*, **6**, 309–316.

Satterthwaite, F.E. (1946). An approximative distribution of estimates of variance components. *Biometrics*, **2**, 110–114.

Satterthwaite, F.E. (1959). Random balance experimentation (with discussion). *Technometrics*, **1**, 111–137.

Saunders, S. (1960). Sequential tolerance regions. *Annals of Mathematical Statistics*, **31**, 198–216.

Savage, I.R. (1961). Probability inequalities of the Tchebyscheff type. *Journal of Research of the National Bureau of Standards*, **65B**, 211–222.

Schach, S. (1969). On a class of nonparametric two-sample tests for circular distributions. *Annals of Mathematical Statistics*, **40**, 1791–1800.

Scheffé, H. (1959). *The analysis of variance*. Wiley, New York.

Scheffé, H. (1963). The simplex-centroid design for experiments with mixtures. *Journal of the Royal Statistical Society. Series B*, **25**, 235–250, Discussion 251–263.

Schneiderman, M.A. and Armitage, P. (1962). A family of closed sequential procedures. *Biometrika*, **49**, 41–56.

Schuster, A. (1898). On the investigation of hidden periodicities with application to a supposed 26 day period of meteorological phenomena. *Terrestrial Magnetism and Atmospheric Electricity*, **3**, 13–41.

Schwarz, H.A. (1885). Über ein die Flächen kleinsten Flächeninhalts betreffendes Problem der Variationsrechnung. *Acta Societatis Scientiarum Fennicae*, **15**, 315–362.

Scott, D.W. (1992). *Multivariate density estimation: Theory, practice, and visualization*. Wiley, New York.

Seber, G.A.F. (1982). *The estimation of animal abundance*. Arnold, London.

Seidel, P.L. (1874). Über ein Verfahren, die Gleichungen, auf welche die Methode der kleinsten Quadrate führt, sowie lineare Gleichungen überhaupt, durch successive Annäherung aufzulösen. *Abhandlungen der Bayer Akademie Wissenschaften*, **11**, 81–108.

Sen, A.R. (1953). On the estimate of the variance in sampling with varying probabilities. *Journal of the Indian Society of Agricultural Statistics*, **5**, 119–127.

Sen, P.K. (1964). Tests for the validity of the fundamental assumption in dilution-direct assays. *Biometrics*, **20**, 770–784.

Shaban, S.A. (1980). Change point problem and two-phase regression: An annotated bibliography. *International Statistical Review*, **48**, 83–93.

Shah, B.V. (1960a). Balanced factorial experiments. *Annals of Mathematical Statistics*, **31**, 502–514.

Shah, B.V. (1960b). A matrix substitution method of constructing partially balanced designs. *Annals of Mathematical Statistics*, **31**, 34–42.

Shah, B.K. and Dave, P.H. (1963). A note on log-logistic distribution. *Journal of Maharaja Sayajirao University of Baroda*, **12**, 15–20.

Shannon, C.E. (1949). Communication in presence of noise. *Proceedings of the IRE*, **37**, 10–21.

Shannon, C.E. and Weaver, W. (1949). *The mathematical theory of communication*. University of Illinois Press, Urbana, IL.

Shapiro, S.S. and Wilk, M.B. (1965). An analysis of variance test for normality (complete samples). *Biometrika*, **52**, 591–611.

Shephard, N. (1996). Statistical aspects of ARCH and stochastic volatility. In *Time series models in econometrics, finance and other fields* (ed. D.R. Cox, D.V. Hinkley and O.E. Barndorff-Nielsen), pp. 1–67. Chapman & Hall, London.

Sheppard, W.F. (1898). On the calculation of the most probable values of frequency constants for data arranged according to equidistant divisions of a scale. *Proceedings of the London Mathematical Society*, **29**, 353–380.

Sherman, B. (1950). A random variable related to the spacing of sample values. *Annals of Mathematical Statistics*, **21**, 339–361.

Sherman, R.E. (1965). Design and evaluation of a repetitive group sampling plan. *Technometrics*, **7**, 11–21.

Shewhart, W.A. (1931). *Economic control of quality of manufactured product.* D. Van Nostrand, New York.

Sibson, R. (1980). The Dirichlet tessellation as an aid in data analysis. *Scandinavian Journal of Statistics*, **7**, 14–20.

Siegel, S. and Tukey, J.W. (1960a). A nonparametric sum of ranks procedure for relative spread in unpaired samples. *Journal of the American Statistical Association*, **55**, 429–445.

Siegel, S. and Tukey, J.W. (1960b). A nonparametric sum of ranks procedure for relative spread in unpaired samples. Corrigenda. *Journal of the American Statistical Association*, **56**, 1005.

Silvey, S.D. (1959). The Lagrangian multiplier test. *Annals of Mathematical Statistics*, **30**, 389–407.

Simon, H.A. (1955). On a class of skew distribution functions. *Biometrika*, **42**, 425–440.

Simon, H.A. (1956). Dynamic programming under uncertainty with a quadratic criterion function. *Econometrica*, **24**, 74–81.

Simonoff, J.S. (1996). *Smoothing methods in statistics.* Springer-Verlag, New York.

Singh, S.N. (1963). Probability models for the variation in the number of births per couple. *Journal of the American Statistical Association*, **58**, 721–727.

Sirken, M.G. (1970). Household surveys with multiplicity. *Journal of the American Statistical Association*, **65**, 257–266.

Siromoney, G. (1964). The general Dirichlet's series distribution. *Journal of the Indian Statistical Association*, **2**, 69–74.

Skellam, J.G. and Shenton, L.R. (1957). Distributions associated with random walk and recurrent events. *Journal of the Royal Statistical Society. Series B*, **19**, 64–118.

Skitovich, V.P. (1954). Linear forms of independent random variables and the normal distribution law. *Izvestiya Akademii Nauk SSSR*, **18**, 185–200.

Slutsky, E.E. (1937). Problems of economic conditions, 3. *Econometrica*, **5**, 105–146.

Smirnov, N.V. (1936). Sur la distribution de ω^2 (Criterium de M. R. von Mises). *Comptes-Rendus de l'Académie des Sciences de Paris*, **202**, 449–452.

Smirnov, N.V. (1939). Estimate of deviation between empirical distribution functions in two independent samples. (in Russian). *Bulletin of Moscow University*, **2**, 3–16.

Smith, W.L. (1955a). Extensions of a renewal theorem. *Proceedings of the Cambridge Philosophical Society*, **51**, 629–638.

Smith, W.L. (1955b). Regenerative stochastic processes. *Proceedings of the Royal Society of London. Series A*, **232**, 6–31.

Smith, W.L. (1958). Renewal theory and its ramifications. *Journal of the Royal Statistical Society. Series B*, **20**, 243–302.

Snedecor, G.W. (1937). *Statistical methods applied to experiments in agriculture and biology*. Collegiate Press, Iowa.

Sobel, M. and Wald, A. (1949). A sequential decision procedure for choosing one of three hypotheses concerning the unknown mean of a normal distribution. *Annals of Mathematical Statistics*, **20**, 502–522.

Sokal, R.R. and Michener, C.D. (1958). A statistical method for evaluating systematic relationships. *University of Kansas Science Bulletin*, **38**, 1409–1438.

Spearman, C. (1904). The proof and measurement of association between two things. *American Journal of Psychology*, **15**, 72–101.

Spearman, C. (1906). Footrule for measuring correlation. *British Journal of Psychology*, **2**, 89–108.

Spearman, C. (1908). The method of right and wrong cases (constant stimuli) without Gauss's formulae. *British Journal of Psychology*, **2**, 227–242.

Spearman, C. (1910). Correlation calculated with faulty data. *British Journal of Psychology*, **3**, 271–295.

Spencer, J. (1904). On the graduation of the rates of sickness and mortality. *Journal of the Institute of Actuaries*, **38**, 334–347.

Spitzer, F. (1956a). A combinatorial lemma and its application to probability theory. *Transactions of the American Mathematical Society*, **82**, 323–339.

Spitzer, F. (1956b). On interval recurrent sums of independent random variables. *Proceedings of the American Mathematical Society*, **7**, 164–171.

Srivastava, R.C. (1964). *A model of convection with entrainment and precipitation*. Ph.D. thesis, McGill University.

Stacy, E.W. (1962). A generalization of the gamma distribution. *Annals of Mathematical Statistics*, **33**, 1187–1192.

Staudte, R.G. and Sheather, S.J. (1990). *Robust estimation and testing*. Wiley,

New York.

Steel, R.G.D. (1960). A rank sum test for comparing all pairs of treatments. *Technometrics*, **2**, 197–207.

Steffensen, J.F. (1930). On Sandsynligheden for at Afkommet uddør. *Mathematik Tidskrift*, **B1**, 19–23.

Stein, C. (1945). A two-sample test for a linear hypothesis whose power is independent of the variance. *Annals of Mathematical Statistics*, **16**, 243–258.

Steiner, J. (1853). Combinatorische Aufgabe. *Journal für die Reine und Angewandte Mathematik*, **45**, 181–182.

Stephan, F.F. (1942). An iterative method of adjusting sample frequency tables when expected marginal totals are known. *Annals of Mathematical Statistics*, **13**, 166–178.

Stevens, S.S. (1968). Measurement, statistics, and the schemapiric view. *Science*, **161**, 849–856.

Stevens, W.L. (1939). Solution to a geometrical problem in probability. *Annals of Eugenics*, **9**, 315–320.

Stigum, B.P. (1963). Dynamic stochastic processes. *Annals of Mathematical Statistics*, **34**, 274–283.

Striebel, C.T. (1961). Efficient estimation of a regression parameter for certain second order processes. *Annals of Mathematical Statistics*, **32**, 1299–1313.

Sturges, H. (1926). The choice of a class-interval. *Journal of the American Statistical Association*, **21**, 65–66.

Sukhatme, B.V. (1957). On certain two-sample nonparametric tests for variances. *Annals of Mathematical Statistics*, **28**, 188–194.

Taguchi, G. (1987). *System of experimental design: Engineering methods to optimize quality and minimize costs*, Volumes I and II. American Supplier Institute, Dearborn, IL.

Takacs, L. (1954). On secondary processes generated by a Poisson process and their applications in physics. *Acta Mathematica Academiae Scientiarum Hungaricae*, **5**, 203–236.

Takacs, L. (1955). On processes of happenings generated by means of a Poisson process. *Acta Mathematica Academiae Scientiarum Hungaricae*, **6**, 81–99.

Takacs, L. (1965). On the distribution of the supremum for stochastic processes with interchangeable increments. *Transactions of the American Mathematical Society*, **119**, 367–379.

Takahasi, K. (1965). Note on the multivariate Burr's distribution. *Annals of the Institute of Statistical Mathematics, Tokyo*, **17**, 257–260.

Tallis, G.M. (1962). The use of generalised multinomial distribution in the estimation of correlation in discrete data. *Journal of the Royal Statistical Society. Series B*, **24**, 530–534.

Tallis, G.M. (1963). Elliptical and radial truncation in normal populations. *Annals of Mathematical Statistics*, **34**, 940–944.

Tang, P.C. (1938). The power function of the analysis of variance tests with tables and illustrations of their use. *Statistical Research Memoirs, University of London*, **2**, 126–149.

Tanner, J.C. (1953). A problem of interference between two queues. *Biometrika*, **40**, 58–69.

Taylor, W.F. (1953). Distance functions and regular best asymptotically normal estimates. *Annals of Mathematical Statistics*, **24**, 85–92.

Tchebychev, P.L. (1867). On mean values. *Journal de Mathématiques Pures et Appliquées*, **12**, 177–184.

Tchebychev, P.L. (1874). Sur les valeurs limites des intégrales. *Journal de Mathématiques Pures et Appliquées*, **19**, 157–160.

Terpstra, T.J. (1952). The asymptotic normality and consistency of Kendall's test against trend, when ties are present in one ranking. *Indagationes Mathematicae*, **14**, 327–333.

Terry, M.E. (1952). Some rank order tests which are most powerful against specific parametric alternatives. *Annals of Mathematical Statistics*, **23**, 343–366.

Tharthare, S.K. (1963). Right angular designs. *Annals of Mathematical Statistics*, **34**, 1057–1067.

Tharthare, S.K. (1965). Generalized right angular designs. *Annals of Mathematical Statistics*, **36**, 1535–1553.

Theil, H. (1957). A note on certainty equivalence in dynamic planning. *Econometrica*, **25**, 346–349.

Theil, H. (1961). *Economic forecasts and policy.* North-Holland, Amsterdam.

Theil, H. (1963). On the use of incomplete prior information in regression analysis. *Journal of the American Statistical Association*, **58**, 401–414.

Theil, H. (1965). The analysis of disturbances in regression analysis. *Journal of the American Statistical Association*, **60**, 1067–1079.

Thiele, T.N. (1889). *Forelæsninger over almindelig Iagttagelseslære: Sandsynlighedsregning og mindste Kvadraters Methode.* Reitzel, Copenhagen.

Thom, R. (1975). *Structural stability and morphogenesis.* Advanced Book Program. XXV. W.A. Benjamin, Reading, MA.

Thompson, H.R. and Seal, K.E. (1964). Serial designs for routine quality control and experimentation. *Technometrics*, **6**, 77–98.

Thompson, M.E. (1997). *Theory of sample surveys.* Chapman & Hall, London.

Thompson, S.K. (1992). *Sampling.* Wiley, New York.

Thompson, W.R. (1935). On a criterion for the rejection of observations and the distribution of the ratio of deviation to sample standard deviation. *Annals of*

Mathematical Statistics, **6**, 214–219.

Thompson, W.R. (1947). Use of moving averages and interpolation to estimate median-effective dose. *BACT Reviews*, **11**, 115–145.

Thompson, W.A. and Wilke, T.A. (1963). On an extreme rank sum test for outliers. *Biometrika*, **50**, 375–383.

Thomson, G.H. (1939). *The factorial analysis of human ability.* University of London Press, London.

Thurstone, L.L. (1931). Multiple factor analysis. *Psychology Review*, **38**, 406–427.

Thurstone, L.L. (1947). *Multiple-factor analysis.* University of Chicago Press, Chicago.

Tiao, G.C. and Guttman, I. (1965). Corrigenda: The inverted Dirichlet distribution with applications. *Journal of the American Statistical Association*, **60**, 1251–1252.

Tippett, L.H.C. (1935). A snap-reading method of making time studies of machines and operatives in factory surveys. *Journal of the Textile Institute*, **26**, 13–51.

Tobin, J. (1958). Estimation of relationships for limited dependent variables. *Econometrica*, **26**, 24–36.

Tong, Y.L. (1970). Some probability inequalities of multivariate normal and multivariate *t*. *Journal of the American Statistical Association*, **65**, 1243–1247.

Tracy, C. and Widom, H. (1994). Level-spacing distributions and the Airy kernel. *Communications in Mathematical Physics*, **159**, 151–174.

Trevan, J.W. (1927). The error of determination of toxicity. *Proceedings of the Royal Society of London. Series B*, **101**, 483–514.

Tukey, J.W. (1947). Non-parametric estimation II. Statistically equivalent blocks and tolerance regions. The continuous case. *Annals of Mathematical Statistics*, **18**, 529–539.

Tukey, J.W. (1949). Comparing individual means in the analysis of variance. *Biometrics*, **5**, 99–114.

Tukey, J.W. (1950). Some sampling simplified. *Journal of the American Statistical Association*, **45**, 501–519.

Tukey, J.W. (1953). *The problem of multiple comparisons.* Unpublished mimeographed notes, Princeton University.

Tukey, J.W. (1956). Keeping moment-like sampling computations simple. *Annals of Mathematical Statistics*, **27**, 37–54.

Tukey, J.W. (1958a). Bias and confidence in not quite large samples. *Annals of Mathematical Statistics*, **29**, 614.

Tukey, J.W. (1958b). A problem of Berkson, and minimum variance orderly estimators. *Annals of Mathematical Statistics*, **29**, 588–592.

Tukey, J.W. (1959). A quick, compact, two sample test to Duckworth's specifications. *Technometrics*, **1**, 31–48.

Tukey, J.W. (1977). *Exploratory data analysis*. Addison-Wesley, Reading, MA.

Tweedie, M.C.K. (1945). Inverse statistical variates. *Nature*, **155**, 453.

Urban, F.M. (1910). The method of constant stimuli and its generalizations. *Psychology Review*, **17**, 229–259.

van der Vaart, A.W. (1998). *Asymptotic statistics*. Cambridge University Press, Cambridge.

van der Vaart, H.R. (1961). Some extensions of the idea of bias. *Annals of Mathematical Statistics*, **32**, 436–447.

van der Waerden, B.L. (1952). Order tests for the two-sample problem and their power. *Indagationes Mathematicae*, **14**, 453–458.

van der Waerden, B.L. (1953). Order tests for the two-sample problem. II, III. *Proceedings of the Koninklijke Nederlandse Akademie van Wetenschappen, Serie A*, **56**, 303–310, 311–316.

van der Waerden, B.L. (1960). Sampling inspection as a minimum loss problem. *Annals of Mathematical Statistics*, **31**, 369–384.

Vartak, M.N. (1959). The non-existence of certain PBIB designs. *Annals of Mathematical Statistics*, **30**, 1051–1062.

Venables, W.N. and Ripley, B.D. (2002). *Modern applied statistics with S* (4th edn). Springer-Verlag, New York, Paris and Berlin.

Venn, J. (1891). On the nature and uses of averages. *Journal of the Royal Statistical Society*, **54**, 429–456.

Volterra, V. (1926). Variations and fluctuations of the number of individuals in animal species living together. In *Animal ecology* (ed. R.N. Chapman), pp. 409–448. McGraw-Hill, New York.

von Bortkiewicz, L. (1898). *Das Gesetz der kleinen Zahlen*. Teubner, Leipzig.

von Mises, R. (1918). Über die Ganzzahligkeit der Atomgewichte und verwandte Fragen. *Physikalische Zeitschrift*, **19**, 490–500.

von Mises, R. (1919). Grundlagen der Wahrscheinlichkeitsrechnung. *Mathematische Zeitschrift*, **5**, 52–99.

von Mises, R. (1931). *Vorlesungen aus dem Gebiete der angewandten Mathematik. Bd. 1. Wahrscheinlichkeitsrechnung und ihre Anwendung in der Statistik und theoretischen Physik*. Franz Deuticke, Leipzig and Wien.

von Neumann, J. (1941). Distribution of the ratio of the mean square successive difference to the variance. *Annals of Mathematical Statistics*, **12**, 367–395.

Wald, A. (1944). On cumulative sums of random variables. *Annals of Mathematical Statistics*, **15**, 283–296.

Wald, A. (1945). Sequential tests of statistical hypotheses. *Annals of Mathematical Statistics*, **16**, 117–186.

Wald, A. (1947). *Sequential analysis.* Wiley, New York.

Wald, A. (1950). *Statistical decision functions.* Wiley, New York.

Wald, A. and Wolfowitz, J. (1939). Confidence limits for continuous distribution functions. *Annals of Mathematical Statistics*, **10**, 105–118.

Wald, A. and Wolfowitz, J. (1940). On a test whether two samples are from the same population. *Annals of Mathematical Statistics*, **11**, 147–162.

Wald, A. and Wolfowitz, J. (1943). An exact test for randomness in the non-parametric case based on serial correlation. *Annals of Mathematical Statistics*, **14**, 378–388.

Walker, A.M. (1950). Note on a generalization of the large sample goodness of fit test for linear autoregressive schemes. *Journal of the Royal Statistical Society. Series B*, **12**, 102–107.

Walker, G. (1931). On periodicity in series of related terms. *Proceedings of the Royal Society of London. Series A*, **131**, 518–532.

Wallman, K. (1993). Enhancing statistical literacy: Enriching our society. *Journal of the American Statistical Association*, **88**, 1–8.

Wand, M.P. and Jones, M.C. (1995). *Kernel smoothing.* Chapman & Hall, London.

Ware, J.H., Muller, J.E, and Braunwald, E. (1985). The futility index: An approach to the cost-effective termination of randomized clinical trials. *The American Journal of Medicine*, **78**, 635–643.

Warner, S.L. (1965). Randomized response: A survey technique for eliminating evasive answer bias. *Journal of the American Statistical Association*, **60**, 63–69.

Watson, G.S. (1961). Goodness-of-fit tests on a circle. *Biometrika*, **48**, 109–114.

Watson, G.S. (1962). Goodness-of-fit tests on a circle. II. *Biometrika*, **49**, 57–63.

Watson, G.S. (1964). Smooth regression analysis. *Sankhyā*, **26**, 359–372.

Watson, G.S. (1965). Equatorial distributions on a sphere. *Biometrika*, **52**, 193–201.

Watson, G.S. (1967). Another test for the uniformity of a circular distribution. *Biometrika*, **54**, 675–677.

Watson, G.S. and Williams, E.J. (1956). On the construction of significance tests on the circle and the sphere. *Biometrika*, **43**, 344–352.

Weaver, W. (1948). Probability, rarity, interest and surprise. *Scientific Monthly*, **67**, 390–392.

Wedderburn, R.W.M. (1974). Quasi-likelihood functions, generalized linear models, and the Gauss–Newton method. *Biometrika*, **61**, 439–447.

Weibull, W. (1939). A statistical theory of the strength of materials. *Ingeniors Vetenskaps Akademiens Handlingar*, **151**, 1–45.

Weiss, L. (1953). Testing one simple hypothesis against another. *Annals of Mathematical Statistics*, **24**, 273–281.

Weiss, L. and Wolfowitz, J. (1966). Generalized maximum probability estimators. *Theory of Probability and its Application*, **11**, 58–81.

Weiss, L. and Wolfowitz, J. (1967). Maximum probability estimators. *Annals of the Institute of Statistical Mathematics, Tokyo*, **19**, 193–206.

Welch, B.L. (1936). Note on an extension of the L_1 test. *Statistical Research Memoirs, University of London*, **1**, 52–56.

Wermuth, N. and Lauritzen, S.L. (1990). On substantive research hypotheses, conditional independence graphs and graphical chain models. *Journal of the Royal Statistical Society. Series B*, **52**, 21–72.

Westenberg, J. (1948). Significance test for median and interquartile range in samples from continuous populations of any form. *Proceedings of the Koninklijke Nederlandse Akademie van Wetenschappen*, **51**, 252–261.

Wetherill, G.B., Chen, H., and Vasudeva, R.B. (1966). Sequential estimation of quantal response curves: A new method of estimation. *Biometrika*, **53**, 439–454.

Wheeler, S. and Watson, G.S. (1964). A distribution-free two-sample test on a circle. *Biometrika*, **51**, 256–257.

Whittaker, E.T. (1943). Chance, freewill and necessity in the scientific conception of the universe. *Proceedings of the Physical Society*, **55**, 459–471.

Whittaker, J. (1990). *Graphical models in applied multivariate statistics*. Wiley, New York.

Whittle, P. (1955). Some distribution and moment formulae for the Markov chain. *Journal of the Royal Statistical Society. Series B*, **17**, 235–242.

Whittle, P. (1965). Some general results in sequential design. *Journal of the Royal Statistical Society. Series B*, **27**, 371–394.

Wicksell, S.D. (1925). The corpuscle problem. A mathematical study of a biometric problem. *Biometrika*, **17**, 84–99.

Wicksell, S.D. (1933). On correlation functions of type III. *Biometrika*, **25**, 121–133.

Wijsman, R.A. (1957). Random orthogonal transformations and their use in some classical distribution problems in multivariate analysis. *Annals of Mathematical Statistics*, **28**, 415–423.

Wilcoxon, F. (1945). Individual comparisons by ranking methods. *Biometrics*, **1**, 80–83.

Wilk, M.B. and Gnanadesikan, R. (1968). Probability plotting methods for the analysis of data. *Biometrika*, **55**, 1–17.

Wilkinson, J.H. (1965). *The algebraic eigenvalue problem*. Clarendon Press, Oxford.

Wilks, S.S. (1932*a*). Certain generalizations in the analysis of variance. *Biometrika*, **24**, 471–494.

Wilks, S.S. (1932*b*). Moments and distributions of estimates of population parameters from fragmentary samples. *Annals of Mathematical Statistics*, **3**, 163–195.

Wilks, S.S. (1942). Statistical prediction with special reference to the problem of tolerance limits. *Annals of Mathematical Statistics*, **13**, 400–409.

Wilks, S.S. (1961). A combinatorial test for the problem of two samples from continuous distributions. *Proceedings of the 4th Berkeley Symposium on Mathematical Statistics and Probability*, **1**, 707–717.

Willcox, W.F. (1897). *The divorce problem: A study in statistics*. Colombia University Press, New York.

Williams, E.J. (1959). The comparison of regression variables. *Journal of the Royal Statistical Society. Series B*, **21**, 396–399.

Wilson, E.B. and Hilferty, M.M. (1931). The distribution of chi-square. *Proceedings of the National Academy of Sciences of the USA*, **17**, 684–688.

Wilson, E.B. and Worcester, J. (1943). The determination of L.D. 50 and its sampling error in bioassay. *Proceedings of the National Academy of Sciences of the USA*, **29**, 79.

Winckler, A. (1866). Allgemeine Sätze zur der Theorie der unregelmässigen Beobachtungsfehler. *Sitzungsberichte der Mathematik-Naturforschung Lk. Königes Akademie Wissenschaften, Wien, Zweite Abt*, **53**, 6–41.

Winston, W.L. (1994). *Operations research: Applications and algorithms*. International Thomson Publishing, Belmont, NY.

Wishart, D. (1969). Numerical classification method for deriving natural classes. *Nature*, **221**, 97–98.

Wishart, J. (1928). The generalized product moment distribution in samples from a normal multivariate population. *Biometrika*, **20A**, 32–52.

Wishart, J. (1932). A note on the distribution of the correlation ratio. *Biometrika*, **24**, 441–456.

Wishart, J. and Bartlett, M.S. (1933). The generalised product moment distribution in a normal system. *Proceedings of the Cambridge Philosophical Society*, **29**, 260–270.

Wold, H. (1938). *A study in the analysis of stationary time series*. Ph.D. thesis, University of Uppsala.

Wold, H. (1948). On stationary point processes and Markov chains. *Skandinavien Aktuarie Tidskrift*, **31**, 229–240.

Wold, H. (1949). A large-sample test for moving averages. *Journal of the Royal Statistical Society. Series B*, **11**, 297–305.

Wold, H. (1953). Etude en économetrie du risque et des situations où le hasard joue un rôle. *Colloques Internationales CNRS*, **40**, 121–126.

Wold, H. (1966). Estimation of principal components and related models by iterative least squares. In *Multivariate Analysis. Proceedings of the International*

Symposium, Dayton 1965, pp. 391–420.

Wolfowitz, J. (1957). The minimum distance method. *Annals of Mathematical Statistics*, **28**, 75–88.

Woodbury, M.A. (1949). On a probability distribution. *Annals of Mathematical Statistics*, **20**, 311–313.

Wright, S. (1921). Correlation and causation. *Journal of Agricultural Research*, **20**, 557–585.

Yates, F. (1933). The analysis of replicated experiments when the field results are incomplete. *Empire Journal of Experimental Agriculture*, **1**, 129–142.

Yates, F. (1934). Contingency tables involving small numbers and the χ^2 test. *Journal of the Royal Statistical Society. Series B*, **1 Suppl**, 217–235.

Yates, F. (1936). Incomplete Latin squares. *Journal of Agricultural Science*, **26**, 301–315.

Yates, F. (1937*a*). *The design and analysis of factorial experiments. Technical Communication of the Commonwealth Bureau of Soils*, Volume 35. Commonwealth Agricultural Bureaux, Farnham Royal.

Yates, F. (1937*b*). A further note on the arrangement of variety trials: Quasi-Latin squares. *Annals of Eugenics*, **7**, 319–331.

Yates, F. (1939). The recovery of inter-block information in balanced incomplete block designs. *Annals of Eugenics*, **9**, 136–156.

Yates, F. (1940). The recovery of inter-block information in balanced incomplete block designs. *Annals of Eugenics*, **10**, 317–325.

Yates, F. and Cochran, W.G. (1938). The analysis of groups of experiments. *Journal of Agricultural Science, Cambridge*, **28**, 556–580.

Yates, F. and Grundy, P.M. (1953). Selection without replacement within strata with probability proportional to size. *Journal of the Royal Statistical Society. Series B*, **15**, 235–261.

Youden, W.J. (1937). Use of incomplete block replications in estimating Tobacco-Mosaic virus. *Contributions Boyce Thompson Institute*, **9**, 41–48.

Youden, W.J. (1951*a*). The Fisherian revolution in methods of experimentation. *Journal of the American Statistical Association*, **46**, 47–50.

Youden, W.J. (1951*b*). *Statistical methods for chemists.* Wiley, New York.

Youden, W.J. (1963). Ranking laboratories by Round-Robin test. *Material Research and Standards*, **3**, 9–13.

Yule, G.U. (1900). The association of attributes in statistics. *Philosophical Transactions of the Royal Society of London. Series A*, **194**, 257–319.

Yule, G.U. (1906). On a property which holds good for all groupings of a normal distribution of frequency for two variables, with applications to the study of contingency-tables for the inheritance of unmeasured qualities. *Proceedings of the Royal Society of London. Series A*, **77**, 324–336.

Yule, G.U. (1911). *An introduction to the theory of statistics.* Griffin, London.

Yule, G.U. (1912). On the methods of measuring association between two attributes. *Journal of the Royal Statistical Society. Series A,* **75**, 107–170.

Yule, G.U. (1921). On the time-correlation problem, with special reference to the variate-difference correlation method. *Journal of the Royal Statistical Society. Series A,* **84**, 497–537.

Yule, G.U. (1925a). The growth of population and the factors which control it. *Journal of the Royal Statistical Society. Series A,* **88**, 1–58.

Yule, G.U. (1925b). A mathematical theory of evolution based on the conclusions of Dr. J.C. Willis. *Philosophical Transactions of the Royal Society of London. Series B,* **213**, 21–87.

Yule, G.U. (1927). On a method of investigating periodicities in disturbed series with special reference to Wolfer's sunspot numbers. *Philosophical Transactions of the Royal Society of London. Series A,* **226**, 267–298.

Yule, G.U. (1944). *The statistical study of literary vocabulary.* Cambridge University Press, Cambridge.

Yule, G.U. (1945). On a method of studying time series based on their internal correlation. *Journal of the Royal Statistical Society. Series A,* **108**, 208–225.

Zadeh, L.A. (1965). Fuzzy sets. *Information and Control,* **8**, 338–353.

Zelen, M. (1954). Bounds on a distribution function that are functions of moments to order four. *Journal of Research of the National Bureau of Standards,* **53**, 377–381.

Zhang, H. and Singer, B. (1999). *Recursive partitioning in the health sciences.* Springer-Verlag, New York.

Zipf, G.K. (1932). *Selected studies of the principle of relative frequency.* Harvard University Press, Cambridge, MA.

Zipf, G.K. (1949). *Human behavior and the principle of least effort.* Addison-Wesley, Cambridge, MA.